Histological & Histochemical Methods

Theory and Practice

SECOND EDITION

Histological & Histochemical Methods

Theory and Practice

SECOND EDITION

by

J. A. Kiernan

Department of Anatomy,
The University of Western Ontario,
London, Ontario, Canada

PERGAMON PRESS
Member of Maxwell Macmillan Pergamon Publishing Corporation

OXFORD · NEW YORK · BEIJING · FRANKFURT
SÃO PAULO · SYDNEY · TOKYO · TORONTO

U.K.	Pergamon Press plc, Headington Hill Hall, Oxford OX3 0BW, England
U.S.A.	Pergamon Press, Inc., Maxwell House, Fairview Park, Elmsford, New York 10523, U.S.A.
PEOPLE'S REPUBLIC OF CHINA	Pergamon Press, Room 4037, Qianmen Hotel, Beijing, People's Republic of China
FEDERAL REPUBLIC OF GERMANY	Pergamon Press GmbH, Hammerweg 6, D-6242 Kronberg, Federal Republic of Germany
BRAZIL	Pergamon Editora Ltda, Rua Eca de Queiros, 346, CEP 04011, Paraiso, São Paulo, Brazil
AUSTRALIA	Pergamon Press Australia Pty Ltd., P.O. Box 544, Potts Point, N.S.W. 2011, Australia
JAPAN	Pergamon Press, 5th Floor, Matsuoka Central Building, 1-7-1 Nishishinjuku, Shinjuku-ku, Tokyo 160, Japan
CANADA	Pergamon Press Canada Ltd., Suite No. 271, 253 College Street, Toronto, Ontario, Canada M5T 1R5

First edition 1981
Second edition 1990

Library of Congress Cataloging-in-Publication Data
Kiernan, J. A. (John Alan)
Histological and histochemical methods : theory and practice / J. A. Kiernan. — 2nd ed.   p. cm.   Includes index.
1. Histology—Technique.   2. Histochemistry—Technique.
I. Title.
QM556.K53   1989   611'.018'028—dc19   89-3843

British Library Cataloguing in Publication Data
Kiernan, John A. (John Alan)
Histological and histochemical methods. – 2nd ed.
1. Biological specimens. Preparation. Histology. Laboratory
I. Title
578'.6

ISBN 0-08-036224-9 Hardcover
ISBN 0-08-036223-0 Flexicover

Printed in Great Britain by BPCC Wheatons Ltd., Exeter

Contents

Each chapter is preceded by a detailed table of contents.

Preface to the Second Edition

If you want to use this book effectively, READ CHAPTER 1 BEFORE ATTEMPTING TO CARRY OUT ANY PRACTICAL INSTRUCTIONS, and look at the list of *Conventions and Abbreviations* on page ix. There is a *Glossary* on page 367.

The purpose of this book is to teach the chemical and physical principles of fixation, staining and histochemistry. In the first edition, the student was urged always to determine the reason for every step in a method before doing it. For this reason, the number of methods for which detailed practical instructions were given was quite small by comparison with the larger compendia of microtechnique. Nevertheless, it soon became apparent that many readers were going straight to the ends of the chapters and following the practical instructions without having read the theoretical parts beforehand. The worst offenders in this respect seem to be research workers! Nearly all the questions I am asked are ones whose answers are to be found a few pages before the technical instructions or are easily accessible through the index. In this edition the theoretical and practical materials have been more closely integrated. Instructions for methods are now given immediately after the explanations of the principles, instead of at the end of each chapter. This should make it more difficult to use the book as a collection of recipes, and encourage a more intelligent approach to microtechnique.

The theoretical Exercises at the end of each chapter have been appreciated by many readers, and have been somewhat expanded for this edition. On the other hand, no teachers of courses in microtechnique or histochemistry have told me they made use of the practical exercises. These have therefore been reduced to occasional suggestions applicable to laboratory teaching.

There have been many advances in histochemistry since the first edition was written, and I have tried to include those that seem likely to become "standard" methods in research, in diagnostic pathology, or in the preparation of materials for teaching purposes. Deletions have been necessary to compensate for the new material, but overall there is more that needs to be said than there was nine years ago.

Methods for Fixing and Processing tissues continue to become more numerous and more diverse, and the first four chapters of this edition contain descriptions of several new reagents and techniques. Ordinary staining methods using dyes are used as much as ever, and greater variety has been introduced into Chapters 6, 7 and 8. More prescribed methods are given than before, but the reader will also find enough theoretical and practical information to make up combinations of staining procedures appropriate to the needs of the moment. The thoughtful use of dyes in this way is encouraged as an alternative to trying out several published procedures in the hope of finding one that is suitable. The chapter on blood stains (7), has been expanded to include methods other than the Romanovsky-Giemsa procedures, and the use of smears and suspensions of cells from tissues other than blood. Dyes and fluorescent compounds that bind specifically to nucleic acids are now widely used, and have been added to Chapter 9. Many ingenious "classical" histochemical methods, especially in the histochemistry of organic functional groups (Chapter 10), have gone out of fashion and have, with regret, been removed or accorded only brief mention. The chemical principles are explained, and the occasional reader needing to use an older technique should look up the appropriate references for technical instructions. Carbohydrate histochemistry, on the other hand, has developed remarkably in the last ten years, and several useful new techniques are discussed in Chapter 11. These include important developments in conventional histochemistry and methods based on the use of lectins. Many labelled lectins are now commercially available, and can be used as easily as simpler compounds. The same is true of all the ordinary reagents required for immunohistochemistry (Chapter 19), which is now as "standard" as any of the older families of staining methods. Other methods based on biological affinity, including nucleic acid hybridization, are also introduced in Chapter 19. Other major changes for this second edition include the treatment of several additional organic compounds with small molecules along with the amines in Chapter 17, and extensive changes in Chapter 18, on neurohistological techniques. The latter chapter is somewhat shorter than in the first edition, but it covers a greater variety of methods.

Another change made for the second edition is the provision of answers to the exercises at the ends of the chapters.

Acknowledgements. Thanks are due to people who have given advice or read the drafts of chapters. A sabbatical year spent in the laboratory of Dr R. W. Horobin, in Sheffield, has improved my understanding of histochemistry in ways that influence most parts of the book. I am also grateful to Richard Horobin for some detailed and constructive criticisms of the first edition. Among colleagues at the University of Western Ontario, I thank Drs P. K. Lala (Chapter 7), B. A. Flumerfelt (Chapter 18), E. A. Heinicke (Chapter 19) and C. C. Naus (Chapter 19). Dr P. E. Reid of the University of British Columbia kindly read the revised draft of Chapter 11, and his advice has resulted in an improved presentation of carbohydrate histochemistry. The comments of students over the years have also prompted corrections and clarification in several places.

London, Ontario, September 1989
J. A. KIERNAN

Conventions and Abbreviations

Conventions

It is important that the reader be familiar with the conventions listed here before attempting to follow the instructions for any practical procedure.

[ ] (i) Enclose a complex, such as $[Ag(NH_3)_2]^+$ or $[PdCl_4]^{2-}$.

(ii) Indicates "concentration of", in molar terms. Thus, $[Ca^{2+}]^3$ = "the cube of the molar concentration of calcium ions".

Accuracy. Solids should be weighed and liquids measured to an accuracy of ± 5%. With quantities less than 10mg or 1.0ml, an accuracy of ±10% is acceptable.

Alcohol. Unqualified, this word is used for methanol, ethanol, isopropanol, or industrial methylated spirit (which is treated as 95% v/v). When the use of a specific alcohol is necessary, this is stated. "Absolute" refers to the commercially available "100%" ethanol, which really contains about 1% water and may also be contaminated with traces of benzene. Absolute ethanol is hygroscopic and should be kept in securely capped bottles, which should be only three-quarters filled to allow for evaporation. In an ordinary covered staining tank, ethanol does not remain acceptably "absolute" for more than about 5 days.

When diluting alcohols for any purpose, use distilled or de-ionized water.

Concentrations expressed as percentages. The symbol % is used in various ways:

(i) For solids in solution, % = grams of solid dissolved in 100ml of the final solution.

(ii) For liquids diluted with other liquids, % = number of millilitres of the principal component present in 100ml of the mixture, the balance being made up by the diluent (usually water). "70% ethanol" means 70ml of absolute ethanol made up to 100ml with water.

(iii) For gases (e.g. formaldehyde), % = grams of the gas contained in 100ml of solution.

(iv) Where doubt may arise, the symbol v/v, w/v, or w/w is appended to the % sign. For dilution of acids and ammonia, see Chapter 20.

Formalin. This word refers to the commercially obtained solution containing 37–40% (w/v) of formaldehyde in water. The shortened form "formal" is used in the names of mixtures such as formal-saline and formal-calcium. The term "formol" will be found in some books, but this is wrong because the ending -ol suggests, incorrectly, that formaldehyde is an alcohol.

Safety precautions. The precautions necessary in any laboratory, especially for prevention of fire, should be observed at all times. Some reagents used in histology and histochemistry have their special hazards. These are mentioned as they arise in the text.

Concentrated mineral acids (especially sulphuric) must be diluted by adding acid to water (**not** water to acid) slowly with stirring.

Formaldehyde and hydrochloric acid should not be thrown down a sink together: their vapours can react together in the air to form *bis*-chloromethyl ether, a potent carcinogen. Each substance should be flushed down the drain separately, with copious running tap water.

Concentrated nitric acid must not be allowed to come into contact with organic liquids, especially alcohol: the strongly exothermic reaction may result in an explosion.

Salts—water of crystallization. The crystalline

forms of salts are shown in instructions for mixing solutions. If the form stated is not available, it will be necessary to calculate the equivalent amount of the alternative material. This is simply done by substitution in the formula

$$\frac{W_1}{M_1} = \frac{W_2}{M_2}$$

in which W = weight, M = molecular weight, and subscripts 1 and 2 refer to the prescribed and the alternative compounds respectively.

For example: 125mg of cupric sulphate ($CuSO_4$) are prescribed, but only the hydrated salt, $CuSO_4.5H_2O$, is available. Molecular weights are 223.14 and 249.68 respectively. Then

$$\frac{125}{223.14} = \frac{W_2}{249.68}$$

$$W_2 = \frac{125 \times 249.68}{223.14}$$

$$= 139.9$$

It will therefore be necessary to use 139.9 (i.e. 140)mg of $CuSO_4.5H_2O$ in place of 125mg of the anhydrous salt.

Solutions. If a solvent is not named (e.g. "1% silver nitrate"), it is assumed to be water. See also under **Water** below.

Structural formulae. Aromatic rings are shown as Kekulé formulae, with alternating double bonds. Thus benzene is

Modern practice favours the designation

which indicates the equivalence of all the bonds in the ring. The Kekulé formulae are used in this book because, with them, it is easier to understand structural changes associated with the formation of coloured compounds,

such as the formation of quinonoid from aromatic rings.

A few deviations from standard chemical notation (e.g. in formulae for lipids) are explained as they arise.

Temperature. Unless otherwise stated, all procedures are carried out at room temperature, which is assumed to be 15–25°C. The other commonly used temperatures are 37°C and about 60°C. A histological laboratory should have ovens or incubators maintained at these temperatures.

If an oven containing melted paraffin wax is used as a 60° incubator, make sure that any aqueous or alcoholic solutions put in it are covered. Water or alcohol vapour may otherwise contaminate the wax.

Water. *When "water" is prescribed in practical instructions, it means distilled or de-ionized water.* When water from the public supply may be used, it is specifically mentioned as "tap water".

Abbreviations

α,β (i) Used to indicate the configuration at position C1 in glycosides (see Chapter 11).

(ii) In aliphatic compounds the α-carbon atom is adjacent to the carbon atom bearing the principal functional group (i.e. α is carbon number 2). The use of numbers and Greek letters is shown below for *n*-hexanol:

$$\overset{\overset{H_2}{}\;\;\overset{H_2}{}\;\;\overset{H_2}{}\;\;\overset{H_2}{}\;\;\overset{H_2}{}}{H_3C-C-C-C-C-C-OH}$$

6	5	4	3	2	1
ϵ	δ	γ	β	α	

(iii) In glycerol and its derivatives, the middle carbon atom is designated as β and the carbons on either side as α and α'.

(iv) In derivatives of naphthalene, indicate the position of a substituent relative to the site of fusion of the rings:

α-naphthol
(= 1-naphthol)

β-naphthol
(= 2-naphthol)

$\triangle$	Symbol used to indicate double bonds in lipids (see Chapter 12).
ε-	Indicates carbon number 6 or a substituent on this atom, as in the case of the amino group at the end of the side-chain of lysine.
μg	microgram (= 10^{-6}g or 10^{-3}mg).
μm	micrometre (= 10^{-6}m or 10^{-3}mm); also sometimes called a "micron".
ACh	Acetylcholine.
AChE	Acetylcholinesterase.
APUD	Amine precursor uptake and decarboxylation.
Ar	An aryl radical (in formulae).
AThCh	Acetylthiocholine.
ATP	Adenosine triphosphate.
ATPase	Adenosine triphosphatase.
bis-	Twice (in names of compounds).
B.P.	British Pharmacopoeia; Boiling point.
BuThCh	Butyrylthiocholine.
B.W. 284C51	1,5-*bis*-(4-diallyldimethylammonium-phenyl)-pentan-3-one. (The initials are for Burroughs Wellcome, the manufacturer).
°C	Degrees Celsius (Centigrade).
ChE	Cholinesterase (= pseudocholinesterase).
C.I.	*Colour Index* (see Chapter 5).
cis-	Indicates a geometrical isomer in which two substituents lie on the same side of the molecule.
CNS	Central nervous system.
Con A	Concanavalin A.
CPC	Cetylpyridinium chloride.
cyt.	Cytochrome (with identifying letter, a, b, c, etc.).
D-	Indicates a compound, usually a

	sugar, of the D-series. The compound itself is not necessarily dextrorotatory.
DAB	3,3′-diaminobenzidine.
dansyl	The 5-(dimethylamino)-1-naphthal-enesulphonyl radical.
DFP	Diisopropylfluorophosphate.
DMAB	*p*-dimethylaminobenzaldehyde (also known as Ehrlich's reagent).
DMP	2,2-dimethoxypropane.
DNA	Deoxyribonucleic acid.
DNase	Deoxyribonuclease.
DOPA	β-3,4,-dihydroxyphenylalanine.
DPN	Diphosphopyridine nucleotide (an obsolete name for NAD^+, q.v.).
DPX	A resinous mounting medium. The initials stand for its three components, **d**istrene (a polystyrene), a **p**lasticizer, and **x**ylene.
E_0, E'_0	Symbols for oxidation-reduction potentials (see Chapter 16).
E600	Diethyl-*p*-nitrophenyl phosphate.
E.C.	Enzyme Commission (see Chapter 14).
EDTA	Ethylenediamine tetraacetic acid. Also known as versene, sequestrene, edetic acid, and (ethylenedinitrilo)-tetraacetic acid. Usually used as its disodium salt.
F_{ab}	Part of the immunoglobulin molecule (see Chapter 19).
FAD	Flavin adenine dinucleotide (see Chapter 16).
FAGLU	A fixative containing formaldehyde and glutaraldehyde (see Chapter 17).
F_c	Part of the immunoglobulin molecule (see Chapter 19).
FITC	Fluorescein isothiocyanate.
FMN	Flavin mononucleotide (see Chapter 16).
GBHA	Glyoxal-*bis*-(2-hydroxyanil).
H-acid	8-amino-1-naphthol-3, 6-disulphonic acid.
H-chain	Part of the immunoglobulin molecule (see Chapter 19).

H.&E. Haematoxylin and eosin (see Chapter 6).

HNAH 2-hydroxy-3-naphthoic acid hydrazide. May equally be named 3-hydroxy-2-naphthoic acid hydrazide.

HRP Horseradish peroxidase.

IgG Immunoglobulin G.

L- Indicates a compound (usually a sugar or an amino acid) of the L-series. The compound itself is not necessarily laevorotatory.

LVN Low viscosity nitrocellulose.

M (as in 0.1M) Molar (= moles per litre).

m- meta- (in names of benzene derivatives).

MBTH 3-methyl-2-benzothiazolone hydrazone.

mole The molecular weight, expressed in grams.

M.W. Molecular weight.

N (as in 0.1N) Normal (= gram-equivalents per litre; see Chapter 20).

N_A Avogadro's number: 6.022×10^{23} molecules per mole.

n- Normal, indicating an unbranched chain, as in n-butanol.

NAD^+ Nicotinamide adenine dinucleotide.

NADI Naphthol-diamine (reaction).

$NADP^+$ Nicotinamide adenine dinucleotide phosphate.

NANA N-acetylneuraminic acid.

Nitro-BT Nitro blue tetrazolium.

nm nanometre (= 10^{-9}m or $10^{-3}\mu$m).

NQS 1,2-naphthoquinone-4-sulphonic acid.

NSM Neurosecretory material.

o- ortho- (in names of benzene derivatives).

OPT o-phthaldialdehyde.

p- para- (in names of benzene derivatives).

PAN Perchloric acid-naphthoquinone (method; see Chapter 12).

PAP Peroxidase-antiperoxidase (reagent; see Chapter 19).

PAS Periodic acid-Schiff (method; see Chapter 11).

PCMB p-chloromercuribenzoate.

pH The logarithm (to base 10) of the reciprocal of the molar concentration of hydrogen ions.

PMA Phosphomolybdic acid.

pg picogram (= 10^{-12} gram).

PMS Phenazine methosulphate.

PNS Peripheral nervous system.

PPD p-phenylenediamine

PTA Phosphotungstic acid.

PVA Polyvinyl alcohol.

PVP Polyvinylpyrollidone (also called "povidone").

R,R' Indicate alkyl or aryl radicals, in formulae.

RNA Ribonucleic acid.

RNase Ribonuclease.

S.G. Specific gravity (also = density, in g per cm^3).

SUSA A fixative mixture introduced by M. Heidenhain: short for *Sublimat-Säure* (see Chapter 2).

t- Tertiary, as in t-butanol: $(CH_3)_3COH$.

TCA Trichloroacetic acid.

THF Tetrahydrofuran.

TMB Tetramethylbenzidine.

TPN Triphosphopyridine nucleotide (an obsolete name for $NADP^+$, q.v.; see Chapter 16).

TPP Thiamine pyrophosphate.

TPPase Thiamine pyrophosphatase.

trans- Indicates a geometrical isomer in which two substituents lie on opposite sides of the molecule.

TRIS *Tris*(hydroxymethyl)aminomethane.

UDPG Uridine-5-diphosphate glucose.

UQ Ubiquinone.

U.S.P. United States Pharmacopoeia.

v/v Volume ÷ volume.

w/v Weight ÷ volume.

w/w Weight ÷ weight.

Specialized abbreviations

The following are used only for special purposes and are explained in the appropriate parts of the text.

Chapter 11. Monosaccharide residues: Fuc, Gal, GalNAc, Glc, GlcNAc, GlcUA, IdUA, Man, Xyl. Lectins: A, DBA, LA , LCA_1, PHA, RCA_1, RCA_{120}, SBA, UEA_1, WGA.

Chapter 16. Tetrazolium salts: BSPT, BT, DS-NBT, INT, MTT, Nitro-BT, NT, TNBT, TTC, TV.

Chapter 17. Biogenic amines: ADR, DA, HIS, 5HT, NA.

1

Introduction to Microtechnique

Many theoretical explanations and practical instructions are contained in this book. **The present chapter concerns various aspects of the making of microscopical preparations that are fundamental to all the techniques described in the later chapters.** It cannot be over-emphasized that unless the student or technician understands the rationale of all that is to be done, he will not do it properly. Some practical information relevant to the manipulations discussed here will be found in Chapter 4 as well as in Section 1.6 of this chapter.

1.1. Thickness and contrast

In order to be examined with the microscope, a specimen must be sufficiently thin to be transparent and must possess sufficient contrast to permit the resolution of structural detail. Thinness may be an intrinsic property of the object to be examined. Thus, small animals and plants, films and smears of cells, macerated or teased tissues, and spread-out sheets of epithelium or connective tissue are all thin enough to be mounted on slides directly as **whole mounts**. In the study of histology and histochemistry, one is more often concerned with the internal structure of more substantial, solid specimens. These must be cut into thin slices or **sections** in order to make them suitable for microscopical examination.

Freehand sections, cut with a razor, are rarely used in animal histology but are still sometimes employed for botanical material. Though some expertise is necessary, sectioning in this way has the advantage of requiring little in the way of time or special equipment. When sections of human or animal tissues are needed in a hurry, **frozen sections** are commonly used. The ordinary **freezing microtome** is used for fixed material, while the **cryostat**, a microtome mounted in a freezing cabinet, may be used for cutting sections of fixed or unfixed tissue. More skill is called for in the operation of a cryostat than for a freezing microtome. Another advantage of cutting frozen sections, aside from speed, is the preservation of some lipid constituents, which are dissolved out during the course of dehydration and embedding in paraffin. A disadvantage of the freezing microtome is that the sections are commonly too thick (e.g., 20–100 µm) for the resolution of fine structural detail within cells. Much thinner sections can be cut with a cryostat. The **vibrating microtome** (Vibratome) can section unfixed, unfrozen specimens. The blade of this instrument passes with a sawing motion through a block of tissue immersed in an isotonic saline solution. The cutting process is much slower than with other types of microtome, so it is not feasible to prepare very large numbers of sections.

When the preservation of lipids and of heat-labile substances such as enzymes is not important, the specimens are **dehydrated, cleared** (which means, in this context, equilibrated with a solvent that is miscible with paraffin), **infiltrated** with molten paraffin wax, and finally, **embedded** (blocked out) in solidified wax. **Paraffin sections**

1

are most commonly cut on a **rotary microtome**, though a **rocking microtome** or a **sledge microtome** can also be used. The sections come off the knife in ribbons, and with sufficient skill it is possible to obtain **serial sections**, which may be as little as 4μm thick, through the whole block of tissue. **Polyester wax** (Steedman, 1960), which is miscible with 95% alcohol, is handled in much the same way as paraffin. Cutting is difficult, but the lower melting point (about 40°C, compared with 55–60°C for paraffin waxes) is an advantage for some tissues and histochemical methods. When sections of large specimens are required, it is convenient to embed in celloidin or low-viscosity nitrocellulose. **Celloidin sections** 50–200μm thick, are usually cut on a sledge microtome. Various **synthetic resins** are also used as embedding media for light microscopy, though their main application is in the cutting of extremely thin sections for examination in the electron microscope. Resin-embedded tissue is usually sectioned with an **ultramicrotome**, using a glass or diamond knife. Sections 0.5–1.0μm thick, suitably stained for optical examination, are valuable for comparison with the much thinner sections used in ultrastructural studies. The light microscope provides greater resolution of detail in plastic-embedded sections than in paraffin sections, but the latter are more easily stained in contrasting colours.

The optical contrast in a thin specimen is determined partly by its intrinsic properties but largely by the way in which it is processed. If the specimen is not stained, contrast will be greatest when the mounting medium has a refractive index substantially different from that of the specimen. Differences of this kind are emphasized in the **phase contrast** and the **differential interference contrast (Nomarski) microscopes**. These instruments are valuable for the study of living cells, such as those grown in tissue culture. In recent years there have been great technical advances in light microscopy, including **video-enhanced contrast microscopy** (Allen, 1987; Breuer *et al.*, 1988), with which otherwise inconspicuous features can be emphasized and filmed, and **confocal microscopy**, in which scanning of the field by a laser or a beam of white light provides images of optical sections through thick specimens (see Fine *et al.*, 1988; Shotton, 1988). Images in different planes can be recorded electronically, and synthesized to provide either three-dimensional pictures or flat pictures of selected objects that are too thick or tortuous to be seen in a single focal plane.

In histology, the natural refractility of a tissue is usually deliberately suppressed by the use of a mounting medium with a refractive index close to that of the anhydrous material constituting the section (approximately 1.53). Almost all the contrast is produced artificially by **staining**.

Fluorescence is the property exhibited by substances which absorb light of short wavelength such as ultraviolet or blue and emit light of longer wavelength, such as green, yellow, or red. The phenomenon can be observed with a **fluorescence microscope** in which arrangements are made for the emitted (long wavelength) light to reach the eye while the exciting (short wavelength) light does not. Fluorescing materials therefore appear as bright objects on a dark background. The fluorescence microscope can be used to observe **autofluorescence** due to substances naturally present and **secondary fluorescence** produced by appropriate chemical treatment of the specimen. The autofluorescence of a tissue, which is due to various endogenous compounds, notably flavoproteins (Benson *et al.*, 1979), frequently interferes with the interpretation of secondary fluorescence. Fluorescent compounds are also formed in tissues by chemical reactions between **fixatives** (see Section 1.4, and Chapter 2) and proteins (Collins & Goldsmith, 1981). This kind of secondary fluorescence is often called autofluorescence.

1.2. Staining and histochemistry

The histologist stains sections in order to facilitate the elucidation of structural details. The histochemist, on the other hand, seeks to determine the locations of known substances within the structural framework. The disciplines of histology and histochemistry overlap to a large extent, but one consequence of the two approaches is that the staining techniques used primarily for morphological purposes are sometimes poorly understood in chemical terms. It is desirable to demonstrate structural components by "staining" for substances they are known to contain, but many valuable empirically derived histological techniques are not based on well understood chemical principles.

1.3. Some physical considerations

The intelligent handling of microscopical preparations requires familiarity with the physical properties of several materials that are used in almost all techniques. All too often, the beginner will ruin a beautifully stained section by forgetting that two solvents are immiscible, or by leaving the slides overnight in a liquid which dissolves the coloured product. The following remarks relate mainly to sections mounted on slides, but are also applicable to blocks of tissue or to smears, films, whole mounts, or free-floating sections.

Water is completely miscible with the common alcohols (methanol, ethanol, isopropanol, methylated ethyl alcohol). Water is immiscible with xylene, benzene, chloroform, and most other clearing agents. These clearing agents are, however, miscible with the alcohols in the absence of water. Melted paraffin wax and the resinous mounting media (Canada balsam, Xam, Permount, DPX, etc.) are miscible with the clearing agents but not with the alcohols or with water. One mounting medium, euparal, is notable for being miscible with absolute alcohol as well as with xylene. Because of these properties of the common solvents, **a specimen must be passed through a series of liquids** during the course of embedding, staining, and mounting for examination.

For example, a piece of tissue removed from an aqueous fixative, such as a formaldehyde solution, must pass through a **dehydrating agent** (such as alcohol) and a **clearing agent** (such as chloroform) before it can be infiltrated with paraffin wax. Ribbons of paraffin sections may be floated on warm water, which will remove wrinkles, when they are being mounted on glass slides. A thin layer of a suitable **adhesive** (see Chapter 4) may be interposed between the slide and the sections, but this is not always necessary. The slides must then be **dried** thoroughly in warm air before being placed in a **clearing agent**, usually xylene, to dissolve and remove the wax. The slides will now bear sections of tissue which are equilibrated with the clearing agent. Passage through **alcohol** (or any other solvent miscible with both xylene and water) must precede immersion of the slides in water. Sudden changes are avoided if possible, so a series of graded mixtures of alcohol with water is used. Most staining solutions and histochemical reagents are aqueous solutions. If a permanent mount in a resinous medium is required, the slides carrying the stained sections must be **dehydrated**, without unintentionally removing the stain, in alcohol or a similar solvent, **cleared** (usually in xylene), and, finally, **mounted** in the resinous medium.

Several synthetic resins are used as embedding media. In most procedures the specimen is first infiltrated with a mixture of monomer and a catalyst at room temperature, and then moved to an oven (60°C) to initiate polymerization. Most monomers are miscible with ethanol or other organic solvents. Some of the polymers are similarly soluble; others can only be removed from the sections by reagents that break covalent bonds in the matrix of resin.

The physical properties of some commonly used solvents are given in Chapter 4, where some practical guidance for their use will also be found.

Resinous mounting media contain clearing agents, so a newly mounted preparation does not become completely transparent for a few hours. The resin has to permeate the section and the solvent has to evaporate at the edges of the coverslip. When these events have taken place, the specimen will be equilibrated with the mounting medium and should have almost the same refractive index as the latter. Consequently, most of the observed contrast will be due to the staining method.

Frozen sections are collected into water or an

aqueous solution. They may be affixed to slides and dried in the air either before or after staining. The frozen section on the slide is, therefore, at first equilibrated with water and must be dehydrated and cleared before mounting in a resinous medium. If the products of staining would dissolve in organic solvents, as is the case with the Sudan dyes and with the end-products of some histochemical reactions, it is necessary to use a water-miscible mounting medium. Several such media are available (e.g. glycerol jelly, Apathy's, Farrant's, polyvinylpyrrolidone) but they usually do not suppress the intrinsic refractility of the specimen as completely as do the resins.

Celloidin sections require special handling owing to the properties of the embedding medium. Nitrocellulose is soluble in a mixture of equal volumes of ethanol and diethyl ether, commonly called ether-alcohol. It also dissolves in absolute methanol and in cellosolve. Celloidin is hardened by 70% ethanol, chloroform, or phenol. Absolute ethanol makes the nitrocellulose swell, but does not dissolve it. Aqueous solutions penetrate freely through the matrix of nitrocellulose, so it is not necessary to remove the latter in order to stain the contained section of tissue. Molten paraffin wax can also permeate a nitrocellulose matrix without dissolving it. Specimens which are expected to be difficult to section are often infiltrated with celloidin, cleared, and then infiltrated with and embedded in wax. This procedure is known as **double embedding**.

Many of the dyes used in histology can be removed from stained sections by alcohol. This property is useful for the extraction of excesses of dye, a process known as **differentiation**, but it can also be a nuisance. Since a stained preparation must be completely dehydrated as well as adequately differentiated, the timing and rate of passage through graded alcohols is often critical. It is one of the arts of histological technique to obtain the correct degree of differentiation.

1.4. Properties of tissues

Freshly removed cells and tissues, especially those of animals, are chemically and physically unstable. The treatments to which they are exposed in the making of microscopical preparations would damage them severely if they were not stabilized in some way. This stabilization is usually accomplished by **fixation**, which is discussed in Chapter 2. For some purposes, especially in enzyme histochemistry, it is necessary to use sections of unfixed tissues. As has already been stated, such sections may be cut with a cryostat or a vibrating microtome. Unfixed sections are stable when dried onto glass slides or coverslips but become labile again when wetted with aqueous liquids that do not produce fixation. Many histological staining methods do not work properly on unfixed tissues.

Most methods of fixation make the tissues harder than they were in the living state. Provided that it is not excessive, hardening is advantageous because it renders the tissues easier to cut into sections. However, some tissues such as bone are too hard to cut even before they have been fixed. These have to be softened after fixation but before dehydration, clearing, and embedding. Calcified tissues are softened by dissolving out the inorganic salts that make them hard, a procedure known as **decalcification** (see Chapter 3). Other hard substances such as cartilage, chitin, and wood require different treatments. Exceptionally robust microtomes equipped with massive chisel-like knives of tungsten-steel have become available in recent years. These will cut sections of undecalcified bone and other hard materials.

Even initially soft specimens sometimes become unduly hard by the time they are embedded in wax. These can be softened by cutting sections to expose the interior of the tissue at the face of the block and then immersing for a few hours in water. Although the solid wax is present in all the interstices of the tissue, materials such as collagen can still imbibe some water and be made much softer. Various proprietary "softening agents" are marketed for the same purpose, but they are, in my experience, no better than plain water. Another important factor in microtomy is the hardness of the embedding mass relative to that of the tissue. This is determined by the composition of the former and by the ambient temperature. Obviously, the proper use of the microtome is also necessary if satisfactory sections are to be cut.

1.5. Books and journals

There is a profusion of books, large and small, that give directions in practical microtechnique. Some of the more modern ones also briefly explain the rationales of the different methods described. Bradbury (1973), Culling (1974), Bancroft & Cook (1984), Humason (1979), Drury & Wallington (1980), and Bancroft & Stevens (1982) can all be recommended, but there are many others equally valuable. The works of Gatenby & Beams (1950) and Gray (1954) were comprehensive in their day, and they contain numerous older recipes and useful technical hints. For histochemistry the major treatise is that of Pearse (1968, 1972, 1980, 1985); other important ones are Barka & Anderson (1963), Ganter & Jollès (1969, 1970; in French), Gabe (1976), and Lillie & Fullmer (1976). In these works it is usually assumed that the reader is familiar with the chemical principles underlying the explanations of how the methods work. The chemical and physical principles of microtechnique and histochemistry are discussed at length by Baker (1958) and Horobin (1982). James (1976) provides a full account of the light microscope and its operation.

Some journals are devoted largely to the publication of papers on methodology. The major ones are *Stain Technology*, the *Journal of Histochemistry and Cytochemistry*, *Histochemistry* (formerly *Histochemie*), the *Histochemical Journal*, *Acta histochemica* and the *Journal of Microscopy* (formerly *Journal of the Royal Microscopical Society*). Relevant papers appear in other journals too, but by scanning the ones listed above it is not difficult to keep up with the major advances in the field.

1.6. On carrying out instructions

THIS IS IMPORTANT. READ THIS SECTION BEFORE ATTEMPTING TO PERFORM ANY OF THE TECHNIQUES DESCRIBED IN LATER CHAPTERS. See also "Conventions and Abbreviations" for methods used to express concentrations of solutions and for the correct interpretation of such terms as "alcohol" and "water" and for guidance on accuracy of measurement of weight, volume, and temperature.

In this and other texts, practical schedules are given for many techniques. Since the number of methods described in this book is relatively small, it has been possible to be quite explicit, so that the methods should all work properly if the instructions are followed exactly. There are, however, some general rules applicable to nearly all staining methods. These will therefore be given now in order to avoid tedious repetition in the following chapters.

1.6.1. De-waxing and hydration of paraffin sections

Place the slides (usually 4–12 of them) in a glass or stainless steel rack and immerse in a rectangular tank containing about 400 ml of xylene. This is the most useful size of tank for most purposes. Smaller ones are available, but when they are used their contents must be renewed more often. A "commercial" or "technical" grade of xylene (mixed isomers) is satisfactory. Agitate the rack, up and down and laterally, three or four times over the course of 2–3 min. If for some reason it is inconvenient to agitate the slides, they should be left to stand in the xylene for at least 5 min. A single slide is de-waxed by moving it slowly back and forth in the tank of xylene for 1 min. Individual slides should be held with stainless steel forceps.

Lift the rack (or individual slide) out of the xylene, shake it four or five times and touch it onto bibulous paper (three or four thicknesses of paper towel, or filter paper) and place in a second tank of xylene. Agitate as described above, but this time 1 min is long enough. The purpose of this second bath of xylene is to remove the wax-laden xylene from the initial bath, thereby reducing the risk of precipitation of wax upon the sections when they are passed into alcohol, in which wax is insoluble. The removal of excess fluid by shaking and blotting is very important and is done every time a rack or slide is passed from one tank to another. If it is not done, the useful life of each tankful of xylene or alcohol will be greatly shortened. **The instruction ''drain slides'' refers to this shaking-off of easily removed excess liquid.**

After the second bath of xylene, drain the rack of slides and place in a tank containing about 400 ml of absolute ethanol. Agitate at intervals of 10–20 s for 1½–2½ min. Drain slides and transfer to 95% ethanol and agitate in this for about 1 min. Drain slides and transfer to 70% alcohol. Agitate for at least 1 min. For an individual slide, it is sufficient to move it about with forceps for about 20 s in each change of alcohol. If the slides have to be left for several hours, or even for a few days, they should be immersed in 70% alcohol. This will prevent the growth of fungi and bacteria on the sections but will not make them come off the slides or become unduly brittle. If some sections do detach from the slides during de-waxing or hydration, more will certainly be lost in later processing. If attachment of the sections appears to be precarious, a film of nitrocellulose should be applied, as described in Chapter 4.

Hydration of the sections is completed by lifting the rack (or individual slide) out of the 70% alcohol, draining it, and immersing in water. Agitation for at least 30 s is necessary for removal of the alcohol. Without agitation, this takes 2–3 min. A second rinse in water is desirable if all traces of alcohol are to be removed.

It is possible to use small volumes of xylene and alcohol by carrying out the above operations in coplin jars (which usually hold up to five slides) or rectangular staining dishes (usually for 8–12 slides). The xylenes and alcohols are poured into these vessels and the slides agitated continuously with forceps. To change the liquid, pour it out (without losing the slides) and replace it with the next one in the series. Working in this way, each

lot of xylene or alcohol should be used only once. When tanks holding 400 ml are used, the liquids can be used repeatedly. They should all be renewed when traces of white sludge (precipitated wax) appear in the absolute ethanol. This commonly occurs after 10–12 racks of slides have been de-waxed and hydrated. In order to minimize evaporation, contamination by water vapour and the risk of fire, all tanks containing alcohol or xylene should have their lids on when not in use.

1.6.2. Staining

An instruction such as ''stain for 5 min'' means that the sections must be in intimate contact with the dye solution for the length of time stated. Slides (alone, or in racks, coplin jars or staining dishes, as convenient) are immersed in the solution, agitated for about 10 s, and then left undisturbed. Free-floating frozen sections are transferred to cavity-blocks, watch-glasses, or the wells of a haemagglutination tray containing the staining solution. Folds and creases in such sections must be straightened out if uniform penetration of the dye is to occur. The best instrument for handling frozen sections is a glass hook or ''hockey stick'', fashioned by drawing out a piece of glass rod in the flame of a Bunsen burner. Cryostat sections are collected onto slides or coverslips, so they are handled in the same way as mounted, hydrated paraffin sections. Sections of nitrocellulose-embedded material, if not affixed to slides, are manipulated in the same way as frozen sections.

In some techniques of enzyme histochemistry and immunocytochemistry, only one drop of a scarce or costly reagent can be applied to each section. This is done with the slide lying horizontally. The slide bearing the section, covered by the drop, is placed on wet filter paper in a closed petri dish: the drop will not evaporate if the air above is saturated with water vapour. Horizontal slides are also used in staining methods for blood films. In this case the reagents are not expensive, so the slides are placed, film upwards, on a pair of glass rods over a sink. The staining solution is poured on to flood the slides and later washed off by a stream of water or of a suitable buffer.

Since sections of tissue take up only minute quantities of dyes and other substances, there is no need for the volume of staining solution or histochemical reagent to be any greater than that required to cover the sections. Exceptions to this general rule are rare and are mentioned in the instructions for the methods concerned.

1.6.3. Washing and rinsing

The excess of unbound dye or other reagent is removed from the stained sections by washing or rinsing, usually with water. A "wash" is a more prolonged and vigorous treatment than a "rinse". Agitation of slides for a minute or more in each of three changes of water constitutes an adequate wash. When tap water is suitable, the slides are placed for about 3 min in a tank through which the water is running quickly enough to produce obvious turbulence. A rack of slides should be lifted out of the running tap water and then replaced every 20–30s in order to ensure that all the slides are thoroughly washed. A rinse, rather than a wash, is prescribed when excessive exposure to water would remove some of the dye specifically bound to the sections. Rinsing is done in the same way as washing, but the slides are agitated continuously and the total time of exposure to water is only about 15s.

With unmounted frozen or nitrocellulose sections, it is more difficult to control the process of washing. The sections are carried through three successive baths (50ml beakers are convenient) of water, and are kept in constant motion for 20–40s in each. Free-floating sections should not be allowed to fold or to crumple into little balls. Stains that are easily extracted by water should not be applied to sections of this kind.

1.6.4. Dehydration and clearing

For stains or histochemical end-products insoluble in water and alcohol, dehydration is a simple matter. The slides are agitated continuously for about 1min in each of the following: 70% alcohol, 95% alcohol; two changes of absolute ethanol, methanol, or isopropanol. There is no objection to taking them straight into 100% alcohol (in which case, three or four changes will be needed), but the use of lower alcohols will protect the more expensive anhydrous liquids from excessive contamination with water. **Slides must be drained** (see Section 1.6.1) as they are transferred from one tank to the next. Clearing is accomplished by passing the slides from absolute alcohol into xylene (two changes, 1min with agitation in each). They may remain in the last change of xylene for several days if mounting in a resinous medium cannot be carried out immediately. Used as described, the alcohols and xylenes, in 400ml tanks, can be used repeatedly for 10–12 racks of slides. The contents of all the tanks must be renewed when the xylene becomes faintly turbid.

Many dyes are extracted by alcohol, especially if water is also present. When this is the case, the instruction will be to "dehydrate rapidly". For **rapid dehydration**, drain off as much water as possible and then transfer the slides directly to absolute alcohol. Agitate very vigorously for 5–10s in each of three tanks of this liquid, draining between changes, and then clear in xylene as described above. The alcohol used for rapid dehydration should be renewed after processing four racks of slides. It is not possible to dehydrate free-floating sections rapidly. These should be mounted onto slides after washing, allowed to dry in the air, and then passed quickly through absolute alcohol into xylene.

1.7. Whole mounts and free cells

The practical exercises at the end of this chapter include the preparation of whole mounts of a thin sheet of tissue and the examination of cells released from epithelial surfaces. Thin specimens of epithelium can be obtained by pressing the surface of an organ on a gelatin-coated slide and carefully peeling it off. The separation of an epithelium from its underlying connective tissue may be enhanced by soaking the fresh material in proteolytic enzymes (collagenase, trypsin), or in other reagents including EDTA and dithiothreitol (Epstein *et al.*, 1979).

The isolation of cells without changing their shapes is called **maceration**. In the method given here, which is that of Goodrich (1942), the forces that hold the cells to their basal lamina and to one another are reduced by soaking in boric acid, which probably works by combining with the carbohydrates of the outside surfaces of cells, displacing calcium and other ions that are involved in intercellular adhesion (Williams & Atalla, 1981). Goodrich's medium for maceration also contains iodine, which kills the cells and preserves their shapes, but does not cause chemical crosslinking. Tissues can also be macerated by prolonged immersion in a dilute solution of any fixative agent that does not form covalent chemical bonds with proteins. Boric acid fails to loosen cells if the tissue has been fixed in glutaraldehyde (see Chapter 2), which stabilizes the structure by forming strong bonds between protein molecules (Vial & Porter, 1975).

Suspensions of living cells, including blood, cultured cell-lines, and cells released from solid organs by gentle enzymatic treatments, may also be deposited on slides and stained (see Chapter 7). Such cells assume spherical shapes, though cilia and other appendages may persist (Kleene & Gesteland, 1983).

1.8. Understanding the methods

There is a reason for everything that is done in making a microscopical preparation. Before trying out a technique for the first time, the student should read about and understand the underlying physics and chemistry. He should then read through all the practical instructions and make sure that he understands the purpose of every stage of the procedure.

Some technical methods can be learned only by practice. These include the cutting of sections, the mounting of sections onto slides, and the application of coverslips to sections. In this book, no attempt is made to teach these skills, which must be acquired under the guidance of more experienced colleagues. Excellent descriptions of

these procedures are to be found in the works of Krajian & Gradwohl (1952), Culling (1974), Gabe (1976), and Bancroft & Stevens (1982), but there is no substitute for practice in the laboratory. Success in microtechnique requires the integration of craftsmanship with intelligent appreciation of scientific principles.

1.9. Exercises

Theoretical

1. What type of preparation would you make in order to investigate the populations of cells in (a) the mesentery of the rat, (b) the articular cartilage of the head of the human femur?

2. A frozen section of skin is stained with an oil-soluble dye in order to demonstrate lipids. What would be an appropriate procedure for making a permanent microscopic preparation of the stained section?

3. If alcohol that has been used to dehydrate a fairly large piece of animal tissue is mixed with water, the resulting liquid is turbid. Why?

4. Some staining methods are applied to whole blocks of tissue which are subsequently embedded, sectioned, cleared, and mounted. To what artifacts would this type of technique be especially prone? How could the artifacts be minimized?

5. It is sometimes necessary to remove the coverslip from a section mounted in a resinous medium. Devise a reasonable procedure by which such a section could be made ready to stain in an aqueous solution of a dye.

Practical

[For Exercises **6**, **7**, and **8**, kill a rat or mouse by overdosage with ether vapour or some other humane method, and make preparations as directed. Other organs may be fixed for subsequent embedding and sectioning. For suitable fixative solutions, see Chapter 2.]

6. Open the abdomen, gently pull out a loop of small intestine, and stretch an area of **mesentery** over the end of a glass microscope slide. Cut away the tissue at the edges of the slide, which will then bear a thinly spread whole mount of mesentery, thickened along the courses of the larger blood vessels. Make several such preparations and air-dry them for 10–15 min. For fixation and removal of fat, immerse the slides in a mixture of absolute or 95% ethanol (95 ml) and glacial acetic acid (5 ml), for 10 min, rinse in absolute followed by 70% ethanol, and then in water. Transfer for 1 min to an aqueous solution of toluidine blue or a similar cationic dye (see Chapter 6, page 92) for 2 min, wash in tap water, apply a coverslip and examine. Nuclei of all cells will be blue. Mast cells will be bright red-purple, and the matrix of the areolar connective tissue pink-violet.

Permanent preparations in a resinous mounting medium can be made if desired. For explanation of the different colours produced by one dye, see Chapter 11.

7. Dissect out the **trachea**, and open it with a longitudinal incision dorsally (where the cartilaginous rings are incomplete). Rinse the specimen in saline (0.9% w/v aqueous sodium chloride) to remove blood and mucus and put it in 0.5ml of Goodrich's (1942) solution (see below) in a capped specimen tube. At 30min intervals, insert a Pasteur pipette and vigorously squirt all surfaces of the specimen with the surrounding fluid. After about 4h, take a drop of fluid, which should now be cloudy, onto a glass slide, and apply a coverslip. Examine, using either a phase-contrast microscope or an ordinary microscope with the condenser defocused. The cells released from the pseudostratified respiratory epithelium are easily recognized by their columnar form and apical cilia.

Macerating solution (Goodrich, 1942)

Boric acid, saturated solution in
saline: 10ml
Iodine, 1% solution in 2% potassium
iodide: 0.1–0.2ml
(Mix before using. Both the stock solutions are stable indefinitely at room temperature. The working solution should be pale yellow.)

8. Remove the skin of the snout and open the **nasal cavities**, using stout scissors or bone cutters. Take a piece of the septum, or a turbinate bone, covered by mucous membrane, from the dorsal half of the nasal cavity. Place it in Goodrich's solution and recover and examine the released epithelial cells, as in Exercise 7. In addition to respiratory epithelial cells, you should be able to see neurosensory cells and supporting cells from the olfactory epithelium. Compare the isolated cells with pictures of the sectioned epithelium in a textbook of Histology.

Why is the volume of macerating solution deliberately kept small in relation to the size of the specimen?

2

Fixation

It is not sufficient for the histologist that a specimen be transparent and that it possess adequate optical contrast. The cells and extracellular materials must be preserved in such a way that there has been as little alteration as possible to the structure and chemical composition of the living tissue. Such preservation is the object of fixation. Without being spatially displaced, the structural proteins and other constituents of the tissue must be rendered insoluble in all the reagents to which they will subsequently be exposed. "Perfect" fixation is, of course, theoretically and practically impossible to attain.

Biological material may be fixed in many ways and some of these are now discussed.

2.1. Physical methods of fixation

2.1.1. Heat

The simplest physical method is the application of heat. This results in the coagulation of proteins and the melting of lipids. The resemblance to the living state is not very close after such treatment, but the method is often used in diagnostic microbiology. The shapes and staining properties of bacteria are preserved well enough to permit identification. Heating of large specimens is accomplished by immersion in boiling saline (usually also containing formaldehyde) or by cooking in a microwave oven (Bernhard, 1974). The latter method compares quite favourably with chemical fixation, and has even been used to prepare specimens for electron microscopy (Chew *et al.*, 1984; Hopwood, 1984). Heating by microwave is also useful for accelerating the crosslinking of proteins by formaldehyde (Hopwood *et al.*, 1988).

2.1.2. Freezing

Animal tissues are sometimes processed by the techniques known as freeze-drying and freeze-substitution, though only the former is purely physical in nature.

A specimen must be frozen in such a way as to minimize the formation of ice crystals, which can make abundant conspicuous holes in the tissue. This artifact is usually seen in large specimens that have been frozen slowly. In preparation for freeze-drying, freeze substitution or the cutting of sections with a cryostat, the piece of tissue, which should be no more than 2 mm thick, is frozen as quickly as possible. Suitable techniques include immersion in isopentane cooled to its freezing point (–170°C) by liquid nitrogen, or placing on a metal block that has already been brought to the temperature of either liquid nitrogen (−196°C) or solid carbon dioxide (dry ice and acetone mixture, −75°C). For practical and theoretical considerations, see Pearse (1980) and Bald (1983).

It is often not feasible to obtain ideal conditions for rapid freezing, but damage due to ice can be greatly reduced by **cryoprotection**, in which the specimen is first equilibrated with a solution of a cryoprotective compound in physiological saline or buffer. Suitable cryoprotective agents are glycerol, dimethylsulphoxide (DMSO) and sucrose, at approximately 10% concentrations (see Terracio & Schwabe, 1981). DMSO is the best understood agent for cryoprotection (Ashwood-Smith, 1971), but the way in which it works is still not properly understood (see Pearse, 1980).

For **freeze-drying**, the frozen specimen is transferred to a special chamber. Here the tissue is maintained *in vacuo* at about −40°C until all the ice it contains has sublimed and has been condensed in a vapour trap maintained at a yet lower temperature. Alternatively, the water vapour may be absorbed into phosphorus pentoxide. The dried specimen may then be infiltrated with paraffin wax and sectioned in the usual way, though the sections cannot be flattened on water. Freeze-drying does not insolubilize proteins, so it is not, strictly speaking, a method of fixation. However, water-soluble substances of low molecular weight are not lost. A freeze-dried block may be chemically fixed in a suitable gas, such as formaldehyde, thus combining the advantages of the two procedures.

In **freeze substitution** the frozen specimen is dehydrated by leaving it in a liquid dehydrating agent, usually ethanol or acetone, at a temperature below −49°C. The liquid dissolves the ice but does not, at the low temperature, coagulate proteins. When dehydration is complete, the temperature is raised to 4°C for a few hours, to allow chemical fixation by the alcohol or acetone to take place. The block is then cleared and embedded in paraffin. Alternatively, *n*-Butanol, which is miscible with wax, can be used for simultaneous freeze-substitution and clearing. Sections of tissue prepared by the latter method are

preferred to cryostat sections for some techniques of enzyme histochemistry (Klaushofer & von Mayersbach, 1979).

2.2. Chemical methods of fixation

For most histological and histochemical purposes, liquid fixatives are used. These substances affect the tissues both physically and chemically. The principal physical changes produced are shrinkage or swelling, and many of the fixatives in common use are mixtures of different agents, formulated so as to balance these two undesirable effects. Most fixatives harden tissues. Moderate hardening is desirable for sectioning with a freezing microtome, or if embedding is to be in nitrocellulose or plastic, but can lead to difficulty in cutting wax-embedded material. Dehydration and infiltration with paraffin always produce some further shrinkage and hardening, whatever the state of the tissue when it came out of the fixative. The volume of a fixed, paraffin-embedded specimen is commonly 60–70% of what it was in life. Another important property of a chemical fixative is its rate of penetration. This rate determines the duration of fixation and the maximum permissible size of the specimen. These physical aspects of fixatives are reviewed by Baker (1958). Many chemical reactions are involved in fixation, and some of these will now be described. The chemistry of fixation has been reviewed by Baker (1958), Pearse (1980), Hayat (1981), Horobin (1982) and Bullock (1984).

2.3. General properties of fixatives

The structure of an animal tissue is determined largely by the configuration of its contained proteins. The main contributors to structure are the **lipoproteins**, which are major components of the plasmalemmae and membranous organelles of cells, the **fibrous glycoproteins** of such extracellular elements as collagen and basement membranes, and the **globular proteins**, which are dissolved in the cytoplasm and extracellular fluid. In some tissues, extracellular **mucosubstances** (e.g. chondroitin sulphates) also contribute substantially to the local architecture. Plant tissues are held together by the **cellulose** and other carbohydrates of the cell walls. All these substances must be stabilized by fixation. The nucleic acids and their associated nucleoproteins should also be preserved, as should the macromolecular carbohydrates (mucosubstances: comprising glycoproteins, proteoglycans, and at least one polysaccharide—glycogen) and, if their histochemical demonstration is required, the lipids. Fortunately, most fixatives render insoluble the proteins, nucleic acids and mucosubstances, though some may be more completely preserved than others by particular agents. Many fixatives do not directly affect lipids, whose preservation depends largely on the avoidance of agents that dissolve them.

2.3.1. Physical considerations

The **rate of penetration** will determine the size of a block to be fixed by immersion. Rapidly penetrating fixatives will usually fix in 24 h a specimen whose least dimension is 3–5 mm. For slowly penetrating fixatives the thickness of the block should not exceed 2 mm. The duration of fixation should not exceed 24 h except in the case of formaldehyde which takes a week to cause full stabilization of histological structure. Distortion due to slow penetration can be offset by perfusion of the fixative through blood-vessels or by injection into thin-walled cavities. **Shrinkage and swelling** are not necessarily detrimental to the quality of fixation, but must be allowed for in quantitative work. The overall change in size is easily determined by measuring appropriate dimensions of the fresh specimen and of the stained, mounted sections. It must not be assumed that all components of an organ or tissue will shrink or swell equally. Empty spaces due to unequal shrinkage of cells or larger regions of tissues are common artifacts, especially in paraffin sections. The consequences of **hardening** due to fixation have already been mentioned. When a specimen contains materials of widely varying hardness (e.g. glands and muscle), the embedding procedure should be chosen to suit the hard-

est component. Double-embedding (Chapter 4) is often the best method for such specimens.

2.3.2. Coagulants and non-coagulants

Fixatives that cause coagulation of cytoplasmic proteins destroy or distort organelles such as mitochondria and secretory granules but they do not seriously disturb the supporting extracellular materials, which are already partly solid before being fixed. It is thought that coagulant fixatives produce a sponge-like proteinaceous reticulum that is easily permeated by the large molecules of melted paraffin, so that sectioning of wax-embedded material is facilitated. Non-coagulant fixatives convert the cytoplasm into an insoluble gel in which the organelles are well preserved but which is thought to be less easily penetrated by paraffin. Artifactual shrinkage spaces and cracks are seen in paraffin sections of specimens fixed by these agents. (Despite the fact that its molecules are large, nitrocellulose causes less distortion of tissues fixed in non-coagulants than does paraffin.) Many fixative mixtures contain both coagulant and non-coagulant compounds and combine the advantages of both. For light microscopy it is necessary to settle for either a mixture that gives adequate cytoplasmic fixation (due mainly to non-coagulant insolubilization of protein) or one that provides superior structural preservation on a larger scale (for which coagulation is necessary). In electron microscopy the cytoplasmic disruption due to coagulant fixation is unacceptable, and non-coagulant agents must be used. Fortunately the plastics used as embedding media for electron microscopy cause much less distortion of the delicate architecture of tissues than does paraffin wax. The latter is still needed, however, for the larger specimens examined with the light microscope.

The nuclei of cells are deliberately stained in most histological preparations. All fixative mixtures should therefore contain a substance that either coagulates the chromatin or renders it resistant to extraction by water and other solvents (see Table 2.1, pp. 26–27). The chromosomes of dividing cells are shown to best advantage after fixation in agents that coagulate nucleoproteins and nucleic acids. Most fixatives do not react chemically with macromolecular carbohydrates and lipids, though these substances are often protected from extraction as a consequence of the insolubilization of associated proteins.

2.3.3. Effects on staining

Another important consideration is the effect of fixation on the subsequent reactivity of the tissues with dyes or histochemical reagents. For example, a glance at Table 2.1 (pp. 26–27) will show that osmium tetroxide should not be used to fix specimens intended for the staining of tissues with anionic (acid) dyes. The fixatives that interfere with staining are those that react chemically with the amino or carboxyl groups of proteins.

2.4. Individual fixative agents

2.4.1. Simple organic coagulants

This group includes liquids such as **acetone**, **ethanol**, and **methanol** that displace water from proteinaceous materials, thereby breaking hydrogen bonds and disturbing the tertiary structure to produce the change known as **denaturation**. Soluble proteins of cytoplasm are coagulated and organelles are destroyed. Nucleic acids are not precipitated and remain soluble in water. At low temperatures (below $-5°C$) ethanol precipitates many proteins without denaturing them. Precipitated protein that has not been denatured retains enzymatic or other biological properties, and remains soluble in water. Alcohols, acetone, and other such solvents extract much lipid from tissues. Carbohydrate-containing components, however, are largely unaffected. Ethanol and methanol make hepatic glycogen insoluble, but acetone does not do this.

Trichloroacetic acid (TCA; CCl_3COOH) is widely used by biochemists to precipitate proteins from solutions. The coagulation is probably due to electrostatic interaction of trichloroacetate anions with positively charged groups ($-NH_3^+$, etc.) of proteins. The highly non-polar Cl_3C- group enables the TCA molecule to penetrate into hydrophobic domains within proteins. The consequent combination of hydrophobic interaction and ionic attraction by the same ion is probably responsible for breaking the hydrogen bonds that hold the protein molecules in their normal conformations. TCA extracts nucleic acids, but only at higher concentrations and temperatures than those used in histological fixation.

Methanol, ethanol and acetone are used alone for fixing films and smears of cells, and previously unfixed cryostat sections. They are not suitable for blocks of tissue (unless very small) because they cause considerable shrinkage and hardening. Brief fixation below 5°C is compatible with some techniques of enzyme histochemistry (Chapter 14), because some proteins are precipitated without being denatured. TCA is a component of fixative mixtures, but is seldom used alone. Simple coagulants distort protein molecules without changing the sequences of their amino acids. Short antigenic sequences (epitopes) that are normally buried in the interior of large protein molecules can thereby be made more accessible to large antibody molecules, which are the reagents used in immunohistochemical methods (Chapter 19).

2.4.2. Mercuric chloride

Solutions of mercuric chloride contain molecules $Cl-Hg-Cl$ and hardly any free Hg^{2+} ions. With water there is slight formation of a hydroxo complex:

$$HgCl_2 + H_2O \rightleftharpoons HO-Hg-Cl + H^+ + Cl^-$$

but this hydrolysis is reversed in acid solutions. In the presence of chloride ions, as from added sodium chloride, a complex anion is produced:

$$HgCl_2 + 2Cl^- \rightleftharpoons [HgCl_4]^{2-}$$

The chemistry of fixation by mercuric chloride is poorly understood, but some inferences can be drawn from known reactions of the compound (see Whitmore, 1921; McAuliffe, 1977). Thus, with ammonium salts or amines:

$$HgCl_2 + 2NH_4^+ \rightleftharpoons Hg(NH_3)_2Cl_2 \downarrow + 2H^+$$

$$HgCl_2 + R\overset{+}{-}NH_3 \rightleftharpoons R\overset{+}{\underset{H_2}{-}}N-Hg-Cl + H^+ + Cl^-$$

Similar mercury-nitrogen bonds are formed with amides and amino acids. Cross-linking of two nitrogens by mercury can also occur. Thus, addition of mercury and cross-linking may account for the insolubilization of proteins by mercuric chloride. However, the carbon-nitrogen bonds are unstable in the presence of halide ions:

$$\rangle N-Hg-N\langle + 2Cl^- \rightleftharpoons 2 \rangle N + 2OH^- + HgCl_2$$

Acids or thiosulphate ions have effects similar to that of halide.

Bonds much more stable than those with nitrogen are formed with the sulphydryl groups of cysteine:

$$HgCl_2 + HS— \boxed{\text{PROTEIN}} \rightarrow$$

$$Cl—Hg—S— \boxed{\text{PROTEIN}} + H^+ + Cl^-$$

This reaction certainly occurs during fixation, but its significance in stabilizing structure is not known. Mercury adds to unsaturated linkages of lipids, but this addition is probably not important in fixation, except in relation to the plasmal reaction (Chapter 12).

Fixative mixtures containing mercuric chloride are notable for enhancing the brightness of subsequent staining with dyes. The reason for this is not known. A crystalline precipitate (of uncertain chemical composition but probably mostly mercurous chloride, Hg_2Cl_2) forms within mercury-fixed tissues and must be removed, before the sections are stained, by treatment with a solution of iodine, followed by sodium thiosulphate (see p. 47). Another important practical point is that $HgCl_2$ penetrates rapidly through the tissue and causes hardening, and the latter effect becomes excessive if the time of fixation is unduly prolonged. In old texts of histological technique, $HgCl_2$ is usually called "corrosive sublimate" or simple "sublimate". It is a very poisonous substance.

2.4.3. Picric acid

Picric acid is trinitrophenol. It is a much stronger acid than unsubstituted phenol in aqueous solution, owing to the electron-withdrawing effect of the three nitro groups on the hydroxyl group:

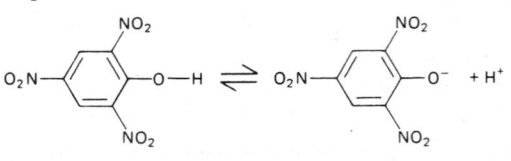

Trinitrophenol is a bright yellow solid, and is also used as a stain (see Chapters 5, 8). It is dangerously explosive when dry and is therefore stored under water. Stock bottles should be inspected from time to time and water added as necessary to give a layer about 2 cm deep on top of the powder.

A near-saturated aqueous solution of picric acid (pH 1.5–2.0) causes coagulation by forming salts (picrates) with the basic groups of proteins. Precipitation does not occur in a neutral solution, and neutralization allows precipitated proteins to redissolve. Tissues fixed in mixtures containing picric acid are usually transferred directly to 70% alcohol, to coagulate the precipitated protein. However, some authorities state (and I agree with them) that it makes no difference whether the specimens are washed in alcohol or water. When other fixatives (e.g. formaldehyde) are mixed with picric acid, it is unlikely that any proteins in the fixed tissue are still soluble in water yet coagulable by alcohol.

Picric acid is sufficiently strong (i.e. low pH due to dissociation into hydrogen ions and picrate ions) to bring about hydrolysis of nucleic acids, so fixatives containing it are avoided when DNA and RNA are to be studied histochemically. The blocks must remain in 70% alcohol (several changes) until as much as possible of the yellow colour is removed. Prolonged contact with picric acid, even in solid paraffin wax, may cause structural deterioration and poor staining, according to Luna (1968). When tissues are sectioned less than a month after fixing, persistence of picric acid does not matter. It is easily washed out of the sections by a dilute (e.g. 0.13% = one-tenth of saturated) aqueous solution of lithium carbonate (Li_2CO_3).

2.4.4. Acetic acid

Acetic acid does not fix proteins, but it coagulates nucleic acids. The mechanism by which this change is brought about is obscure. Like the rapid penetration and production of swelling (Table 2.1), it is a property of the undissociated acid, and not of the acetate ion. These properties are shared by other carboxylic acids that are miscible with both water and oils (Zirkle, 1933). Acetic acid is included in fixative mixtures to preserve chromosomes, to precipitate the chromatin of interphase nuclei, and to oppose the shrinking actions of other agents such as ethanol and picric acid.

2.4.5. Chromium compounds

The compounds of chromium used in fixation are chromium trioxide and potassium dichromate. These contain the metal in its highest oxidation state, + 6 (see Cotton & Wilkinson, 1980). Chromium trioxide dissolves in water to form the deep red-orange "chromic acid", which is completely ionized. The anions are $HCrO_4^-$ (hydrogen chromate) and $Cr_2O_7^{2-}$ (dichromate):

$$CrO_3 + H_2O \rightarrow H^+ + HCrO_4^-$$

$$2HCrO_4^- \rightleftharpoons Cr_2O_7^{2-} + H_2O$$

The position of the equilibrium of the second reaction is influenced by hydrogen ions, so that a high [H+] (= low pH) favours the formation of $HCrO_4^-$ (red), whereas $Cr_2O_7^{2-}$ (orange) predominates in moderately acidic solutions. If a dichromate solution is made alkaline (low [H+], high [OH−], high pH) the yellow chromate ion is generated:

$$Cr_2O_7^{2-} + 2OH^- \rightleftharpoons 2CrO_4^{2-} + H_2O$$

The strength of these Cr(VI) anions as oxidizing agents varies inversely with the pH (Waters, 1958). Acidic solutions oxidize many organic compounds, with concomitant reduction of Cr(VI) to Cr(III). The latter occurs as the chromic (Cr^{3+}) cation.

Strongly acid (pH < 3.5) solutions of chromium trioxide or potassium dichromate coagulate proteins, and chromatin. A reticulated texture is produced in the cytoplasm, and the chromosomes of dividing cells are well preserved, as is the mitotic spindle. DNA is partly hydrolysed by chromic acid so that it gives a positive reaction in histochemical tests for aldehydes. Any strong acid will hydrolyse DNA; the reaction is discussed in Chapter 9.

A change in the fixative properties of the dichromate ion occurs when the pH is higher than 3.5. The less acid solutions insolubilize proteins without coagulation, but do not fix nucleic acids (see Table 2.1). Alkaline solutions containing Cr(VI) (chromates) are not used as fixatives.

The chemistry of fixation of proteins by dichromate is poorly understood. Zirkle (1928) and Casselmann (1955) suggested that reduction of Cr(VI) by components of the tissue was followed by the formation of coordination compounds of Cr(III) with oxygen and nitrogen atoms of structural macromolecules. The chromic ion is notable for its ability to form six coordinate bonds with a great variety of ligands. The complexes are formed and broken much more slowly than are similar bonds with most other metals (Cotton & Wilkinson, 1980). They have such forms as

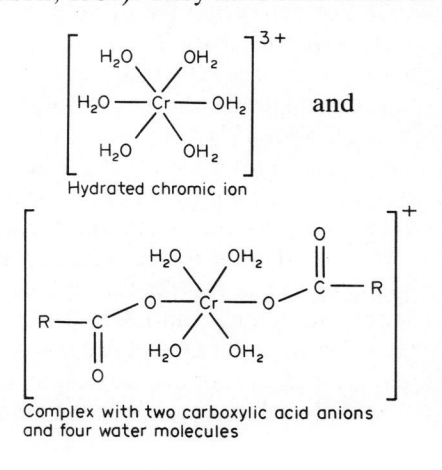

Hydrated chromic ion

and

Complex with two carboxylic acid anions and four water molecules

Chromic salts are used in chrome tanning (Gustavson, 1956), which is one of the industrial methods for conversion of collagen into leather, and in the hardening of gelatin for photographic emulsions (Burness & Pouradier, 1977; Pouradier, 1977). Hardened gelatin is insoluble in water. Both processes are comparable to fixation, and it is known that the greatest numbers of coordinate bonds are formed between chromium atoms and the carboxyl groups of amino acids of collagen:

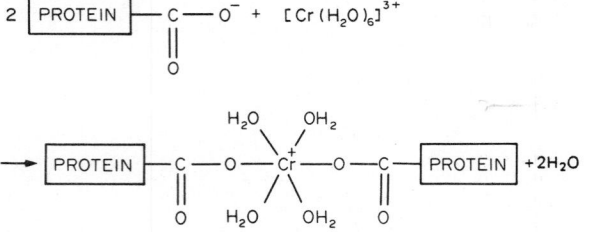

The pH of a chrome tanning solution is critical, because excessive acidity inhibits the formation of complexes, whereas neutrality favours the formation of large ions in which water molecules are cross-linked through chromium. Macromolecular complex ions cannot permeate the collagenous matrix (Britton, 1956; Thorstensen, 1969; Heidemann, 1988).

It might be expected that chromic salts would have the same fixative properties as dichromates, but this is not so (Kiernan, 1985). Solutions of chromic sulphate or acetate are destructive to animal tissues, whatever the pH. However, chromic ions do contribute significantly to the properties of a few excellent fixative mixtures. I have suggested (1985) that the formation of Cr(III) complexes occurs too slowly to be useful for primary fixation, but that cross-links are formed after primary stabilization by other agents such as formaldehyde. The protein-chromium-protein linkages protect the fixed tissue from further damage during the course of embedding in paraffin.

Oxidation by the dichromate ion is exploited in the chromaffin reaction, whereby catecholamines (adrenaline and noradrenaline) are transformed into brown compounds (Chapter 17). Gelatin can be hardened by oxidation in an alka-line medium (Tull, 1972), but there is no reason to believe that oxidation contributes to the fixation of tissues by chromium compounds. As well as fixing proteins, the dichromate ion can react with phospholipids in such a way as to make them insoluble in non-polar solvents (see Chapter 12). This lipid-stabilizing action does not occur with the usual 1 or 2 days of fixation at room temperature.

Material fixed in dichromate must be washed for 12–48 hours in running tap water before being transferred to the dehydrating alcohol, in order to avoid the reaction:

$$Cr_2O_7^{2-} + 3C_2H_5OH + 2H^+ \rightarrow$$
$$2Cr(OH)_3\downarrow + 3CH_3CHO + H_2O$$

which would produce an insoluble green precipitate in the tissue. Potassium dichromate is used in mixtures such as Zenker's and Helly's fluids. The first of these contains enough acetic acid to cause the fixative action to be that of a strongly acid solution. Helly's fluid is less acid, and the dichromate in it acts as a non-coagulant fixative. Both these mixtures also contain mercuric chloride, which is a coagulant. Helly's fluid contains formaldehyde too, and is therefore unstable owing to the reaction:

$$Cr_2O_7^{2-} + 3HCHO + 5H^+ \rightarrow$$
$$2Cr^{3+} + 3HCOO^- + 4H_2O$$

The reaction proceeds quite slowly, however, because [H+] in Helly's fluid is not very high. Both the dichromate and the formaldehyde have time to act upon the tissue before they are themselves changed into chromic and formate ions.

2.4.6. Osmium tetroxide

This substance, sometimes wrongly called "osmic acid", is a non-ionic solid, OsO_4, which is volatile at room temperature. The vapour is irritating and can cause corneal opacities. OsO_4 is soluble (without ionization) in water but much more soluble in non-polar organic solvents. It is somewhat unstable in solution, being reduced by traces of organic matter to the dioxide, $OsO_2.2H_2O$. This reduction is also brought about by alcohols, but not by carbonyl compounds such as formaldehyde, glutaraldehyde, and acetone, provided that these substances are pure. Osmium tetroxide is one of the oldest chemical fixatives, having been used for this purpose since 1865 (Maxwell, 1988).

OsO_4 is toxic, but it is not an environmental hazard because it is so quickly reduced (Smith *et al.*, 1978). However, it is wasteful to throw used solutions of this expensive substance down the sink. They should be stored and then chemically processed to recover the osmium (for methods, see Jacobs & Liggett, 1971 or Kiernan, 1978).

Although OsO_4 can react in the test-tube with proteins (especially sulphydryl groups) and carbohydrates (Bahr, 1954), it extracts quite large amounts of these substances from tissues during the course of fixation. Protein solutions are gelated but not coagulated by OsO_4, and some cross-linking occurs. The chemical reactions involved in these changes are not yet understood (Hopwood, 1977; Nielson & Griffith, 1979). The best understood fixative action is with the unsaturated linkages of lipids (see Adams *et al.*, 1967; Schroder, 1980). A cyclic ester is first formed, by addition:

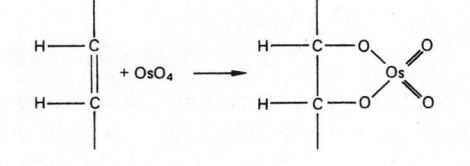

In this reaction, oxidation occurs at the carbon atoms (2 electrons lost), while the osmium is reduced, by gaining 2 electrons, so that its oxidation number changes from +8 to +6.

When two unsaturated linkages are suitably positioned they may be cross-linked (Wigglesworth, 1957). Chemical studies by Korn (1967) indicate that a diester is formed by the reaction

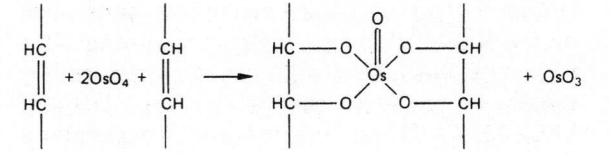

The oxide OsO_3 is unstable and disproportionates:

$$2OsO_3 \rightarrow OsO_2\downarrow + OsO_4$$

The osmium (VI) addition compounds and diesters are colourless or brown and may be soluble in organic solvents. The precipitated osmium dioxide is black and insoluble. Blackening of tissue fixed or stained with osmium tetroxide is increased in some circumstances with passage through alcohol, which may effect a reaction of the type:

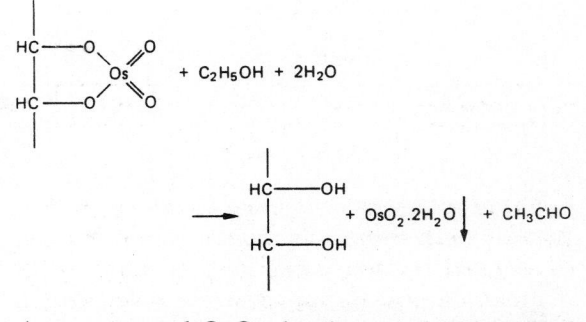

Any unreacted OsO_4 that has not been washed out of the tissue is similarly reduced:

$$OsO_4 + 2C_2H_5OH \rightarrow OsO_2.2H_2O \downarrow + 2CH_3CHO$$

Ordered arrays of lipid molecules are present predominantly in biological membranes, and OsO_4 is most valuable as a fixative for these structures, which are rendered insoluble, black and electron dense. Although individual membranes cannot be resolved with the light microscope, it is possible to see structures that are largely composed of membranous material, such as mitochondria and the myelin sheaths of nerve fibres. Solution of OsO_4 in non-polar substances, followed by its reduction, causes blackening of the contents of fat cells.

Treatment with OsO_4 largely abolishes the affinity of tissue proteins for anionic (acid) dyes, and normally acidophilic elements become stainable by cationic (basic) dyes. This change, which is in need of investigation, may be due to oxidation of terminal and side-chain amino groups of proteins with concomitant formation of carboxyl groups. By analogy with other oxidative deaminations the reaction would be expected to occur in three stages:

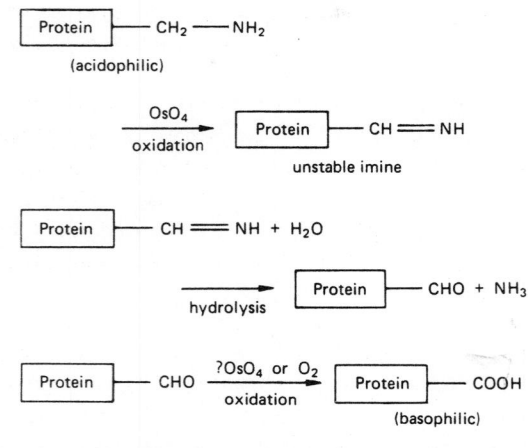

The product of reduction of OsO_4 is probably not $OsO_2.2H_2O$ because, if it were, proteins would be blackened. A soluble osmate, $[OsO_2(OH)_4]^{2-}$, or osmiamate, $[OsO_3N]^-$, may be the principal by-product of the reaction.

Osmium tetroxide penetrates blocks of tissue to a depth of only 0.5–1.0 mm, so its use as a fixative is limited to small blocks. By stabilizing membranes and gelating dissolved proteins, OsO_4 provides lifelike fixation of the internal structures of cells. The absence of morphological artifact has been demonstrated by microscopic observation of living cells during the course of fixation by OsO_4 (Strangeways & Canti, 1927; Policard, Bessis & Bricka, 1952). Unfortunately, however, pieces of tissue fixed in this agent have a crumbly consistency which is made worse by embedding in wax, and leads to the formation of cracks and shrinkage spaces. Osmium tetroxide may be used as a secondary fixative (post-fixation) after formaldehyde or glutaraldehyde, as in electron microscopy, and it may be used to stain frozen sections for unsaturated lipids. The vapour of OsO_4 is as effective as an aqueous solution, both as a fixative and as a stain. In several solutions for fixing and post-fixing specimens for electron microscopy, OsO_4 is mixed with other substances (see White *et al.*, 1979; Goldfischer *et al.*, 1981; Hayat, 1981; Emerman & Behrman, 1982; Carrapico *et al.*, 1984; De Bruijn *et al.*,

1984; Neiss, 1984), for the purposes of increasing electron-density and reducing the extraction of proteins. Apart from osmium-iodide mixtures (Chapter 18), such solutions are not used as fixatives for light microscopy.

(The uses of OsO_4 in the histochemical study of lipids are discussed in Chapter 12.)

2.4.7. Formaldehyde

Formaldehyde is a gas (B.P. $-21°C$) with the structural formula:

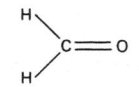

It is available to the histologist as a solution (**formalin**) containing 37–40% by weight of the gas in water, and as a solid polymer, **paraformaldehyde**, which is $HO(CH_2O)_nH$, n being 6–100. Paraformaldehyde is also seen as the white precipitate that forms in old bottles of formalin. Old, milky solutions can be clarified by heating for 30 minutes in a Kilner jar in an autoclave (Cares, 1945), but this is rarely done because formalin is not expensive. Although formaldehyde is the simplest of the aldehydes, its chemistry is quite complicated (see Walker, 1964; Fox *et al.*, 1985).

In aqueous solutions formaldehyde is present as methylene hydrate (= methylene glycol), the product of the reaction:

$$H_2C = O + H_2O \rightleftharpoons HOCH_2OH$$

Formaldehyde   Methylene hydrate

The equilibrium lies far to the right, and hardly any true formaldehyde is present in the solution. However, the chemical reactions of methylene hydrate are those of formaldehyde in the presence of water, so it is usual to speak and write of aqueous solutions as if they contained formaldehyde. Formalin also contains soluble polymers of the form $HO(CH_2O)_nH$ (where $n = 2$–8), known as lower polyoxymethylene glycols. Continued addition of methylene hydrate molecules occurs spontaneously, and this is the reason for the precipitation of paraformaldehyde in formalin that has been stored for a long time. Formalin also contains methanol (commonly about 10% v/v), which is added as a stabilizer to inhibit polymerization. Methanol does this by forming with formaldehyde a hemiacetal (methylal), which is more stable than methylene hydrate.

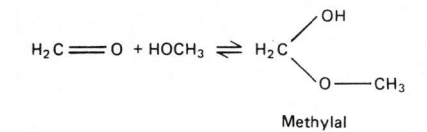

Methylal

The polymers are hydrolysed when formalin is diluted with an excess of water. Thus, for trioxymethylene glycol, $HO(CH_2O)_3H$,

$$HO-\underset{H_2}{C}-O-\underset{H_2}{C}-O-\underset{H_2}{C}-OH + 2H_2O \rightleftharpoons 3HO\underset{H_2}{C}OH$$

The equilibrium is displaced to the right because of the high concentration of water, but the reaction is very slow (taking some weeks for completion) between pH 2 and 5, as in non-neutralized formalin. The depolymerization occurs quite rapidly, however, in neutral media. If formalin is to be used as a fixative in simple aqueous solution it should be diluted several days in advance, but when buffered to approximate neutrality it may be used immediately. Neutrality may also be achieved by keeping some marble chips (calcium carbonate) in the bottom of the bottle of diluted formalin.

Formalin deteriorates during storage as the result of a Cannizzaro reaction:

$$2HOCH_2OH \longrightarrow CH_3OH + HCOOH + H_2O$$

The pH of the solution falls, though very slowly, as more formic acid is produced (Fox *et al.*, 1985). With other aldehydes, Cannizzaro reactions are significant only in alkaline media. Oxidation of formaldehyde by atmospheric oxygen (which would also generate formic acid) is exceedingly slow, and is not a significant cause of deterioration.

Formaldehyde solutions for use as fixatives are also made by depolymerizing paraformaldehyde. This substance dissolves very slowly in water but more quickly in near-neutral buffer solutions. Methylene hydrate made in this way does not contain methanol or formic acid, and is preferred to diluted formalin as a fixative for histochemistry and electron microscopy. Paraformaldehyde is also depolymerized by heating. Small specimens, including those of freeze-dried tissues, can be fixed by exposure to the vapour, which is gaseous formaldehyde, at 50–80°C.

The content of formaldehyde in a fixative is best denoted by stating the percentage by weight of the gas rather than the amount of formalin used in preparing the mixture. Thus, "4% formaldehyde" is preferred to "10% formalin" (for the same solution), though the latter designation is in common use.

Formaldehyde reacts with several parts of protein molecules (see Walker, 1964; Hopwood, 1969; Pearse, 1980, for more information). The methylene glycol molecule adds to many functional groups to form hemiacetals and related adducts. For example, with primary amines (N-terminal amino acids and lysine side-chains):

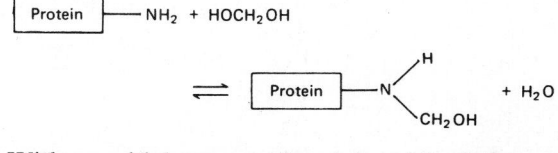

With guanidyl groups of arginine side-chains:

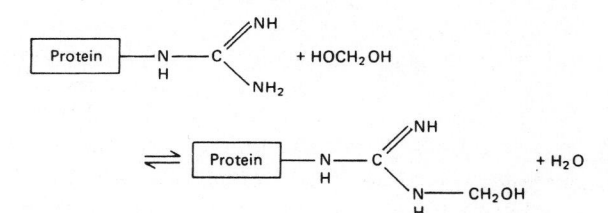

With sulphydryl groups of cysteine:

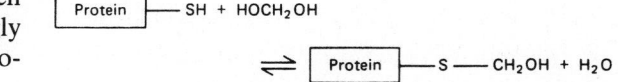

With aliphatic hydroxyl groups (serine, threonine):

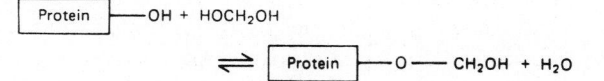

With amide nitrogen (at accessible peptide linkages):

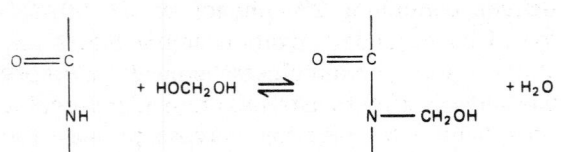

These reactions are all readily reversible by washing in water or alcohol, so the simple addition of formaldehyde to —NH$_2$, —NHC(NH)NH$_2$, and —SH groups of proteins does not contribute significantly to fixation. However, the hemiacetal-like adducts all have free hydroxymethyl groups and these are capable of further reaction with suitably positioned functional groups of proteins:

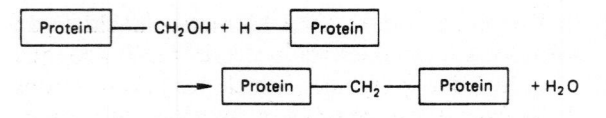

Thus, different protein molecules can be joined together by methylene bridges, which are chemically stable. Formaldehyde is used in the tanning of collagen to produce leather. Investigations of this industrial process (see Gustavson, 1956) have led to the conclusion that although cross-links of many kinds are possible, the great majority are formed between ε-amino groups of lysine and the amide nitrogen atoms of peptide linkages:

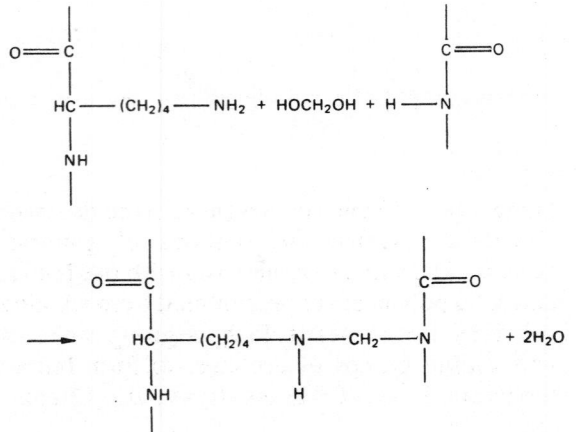

The formation of methylene bridges in this reaction is probably largely responsible for the cross-linking of protein molecules that constitutes fixation by formaldehyde. Many ε-amino groups will not be close to peptide linkages and will therefore be able to form only the unstable hemiacetal-like adducts. Thus, the histochemical reactivity of primary amines (including the binding of anionic dyes) will be only slightly depressed in the fixed tissue. After prolonged storage (e.g. 6 months to 10 years) in formalin, the free —NHCH$_2$OH groups may be oxidized by the atmosphere to the more stable —NHCOOH. This change would account for the eventual loss of stainability by anionic dyes that occurs in old specimens.

Cross-linking of protein molecules by formaldehyde is much slower than the chemical reactions of other fixative agents, and requires 1–2 weeks for completion at room temperature. For histochemical purposes, tissues are commonly fixed for 12–24h at 4°C, but many non-histochemical methods especially for the nervous system, work better after complete fixation. Sufficient hardening for the cutting of frozen sections is usually attained after 24h. Very long periods of storage in formaldehyde solutions result in excessive hardening, loss of stainability of nuclei, and (with acidic solutions) deposition of brown "formalin pigment". This is a haematin formed by acid degradation of haemoglobin. It can be removed by treating the sections with an alcoholic solution of picric acid or with any of a variety of oxidizing agents or alkalis (see Lillie & Fullmer, 1976). The "pigment" does not form if the formaldehyde solution is buffered to approximate neutrality.

Formaldehyde preserves most lipids, especially if the fixing solution contains calcium ions which, for ill-understood reasons, reduce the solubilities of some phospholipids in water. The only chemical reactions of formaldehyde with lipids under ordinary conditions of fixation are (1) addition to the amino groups of phosphatidyl ethanolamines, which is probably reversible by washing in water, and (2) prevention of the histochemical reactivity of plasmalogens owing to oxidation, probably to a glycol, of the ethylenic linkage next to the ether group. The latter reaction may be brought about by atmospheric oxygen rather than by formaldehyde. With prolonged fixation in formaldehyde solutions (3 months to 2 years), other double bonds are attacked, and several products are formed, all of which are more soluble in water than the original lipids (Jones, 1972).

Formaldehyde does not react significantly with carbohydrates. All the common mucosubstances can be demonstrated after fixation with formaldehyde, though appreciable quantities of glycogen are lost.

A neutral, buffered aqueous solution (pH 7.2–7.4) containing 2–5% of formaldehyde, is the most generally useful (though not the best) fixative for most histological and histochemical purposes. When histochemical methods are not to be used, mixtures such as Bouin's fluid and SUSA (see Section 2.5.3.) are preferable to formaldehyde because (a) they protect the tissue against damaging effects of embedding in wax, and (b) staining with almost all dyes is brighter than after formaldehyde alone.

The speed and quality of fixation with neutral formaldehyde solution can be improved by

2.4.8 and 2.4.9(g)). Hopwood *et al.* (1989) recommend primary fixation in 4% formaldehyde containing 2% phenol at pH 7.0, followed by secondary fixation at pH 5.5. Many stains and immunohistochemical methods worked well after such treatment, and reasonable preservation for electron microscopy was also obtained.

2.4.8. Glutaraldehyde and other aldehydes

Glutaraldehyde,

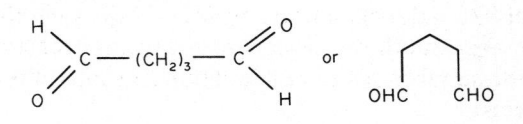

is the most widely used bifunctional aldehyde fixative. Glutaraldehyde polymerizes in aqueous solution. The polymers, which are of a different type from those present in solutions of formaldehyde, are formed by aldol condensation. This reaction yields a product in which an olefinic double bond (C=C) is conjugated with the carbonyl (C=O) double bond of the aldehyde group (Monsan, Puzo & Marzarguil, 1975):

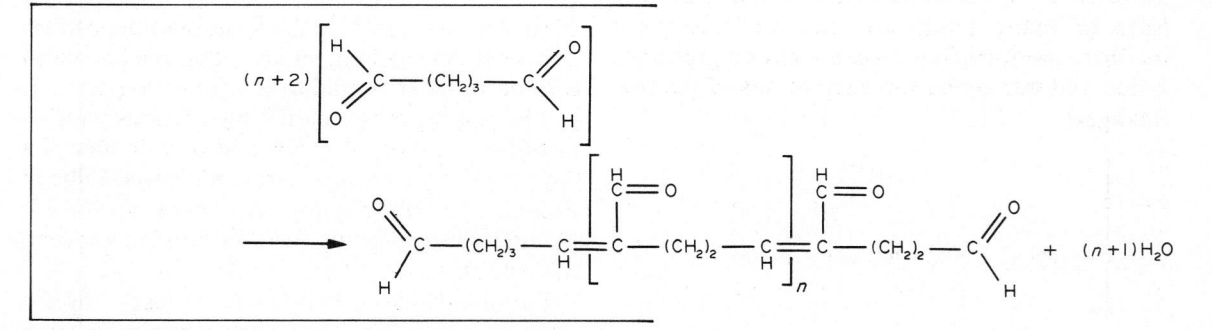

addition of an aliphatic amine such as cyclohexylamine or lysine (Luther & Bloch, 1989), or of phenol (Hopwood & Slidders, 1989). These compounds react with formaldehyde and their presence probably results in the formation of polymeric cross-links of variable length within the tissue, providing fixation comparable to that obtainable with glutaraldehyde (see Sections

Dimers ($n = 0$) and trimers ($n = 1$) are the most abundant polymers in solutions of glutaraldehyde. The value of n increases with rise in pH, and solid polymers are precipitated from alkaline solutions. Both types of aldehyde group can react with amino groups of proteins, to form **imines** (compounds with C=N bonds; see also Chapter 10):

$$R\text{---}CHO + H_2N\text{---}R \rightleftharpoons RC{=}\underset{H}{N}\text{---}R + H_2O$$

The imines that form from the terminal aldehyde groups of poly(glutaraldehyde) are, like most aliphatic imines, unstable. However, the aldehyde groups within the repeating units of the polymer can form imines that are stable, owing to conjugation of their C=N bonds with the C=C bonds of the polymer. Consequently, the only significant reaction of poly(glutaraldehyde) with amino groups of protein is:

chemical production of aldehydes. Free aldehyde groups can also bind, non-specifically, any reagents that are proteins, including enzymes and antibodies. Furthermore, the reducing properties of the aldehyde group can lead to false-positive results in autoradiography. **Aldehyde groups introduced by the fixative must be irreversibly chemically blocked before any of the above methods can be used.** Suitable blocking procedures are described in Chapter 10.

Glutaraldehyde penetrates tissues slowly, so it

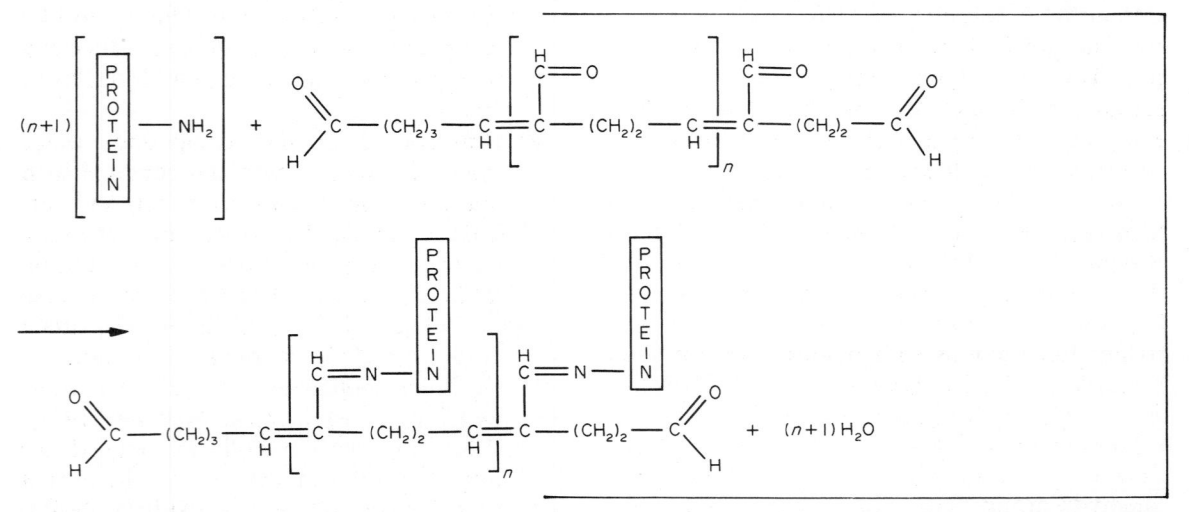

Like formaldehyde, glutaraldehyde probably also combines with reactive groups of proteins other than the ε-amino of lysine, but the reactions have not been studied in the context of histological fixation.

Mid-chain aldehyde groups of the poly(glutaraldehyde) molecule are the only ones involved in fixation (Monsan *et al.*, 1975), so the terminal aldehyde groups remain free, as do any mid-chain groups that have not combined with reactive side-chains of proteins. Consequently, **glutaraldehyde-fixed tissues are full of artificially introduced aldehyde groups, which react with many histochemical reagents.** This phenomenon must be taken into account when sections of glutaraldehyde-fixed material are to be treated by the Feulgen or periodic acid-Schiff procedures, in which the detection of DNA or of carbohydrate-containing substances is made possible by the

should be perfused through the vascular system if possible. Cross-linking of proteins occurs much more rapidly than with formaldehyde: fixation is complete after only a few hours. Tissues fixed in glutaraldehyde are more strongly stabilized by cross-linking than those fixed in formaldehyde, and this is probably the reason why ultrastructural features are so well preserved by the former substance. However, glutaraldehyde causes even more difficulty than does formaldehyde with the sectioning paraffin-embedded material.

The activities of some enzymes are preserved after brief fixation in glutaraldehyde, but formaldehyde usually causes less inhibition, probably because it forms fewer cross-links between protein molecules. Cross-linking interferes chemically with the antigenicity of proteins, and also retards the passage of large antibody molecules into sections of fixed tissue. Consequently, glutaraldehyde should be avoided if possible when immunohistochemical staining methods are to be used.

A solution containing glutaraldehyde and formaldehyde, buffered to pH 7.2–7.4, is the most widely used fixative for electron microscopy. The mixture produces much more chemical modification of proteins than either of the aldehydes acting alone (Kirkenby & Moe, 1986), and can even cause insolubilization of free amines and amino acid ions (Conger, Garcia, Lossinsky & Kaufmann, 1978). Aldehyde fixatives do not add electron-opacity to tissues, so post-fixation in osmium tetroxide and contrast-staining with salts of other heavy metals are usually practised when ultrathin sections are to be examined by electron microscopy.

Several other aldehydes have been used as fixatives. A few examples follow. **Chloral hydrate**, Cl_3C—$CH(OH)_2$, is the stable hydrated derivate of trichloroacetaldehyde. It is included in several older fixatives, especially for nervous tissue. Its reactions with tissues have not been critically examined. **Acrolein**, H_2C=$CHCHO$, is a very toxic unsaturated aldehyde occasionally used as a fixative for electron microscopy. Reaction occurs with the olefinic as well as with the aldehyde group, with resultant cross-linking of proteins. The actions of acrolein on tissues closely resemble those of glutaraldehyde. **Hydroxyadipaldehyde**, OHC—$CH(OH)$—$(CH_2)_3$—CHO, and **crotonaldehyde**, H_3C—CH=CH—CHO, were introduced as fixatives for electron microscopy by Sabatini, Bensch & Barnett (1963), at the same time as glutaraldehyde. They cause less inhibition of enzymes than glutaraldehyde or acrolein, but this is associated with inferior ultrastructural preservation.

2.4.9. Other fixative agents

The major individual fixatives have now been described, but many other substances have also been used, some in mixtures that are now obsolete or at least out of fashion, and others in more recent years for special purposes. These minor fixative agents include:

(a) Mineral acids, which also serve as coagulants, e.g. H_2SO_4, HNO_3.

(b) Organic protein coagulants, e.g. tannic acid (Chaplin, 1985), *p*-toluenesulphonic acid (Malm, 1962).

(c) Several metal ions and complexes which cause precipitation of proteins, e.g. Cu^{2+}, Cd^{2+}, $[PtCl_6]^{4-}$, Pb^{2+}, $[UO_2]^{2+}$, Co^{2+}. See Osterberg (1974) for a review of metal ion-protein interactions.

(d) Cationic surfactants (detergents) and cationic dyes, which form insoluble salt-like complexes with polyanions, especially with proteoglycans (Williams & Jackson, 1956).

(e) Bifunctional organic compounds other than aldehydes. These can combine with and cross-link amino, hydroxyl, and carboxyl groups. Examples are: cyanuric chloride, carbodiimides, imido esters, quinones, diethyl pyrocarbonate (Pearse & Polak, 1975; Bu'Lock *et al.*, 1982; Robinson, 1987; Panula *et al.*, 1988).

(f) Sodium periodate with lysine and paraformaldehyde (McLean & Nakane, 1974). With this combination, introduced for immunohistochemical studies, the idea is that the periodate ion oxidizes carbohydrate side-chains of glycoproteins to aldehydes (see Chapter 11), which then combine with the two amino groups of lysine (see Chapter 10). Thus, glycoprotein molecules, especially those on the surfaces of cells, should be cross-linked. However, Hixson *et al.* (1981) have shown that the lysine in the mixture reacts immediately with the formaldehyde, forming polymeric products, and that glycoproteins are cross-linked after treatment with periodate alone. They suggest that periodate-generated aldehyde groups combine with the amino groups of nearby protein molecules, and that the formaldehyde-lysine compounds form longer cross-links than does formaldehyde alone, thus favouring the permeation of the tissue by large antibody molecules.

None of these fixatives has yet acquired great importance, so they will not be discussed further. Recent research has been directed towards fixatives that would give adequate structural preservation at the level of the electron microscope without unduly disturbing the activities of enzymes and the antigenic properties of substances to be studied by immunohistochemical techniques.

2.4.10. Chemically inactive ingredients

In addition to one or more of the components discussed above, an aqueous fixative mixture commonly includes "indifferent" substances which influence its osmotic pressure and its pH.

The most rapidly penetrating component of an aqueous fixative is water, so the central parts of a specimen are likely to be bathed in a hypotonic medium before they are fixed. This would lead to swelling or rupture of cells and disorganization of their surrounding connective tissue. To prevent such damage, unreactive salts with small, rapidly diffusing ions (e.g. sodium chloride or sulphate) are incorporated into the fixative mixture. Sucrose is also used to raise the osmotic pressure, but its molecules are fairly large, so it should probably be employed only in fixatives for vascular perfusion or for immersion of very small specimens. Most histologists and electron microscopists agree that the best results are obtained when the osmotic pressure due to the chemically inactive components is slightly higher than that of the extracellular fluid (Schook, 1980; Hayat, 1981). The additional osmotic effect of the active ingredients is not taken into account, because the osmotic properties of cell membranes are changed by contact with such substances.

Non-aqueous fixatives coagulate the tissue as they penetrate, so for these the control of osmotic pressure is unnecessary. The solvent penetrates more rapidly than any dissolved ingredients, however. For example, the centre of a large specimen immersed in an alcoholic solution of picric acid will be fixed primarily by the alcohol alone.

The pH of a fixative must be appropriate for the chemical reactions of fixation. Formaldehyde, glutaraldehyde, and OsO_4 are all fully active at and around neutrality, so their solutions may, with advantage, be buffered to approximately the same pH as the extracellular fluid. This is pH 7.2–7.6 for mammals. The prevention of an abrupt change in pH during fixation probably minimizes fine structural disturbance within dying cells. Some of that disruption may be due to spontaneous acidification by metabolites such as carbon dioxide and lactic acid. Buffering is of greatest value in the case of formaldehyde, which reacts with proteins much more slowly than do other fixatives.

Most fixatives other than the three mentioned in the preceding paragraph are active only when their solutions are considerably more acid than the extracellular fluid. The correct pH is produced when the fixative mixture is made and a buffer is not needed. In the design of new fixatives, care must be taken to avoid irrational combinations. For example, the advantages of mixing acetic acid with formaldehyde or osmium tetroxide would be lost if the mixture were then carefully buffered to pH 7.4. The effect of pH on the type of fixation produced by the dichromate ion has already been described.

Solutions of indifferent salts, buffers, and sucrose are often used to wash fixed specimens, especially in processing for electron microscopy. Completely fixed tissues are, however, no longer responsive to changes in the osmotic pressures of the solutions in which they are immersed. Plain water is satisfactory for washing all specimens intended for ordinary histological or cytological examination with the light microscope. Isotonic buffers are recommended, however, for washing specimens that have received only partial fixation preparatory to the application of histochemical methods for enzymes or to subsequent processing for electron microscopy. Paljarvi et al. (1979) have shown that tissues are still susceptible to damage by osmotic stress after fixation for 24h in buffered 4% formaldehyde, though 3% glutaraldehyde used for the same length of time confers complete stabilization.

2.4.11. Summary of some properties of individual fixatives

The more important effects of the major fixative agents upon tissues are presented briefly in Table 2.1. pp. 26–27.

2.5. Choice of fixative

A fixative is chosen according to the structural or chemical components of the tissue that are to be demonstrated. Often a mixture of different agents is employed in order to offset undesirable effects of individual substances and to obtain more than one type of chemical fixation. A few commonly used liquids will now be described. Fixatives for specialized purposes are described in other chapters.

2.5.1. Non-aqueous fixatives

The following mixtures are used for general histology and for the preservation of nucleic acids and macromolecular carbohydrates.

Clarke's fluid

Absolute ethanol: 3 volumes ⎫ Mix just
Glacial acetic acid: 1 volume ⎭ before using

One of the oldest fixatives, but notable for excellent results with subsequent paraffin embedding. In semi-quantitative comparisons of fixatives for light microscopy, this crude mixture always gets a high score for micro-anatomical preservation (see Baker, 1958; Kiernan, 1985).

Carnoy's fluid

Absolute ethanol: 60 ml ⎫ Mix just before
Chloroform: 30 ml ⎬ using
Glacial acetic acid: 10 ml ⎭

Like Clarke's, this is a rapidly penetrating fixative, which coagulates protein and nucleic acids

and extracts lipids. Many carbohydrate components are also preserved. Blocks of tissue up to 5 mm thick are fixed for 6–8 h. Fixation for more than 18 h can result in hydrolysis of nucleic acids, with loss of RNA. This effect can be suppressed by using 5 ml instead of 10 ml of acetic acid. The fixative is then called **modified Carnoy** (James & Tas, 1984).

Alcohol–formalin–acetic mixtures

Ethanol (95–100%):   85 ml ⎫
Formalin (37–40%   ⎪ Mix before
  HCHO):   10 ml ⎬ using
Glacial acetic acid: 5 ml ⎭

This is one of many similar "A.F.A." mixtures. The amount of acetic acid or formalin may be varied, and methanol is sometimes used instead of ethanol. Penetration is rapid; specimens are fixed for 4 to 48 h. Smears or whole mounts on slides are fixed for 10–30 min. Probably the formaldehyde does not contribute much to fixation when the duration is short. As with Carnoy, A.F.A. preserves general morphology, nucleic acids and carbohydrates well. Most lipids are extracted. After fixation the specimens are transferred to 95% alcohol, and then processed for embedding in wax.

2.5.2. Aqueous formaldehyde solutions

Neutral buffered formalin

This solution of formaldehyde in a 0.075 M phosphate buffer was introduced by R. D. Lillie, after whom it is sometimes named.

Sodium phosphate, monobasic
  ($NaH_2PO_4.H_2O$):   4.0 g
Sodium phosphate, dibasic
  (Na_2HPO_4):   6.5 g
Either,
  dissolve in 750 ml of water, add 100 ml of formalin (37–40% HCHO), and make up to 1000 ml with water;

TABLE 2.1. *Properties of individual fixative agents* (Arrows indicate increase (↑) or decrease (↓). Plus signs

	Ethanol, methanol, acetone	Acetic acid	Trichloracetic acid	Picric acid
Usual concentration (alone or in mixtures) (%)	70–100	5–35	2–5	0.5–5
Penetration	Rapid	Rapid	Rapid	Slow
Change in volume of tissue	↓ + + +	↑ + + +	↑ + +	↓ +
Hardening	+ + +	Nil	Nil (?)	+
Fixative effect on proteins	Non-additive; coagulant	Nil	Non-additive; coagulant	Additive; coagulant
Action on nucleic acids	Nil	Precipitation	Some extraction	Partial hydrolysis
Effects on carbohydrates	Nil	Nil	Nil	Nil
Effects on lipids	Some extraction	Nil	Nil	Nil
Effects on enzyme activities	Some preserved if cold	Inhibition (?)	Inhibition	Inhibition
Effects on organelles (especially mitochondria)	Destroyed	Destroyed	Preserved	Distorted
Staining with: anionic dyes	Satisfactory	Poor	Satisfactory	Good
cationic dyes	Satisfactory	Good	Good	Satisfactory
Value as sole fixative agent	For small blocks (for glycogen) or cryostat sections	Nil	Not used alone (but see Gabe, 1976)	Poor

or,

dissolve in water to 1000 ml, heat to 60–70°C (in fume hood), add 40 g of paraformaldehyde powder, stir well, cool and filter.
(Can be kept for several weeks. The pH is 7.2–7.4.)

Some batches of paraformaldehyde are unsuitable because they dissolve too slowly. The concentration of formaldehyde is not critical, and may range from 2.5% to 10%. The mixture made from formalin is used when the presence of traces of methanol and formate can be tolerated (as in most histological work). Paraformaldehyde is always used when the fixative is for electron microscopy, and often when it is for enzyme histochemistry or immunohistochemistry. Fix for 12–24 h at 4°C to preserve enzymes and antigens. For general histology and traditional neurological staining, fix for at least one week at room temperature. Before dehydrating wash in water to remove the buffer salts, which are insoluble in alcohol.

The addition of 10 ml of 25% aqueous **glutaraldehyde** to 90 ml of the above mixture gives a solution similar to Karnovsky's (1965) fixative that preserves many intracytoplasmic structures such show relative magnitudes of effects.   ? indicates uncertainty)

Formaldehyde	Glutaraldehyde	Mercuric chloride	Dichromate ion pH < 3.5 (CrO$_3$)	Dichromate ion pH > 3.5 (K$_2$Cr$_2$O$_7$)	Osmium tetroxide
2–4	0.25–4	3–6	0.2–0.8	1–5	0.5–2
Fairly rapid	Slow	Fairly rapid	Slow	Fairly rapid	Slow
Nil	Nil	↓ +	↓ +	Nil	Nil
+ +	+ +	+ +	+ +	+	+
Additive; non-coagulant	Additive; non-coagulant	Additive; coagulant	Additive; coagulant	Additive; non-coagulant ·	Non-coagulant (see p. 18); some extraction
Slight extraction	Slight extraction	Coagulation	Coagulation; some hydrolysis	Some extraction	Slight extraction
Nil	Nil (?)	Nil	Oxidation	Nil	Some oxidation (?)
Nil (but see p. 21)	Similar to formaldehyde	Plasmal reaction	Oxidation of C=C	(see p. 16)	Addition to and oxidation of C=C
Some preserved if cold and short time	Most are inhibited	Inhibition	Inhibition	Inhibition	Inhibition
Preserved	Well preserved	Preserved	Considerable distortion	Very slight distortion	Well preserved
Rather poor	Rather poor	Good	Satisfactory	Good	Acidophilia changed to basophilia
Good	Satisfactory	Good	Satisfactory	Satisfactory	Satisfactory
Useful	Rarely used alone	Poor	Poor	Poor	Specialized applications only

as mitochondria. These organelles may be further stabilized for paraffin embedding by post-fixing small pieces of the tissue in osmium tetroxide (1% aqueous, 6h) or potassium dichromate (3% aqueous, 7 days).

This is devised to insolubilize phospholipids prior to cutting frozen sections. Specimens are fixed for 1–3 days. (See also Chapter 12.)

2.5.3. *Aqueous fixatives for general histology*

Formal-calcium

Formalin (37–40% HCHO):	100ml	Prepare just before using
Calcium acetate, (CH$_3$COO)$_2$Ca.H$_2$O	20 g	
Water:	to 1000 ml	

Bouin's fluid

Saturated aqueous picric acid:	750ml	Keeps indefinitely
Formalin (40% HCHO):	250ml	
Glacial acetic acid:	50ml	

Not used in histochemistry but preserves morphological features, especially of connective tissue, well. This fixative is valuable for purely histological work, because physical distortion of tissues is minimal, but intracellular structures other than nuclei are poorly preserved. Specimens are usually fixed in Bouin for 24h, but material stored in it for several months is sometimes still usable.

Gendre's fluid (alcoholic Bouin)

Saturated alcoholic picric acid:	800ml	Keeps
Formalin (40% HCHO):	150ml	indefinitely (but see below)
Glacial acetic acid:	50ml	

(The alcoholic picric acid solution is made up in 90–95% ethanol or industrial methylated spirit.)

This mixture can be kept indefinitely, but its chemical composition changes considerably. The concentrations of ethanol and formaldehyde decline, and there is concomitant improvement of the fixative properties, at least for some invertebrate tissues (Gregory, Greenway & Lord, 1980). Gregory (1980) recommends an "**artificially aged alcoholic Bouin**" with the following composition:

Formalin (40% HCHO):	0–15ml
(Stainability of tissue can be modified by varying the amount of formalin in the mixture)	
Ethanol:	25ml
Glacial acetic acid:	5ml
Ethyl acetate:	5ml
Dioxymethane:	15ml
(If not available, increase ethyl acetate to 25ml)	
Picric acid:	0.5g
Water:	to 100ml

These non-aqueous fixatives are similar to Bouin, but also immobilize some water-soluble carbohydrates (glycogen; mast cell granules of some species) that dissolve in aqueous fixatives. Specimens are fixed overnight (16–20h) and then washed in several changes of 95% alcohol.

Heidenhain's "SUSA"

Mercuric chloride (HgCl$_2$):	45g	
Sodium chloride (NaCl):	5g	
Trichloroacetic acid:	20g	Keeps indefinitely
Glacial acetic acid:	40ml	
Formalin (40% HCHO):	200ml	
Water:	to 100ml	

Pieces of tissues are fixed for no more than 24h and then transferred directly to 95% alcohol, supposedly to avoid swelling. Stains for nuclei, cytoplasm, and connective tissue all work brightly after fixation in SUSA, as do some histochemical methods for carbohydrates. **Mercurial deposits must be removed (see Chapter 4) prior to staining.**

Helly's fluid

Stock solution:

Mercuric chloride (HgCl$_2$):	50g	
Potassium dichromate (K$_2$Cr$_2$O$_7$):	25g	Keeps indefinitely
Sodium sulphate (Na$_2$SO$_4$.10H$_2$O):	10g	
Water:	to 1000ml	

Working solution:

Immediately before use, mix 5ml of formalin (40% HCHO) with 100ml of the stock solution.

Helly's fluid is used mainly to fix cytoplasmic elements such as mitochondria and secretory granules, as in endocrine organs and haemopoietic tissue. The non-coagulant components (formaldehyde, dichromate) offset the coagulant action of the mercuric chloride so that cytoplasm is not coarsely coagulated. The mixture goes green and murky on standing, owing to reduction of dichromate to chromic salts by formaldehyde. Fixation should not be for more than 12–24h. **It is important to wash out all the dichromate before dehydrating, and to remove mercury deposits before staining.**

Zenker's fluid

This fixative is similar to Helly's, except that the formalin is replaced by 5 ml of glacial acetic acid. This mixture is stable indefinitely. Because of the low pH at which the dichromate is used, cytoplasmic proteins are rather coarsely coagulated, and most organelles are not preserved. Helly and Zenker are excellent for morphological work, but are not compatible with many histochemical techniques.

For embryonic tissues, structural preservation by Zenker (6 hours) can be improved by postfixing for 7 days in neutral, buffered formaldehyde (Howard *et al.*, 1989). Possibly the formaldehyde reduces residual Cr(VI) in the tissue to Cr(III), which then causes additional cross-linking of proteins (see Section 2.4.5).

2.5.4. Traditional cytological fixatives

Although intracellular structures are best studied with the electron microscope, most of the components of the nucleus and cytoplasm were discovered long before that instrument was invented. Different fixatives must be used for the examination of chromosomes or of cytoplasmic organelles with the light microscope. The nucleoprotein and DNA of chromosomes are rendered insoluble by either acetic acid or chromic acid (pH $<$ 3.5). The latter also preserves the mitotic spindle, but largely destroys the cytoplasm. Neutral dichromate (pH $>$ 3.5) is included in solutions for the fixation of organelles. Osmium tetroxide is a common component of both types of cytological fixative. The mixtures are made up from stock aqueous solutions of chromium trioxide (5%), potassium dichromate (5%) and osmium tetroxide (2%). The pieces of tissue must be tiny (1mm or less in thickness). Fixation for 6–18h is followed by thorough washing in water to remove unbound chemicals. Very clean glass (not plastic) containers must be used for all solutions containing osmium tetroxide.

Altmann's fixative

2% osmium tetroxide:	5.0ml
5% potassium dichromate:	5.0ml

Mix before using, though it keeps indefinitely in a clean, tightly capped bottle. This cytoplasmic fixative can also be used to post-fix material fixed in a buffered glutaraldehyde or formaldehyde-glutaraldehyde mixture.

Flemming's strong fluid

2% osmium tetroxide:	2.0ml
5% chromium trioxide:	1.5ml
Glacial acetic acid:	0.5ml
Water:	to 10.0ml

Mix before using, but it is stable for several days. This fixative is for mitotic and meiotic nuclei.

2.5.5. Fixatives for immunohistochemistry

These two mixtures are used principally for specimens to be examined by immunohistochemical methods. Their possible mechanisms of action have been discussed earlier in this chapter.

Buffered formaldehyde with picrate

Paraformaldehyde:	20g
Picric acid (saturated aqueous solution, filtered):	150ml

Heat to 60°C. Add 2–5 drops of 1.0 M (4%) aqueous sodium hydroxide, if necessary, to dissolve the paraformaldehyde. Filter, cool, and make up to 1000ml with the following phosphate buffer:

Sodium phosphate, monobasic
(NaH$_2$PO$_4$.H$_2$O):   3.31g
Sodium phosphate, dibasic
(Na$_2$HPO$_4$):   17.89g
Water: to make 1000ml

The pH is 7.3. This fixative (Stefanini *et al.*, 1967) is stable for at least 12 months at room temperature. Picrate does not precipitate proteins in neutral solution, and the reasons why it improves the preservation of structure and antigenicity (Accini *et al.*, 1974) are not known.

Periodate-lysine-paraformaldehyde

A 0.1 M solution of dibasic sodium phosphate and a 0.1 M sodium phosphate buffer, pH 7.4 are needed (see Chapter 20). A stock 8% solution of formaldehyde should also be available. It is made by heating 750ml of water to 60–70°C, then adding 80g of paraformaldehyde followed by a few drops of 1.0M (4%) sodium hydroxide, and making up to 1000ml with water when the solution is transparent. The working solution is prepared as before using, as follows:

Lysine monohydrochloride: 1.83g
Water: 50ml

(This is a 0.2 M solution; adjust weight of lysine if a form other than the monohydrochloride is used.)

0.1 M Na$_2$HPO$_4$:   until pH is 7.4
0.1 M sodium phosphate buffer,
pH 7.4: to bring volume to   100ml
8% paraformaldehyde solution: 33ml
Sodium metaperiodate (NaIO$_4$): 284mg
(Alternatively, Sodium
paraperiodate, Na$_3$H$_2$IO$_6$: 391mg)

The pH of the final mixture is not adjusted when it falls following addition of the sodium periodate.

2.6. Methods of fixation

Films, smears, and cryostat sections of unfixed material are immersed in buffered formal-dehyde, cold (0–4°C) acetone, or absolute ethanol or methanol, either before or after the application of blood stains (Chapter 7) or histochemical methods for enzymes (Chapters 14–16), according to the requirements of the particular techniques. Solid specimens are fixed by **immersion** in at least twenty times their own volumes of the appropriate solution, or by **perfusion** of the fixative through the vascular system. The latter technique has the advantage of ensuring that the fixing agent is rapidly brought into contact with all parts of all organs. Usually a whole animal is perfused (Fig. 2.1), but it is also feasible to

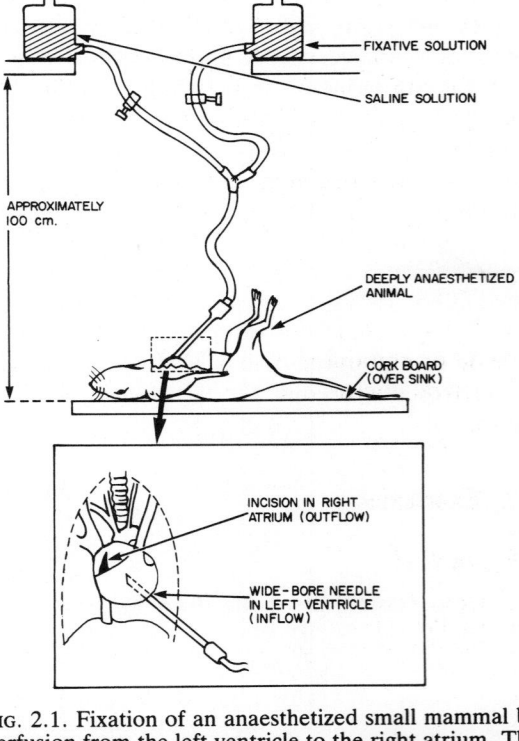

FIG. 2.1. Fixation of an anaesthetized small mammal by perfusion from the left ventricle to the right atrium. The saline is run through first, until the effluent fluid is clear. The fixative (usually a formaldehyde or glutaraldehyde solution) is then perfused until the whole body of the animal is hard and inflexible. About 100ml of fixative is required for a 200–300g rat, and the whole procedure takes about 15 minutes. (According to Thorball & Tranum-Jensen (1983), it takes 5–10 minutes to saturate all the glutaraldehyde binding sites in the tissues of a well perfused rat.)

cannulate and perfuse the vasculature of a single organ. Fixation of material removed from a perfused animal is usually continued by immersion in the same fluid.

Very small or thin pieces may be fixed by leaving them for 2–24h in the **vapour** above an aqueous solution of osmium tetroxide, and freeze-dried blocks are fixed in the vapour of formaldehyde or glutaraldehyde. Despite the excellence of the latter procedures, they are rarely used except in histochemical studies of monoamines.

Another method for small pieces is **phase partition fixation**, in which the fixative agent is dissolved in a liquid that does not mix with water. A solution of glutaraldehyde in heptane (McAuliffe & Nettleton, 1984) is suitable. When the specimen is immersed, the glutaraldehyde passes from the non-aqueous phase (heptane) into the specimen's aqueous phase (water, of extra- and intracellular fluids). The ionic compositions of fluids in the specimen are not disturbed by dilution with an unnatural aqueous solvent, as they are with conventional immersion or perfusion, and the solvent extracts less protein and free amino acids than an aqueous fixative (Mays *et al.*, 1984). Phase partition fixation in formaldehyde or glutaraldehyde is acceptable for light or electron microscopy (Nettleton & McAuliffe, 1986).

2.7. Exercises

Theoretical

1. Deduce the differences that you would expect to see between the appearances of stained sections of animal tissue fixed in: (a) formalin, diluted ten times with water immediately before use, (b) an isotonic solution containing 4% formaldehyde and buffered to pH 7.4.

2. Why is it preferable to use Bouin's fluid rather than an aqueous solution of picric acid alone as a fixative?

3. Before the discovery of the fixative properties of glutaraldehyde, osmium tetroxide (in buffered, isotonic solutions) was the best available fixative for specimens to be examined in the electron microscope. In light microscopy, however, OsO_4 is of limited value as a fixative. Why should this be? Why is OsO_4 useful as a fixative for peripheral nerves but almost useless for central nervous tissue in light microscopy?

4. Glutaraldehyde, although it is an excellent fixative for electron microscopy, is not often used to fix tissues that are to be examined in the light microscope. Why? For which purposes would you expect glutaraldehyde to be the fixative of first choice in light microscopy?

5. What fixative procedure would you use in order to preserve: (a) RNA, (b) neutral fat, (c) the Golgi apparatus, (d) myelin, (e) collagen and reticulin, (f) free amino groups of protein, (g) free carboxyl groups of mucosubstances, (h) an enzyme that was inactivated by aldehydes and by any protein-precipitating reagent?

6. Cetylpyridinium chloride (CPC)

$$CH_3(CH_2)_4CH_2 \overset{+}{N} \quad Cl^- . H_2O$$

has been used as a fixative for a specific histochemical purpose. What type of substance would you expect to be insolubilized by CPC? Many morphological features are distorted after fixation in CPC. Explain why.

Practical

(See also practical exercises at end of Chapter 1.)

7. Kill a rat or mouse and remove and fix pieces of organs that will be needed for the methods described in later chapters.

A list of materials suitable for different purposes is given in the last section of Chapter 20. For enzyme histochemistry, unfixed tissues (frozen, for sectioning in cryostat) are often needed, and when fixation is feasible, the specimens must usually be used within 24–48 hours. For general and connective tissue methods, fix specimens in different fluids so that comparisons can be made.

For instructional purposes, it is useful to have paraffin-embedded material available, as blocks and as mounted sections.

3

Decalcification and Other Treatments for Hard Tissues

Tissues such as bone, dentine or wood can be cut with a particularly rugged microtome, equipped with a chisel-shaped knife of hardened steel. Hard specimens cannot, however, be sectioned with an ordinary microtome. Such materials can be softened after fixation, usually by removing the substances responsible for the hardness of the tissue. **Decalcification** is the chemical dissolution of insoluble calcium salts with a suitable acid or chelating agent.

3.1. Decalcification by acids

The principal mineral component of the calcified tissues of vertebrate animals is a hydroxyapatite formally designated as $Ca_{10}(PO_4)_6(OH)_2$. Like other "insoluble" salts this exists, when wet, in equilibrium with its saturated solution, which contains very low concentrations of calcium, phosphate, and hydroxide ions:

$$Ca_{10}(PO_4)_6(OH)_2 \rightleftharpoons 10Ca^{2+} + 6PO_4^{3-} + 2OH^-$$

Continuous removal of calcium, phosphate, or hydroxide ions from the solution will prevent the system from reaching equilibrium. The reaction will therefore proceed from left to right until all the hydroxyapatite has dissolved. If the liquid surrounding the specimen has a high concentration of hydrogen ions, the reaction

$$H^+ + OH^- \rightleftharpoons H_2O$$

will be driven from left to right, removing the hydroxide ions liberated as a result of dissolution of the hydroxyapatite. Any strong acid could serve as a source of hydrogen ions, but those that form sparingly soluble calcium salts (e.g. sulphuric acid) are, for obvious reasons, unsuitable. Hydrochloric, nitric and formic acids are the ones most often used. The overall reactions of these acids with hydroxyapatite are respectively:

$$Ca_{10}(PO_4)_6(OH)_2 + 20H^+ + 20Cl^- \rightarrow$$
$$10Ca^{2+} + 20Cl^- + 6H_3PO_4 + 2H_2O$$

$$Ca_{10}(PO_4)_6(OH)_2 + 20H^+ + 20NO_3^- \rightarrow$$
$$10Ca^{2+} + 20NO_3^- + 6H_3PO_4 + 2H_2O$$

$$Ca_{10}(PO_4)_6(OH)_2 + 20HCOOH \rightarrow$$
$$10Ca^{2+} + 20HCOO^- + 6H_3PO_4 + 2H_2O$$

The calcium from the tissue ends up as calcium ions dissolved in the decalcifying fluid. If the latter is changed frequently the completion of decalcification can be recognized when extracted calcium ions are no longer detectable by a simple chemical test.

Decalcification by strong acids can result in hydrolysis of nucleic acids. This occurs less completely than in a deliberate Feulgen hydrolysis (see Chapter 9), but may still interfere with the interpretation of the results obtained with histochemical techniques. Most enzymes are put out of action by acid decalcifying agents, but the structure of the tissue is only slightly disrupted *provided that fixation has been adequate*. The crystal lattice of the hydroxyapatite of bones and teeth incorporates small numbers of carbonate ions in addition to the more abundant phosphates and hydroxides. The mineralized tissue contains, in effect, a small percentage of calcium carbonate. This is dissolved by acids:

$$CaCO_3 + 2H^+ \rightarrow Ca^{2+} + H_2O + CO_2 \uparrow$$

Minute bubbles of carbon dioxide are formed within and on the surfaces of specimens being decalcified in acids, but they do not usually produce signs of damage visible under the microscope.

3.2. Decalcification by chelating agents

A chelating agent is an organic compound or ion which is able to combine with a metal ion to form a compound known as a metal chelate. In the chelate, the metal atom is covalently bound as part of a 5- or 6-membered ring. Chelates are stable compounds and do not readily decompose to liberate metal ions. Consequently, the reaction of a metal ion with a chelating agent, although reversible, proceeds almost to completion. If the chelating agent is present in excess, virtually all the metal ions will be removed from the solution.

Ethylenediamine tetraacetic acid (EDTA) forms ordinary salts with sodium and other alkali metals, but the ethylenediamine tetraacetate ions combine with most other metal ions to form stable, soluble chelates. If a piece of calcified tissue is immersed in liquid containing EDTA anions, calcium ions will be removed from the solution by chelation. Hydroxyapatite will therefore dissolve because it will be unable to attain equilibrium with a saturated solution.

For a full account of the chemistry of chelation, see Chaberek & Martell (1959). The chelation of metal ions by dye molecules is described in Chapter 5 of this book.

Decalcification by EDTA differs from decalcification by acids in that hydrogen ions play no part in the chemical reaction involved. The chelating agent is used on the alkaline side of neutrality, so the deleterious effects of strong acids on labile substances such as nucleic acids and enzymes are avoided. The main disadvantage of EDTA is that it acts much more slowly than the acids.

3.3. Decalcification in practice

Specimens that are to be decalcified must be properly fixed. The fixative must be thoroughly washed out of the tissue prior to decalcification in order to avoid undesirable chemical reactions. For example, dichromate ions would be reduced by formic acid; mercury would form a chelate with EDTA. The volume of decalcifying fluid should be at least twenty times that of the specimen. An acid mixture is changed every 24–48h; an EDTA solution every 3–5 days. If the ammonium oxalate test (Section 3.3.3) is to be used, the anticipated last change of acid should have only five times the volume of the specimen, so that any calcium in the liquid will be present at a higher concentration than in a larger volume.

3.3.1. Acid decalcifiers

Formic acid, at pH 1.5–3.5 is the decalcifier of choice for most purposes. Several mixtures similar to the one below have been described. Some of them also contain formaldehyde. Other organic acids used for decalcification include acetic and citric, which are slow, and lactic, which is nearly as fast as formic (Eggert & Germain, 1979). De Castro's fluid is one of many older decalcifying fluids that contain a mineral acid.

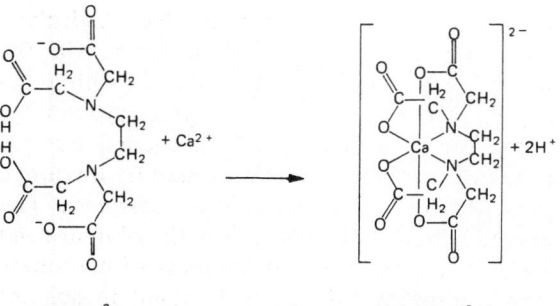

(EDTA)$^{2-}$
As in a solution of the disodium salt of EDTA

[CaEDTA]$^{2-}$
The bonds to the Ca atom are of equal length and mutually at right angles, directed as if to the vertices of a regular octahedron

Buffered formic acid (Clark, 1954)

> Formic acid (90%): 250 ml ⎫
> Water: 750 ml ⎬ Keeps
> Sodium formate   ⎪ indefinitely
> (HCOONa):   34 g ⎭

Alternatively, a solution with the same composition can be made by dissolving 19.8 g of sodium hydroxide in 729 ml water and adding 271 ml of 90% formic acid.

This solution, whose pH is 2.0, produces only minimal suppression of the stainability of nucleic acids. Ippolito, LaVelle & Pedrini (1981) found that formic acid at pH 2.0 extracted less proteoglycan from cartilage than 5 other decalcifying agents tested.

De Castro's fluid

> Absolute ethanol: 300 ml ⎫
> Chloral hydrate: 50 g ⎬ **Mix in**
> Water: 670 ml ⎪ **order**
>   ⎪ **stated.**
> Concentrated nitric   ⎪ **Keeps**
> acid (70% HNO₃): 30 ml ⎭ **indefinitely**

Caution. The nitric acid must be added last. Concentrated HNO_3 reacts explosively with absolute ethanol.

De Castro's fluid is traditionally used in conjunction with neurological staining methods. It is strongly acidic (pH 1.0) and decalcifies rapidly. The alcohol and chloral hydrate in the mixture are supposed to prevent swelling of the tissue, presumably by increasing the osmotic pressure of the fluid. This is probably unimportant when adequately fixed specimens are being decalcified. However, de Castro's mixture is occasionally used as a simultaneous fixing and decalcifying agent. In this circumstance the osmotic effects, as well as the fixative properties of all three ingredients, assume greater significance.

3.3.2. Chelation with EDTA

Decalcification with EDTA proceeds much more slowly than with acids, several weeks often being required, but this is not injurious to the tissues. Some histochemical methods for enzymes can subsequently be performed upon frozen sections.

Either 5 or 10 g of disodium ethylenediamine tetraacetate, $[CH_2N(CH_2COOH)CH_2COONa]_2.2H_2O$, is dissolved in 100 ml of water, and 4% NaOH added until the pH is between 7 and 8. More strongly alkaline solutions are avoided because they extract proteoglycans (Ippolito *et al.*, 1981). The solution is changed every fourth day. Tissues treated with EDTA should be washed in water before dehydration, because the chelating agent is insoluble in alcohol. Alcohol-soluble salts of EDTA are available, but have no special advantages as histological decalcifying agents (Eggert *et al.*, 1981).

3.3.3. End-point of decalcification

If the specimen is larger than necessary and contains calcified tissue throughout, it may be trimmed with a scalpel at intervals during decalcification. When it can be cut easily with the scalpel blade it will also be soft enough to be sectioned on a microtome. Sometimes the specimen is too small to be trimmed, or the calcified material is isolated in the middle of the block. An otherwise unwanted piece of bone of approximately the same size can then be processed alongside the specimen. When this piece of bone is fully softened the specimen for histological study should also be decalcified. Specimens should never be tested by poking needles into them; the holes will persist in the sections.

A more exact test for completeness of decalcification makes use of the fact that calcium oxalate, though soluble in mineral acids, is insoluble in water and in aqueous solutions of alkalis.

$$Ca^{2+} + C_2O_4^{2-} + H_2O \rightarrow CaC_2O_4.H_2O$$

The test is conducted as follows:

(1) Add drops of strong ammonia solution (ammonium hydroxide, S.G. 0.9) to about 5 ml of used decalcifying fluid until the mixture becomes alkaline to litmus (pH > 7).

(2) Add 5 ml of a saturated aqueous solution

of ammonium oxalate (approximately 3% $(NH_4)_2C_2O_4.H_2O$) and leave to stand for 30 min.

If no white precipitate has formed after this time, the fluid contains no calcium ions. This test can also be used to determine the end-point of decalcification by EDTA, even though solutions of the latter do not contain free calcium ions (Eggert & Germain, 1979). The $[CaEDTA]^{2-}$ anion presumably dissociates as the highly insoluble oxalate is formed. Rosen (1981) recommends adjusting EDTA solutions to pH 3.2–3.6 for maximum sensitivity of the oxalate test.

If an X-ray machine is available, the presence or absence of calcified deposits in a specimen can be demonstrated by radiography. Control specimens known to contain and not to contain calcified material should be X-rayed alongside the specimen being tested in order to enable the amount of radio-opacity due to soft tissues to be assessed.

3.4. Softening of non-calcareous materials

Cartilage in vertebrates and **chitin** in arthropods are composed largely of macromolecular carbohydrates, but they often also contain insoluble calcium salts. These materials can be softened to some extent by demineralization in acids or chelating agents. **Wood** is cellulose, reinforced by lignin (see Chapter 10), and often containing deposits of silica (SiO_2). Hydrofluoric acid is sometimes used as a softening agent for hard plant tissues, but Carlquist (1982) preferred a solution containing ethylene diamine. Its mechanism of action is uncertain.

Ethylene diamine solution for softening wood

Ethylene diamine
($H_2N(CH_2)_2NH_2$):   10 ml
Water: to make 100 ml

Caution. Ethylene diamine is a caustic alkali. Keep stoppered to reduce absorption of atmospheric CO_2.

Put boiled or alcohol-fixed wood in the ethylene diamine solution for 3 days. Dehydrate and clear using *t*-butanol (see Chapter 4) and embed in paraffin wax. The paper by Carlquist (1982) contains many other useful practical instructions for processing hard plant tissues.

Softening specimens already embedded in wax

Incomplete dehydration or clearing is one cause of excessive hardening of normally soft tissues embedded in paraffin, but specimens are sometimes unexpectedly difficult to cut for no discernible reason. Hard material encountered at the time of sectioning can be softened by immersing the face of the wax block in water for 15–30 min. Water enters the exposed tissue, and often softens it for a depth of up to 500 µm into the block. Proprietary softening agents (of undisclosed composition; probably aqueous phenol) do not seem to be any better than water for this purpose.

3.5. Exercises

1. Following decalcification in acidic mixtures, tissues are often placed overnight in 5% sodium sulphate. This is often said to be done in order to neutralize any acid remaining in the specimen, with the effect of preventing swelling during subsequent processing. Is this justification of the use of sodium sulphate correct?

2. Sections are required of a bone-containing specimen fixed in Bouin's fluid. What would you do before embedding the specimen in paraffin wax? Give reasons for all stages of the procedure. How would the procedure differ for a piece of Zenker-fixed tissue?

4

Processing and Mounting

This chapter contains some practical instructions for the processing of specimens and the handling of sections and slides. The procedures described are basic to nearly all the staining and histochemical techniques presented elsewhere in the book. Much of the underlying theory has already been explained in Chapter 1.

4.1. Processing

For comprehensive accounts of dehydration, clearing, and embedding the reader is referred to Gabe (1976) and Humason (1979). The properties of several of the solvents used in these procedures are summarized in Table 4.1 (pp. 39–40).

4.1.1. Gelatin embedding

Gelatin does not penetrate the minute interstices of a specimen in the same way as paraffin or nitrocellulose. It surrounds the block of tissue and fills in the larger cavities and cracks. This amount of support permits the cutting of frozen sections of objects that would otherwise disperse into fragments when placed into water. Prolonged infiltration with gelatin, followed by drying, gives a hard block from which thin (2 μm) sections may be cut. This classical technique (see Gabe, 1976) has been replaced by the use of synthetic resins (Section 4.1.5), which are easier to use.

1. Wash formaldehyde-fixed specimens in a large excess of water for 30–60 min. (If fixation has been brief, as for enzyme histochemistry, use a suitable buffered saline instead of water.)
2. Infiltrate with the following solution (which has to be melted before use) for 1–2 h at 37°C, with occasional turning of the specimen.

Gelatin powder: 30 g
Glycerol: 30 ml
Water: 140 ml
Thymol (as a bacteriostatic): one small crystal

} Keeps for a few weeks at 4°C

(For enzyme histochemistry, use freshly dissolved 10% gelatin in buffered saline, for 30 min.)

3. Orientate the specimen in the gelatin in a small petri dish, and place in refrigerator until set: about 1 h.

4. Cut out a square block, leaving about 3mm of gelatin around the specimen, and place the block in 4% neutral, buffered formaldehyde (see Chapter 22) at 4°C overnight. This will make the gelatin insoluble. (For enzyme histochemistry, the treatment with formalin should be as brief as possible. 2h at 4°C is recommended.)

5. Cut frozen sections in the usual way, and mount on to slides. Albumen or chromegelatin may be used as an adhesive. Dry by draining, and then place the slides on a hotplate (45–50°C) for 10–15min, or until gelatin begins to soften. (For enzyme histochemistry, omit the heating.)

The gelatin embedding mass, which cannot be removed, is coloured quite strongly by anionic dyes, but this does not interfere with the interpretation of the appearance of the stained section.

Other water-miscible embedding masses are agar and polyacrylamide (Hausen & Dreyer, 1981). The latter is suitable only for specimens to be cut with a cryostat. Neither is as easy to use as gelatin. There are also water-soluble waxes, but they are used only for specialized purposes.

4.1.2. Dehydration, clearing, and paraffin embedding

Many forms of wax are available, both commercially and as extempore mixtures. Satisfactory results are obtainable with almost any proprietary paraffin wax that melts at 56–58°C and contains a little synthetic polymer (whose identity is unfortunately not usually disclosed by the manufacturer). The wax should be kept about 2°C above its melting point; higher temperatures (above 60°C) can cause chemical degradation of the synthetic additives. The melting point can be varied by mixing different waxes; a higher melting point gives a harder block. Optional extra ingredients can include beeswax or crepe rubber to promote ribboning of sections, and dimethylsulphoxide (DMSO) to accelerate the penetration of the tissue. The virtues and

vices of different waxes are discussed at length by Steedman (1960).

It is often said that prolonged immersion in molten wax, especially if the temperature is too high, causes excessive hardening of the tissue. Gabe (1976) maintained that hardening occurred only if the specimens had not been completely dehydrated. I agree with Gabe; the effects of insufficient infiltration are much more serious than those of unnecessarily long exposure to hot paraffin.

4.1.2.1. Standard procedure

This "standard procedure" is suitable for specimens 3–5mm thick. The times for washing, dehydration, clearing, and infiltration should be shorter for small specimens and longer for larger ones.

The starting point given below is water. If alcoholic fixatives have been used, the earlier stages of dehydration are omitted. Throughout the procedure, the volume of liquid should be 10–20 times that of the specimen. The "alcohol" may be ethanol, methanol, industrial methylated spirit (up to 95%), or isopropanol. The "oily clearing agent" is one of the more viscous and less volatile liquids from Group II of Table 4.1: amyl acetate, benzyl benzoate, cedarwood oil, methyl salicylate or terpineol.

Pass the specimen through:

1. (Delicate specimens only; most solid pieces can go directly to stage 2). Two hours in each of: 15%, 25%, and 50% alcohol
2. 70% alcohol: 2h (or overnight)
3. 95% alcohol: 2h
4. 100% alcohol (1):   2h
5. 100% alcohol (2):   1h
6. An oily clearing agent (1):   2–24h
7. An oily clearing agent (2):   2–24h
8. Benzene or toluene: 15–30 min (to remove oily material from surface of specimen)

9. Wax (1)   1h
10. Wax (2)   1h
11. Wax (3)   1h
12. Wax (4)   1h

The infiltration with wax may be carried out in the oven or in a vacuum-embedding chamber. The latter has the advantages of accelerating removal of volatile solvents and of extracting air bubbles.

13. Block out in wax in a suitable mould.

Alcohols are used only once. New solvents must be used for Stage 7, but once-used solvents are permissible for stage 6. The first two lots of wax must be discarded after they have been used to infiltrate specimens cleared in non-volatile solvents. The wax from the third or fourth change may be used once again for changes 9 and 10 with subsequent specimens. Quantities of inflammable liquids greater than 100ml should not be poured down the sink, especially if immiscible with water. They should be collected in metal solvent drums for eventual safe disposal.

Less viscous clearing agents include chloroform, and a mixture of chloroform with benzene or toluene (equal volumes). Times are about half those for the more viscous ("oily") liquids, and stage 8 is not needed.

Some liquids (Group III in Table 4.1) are miscible with water and with melted wax, and can therefore be used as combined dehydrating and clearing agents. These solvents are usually rather expensive, but processing is quicker when they are used. Stages 2–8 of the standard procedure are replaced by 4 changes (each 1–2h) of dioxane, tetrahydrofuran or *t*-butanol. Specimens that are already in 70% alcohol can be cleared in 3 changes (each 1–2h) of *n*-butanol.

Clearing agents of high refractive index (e.g. methyl salicylate, cedarwood oil) render pieces of tissue transparent. This change indicates that "clearing" is literally complete and is responsible for the customary but rather illogical use of the term. Several authors have advocated such expressions as "intermediate solvent", but "clearing agent" remains fixed in the practising histologist's vocabulary.

4.1.2.2. Chemical dehydration

Instead of replacing the water in a specimen with alcohol, dioxane, or the like, it is possible to use a reagent that reacts chemically with water to give liquid products which are miscible with the clearing agent. **2,2-Dimethoxypropane (DMP)** is such a reagent. It is a ketal and is hydrolysed by water in the presence of an acid as catalyst, to yield methanol and acetone:

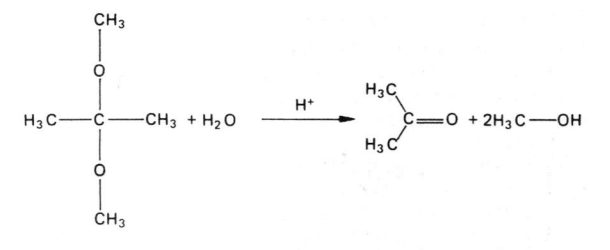

If a reasonable excess of DMP is provided, almost all the water in a specimen will react (Erley, 1957). The hydrolysis is a strongly endothermic reaction, so it is advisable to warm the specimen and the acidified DMP to about 30°C to minimize the risk of formation of ice within the specimen while it is being penetrated by the reagent (Prentø, 1978). When dehydration is complete, the specimen is equilibrated with a mixture of DMP, methanol, and acetone. DMP itself is immiscible with water but it is fully miscible with most organic solvents and with paraffin wax, and it extracts most lipids from the tissue (Beckmann & Dierichs, 1982). Acetone and methanol do not mix with melted paraffin however, so the specimens are cleared before infiltration.

Several practical procedures for chemical dehydration with DMP have been described. The following method (based on the present author's experiences) is suitable for fixed pieces of tissue 1–5mm in thickness.

1. Add 0.02ml (1 tiny drop) of concentrated hydrochloric acid to 50ml of 2,2-dimethoxypropane (DMP) and stir thoroughly. 5–10ml is needed for each specimen. (Any left-over acidified DMP can be kept, but it evaporates quickly if not covered.)

2. Warm the acidified DMP and the specimen (which may be in water or 70% alcohol) to 25–35°C. This is conveniently done by standing the vials on top of an oven or hot-plate.

3. Place each specimen in 5–10ml of the warm

TABLE 4.1. *Properties of solvents used in histology*

Name	Refractive index (at 20°C)	Boiling point(°C)	Fire hazard[a]	Remarks
GROUP I. *Dehydrating agents*				
Miscible with water and with clearing agents (Group II), but not miscible with melted paraffin wax or with resinous mounting media.				
ACETONE	1.36	56	++	The "absolute" liquids usually
ETHANOL (ethyl alcohol)	1.36	78	+	contain about 1% by volume of
ISOPROPANOL (isopropyl alcohol; propan-2-ol)	1.38	82	+	water, but this can usually be ignored. Ethanol and
METHANOL (methyl alcohol)	1.33	65	+	isopropanol are preferred to the more volatile solvents for most purposes
GROUP II. *Clearing agents*				
Miscible with dehydrating agents (Group I), melted paraffin wax, and resinous mounting media. Not miscible with water.				Tissues become transparent only when the refractive index of the clearing agent is higher than 1.47. Replacement of clearing agent by wax is notably slower when the boiling point of the former is above 150°C
AMYL ACETATE (isoamyl acetate)	1.40	142	+	Strong odour of "pear drops"
BENZENE	1.50	80	+	Avoid inhalation of vapour
n-BUTANOL (*n*-butyl alcohol; butan-1-ol)	1.40	118	+	Partially miscible with water. May be used for dehydration of blotted sections. Less extraction of dyes than with ethanol. Irritating vapour
CARBON TETRACHLORIDE	1.46	77		Toxic vapour
CHLOROFORM	1.45	61		S.G. 1.49. Cleared tissues do not sink to bottom
CEDARWOOD OIL	1.51	260(approx.)		Only partially miscible with methanol
BENZYL BENZOATE	1.57	323		Almost odourless
METHYL BENZOATE (oil of niobe)	1.51	200		Unpleasant odour
METHYL SALICYLATE (oil of wintergreen)	1.54	223		Penetrating odour, not unpleasant
TERPINEOL (mixed isomers; synthetic oil of lilac)	1.48	200(approx)		Miscible with ethanol containing 15% water
TOLUENE	1.50	111	+	Avoid inhalation of vapour
XYLENE (mixed isomers)	1.50	140(approx.)	+	Clearing agent of choice for sections. Best avoided for blocks because it causes more hardening than any other clearing agent

(Continued on next page)

TABLE 4.1. *Properties of solvents used in histology (continued)*

Name	Refractive index (at 20°C)	Boiling point(°C)	Fire hazard[a]	Remarks
GROUP III. *Solvents for combined dehydration and clearing*				
Miscible with water and with melted paraffin wax. Not necessarily miscible with resinous mounting media.				
t-BUTANOL (tertiary butyl alcohol; 2-methyl-propan-2-ol)	1.39	83	+	Freezes at 26°C
DIOXANE (1,4-dioxane; diethylene oxide)	1.42	101	+	⎫ Toxic vapours. Explosive peroxides can form after long storage
TETRAHYDROFURAN (THF) (tetramethylene oxide)	1.41	66	++	⎭
TETRAHYDROFURFURYL ALCOHOL	1.45	177		Rarely used
GROUP IV. *Miscellaneous liquids*				
WATER	1.33	100		
ETHYLENE GLYCOL	1.43	198		Miscible with water and dehydrating agents (Group I) but not with clearing agents (Group II)
GLYCEROL (glycerin)	1.47	290		
PROPYLENE GLYCOL	1.43	188		
ETHER (diethyl ether)	1.35	35	++	(Mixed with ethanol.) Used in nitrocellulose embedding. Slight miscibility with water. The vapour softens nitrocellulose and enhances adhesion of celloidin sections to slides
LIGROIN (low boiling)		60–120	++	⎫ Sometimes used as clearing agents for specimens or slides. Only partially miscible with alcohols
PETROLEUM ETHER (mainly hexanes)	1.37 (*n*-hexane)	60(approx)	++	⎭
ISOPENTANE (2-methyl butane)	1.42 (at −150°C)	28	++	Freezes at −160°C. The cooled liquid is used for rapid freezing of tissues.
LIQUID NITROGEN		−196		Used in rapid freezing procedures
ACETIC ACID	1.37	118		Miscible with water and all solvents in Groups I–III. Freezes at 16.6°C. Used in fixatives, buffer solutions, and staining solutions

[a] + means the liquid is inflammable and should not be used near a naked flame. + + means the liquid is very dangerously inflammable. When these liquids are in use, all flames must be extinguished and no smoking allowed anywhere in the room. No symbol in this column means the liquid, if inflammable, is not volatile enough to constitute a serious hazard. Water, chloroform, and carbon tetrachloride are the only volatile liquids in the list whose vapours are non-flammable.

acidified DMP. Leave for 30–60 min, with agitation at 10–15 min intervals.

4. Clear in benzene or toluene (15ml per specimen) for 15 min, with agitation every 5 min. Only one change of the clearing agent is needed. Oily clearing agents may also be used.

5. Infiltrate with molten wax (3 changes, each 45 min), and make blocks in the usual way.

DMP is not a viscous liquid, so its rather slow penetration is probably due to its immiscibility with water. Specimens more than 6–8 mm thick cannot be dehydrated with DMP by the method given above.

4.1.3. Nitrocellulose embedding

Suitable types of nitrocellulose are sold under such names as celloidin, Parlodion, Necolloidin, and low-viscosity nitrocellulose (LVN). The solutions of nitrocellulose are made up well in advance and are stored in tightly screw-capped bottles. **They are dangerously inflammable**. Those required are: 8%, 4%, and 2% nitrocellulose, dissolved in a mixture of equal volumes of ethanol and diethyl ether. The times given below are for specimens approximately 15 mm thick. They may be halved for specimens 5 mm thick. The volume of liquid should be 5–10 times of that of the specimen. The tissue is equilibrated with absolute ethanol and then transferred to:

Ether-alcohol
  mixture: overnight   ⎫ In tightly
2% nitrocellulose: 1 week   ⎬ stoppered
4% nitrocellulose: 1 week   ⎪ specimen
8% nitrocellulose: 1 week   ⎭ tubes

Orient the tissue in a suitable mould, which should be at least three times as deep as the specimen. Mark the level of the surface of the 8% nitrocellulose solution on the outside of the vessel. Leave covered for 12h or until free of bubbles, then remove the lid from the mould and place in a desiccator, with the lid partly open, in a fume hood. Leave until the depth has halved as a result of **slow** evaporation of the solvent. This may take 2–10 days. Put some chloroform in the bottom of the desiccator, close the lid, and leave for a further 48h. The nitrocellulose is hardened by the action of the chloroform vapour. Cut out a block of celloidin containing the specimen and glue it with 4% nitrocellulose to a wooden block. The block is stored in 70% ethanol, which causes further hardening of the embedding medium. The hardened nitrocellulose blocks must not be allowed to dry out.

If LVN is used, 5%, 10%, and 20% solutions are used (in ether-alcohol containing 0.5% of castor oil), the times of infiltration given for celloidin may be halved, and evaporation is allowed to proceed only until a hard crust has formed. Culling (1974) gives a full account of the use of LVN.

Other embedding media that can be concentrated by evaporation of their solvents include polystyrene (Frangioni & Borgioli, 1979) and fully polymerized methyl methacrylate (Gorbsky & Borisy, 1986), but these are not often used.

4.1.4. Double embedding

Hardened nitrocellulose can itself be infiltrated with melted wax in the procedure known as double embedding. This should be used for specimens that contain both hard and soft tissues. Some published methods probably do not introduce enough nitrocellulose into the specimen to make much difference to the hardness of the softer tissues. The following procedure (based on Pfühl's method; see Gabe, 1976) is slow, but it allows thorough infiltration with nitrocellulose, which is then hardened by chloroform and by the action of the phenol dissolved in the clearing agent.

Nitrocellulose solutions

2% and 4% nitrocellulose (celloidin, Parlodion, LVN) in *either* ether-alcohol (50:50) *or* absolute methanol. These solutions are stable indefinitely. Evaporative losses can be made good by replacing the solvent. *Caution*. Dangerously flammable.

Clearing agent (phenolic benzene)

Weigh out 100 (±10)g of solid phenol into a beaker. Stand in an oven at 60°C until all melted, then pour it into 800ml of benzene in a 1 litre bottle. Add benzene to make 1000–1100ml. Keeps for a year in a dark cupboard. Light brown discoloration does not matter; discard when dark brown.

Procedure

1. Dehydrate specimens into *either* ether-alcohol *or* methanol: 4–8h.
2. 2% nitrocellulose: 2–4 days.
3. 4% nitrocellulose: 3–6 days.

4. Wipe off excess nitrocellulose, and transfer to chloroform: 1–2 days.
5. Clear in phenolic benzene: 12–24 h.
6. Infiltrate with wax (4 changes, each 2h), and make blocks.

4.1.5. Embedding in plastic

Synthetic resins ("plastics") are the only acceptable embedding media for electron microscopy. The principles involved in their use are simple: the specimen is infiltrated with a reactive monomer (small molecules), which is then polymerized to form the plastic (large molecules or a cross-linked matrix). The resin is harder than wax or nitrocellulose, making possible the cutting of the ultrathin sections needed for transmission electron microscopy. It is also possible to cut sections at 0.5–3.0μm for light microscopy. Much more fine detail is visible than in sections of wax-embedded material, which cannot usually be sectioned below 4μm. Specimens for electron microscopy, which are less than 1mm across, are cut with glass or diamond knives. Larger specimens for light microscopy must be cut with a steel knife or with a special long glass knife (known as a Ralph knife), and the plastic must not be as hard as it is for cutting ultrathin sections of tiny pieces of tissue. The desired hardness is attained by mixing the monomer with wax or a wax-like plasticizer. This is more conveniently done with methacrylate resins than with the epoxy resins favoured by electron microscopists.

Methacrylates are esters with the general formula $H_3C=C(CH_3)COOR$, in which the C=O bond is conjugated with the C=C bond. R is methyl in methyl methacrylate, n-butyl in n-butyl methacrylate, or 2-hydroxyethyl in glycol methacrylate. A methacrylate polymerizes thus:

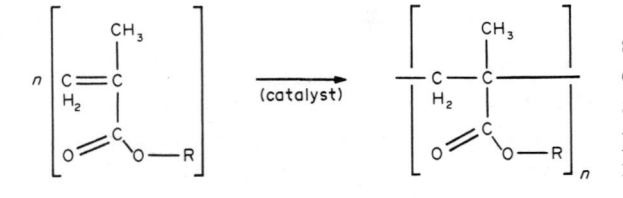

(R is CH_3— in methyl methacrylate, n-C_4H_9— in butyl methacrylate, or $HOCH_2CH_2$—in glycol methacrylate.)

The monomer supplied by the manufacturer contains a stabilizer, usually hydroquinone, to retard spontaneous polymerization. The stabilizer can be removed, but an easier way to overcome its action is to add more of the polymerization catalyst (usually benzoyl peroxide) than would otherwise be needed. Butyl methacrylate was the first synthetic resin to be used in histology (Newman, Borysko & Swerdlow, 1949).

Two procedures are briefly described below. Techniques for cutting the sections are outside the scope of this book, and the original references should be consulted for technical details.

4.1.5.1. Butyl methacrylate with paraffin

Specimens up to 5 mm thick, fixed in formaldehyde or glutaraldehyde, should be post-fixed in Bouin, which was found by Engen & Wheeler (1978) to improve staining and also the handling properties of the sections. Dehydrate, transfer from absolute alcohol into two changes of n-butyl methacrylate (each 2 h), and then into the following mixture:

Butyl methacrylate: 10ml
Paraffin wax: 3.5g
  Mix and put in an oven at 60°C, 2–3 hours before using. Then add:
Benzoyl peroxide: 120–460mg
  The optimum amount of catalyst should be determined by trial and error (McMillan et al., 1983). It varies among batches of butyl methacrylate. 200mg is suggested for the first attempt.

Leave at 50°C for 18–24h (or longer if necessary, until hard). The container may be a gelatin capsule, a disposable plastic ice-cube holder, or a peel-away embedding mould. Oxygen inhibits polymerization and should be excluded by covering the surface of the liquid with a sheet of plastic or (preferably) by having an atmosphere of nitrogen in the oven.

Allow the polymerized block to cool, then trim it (using a small hacksaw) to within 1–2 mm of the tissue. Place the block in melted paraffin for 30 min, and cast a new block, so that the plastic-embedded specimen is surrounded by wax. Mount this block on the microtome and section at 1–4 μm. (See papers cited above for cutting techniques.)

The embedding media are removed from the sections with xylene before staining.

4.1.5.2. Glycol methacrylate

Glycol methacrylate is miscible with water or alcohol, but its polymer is insoluble in all common solvents, so the sections have to be stained in the presence of the embedding mass. Fixed specimens may be dehydrated in ethanol or in glycol methacrylate. They are then placed in the embedding medium:

Glycol methacrylate:	95 g	Can be stored in presence of oxygen (air bubbles)
Polyethylene glycol 400:	5 g	
Benzoyl peroxide:	1.5 g	

Two changes, each 1 h, are recommended. The specimens are then transferred to a third change, in embedding blocks, with exclusion of oxygen (see previous method) at 38–40°C. Polymerization may take as long as 4 days. See Bennett *et al.* (1976), Murgatroyd (1976) and Hayat (1981) for more information, including alternative embedding mixtures. The polymerized resin is hydrophilic, but cannot be removed. It is penetrated by aqueous stains and other reagents, but longer times are usually needed than for dewaxed paraffin sections.

4.2. Keeping sections on slides

4.2.1. Adhesives for sections

Properly flattened sections will usually stick to grease-free glass slides without the assistance of an adhesive. Thin sections (<7 μm) generally stick better than thicker ones. For most purposes, however, the use of an adhesive is strongly advised. It is essential when the sections are to be exposed to solutions more alkaline than about pH8. The two adhesives given below are suitable for sections to be stained by almost any techniques. Mayer's albumen has better keeping properties than chrome-gelatin, but the latter is much more efficacious, especially for protection against alkaline reagents. Albumen may be preferable to gelatin if strong acids are used, as in the hydrolysis of nucleic (Chapter 9) or sialic (Chapter 11) acids. Other adhesives have been proposed in recent years, including glues and other bonding agents (Edwards & Price, 1982; Järvinen & Rinne, 1983; Fink, 1987a,b), and making the glass surface cationic by staining it with alcian blue (Archimbaud, Islam & Preisler, 1986).

4.2.1.1. Mayer's albumen

Preparation (three possibilities are available)

(a) Collect the whites of one or two eggs into a graduated 250 ml beaker. Add an equal volume of glycerol and mix thoroughly. Filter through cloth and then through cotton wool or coarse filter paper. Filtration is accelerated if carried out in an oven at 55–60°C. Add a small crystal of thymol to inhibit growth of micro-organisms.

(b) Make a 5% solution of dried egg-white (commonly called "albumen, egg" in catalogues) in 0.5% aqueous sodium chloride. It takes a day to dissolve, with occasional stirring. Filter through coarse filter paper and add an equal volume of glycerol to the filtrate. Add a crystal of thymol.

(c) Buy a ready-made Mayer's albumen solution.

These solutions will keep for several years at 4°C.

Application

Place a *small* drop on a slide and distribute it evenly over the surface with the tip of the finger.

Float out the sections on water, collect onto the slide, drain, and flatten in the usual manner. Alternatively, do not put Mayer's albumen solution on the slides but add about 20ml of it to each litre of the water used for floating out the sections.

When the slides are dry, put them in the wax oven at about 60°C for 30 min or turn up the temperature of the hotplate used for drying the slides. The heat coagulates the egg albumen. The melting of the wax probably also promotes closer contact between the sections and the glass. The albumen is lightly stained by most dyes and is most conspicuous around the sites of the edges of the wax ribbon. The layer between the sections and the slide must be extremely thin.

4.2.1.2. Chrome-gelatin

Preparation

Dissolve 1.0g gelatin powder (a high quality bacteriological grade gelatin is advised) in 80ml of warm water and allow to cool. Dissolve 0.1g chrome alum (chromic potassium sulphate: $CrK(SO_4)_2.12H_2O$) in 20ml of water. Mix the two solutions. This mixture quickly becomes infected and should not be used if it is more than 3 days old. The solution cannot be diluted after mixing the two ingredients.

Application

Place a *large* drop on the slide, spread it over the surface with a finger, and leave for about 10 min to dry. Another method is to make a more dilute solution (0.1% gelatin, 0.01% chrome alum), immerse the slides in it, drain, and dry. Treatment of slides with chrome-gelatin is often called "subbing". Subbed slides may be kept for at least 3 months if protected from dust. Float the sections onto the slides, flatten, drain and dry in the usual way.

4.2.2. Coating slides with nitrocellulose

When mounted sections have to be subjected to rough treatment, especially immersion in alkalis or hot acids, it is advisable to encase the slide in a thin film of nitrocellulose. This will hold the sections in place if there is failure of the adhesion between tissue and glass. The film should cover the whole slide (both surfaces and all four edges) if it is to be effective. The procedure is as follows:

1. Take slides to absolute alcohol.
2. Immerse in ether–alcohol (equal volumes of ethanol and diethyl ether) for 30–60s.
3. Immerse in a 0.2–0.4% solution of nitrocellulose in ether–alcohol for 30–60s. (This solution is made by diluting one of the stock solutions kept for nitrocellulose embedding. Remember that ether is volatile and dangerously flammable.)
4. Lift out the slides, drain, and allow them to become almost dry. A change in the reflection of light from the glass surfaces indicates the moment at which to move on to stage 5. This end-point is easily learned with a little practice.
5. Place slides in 70% ethanol for 2 min to harden the film of nitrocellulose.
6. Carry out the staining procedure. Dehydrate as far as the 95% alcohol stage.
7. Transfer the slides from 95% alcohol into ether–alcohol, with minimum agitation. Leave in ether–alcohol for 2–3 min to dissolve the nitrocellulose film.
8. Carefully take the slides from ether–alcohol into xylene (1 min) and then into a second change of xylene (at least 1 min). Try not to agitate the slides: this could loosen the sections.
9. Apply coverslip, using a resinous mounting medium.

4.3. Mounting media

4.3.1. Resinous media

Resinous mounting media are of three types: natural, semi-synthetic, and wholly synthetic. They are miscible with xylene, but not with alcohol or water. Instructions for making several resinous mounting media are given by Lillie & Fullmer (1976), but it is generally best to buy them ready-made. Unfortunately, many useful media are obtained as proprietary products of undisclosed composition.

4.3.1.1. Canada balsam

This, the traditional natural mounting medium, is the resin of a fir, *Abies balsamea*. "Natural" Canada balsam, which contains the volatile components of the resin, is a viscous yellow liquid that softens when warmed. Dried balsam is a solid, and must be mixed with xylene to give a workable mounting medium. 50–60 g dry balsam is dissolved in 100 ml xylene. It takes a few days to dissolve. Unsaturated compounds in the resin make Canada balsam a mild reducing agent. Consequently, preparations stained with dyes fade after months or years in this medium. On the other hand, the black colour of cobalt sulphide (a product of some histochemical reactions which is slowly decolorized by atmospheric oxidation) lasts longer in Canada balsam than in synthetic resins. The acid components of Canada balsam, notably abietic acid, $C_{19}H_{29}COOH$, may contribute to the deterioration of specimens stained by cationic dyes. Other natural resins used in mounting media include dammar and sandarac. Their effects on stained material are similar to those of Canada balsam.

4.3.1.2. Synthetic resins

A polystyrene-based medium, DPX, is recommended for all purposes except the preservation of cobalt sulphide deposits. It contains polystyrene (M.W. 80,000), a plasticizer (tri-*o*-cresyl phosphate or dibutyl phthalate), and xylene. It contains no acids or reducing agents, and is non-fluorescent. The addition of 0.1 ml of 2-mercaptoethanol (foul smell!) to each 10 ml of DPX gives a medium in which the fading of fluorescent dyes is retarded (Franklin & Filion, 1985).

The initials stand for "Distrene-80" (a brand of polystyrene), Plasticizer, and Xylene. Add polystyrene beads (25 g) to dibutyl phthalate (5 ml) and xylene (65–75 ml). Dissolution takes a few days. Unsatisfactory batches of DPX are occasionally encountered. Other synthetic media such as "Entellan" or "Permount" can then be used instead.

4.3.2. Aqueous media

Water-miscible media are easily made in the laboratory. Five are listed below. They should be stored in screw-capped bottles, and care is necessary to ensure that the lids do not become too firmly cemented on.

4.3.2.1. Glycerol jelly

Gelatin powder:	10 g
Water:	60 ml

   Dissolve by warming and add:

Glycerol:	70 ml

Add **either** one drop of saturated aqueous solution of phenol **or** 15 mg of sodium merthiolate as an antibacterial agent

Keeps for a few weeks at 4°C. Discard when turbid or mouldy.

Glycerol jelly must be melted and freed of air bubbles before use. This is conveniently done in a vacuum-embedding chamber. Because of the low refractive index (1.42), many unstained structures remain visible in this medium.

4.3.2.2. Buffered glycerol with PPD

This does not solidify, but the coverslip can be held in position by applying a little nail varnish or DPX to its edges. Buffered glycerol is used for fluorescent immunohistochemical preparations (see Chapter 19). The high pH provides for optimally efficient fluorescence, and the added *p*-phenylenediamine (PPD) retards fading (Platt & Michael, 1983).

> *Either*
> 0.1M Sodium phosphate buffer
> (pH 7.4):   10ml
> *or*
> 0.1M TRIS buffer (pH 9.0):   10ml
> *p*-Phenylenediamine
>   hydrochloride: 100mg
> Glycerol: 90ml
>
> Keeps for 3 months, in dark at −20°C

4.3.2.3. Fructose syrup

> Fructose (=laevulose):   15g
> Water: 5ml
>
> Put together in a securely capped bottle and leave at 60°C for 1–2 days, until all the sugar has dissolved to form a clear syrup. Keeps for several weeks.

This medium is sticky enough to hold a coverslip in position, and it does not evaporate and crystallize. It is very easy to use, but too acid for preserving cationic dyes.

4.3.2.4. Apathy's medium

> Gum arabic (=gum acacia):   50g
> Sucrose: 50g
> Water: 50ml
> Thymol: One small crystal
>
> Keeps for a few months at room temperature. Discard if it becomes infected or if the sugar crystallizes.

Dissolve the ingredients, with frequent stirring and occasional heating on a water-bath. The final volume should be approximately 100ml. Apathy's medium has a refractive index (about 1.5) higher than that of glycerol jelly, so it provides more transparent preparations.

4.3.2.5. Polyvinylpyrollidone (PVP) medium

> Polyvinylpyrollidone (M.W.
>   10,000):   25g
> Water (*or* phosphate or TRIS
>   buffer, as in Section 4.3.2.2
>   above):   25ml
>
> Dissolve the PVP by leaving for several hours on a magnetic stirrer. Then add:
> Glycerol: 1.0ml
> Thymol: one small crystal
>
> Usually keeps for 2 to 3 years. Discard if it looks infected.

This mounting medium is less viscous than glycerol jelly or Apathy's and is very easy to handle. The refractive index is 1.46 (Pearse, 1968), but increases as the water evaporates at the edges of the coverslip until unstained structures are barely visible. Polyvinyl alcohol can be used instead of PVP, and either PPD (10mg per 100ml) or *n*-propyl gallate (600mg per 100ml) may be added, if desired, to inhibit fading of fluorescence (Valnes & Brandtzaeg, 1985).

4.4. Treatments before staining

The following treatments are for the eradication of artifacts induced by fixation. The rationales of the methods are explained in Chapter 2.

4.4.1. Removal of mercury precipitates

Tissues fixed in mixtures containing mercuric chloride contain randomly distributed black particles. The nature of this deposit is not known, but is generally assumed to be mercurous chloride (Hg_2Cl_2), possibly with some metallic mercury. The precipitate does not disturb the structure of the tissue and it is easily removed by the following method. **Sections of all mercuric-chloride-fixed specimens must be subjected to this procedure before staining.**

Solutions required

A. Alcoholic iodine

Iodine:	2.5g
70% ethanol:	500ml

Keeps indefinitely and can be re-used many times (see *Note* below)

B. Sodium thiosulphate solution

Sodium thiosulphate ($Na_2S_2O_3.5H_2O$):	15g
Dissolve in water and make up to:	250ml

Keeps indefinitely. May be re-used three or four times

Procedure

1. Take sections to 70% ethanol.
2. Immerse in alcoholic iodine (solution A) for approximately 3 min. With continuous agitation, 1 min is sufficient.
3. Rinse in water.
4. Immerse in 5% sodium thiosulphate (solution B) until the yellow staining due to iodine has all been removed. This usually takes about 30s.
5. Wash in running tap water for 2 min, then rinse in distilled water.

Note: Alternatively, a 1% solution of iodine in 2% aqueous potassium iodide may be used. This stains the sections deep brown and the decolorization at stage 4 takes longer than after 0.5% alcoholic iodine.

4.4.2. Removal of picric acid

The yellow colour of picric acid (from Bouin's and similar fixatives) rarely interferes with staining, but it can be removed by immersing the hydrated slides for about 30 seconds in a dilute alkali. Lithium carbonate is traditional:

Lithium carbonate ($Li_2CO)_3$:	1g
Water:	500ml

Keeps indefinitely

(This may be used directly, or diluted 10–20 times with water. Discard after use.)

4.4.3. Removal of "formalin pigment"

The dark, crystalline deposits of "acid haematin" that form after fixation of blood-rich tissues in formaldehyde at pH<6.0 are removed by immersion of the hydrated sections for 5 minutes in a saturated (about 8%) solution of picric acid in 95% alcohol.

4.4.4. Removal of osmium dioxide

Black deposits are present in tissues that have been fixed or post-fixed in osmium tetroxide. The dark colour can be removed from sections by immersion in 1 or 2% aqueous hydrogen peroxide for 2–30 minutes. The solution should be freshly diluted from a stock bottle of 30% ("100 volumes available oxygen") H_2O_2. (*Caution*. The concentrated solution is injurious to skin and clothing; the diluted solution is harmless.)

When the sections are decolorized, they must be washed thoroughly to remove the OsO_4 formed by oxidation of the black material: 30 minutes in running tap water for mounted sections; 10 changes of 50ml water, each 1–2 min, with agitation, for free-floating sections.

4.4.5. Blocking free aldehyde groups

The following procedure will prevent non-specific histochemical reactions and binding of proteins by the free aldehyde groups introduced by fixation with glutaraldehyde. (For rationale, see Chapter 10.)

Sodium borohydride solution

0.07M sodium phosphate (dibasic) (=disodium hydrogen phosphate; 1.0% (w/v) aqueous Na_2HPO_4):	100ml
Sodium borohydride ($NaBH_4$):	50mg

This solution must be fresh. *Caution*. Large volumes of hydrogen are evolved if borohydride solutions are acidified, so do not acidify. Wash down sink with copious running water after use.

Procedure

Immerse sections in the borohydride solution for 10 min, with gentle agitation every 2 min to shake off bubbles. Wash in 4 changes of water.

Note. The solution is alkaline, and therefore likely to cause losses of mounted sections. Coating with nitrocellulose (Section 4.2.2) is often needed.

4.5. Exercises

Theoretical

1. Sections of a formaldehyde-fixed specimen are required for a histochemical study of carbohydrate-containing lipids. Which embedding procedures would and would not be satisfactory, and why?

2. Explain how a piece of tissue is dehydrated by (a) dioxane, (b) 2,2-dimethoxypropane. Devise a reasonable practical procedure for the use of dioxane in taking a specimen 5mm³ from Helly's fixative into paraffin wax.

3. Paraffin sections of material fixed in Heidenhain's SUSA are to be stained by a histochemical method for arginyl residues which involves the use of a strongly alkaline reagent. How would you handle the slides between removal of the wax and immersion in the alkaline solution?

4. Euparal is a semi-synthetic resinous mounting medium containing gum sandarac and paraldehyde. It is miscible with absolute ethanol or xylene. Its refractive index is 1.48. For which of the following purposes would it be reasonable to use euparal?

(a) Sections stained wth a cationic dye.
(b) Sections in which occasional cells contain a stable, insoluble, coloured histochemical end-product.
(c) Unstained whole mounts of small multicellular animals or plants.
(d) Sections stained with fat-soluble dyes.

Practical

5. Take two slides bearing paraffin sections of a tissue fixed in a mercuric-chloride-containing mixture. Process one but not the other as described in Section 4.4.1. Stain both slides by one of the methods described in Chapter 6. Observe the effect of failing to remove mercurial deposits. This should be the first and last time that you will see this artifact.

6. Stain some paraffin sections of any tissue with (a) polychrome methylene blue (p. 93), (b) haematoxylin and eosin (p. 96); and (c) nothing. Mount in DPX, glycerol jelly, and Apathy's medium (nine slides altogether). Examine the slides immediately, after 1 or 2h and after 1 or 2 weeks. Account for the changes that occur. If you do anything foolish, such as mounting from xylene into an aqueous medium, take note of the results and do not do it again.

Many compounds are coloured but not all of them are dyes. A dye is a coloured compound that can be bound by a substrate. In histological staining dyes are used to impart colours to the various components of tissues. Sometimes the colouring process has a high degree of chemical specificity, so that the dye can be used as a histochemical reagent. In many other histochemical techniques the reagents are colourless, but coloured substances are formed by reactions involving components of the tissue. Often these end-products are insoluble compounds chemically related to dyestuffs.

An important work of reference for biologists who use dyes is *Conn's Biological Stains* (9th ed., ed. Lillie, 1977), which includes descriptions of many dyes and much other information. Abrahart (1968), Allen (1971) and Gordon & Gregory (1983) are textbooks of dye chemistry. More advanced treatises include Venkataraman (1952–1978) and the *Colour Index* (see p. 58).

5.1. General structure of dye molecules

A molecule of a dye has two characteristic parts. These are the **chromogen**, which is the coloured part, and the **auxochrome**†, which is the part of the molecule that attaches to the substrate. The auxochrome is an ionizable substituent, or a substituent that reacts to form a covalent bond with a substrate or a coordinate bond with a metal ("mordant") ion. A **chromophore** is an arrangement of atoms within the chromogen that is responsible for the absorption of light in the

---

†The word "auxochrome" is used here as it is by dye chemists and histologists. Organic chemists use the same word to mean an atom or group of atoms which changes the wavelength at which a chromophore absorbs maximally. Gurr (1971) used "colligator" to denote the part of a dye molecule involved in attachment to the substrate, but this term has not found wide acceptance.

visible part of the spectrum. Many dyes contain more than one auxochrome and most contain additional radicals that modify the colour of the complete compound. The colour is also influenced by the number and types of auxochromic groups. A compound that includes a chromophore but is unable to ionize will be only weakly coloured. The ionic charge of an auxochrome is balanced by an oppositely charged ion, most commonly H^+, Na^+, or Cl^-. The old term **gegen-ion** conveniently describes the balancing ion.

A **fluorochrome** absorbs ultraviolet or blue (or occasionally green) light and emits light of longer wavelength. Many fluorochromes are not or scarcely coloured. They are used in fluorescence microscopy. Some compounds become fluorescent only after they have united with components of tissues.

5.2. Colour and chromophores

Compounds are coloured because their molecules absorb quanta of electromagnetic radiation in the visible part of the spectrum. The energy absorbed by the molecule causes changes in the energy levels of electrons—those involved in covalent bonding as well as lone pairs associated with some atoms. With organic compounds such as dyes the absorption of visible light is largely attributable to the electrons associated with those combinations of atoms known as **chromophores**. Chromophores are arrangements of the following atomic linkages:

$$C=C, \ C=O, \ C=S, \ C=N, \ N=N, \ N=O, \ -NO_2$$

The double bonds always alternate with single bonds to form what are known as **conjugated systems.** The bonding electrons are able to move from one atom to another along a conjugated system, with the effect of exchanging the positions of the atomic linkages formally designated as double and single bonds. This phenomenon is known as **resonance**. Benzene is a resonance hybrid of the extreme structures

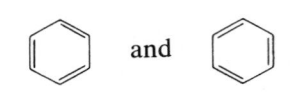 and

Intermediate forms cannot be represented by simple structural formulae. The equivalence of the bonds in a ring such as that of benzene confers a stability that would not otherwise be expected in a cyclic unsaturated compound. Rings stabilized by resonance are said to display **aromatic** character. The conjugated chromophoric system of a dye usually includes double bonds embodied in the formal structure of one or more aromatic rings. The resonance of the bonds in such rings commonly permits the existence of both aromatic and non-aromatic configurations in different rings within the same dye molecule.

The following chromophoric systems are those most frequently encountered in biological stains:

(a) The **nitro** group:

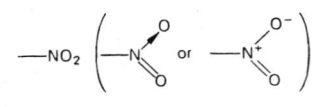

The single (dative) and double bonds between N and O are equivalent because of resonance.

(b) The **nitroso** group: $-N=O$.

(c) The **indamine** group: $-N=$. This always forms part of a larger chromophoric system.

(d) The **azo** group: $-N=N-$. This will be discussed at some length later (see Section 5.9.3).

(e) The **quinonoid** configuration:

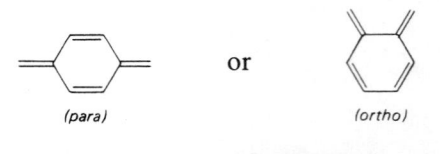

Note that the quinonoid ring is not aromatic. In dyes, resonance often exists among quinonoid and aromatic rings in the same molecule.

Often the common chromophores (a)-(e) shown above occur in combination. Thus, the **quinone-imine** configuration

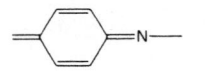

is present in many dyes, such as the triphenylmethanes, the azines, the oxazines and the thiazines. A few less common chromophoric systems will be mentioned in connection with certain dyes.

5.3. Auxochromic groups

The auxochrome or "colligator" (Gurr, 1971) is traditionally held to be responsible for attaching the chromogen to the substrate. Other factors, however, are also involved in the binding of dyes; these will be discussed later.

The **basic** auxochromes are amines, which ionize thus:

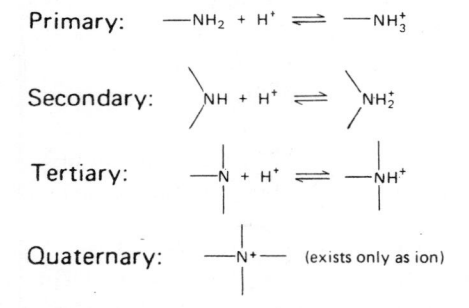

A dye with a net positive charge is called a **basic** or (preferably) a **cationic dye**. On treatment with strong alkali, the free bases (typically unionized amines) of such dyes will be liberated. Acid solutions will favour the formation of cations. The gegen-ions of such dyes are usually chloride or sulphate ions.

The **acid** auxochromes are derived from carboxylic or sulphonic acids or from phenolic hydroxyl groups:

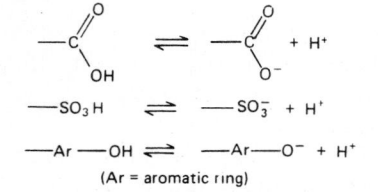

(Ar = aromatic ring)

They occur in **acid (anionic)** dyes, which may be encountered as the free acids or phenols, or as salts, usually of sodium. The carboxylic acids are weak acids, being only partially ionized in aqueous media. Most phenols are even weaker acids. Consequently, dyes with carboxylic acid or phenolic auxochromes cannot exist as anions in the presence of high concentrations of hydrogen ions, because the equilibrium

$$Dye\text{---}H \rightleftharpoons Dye^- + H^+$$

will be pushed over to the left. On the other hand, sulphonic acids are strong, being fully ionized even at very low pH. Sulphonated dyes will therefore behave as anions even in strongly acid media.

From the foregoing paragraphs it will be seen that the pH of the solution will have profound effects on the colouring abilities of anionic and cationic dyes.

Reactive auxochromes are able to combine with hydroxyl and amino groups of the substrate to form covalent bonds. The most important reactive auxochromes are the halogenated triazinyl groups, such as dichlorotriazinyl,

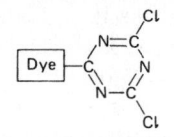

derived from cyanuric chloride. These auxochromes are usually attached to a nitrogen atom of the chromogen. In the monochlorotriazinyl group, one of the chlorine atoms in the above formula is replaced by an alkyl or amino group. The chlorine atoms confer high chemical reactivity, similar to that of the acyl halides. Condensation with hydroxyl or amino groups of the substrate, with elimination of HCl, occurs most readily in alkaline conditions:

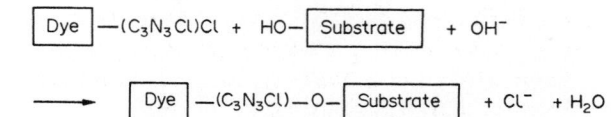

The dichlorotriazinyl group is more reactive than monochlorotriazinyl, and will combine with —OH or —NH$_2$ at room temperature. Other reactive auxochromes used in textile dyeing include pyrimidines and vinyl sulphone groups, which also react in alkaline conditions.

5.4. Fluorescent compounds

The chemical features that enable an organic compound to fluoresce are less easily defined than those responsible for colour. In order to be fluorescent, a molecule must contain a system of conjugated double bonds in its hydrocarbon skeleton. The double bonds may include those of aromatic rings; indeed, cyclic conjugated systems are associated with stronger fluorescence than are linear ones. Strongly fluorescent compounds, the only ones of interest in microscopy, usually have rigidly coplanar molecules. Thus, fluorene, in which planarity is stabilized by a methylene bridge, is more strongly fluorescent than biphenyl in which the two phenyl radicals are free to rotate about the single bond that joins them:

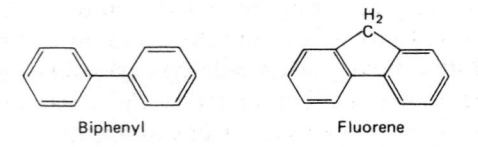

Biphenyl   Fluorene

Compounds with coplanar fused ring systems are more strongly fluorescent when the rings are all joined side to side than when there is a bend in the structure. Anthracene therefore fluoresces more brightly than phenanthrene in response to the same level of exciting illumination. That is, anthracene has the higher **fluorescence efficiency** of the two compounds:

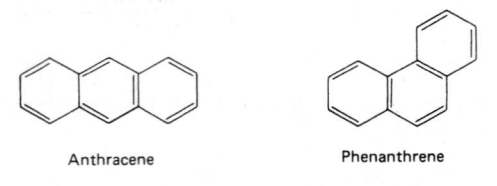

Anthracene   Phenanthrene

Fluorescence is also affected by substituents on the conjugated hydrocarbon skeleton of the molecule. When a single substituent is present the fluorescence efficiency is usually increased by —OH, —OCH₃, —F, —CN, —NH₂, —NHCH₃, and —N(CH₃)₂, and reduced by $\geq$C=O, —COOH, —Cl, —Br, —I, —NO₂, or —SO₃H. Alkyl side-chains generally have no effect unless they sterically hinder the assumption of a planar configuration. The above generalizations do not always apply when more than one substituent is present. Salicylic acid (*o*-hydroxybenzoic acid), for example, is fluorescent despite its carboxyl group. With heterocyclic compounds, including many dyes, the situation is complex and not fully understood. The fluorescent properties of such substances cannot be predicted by simple inspection of their structural formulae. Juarranz *et al*. (1986) have found that fluorescence of dyes is correlated with the number of conjugated bonds in the molecule. A dye is most likely to be fluorescent if this number is in the range 28–30.

The fluorescence of a compound can be influenced by other substances with which it is mixed and by the pH of an aqueous solvent or histological mounting medium. For a more detailed but still introductory account of the fluorescence of organic compounds, see Bridges (1968).

5.5. Combination of dyes with substrates

The histologist uses a large excess of the staining solution, usually at room temperature, and the amount of dye taken up by the tissue is a negligible proportion of that present in the vessel. Staining is said to be **progressive** when a solution of a dye is allowed to act slowly until the desired effect is obtained. In **regressive** staining, the tissue is deliberately overstained, and then placed in a reagent (often water, alcohol, or acidified alcohol) that slowly removes the dye until the latter is left behind only in the components in which colour is wanted. This process of controlled removal of a dye is known as **differentiation** or destaining. The success of both the progressive and the regressive modes of staining depends on the fact that dyes generally have higher affinities for some components of tissues than for others.

5.5.1. General physical considerations

The binding of dyes to textiles has been studied for many years, but is still not fully understood (see Giles, 1975). Less is known about the interactions between dyes and tissues. This subject has been reviewed in great detail by Horobin (1982). The occurrence or non-occurrence of staining by a dye is determined by thermodynamic principles that apply to all dyes and substrates (see Goldstein, 1963), and by formation of chemical bonds that vary with the different types of dye and substrate. Physical prerequisites for the dyeing of textile fibres or tissues include (*a*) the **size of the dye particles**, (*b*) the **porosity** of the substrate, and (*c*) the **interfacial effect**.

An approximate indication of the particle size is given by the molecular weight of the dye. However, most dye molecules have a strong tendency to stick together and form **aggregates**. Aggregation, which is a reversible process, occurs most readily with the larger dye molecules, in concentrated solutions, and at low temperatures. Large particles diffuse in and out of a fibre or a component of a tissue more slowly than small ones.

Dye molecules or aggregates diffuse through submicroscopic spaces or pores within the substrate. These channels cannot usually be directly observed by electron microscopy, but their properties can be determined by studying the rates of permeation of molecules of known size. Thus, it is found that much of the volume of cotton (cellulose) fibres consists of pores about 3 nm in diameter. In sections of fixed biological specimens, the refractive index is an indicator of porosity (Goldstein, 1964). Cellulose is observed to be more permeable than most of the other substances that contribute to the structure of plant and animal tissues.

The interfacial effect is a consequence of the lowering of the surface tension of water by many solutes, including most dyes: the dye has a higher concentration at the surface than in the bulk of the solution. When a dye solution permeates a porous substrate, the solute becomes concentrated at the interface between the solution and the walls of the pores. In cotton fibres, for example, dyes are concentrated some 50–60 times, without the formation of any chemical bonds.

5.5.2. Electrovalent attraction

The simplest mode of dyeing is that whereby there is simple electrostatic attraction of oppositely charge ionized groups in the dye and substrate. For example:

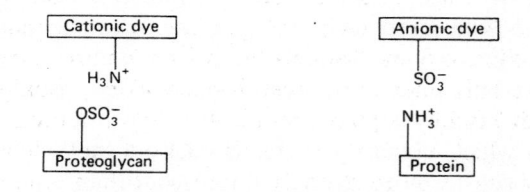

The electrovalent bonds are readily disrupted by acids or by high concentrations of electrolytes, which may therefore be included in the staining mixture to retard or limit the extent of dyeing.

Both cationic and anionic dyes are usually used in acid solutions. This enhances the ionization of amino and related groups of cationic dyes:

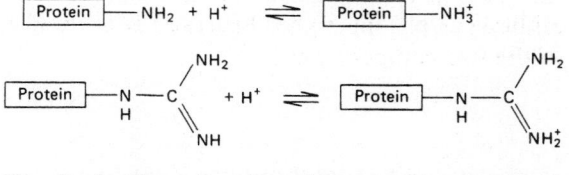

and of basic groups of proteins:

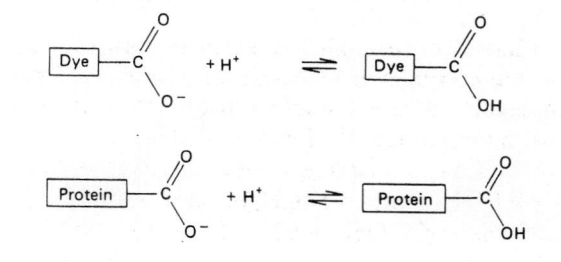

The ionization of carboxyl groups is suppressed in acid media:

and so is that of the phosphoric acids of DNA and RNA, though in this case a higher $[H^+]$ is needed, since phosphoric acid is stronger than carboxylic acids:

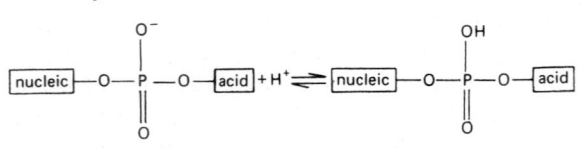

These effects of acidification are simple applications of the law of mass action to the reversible chemical reactions of protonation of bases and dissociation of weak acids. Sulphonic acids are strong, so sulphonated dyes exist as anions even at very low pH. The half-sulphate ester groups of proteoglycans (see Chapter 11) are also strong acids, as are any sulphate or sulphonate radicals introduced into tissues by treatment with histochemical reagents.

The order of acidic strength of the acid (potentially anionic) groups normally present in tissues is sulphate > phosphate > carboxylate > phenol. Cationic dyes are ordinarily used at the pH sufficiently low to suppress the ionization of phenolic hydroxyl groups and carboxyl groups but high enough to allow the esters of phosphoric and sulphuric acids (i.e. nucleic acids and proteoglycans respectively) to exist as anions, and therefore to be stained. The uses of cationic dyes will be discussed further in Chapters 6, 9, and 11.

Electrostatic attractions operate over longer distances than the other forces involved in dye binding. They are probably important in pulling dye molecules towards oppositely charged parts of tissues. Ideally, an insoluble ion-pair or "salt" should be formed if the dyeing is to be permanent, but this does not usually happen in histological staining. Ionic bonds may be the only forces holding dye to substrate when staining is by a dilute solution of a cationic or anionic dye with small molecules. Other types of binding are also involved when dyes with larger molecules are used, as will be seen below.

5.5.3. Covalent combination

Covalent bonds are probably not formed as the result of reactions between ordinary anionic or cationic dyes with substrates. Many histochemical reactions, on the other hand, certainly produce covalent bonds between chromogens and reactive groups in the tissue. Covalently bound coloured material cannot be washed out of a section by simple solvents or salt solutions, but resistance to extraction does not prove the existence of covalent binding; the coloured product might simply be insoluble.

Reactive dyes (see p. 51) are, to date, not used as ordinary histological stains, probably because they are not selective enough in their reactions. Auxochromes like the dichlorotriazinyl group combine with free hydroxyl, amino and other groups, so dyes containing them would stain everything in the tissue, with no selectivity.

5.5.4. Hydrogen bonding

Hydrogen bonds form between hydrogen atoms (which must be bonded covalently to oxygen or nitrogen) and bicovalent oxygen or tricovalent nitrogen. The bond is represented by a dotted line thus:

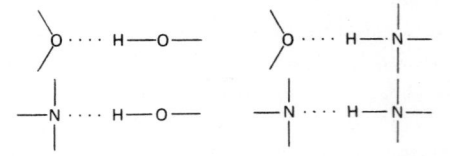

The bond is due to attraction between the lone pair of electrons (i.e. the pair not used in the covalent bond) on the oxygen or nitrogen atom, and the nucleus of the hydrogen atom. This nucleus is partially exposed because the electron orbital around it is drawn towards the strongly electronegative oxygen or nitrogen atom to which the hydrogen is covalently bonded. Another type of hydrogen bond can join the delocalized π-electrons of an olefinic double bond or of an aromatic ring to the hydroxyl group of an alcohol or phenol:

C
‖ · · · H — O — R   ⬡ · · · · H — O — R
C

Each of the electrical charges resulting in the hydrogen bond is of lower magnitude than the full charge of an electron or proton, so the bond is weaker than a covalent bond, though it is stronger than a van der Waals attraction.

The hydrogen atoms of water are hydrogen bonded to the oxygen atoms of other water molecules. Many dyes and components of tissues have hydrogen, oxygen, and nitrogen atoms capable of forming hydrogen bonds with water. It is therefore unlikely that bonds of this kind contribute significantly to the attachment of dyes to substrates when staining is carried out in aqueous media. The molecules of water, being much more numerous than those of the dye, would successfully compete for the available hydrogen bonding sites of the substrate. Horobin (1982) suggests that hydrogen bonding occurs when dyes are applied from non-aqueous solvents for staining glycogen, which has numerous —OH groups.

5.5.5. Charge-transfer bonding

This is another type of rather weak bonding which results in the formation of non-covalent "addition compounds". On one side of the charge-transfer bond is a molecule with electrons available for donation. These are commonly the delocalized π-electrons of systems of double bonds or of aromatic rings, though they may be lone-pair electrons of amine nitrogen atoms. On the other side of the bond is an organic molecule capable of acting as an electron-acceptor, though the exact mechanism of acceptance is not fully understood (see March, 1977, for more information). Picric acid, which is used as both a fixative and a dye, is well known for its ability to form charge-transfer complexes with aromatic hydrocarbons, olefins, and amines. Some other aromatic nitro compounds share this property.

It is conceivable that charge-transfer bonding could be involved in the binding of picric acid and other nitro-dyes to stained tissues, but this possibility has not been investigated.

5.5.6. van der Waals forces

The van der Waals forces are the electrostatic attractions that always exist between the electrons of one atom and the nucleus of another. They are weak forces, and can act only between molecules that are already very close together. Two types of force are recognized. **Polarization forces** are between dipoles; a dipole is a molecule in which the electrons are unsymmetrically distributed, so that one end carries a fractional electric charge relative to the other end. Water is a dipole, and so are most organic compounds that are not hydrocarbons. The strength of a polarization force varies inversely with the sixth power of the distance separating the dipoles. **Dispersion or London forces** exist between symmetrical molecules, because of the transient momentary existence of regions of uneven electron distribution. Dispersion forces are most effective between large molecules, and can cause long dye molecules to become aligned on long molecules of a fibre (Giles, 1975).

The van der Waals forces may be important in the "direct" dyeing of cellulose, discussed later in this chapter, in the staining of elastin (see Chapter 8), and in the attachment of some mordant-dye complexes (Section 5.5.8) to their substrates.

5.5.7. Hydrophobic interaction

The affinity of non-polar molecules or parts of molecules for one another is due to van der Waals forces between the hydrophobic groups, and to hydrogen bonding among nearby water molecules. The polar parts of otherwise hydrophobic molecules are hydrogen-bonded to clumps of water molecules, so the non-polar regions are the only parts that can approach one another closely enough to be held together by van der Waals forces. Hydrophobic interaction is important when a hydrophobic substrate such as cellulose acetate or polyester is dyed. The dye molecules can enter pores that are inaccessible to water, and stay there.

In histology, hydrophobic interactions are

responsible for the penetration of oil-soluble dyes into lipids and for their retention at hydrophobic sites in the tissue. Dyes of this type (solvent dyes; see p. 60) are applied as solutions in moderately polar organic solvents (e.g. 70% alcohol), and the stained preparations must be mounted in aqueous media. For staining to occur it is necessary for the dye to be less soluble in its solvent than in the hydrophobic substrate. The solvent should therefore be as polar as possible and saturated with the dye. Coloured, non-polar compounds are often formed as end-products in histochemical methods for enzymes (see Chapter 14). Such products may become attached by hydrophobic interaction to lipids, and thus lead to faulty localization of the enzymes. The uses of solvent dyes in lipid histochemistry are discussed further in Chapter 12.

5.5.8. Coordinate bonds, chelation and mordants

A mordant is a substance that serves to bind a dye to a substrate. Thus cotton can be impregnated with tannic acid, with which cationic dyes form insoluble salts. The tannic acid serves as a mordant for the dye, which would not adhere to the uncharged cellulose molecules of the cotton. In histological parlance, however, the term is restricted to metal ions that are able to bind covalently to suitable dye molecules, forming **complexes** (also known as "coordination compounds" or "dye-metal complexes"). An ion or molecule that combines with a metal ion is called a **ligand**. A ligand must contain at least one atom, the **donor**, with an unshared pair of electrons in its outermost shell. These electrons become shared by the metal ion and the donor to produce the covalent bond between the two atoms.

A bond of this type resembles an ordinary covalent bond, in which each atom contributes one of the electrons of the shared pair, in that it may be very strong and resistant to harsh physical and chemical treatments, or it may be weaker, having some ionic character. Bonds with two electrons donated by the same atom are often called **coordinate**, **dative**, or **semi-polar** bonds,

and several conventional representations are used in structural formulae. These include such symbols as: $Cu—NH_3$ (indistinguishable from any other covalent bond). $Cu^{\ominus}—^{\oplus}NH_3$ (indicating the acquisition of the electron by the copper atom, which consequently has a higher density of negative charge surrounding its nucleus than does nitrogen), $Cu \leftarrow NH_3$ (showing clearly that the nitrogen atom was the donor; relative electronegativity at the head of the arrow) and $Cu...NH_3$ (in which the covalent or ionic character of the bond is left deliberately vague. This should not be confused with a hydrogen bond). None of these notations are entirely satisfactory. The first one shown (representation as a simple covalency) is used in this book.

The total number of atoms covalently bonded to a metal atom or ion is the **coordination number** of the latter. The charge remaining on an atom stripped of all its ligands is its **oxidation number**. For the rules governing the assigning of oxidation numbers to elements, the reader should consult a textbook of general and inorganic chemistry. It is important to remember that the oxygen atom of the water molecule has an unshared pair of electrons and therefore serves as a donor. Thus the copper(II) or cupric ions in an aqueous solution of $CuSO_4$ exist as $[Cu(H_2O)_4]^{2+}$, or tetraaquocopper(II) ions. The water molecules of this complex ion can be displaced by other ligands, such as ammonia in the copper(II) tetrammine ("cuprammonium") complex $[Cu(NH_3)_4]^{2+}$. In these complexes the coordination number of the cupric ion is 4 and the oxidation number, represented by the roman numeral, is +2.

When a ligand has two or more atoms suitably positioned to form coordinate bonds with the same metal ion, the latter will be incorporated into a ring. This arrangement commonly confers stability upon the complex, which is then known as a **chelate**. Thus, cupric ions combine with sulphosalicylate ions:

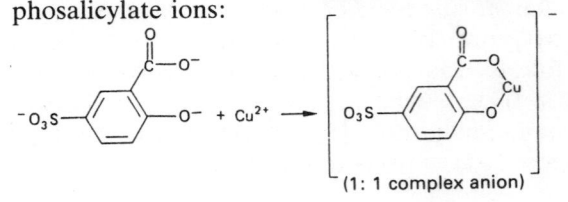

(1: 1 complex anion)

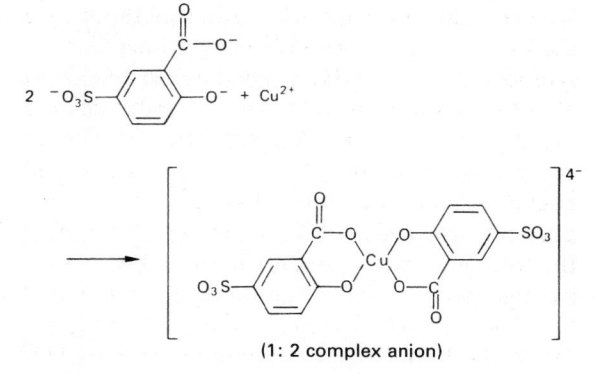

(1: 2 complex anion)

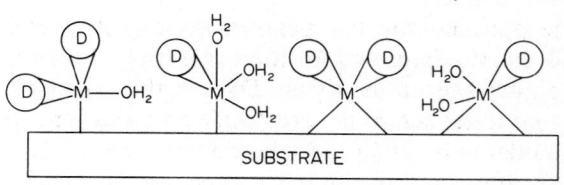

SUBSTRATE

Complexes of the chelate type are involved in the interactions of mordants with dyes. The cations most often encountered in conjunction with histological stains are those of aluminium, iron, and chromium. Chelation of many other metals is important in a wide variety of histochemical techniques.

The most important donor atoms of dyes are oxygen (of phenolic hydroxyl and carboxyl groups and of quinones) and nitrogen (of azo, amine, and nitroso groups). These donor atoms must, of course, be suitably positioned on the dye molecule for the latter to be able to serve as a chelating ligand. Substituents *ortho* to one another on a benzenoid ring or at the 1 and 8 positions of naphthalene, or at sterically equivalent positions on other fused aromatic rings, are usually present in dyes that form complexes with metals.

For a more thorough account of the chemistry of dye-metal complexes, with special reference to those of azo dyes, the reader is referred to Zollinger (1961).

When a metal ion acts as a mordant, coordinate bonds are formed with both the dye molecule and the substrate. The donor atoms in tissues stained by mordant dyes cannot be identified with certainty. It may reasonably be presumed, however, that any suitably positioned oxygen or nitrogen atoms that are not already fully coordinated with parts of other molecules in the tissue will be potentially available for binding a mordant metal ion. Dye-mordant-substrate complexes have such forms as

(D = dye molecule; M = metal ion capable of forming six covalent bonds; H_2O, OH_2, coordinately bound water molecules).

The mordant is sometimes applied before the dye, but is usually mixed with it. Occasionally a metal salt is applied after the dye (see also p. 60). In a mixture of dye and mordant there will be competition among the various potential ligands (dye, water and substrate) for the metal ions. Such a mixture also contains fully coordinated dye-metal (often 2:1) complex molecules, and these may have their own characteristic staining properties. A complex in which the metal ions are outnumbered by dye molecules is known as a **lake**. In practical staining mixtures, an excess of mordant ions over dye molecules is nearly always present in order to ensure that the former will be in sufficient abundance to combine with the tissue as well as with the dye.

True mordant dyeing may occur in the staining of nuclei with iron- or aluminium-haematein (see Chapter 6). It is known that nucleic acids are not the only substances involved in the binding of these dyes, and experimental evidence supports the contention that metal ions are interposed between the molecules of the dye and those of the tissue. For example, stained preparations can be differentiated by treating with aqueous solutions of the salts and used as mordants. Nuclear staining by mordant dyes can also be differentiated by treatment with acids, which work by cleaving the bond between mordant and tissue (see Baker, 1962).

Other dye-metal complexes, including chromium-gallocyanine, are simply large coloured cations (Berube *et al.*, 1966; Marshall & Horobin, 1972a), and there is no evidence for the interposition of a "mordant" atom between the tissue and the chromogen (see Lillie, 1977). It has also been suggested (Marshall & Horobin, 1973) that preformed dye-metal complexes may be held to their substrates by van der Waals forces rather than by electrovalent attraction or by covalent bonds. For such a mechanism to be effective it is necessary that the molecules of the complex be large. Since in most lakes two dye molecules are bound to each metal ion, this condition is likely to be fulfilled.

It is generally characteristic of histological staining by mordant dyes that the colours imparted to the tissues are not extracted by water, alcohols, or weakly acidic solutions used in the counterstaining and later processing of the preparations. Stronger acids decolorize the stained sections and are used for differentiation. Solutions of the mordant salt alone may also be employed for this purpose.

5.6. Nomenclature and availability of dyes

The formal chemical names of most dyes are so long and cumbersome that it would be inconvenient to have to say or write them frequently. Trivial names are used instead. Some of these have been in use long enough to be considered "traditional" (e.g. haematoxylin, methylene blue, acid fuchsine). Others are trade names of manufacturers (e.g. alcian blue 8GX, procion brilliant red M2B). Letters or numbers following names are called "shade designations" and form integral parts of the names. Thus, eosin and eosin B (p. 79) are different, though related, compounds. Unfortunately, the trivial names of dyes give no clues to their chemical natures or their uses.

In the *Colour Index* (3rd edn., 1971), published by the Society of Dyers and Colourists, dyes and other useful coloured compounds are given numbers (the "C.I. numbers") and also names that indicate their modes of industrial application and their colours. For example, methyl blue (p. 77) is C.I.42780, Acid blue 93. Chromoxane cyanine R (p. 77) is C.I.43820, Mordant blue 3. Natural dyes are named as a separate class, so that haematoxylin is C.I.75290, Natural black 1, despite the fact that it is a mordant dye. The C.I. numbers are used in other

works as well as the *Colour Index* itself and they should always be specified when ordering dyes from suppliers. Some of the dyes and related compounds used in biological work are not included in the *Colour Index*, but are described and discussed in *Conn's Biological Stains* (9th edn., Lillie, 1977). The preferred names given in the latter work are used in this book, but C.I. numbers and commonly encountered synonyms are also given.

Authors have been inconsistent in their spelling of the names of dyes and in the use of initial capital letters. In this book, initial capitals are used in the preferred names only for words that are also proper names of people or places. Other words, including trade names, are printed entirely in lower case letters, but shade designations are capitalized (e.g., Victoria blue 4R; methylene violet, Bernthsen; toluidine blue O). The C.I. applicational names are treated as a botanist or zoologist would treat the generic and specific names of a species, with an initial capital for the class and lower case for the colour (e.g. Basic violet 10; Mordant black 37). The policy of Baker (1958) is followed with respect to the spelling of words in traditional preferred names that end in -in or -ine. Thus the ending -ine is used when the dye is an amine or a derivative of an amine (e.g. acid fuchsine, **not** acid fuchsin; or safranine, **not** safranin). For a dye that is not an amine, there is no terminal e (e.g. eosin, **not** eosine; haematein, **not** haemateine). This convention results in differences between some of the spellings in this book and those used by Lillie (1977). Other differences arise from the use of diphthongal ae (e.g. haematoxylin rather than hematoxylin), and that of ph rather than f in such words as "sulphur" and 'sulphonic". The latter discrepancies merely reflect differences between English and American spellings.

Dyes for use as stains and histochemical reagents can be purchased from many suppliers of laboratory chemicals. A few firms specialize in biological stains, but their products are not necessarily superior to those obtained from more ordinary sources. A few dyes (e.g. aldehyde-fuchsine, p. 128) are easy to synthesize in the laboratory. Dyes are not usually manufactured by the firms that sell them. They are made in factories for the primary purpose of satisfying the requirements of the textile industry, though some stains are obsolete as industrial dyes and are now produced solely for use by microscopists.

Because of the different requirements of the dyer and the histologist, **most dyes are not marketed as pure compounds**. The dye is nearly always mixed with a substantial proportion of inactive filler (often sodium chloride) and several different coloured substances are often present in a sample sold as a single, pure dyestuff. Inert fillers are added in order to standardize the potency of different batches of dye so that they will behave similarly towards textiles. Some salt is often inevitably present as a result of the salting-out process used to recover the dye in solid form from the solution in which its synthesis was accomplished. The chemical reactions by which dyes are generated are usually accompanied by side reactions. These often lead to the production of other dyes, which end up as contaminants in the final products. Changes occur after dissolving some dyes, and this process of "maturation" may lead to deterioration of a staining solution (as with gallocyanine, see p. 81) or to improvement in its properties (as with haematoxylin as explained on p. 72 and with methylene blue as explained on p. 83). Occasionally a dye with a single name is a deliberate mixture of two quite different substances (e.g. alcian green; see p. 87). In recent years there have been many investigations of the purity of dyes using paper and thin-layer chromatography. Nearly every histological stain tested has been shown to contain several coloured components. Often one of these predominates and this usually corresponds to the name on the bottle, but some commonly used dyes have been shown to be mixtures of two or three major components with ten or more minor contaminants. It is hardly surprising that the properties of a biological stain vary from one batch to another of what is supposed to be the same single dyestuff.

It is necessary for the histologist to have dyes that can be relied upon to give satisfactory and repeatable results. Such constancy of performance has assumed even greater importance in recent years than in the past with the advent of automated screening devices for the examination of blood and exfoliated cells in pathology laboratories. In the early days of staining, microscopists depended upon a few suppliers, notably Dr. Georg Grübler & Co. of Berlin, who tested the products of various manufacturers and then bought up and packaged batches of dyes that proved satisfactory as stains. Grübler's dyes ceased to be available outside Germany in 1914, and most biologists became dependent upon the dye industries of other countries. The difficulty in obtaining usable stain led to

the founding of the Biological Stains Commission in the United States in 1920. This body tests batches of dyes submitted by manufacturers and suppliers and certifies those suitable for their intended uses. The label on a bottle of a certified stain shows the percentage by weight of dye in the material, the C.I. number, and a batch number. Newly certified batches of dyes are reported in the Commission's journal, *Stain Technology*. The main purpose of certification is to serve pathologists, so many dyes used in other biological sciences are not tested. All samples of such dyes should be viewed with suspicion and carefully tested by the user before applying them to valuable specimens. The history of the availability and standardization of biological stains is recounted by Lillie (1977), who also provides practical instructions for assaying and testing many dyes. Satisfactory assay methods are not available for some dyes. These must be tested empirically by standardized staining procedures.

When a weight or concentration of any dye is specified in the instructions for a method, in this or any other text, it refers to the material constituting an acceptable sample of the dye—not to the pure dyestuff. For the more important stains the acceptable percentages of active substances present in the commercial products are given with the descriptions of dyes in Section 5.9. The data are mostly from Lillie (1977). If the content of dye stated on the label of a bottle differs by more than a few percent from the recommended acceptable value, appropriate correction must be made in preparing a working solution. With samples of dyes not certified by the Biological Stains Commission, the dye content is not usually stated on the label, and must be guessed. Most dyes can be stored as dry powders for many years without deterioration (Emmel & Stotz, 1986), but old stock was not necessarily good when it was manufactured, especially if that was before the days of standardized testing and certification.

5.7. Classification of dyes by methods of application

Dyes and related colouring agents may be arranged in sixteen groups according to the different ways in which they are applied to textile fibres and other substances. These techniques are not all used in histological practice but they are briefly set out below to give the reader some idea of the variety of colouring methods employed in industry. For more information, see Allen (1971), and Bird & Boston (1975).

Basic dyes. These are coloured cations, which attach by electrostatic forces to anionic groups in the substrate. They are used on proteinaceous fibres, cellulosic fibres that have been mordanted with tannic acid and polyacrylonitrile fibres. Cationic dyes form an important group of histological stains and histochemical reagents.

Acid dyes. These are coloured anions, with sulphonic or carboxylic acid auxochromes. Many dyes in this group are amphoteric, having amine as well as acid groups. An excess of anionic substituents confers a net negative charge on the molecule. The sulphonic acids are the most important industrial dyes in this group because they are strong acids, fully ionized even at low pH. They are applied from acidic dyebaths, mainly to proteinaceous fibres such as wool and silk with free amino and guanidino groups. As will be seen later, ionic interactions are not the only forces involved in the binding of acid dyes to their substrates. Anionic dyes are important as stains for cytoplasm and extracellular structures.

Direct dyes. These are anionic dyes with large molecules. Nearly all are azo dyes. They bind to cellulosic fibres directly (i.e. without the need for a mordant). Some are used as histological stains. The mode of dyeing by direct dyes will be discussed later (p. 67).

Mordant dyes. Mordants and their functions have already been discussed (pp. 56–59). By convention (*Colour Index*), the mordant dyes are defined as those used in conjunction with metal salts. The mordant may be applied **before** the dye, together **with** the dye as a soluble dye-metal complex (the "metachrome" process) or **after** the dye (the "afterchrome" process). In some instances a mordant dye is also an anionic or, rarely, a cationic dye in its own right. Mordant dyes have many uses in histology and histochemistry.

Reactive dyes. These dyes combine covalently with their substrates. The most important are those with auxochromes capable of uniting with the hydroxyl groups of cellulose. Isothiocyanates, sulphonyl chlorides, diazonium salts, and a few other derivatives of ordinary dyes and fluorochromes are used for making covalently labelled proteins for use as tracers and as reagents in immunohistochemical techniques (Chapter 19). This labelling process is chemically akin to the industrial use of reactive dyes. A reactive dye bound to an insoluble hydrophilic polymer can be used for purification of enzymes by affinity chromatography. The active sites that bind coenzymes are the only parts of the apoenzyme molecules flat enough to have strong non-ionic affinity for the bound dye. The enzyme is eluted with a solution of the coenzyme (see Clonis *et al.*, 1987).

Solvent dyes. The simple solvent dyes are coloured substances that dissolve in hydrophobic materials but not in water. Solvent dyes are used in wood stains, lacquers, polishes, printing inks, inks for ball-point pens, and in the mass colouring of wax, some plastics, and soap. Other solvent dyes are salts formed by the interaction of anionic chromogens with hydrophobic cations. The latter are released during dyeing, leaving the water-insoluble coloured anions behind in the substrate. Biologists sometimes call solvent dyes **lysochromes**, a term introduced by Baker (see Baker, 1958). Several lysochromes are used in the histochemical study of lipids (see Chapter 12).

Vat dyes. These are important in the textile industry but are not used by biologists. A vat dye is applied to cloth as its leuco compound (which is not usually colourless in practice). This is then oxidized, commonly by treatment with air and steam, to the fully coloured form of the dye, which is insoluble. Vat dyes are used mainly on cotton. The oldest vat dye is indigo (see page 83), derivatives of which are produced in the indigogenic techniques of enzyme histochemistry (see Chapter 15).

Sulphur dyes. These are mixtures of uncertain composition made by heating a variety of organic compounds with sulphur or with alkali metal polysulphides. A sulphur dye is applied (mainly to cellulosic fabrics) mixed with aqueous sodium sulphide, which causes reduction to the leuco compound. The insoluble coloured dye is regenerated by exposure of the dyed material to air. The process is similar in principle to vat dyeing. Sulphur dyes are not used in histology.

Ingrain dyes. Originally this term included all dyes generated within the substrate, but it is now used specifically for temporarily solubilized phthalocyanines that change into insoluble pigments after application to their substrates. The phthalocyanines are discussed on p. 86). Alcian blue, used in carbohydrate histochemistry, is the most important biological stain in this class.

Polycondensation dyes (condense dyes). This category includes dyes with a variety of chromophoric systems. They are applied as soluble monomers, the molecules of which then react with one another to form insoluble coloured polymers. Such dyes are not used as biological stains, but some histochemical methods for enzymes result in the production of coloured polymeric end-products.

Azoic dyes. These are formed within the substrate by reaction of diazonium salts (**azoic diazo components**) with suitable phenols or aromatic amines (**azoic coupling components**). They are important to the histochemist and are discussed later in this chapter.

Oxidation bases. Certain colourless bifunctional amines and aminophenols are used as dyes, especially for fur and hair. Oxidation, usually by hydrogen peroxide, produces insoluble brown or black polymers containing the quinonoid chromophore. The colour that develops may be modified by adding metal salts or phenolic compounds to the oxidation base. Similar chemical reactions are encountered in several histochemical techniques.

Disperse dyes. These are insoluble coloured compounds used as aqueous suspensions (particle size 1–10μm) for dyeing cellulose acetates and polyesters. The colouring process is enhanced by the presence in the dyebath of suitable oily agents (**carriers**), which are believed to promote the entry of the particles of dye into the interstices of the hydrophobic fibres, where water cannot penetrate. Disperse dyes are not used in histology.

Pigments. These are finely divided white, black, or coloured materials, insoluble in water and organic solvents. They are used as suspensions or emulsions in liquid media, as paints, and for the mass coloration of plastics, rubber, and synthetic fibres. There are many inorganic and organic pigments. They are often formed as the visible end-products of histochemical reactions.

Fluorescent brighteners. These compounds (see review by Gold, 1971) serve to increase the amount of visible light reflected from white or coloured objects. Invisible ultraviolet radiation absorbed by the brightening agent is emitted at a higher, visible wavelength. Most fluorescent brighteners contain the stilbene configuration (see p.71). Some of them are used in microscopy as fluorochromes. Their staining properties are those of the direct cotton dyes.

Food colours. This group includes dyes of various kinds whose principal industrial applications are in the manufacture of foodstuffs. Fast green FCF, an anionic triphenylmethane dye (p.76), is the only one commonly used as a histological stain.

5.8. Classification of stains by microscopists

Biological stains have traditionally been designated by their users as basic, acid, or mordant dyes or as "fat stains", the last-named being the lysochromes or solvent dyes. These four groups reflect in a rough and ready way the purposes for which the dyes are used. Thus basic (cationic) dyes are used to stain nuclei and the major anionic compounds of cytoplasm and extracellular structures. Acid (anionic) dyes demonstrate cytoplasmic and extracellular proteinaceous material. Mordant dyes are used mainly as nuclear stains, and lysochromes are for fats and other lipids.

This classification has the advantages of simplicity and widespread usage, but it is not adequate for the student trying to understand the reasons why different dyes are used, alone or in combination, for the demonstration of structural and chemical features of tissues. The uses of the dyes are determined by their physical and chemical properties, and these cannot be made to fit into any simple scheme of classification. Any dye used as a stain should be thought of in terms of its colour, its molecular size and shape, its solubility, and its potential chemical reactivity. All these properties affect the method of use and the results obtained.

5.9. Classification of dyes by chromophoric systems

We shall now consider the major groups of dyes following a chemical scheme of classification based on the structures of the major chromophoric systems. The more important members of each group, including all the dyes mentioned in later parts of this book, will be described. Some of the groups of dyes have little or no importance as biological stains, so for these only single representative examples are given. The number of dyes used in histology and histochemistry is likely to increase in the future, so the potential user should be aware of the variety available. For more information about individual dyes, see Venkataraman (1952–1978) and Lillie (1977). For a brief introduction to the literature of dye chemistry and usage, see Horobin (1983).

5.9.1. Nitroso dyes

Nitrous acid reacts with phenols to give nitrosophenols in which the radical —N=O replaces hydrogen *ortho* or *para* to the phenolic hydroxyl group. There is tautomerism between the nitroso compounds and the corresponding quinone oximes. Thus for *p*-nitrosophenol

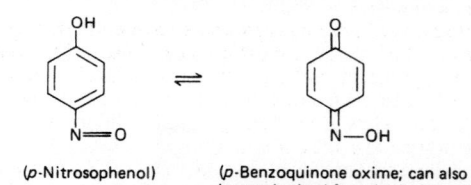

(*p*-Nitrosophenol)   (*p*-Benzoquinone oxime; can also be synthesized from hydroxylamine and *p*-benzoquinone)

The *o*-nitroso compounds can form coloured chelates with metal ions and so function as mordant dyes. Nitroso dyes have found hardly any uses in histology, though they have properties that might well be exploited.

Examples

Naphthol green Y (C.I. 10005; Mordant green 4; M.W. 173)

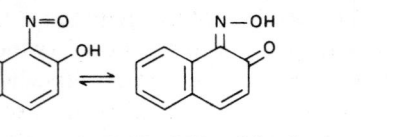

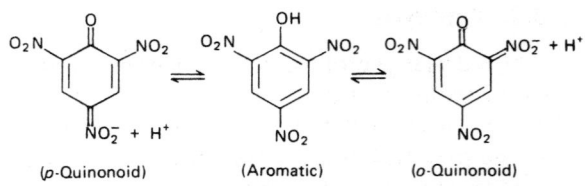

(*p*-Quinonoid)   (Aromatic)   (*o*-Quinonoid)

With ferrous ions and NaOH, this forms an important pigment, **pigment green B** (C.I. 10006; Pigment green 8), with the formula Na$^+$ [(Dye)$_3$Fe]$^-$.

NAPHTHOL GREEN B (C.I. 10020; Acid green 1; M.W. 901)

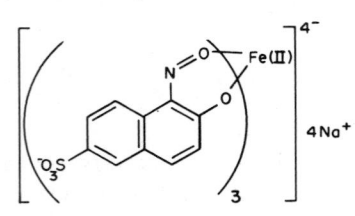

This stable dye-metal complex is used as the stain for collagen in some mixtures of anionic dyes.

5.9.2. Nitro dyes

The nitro group may be present as a substituent in dyes with other chromophoric systems, but there are some in which it is the only chromophore. Hydroxyl or amino groups are also present in such dyes, and aromatic and quinonoid tautomers exist as is shown below for picric acid. The nitro group, formally represented as —NO$_2$, is a resonance hybrid of the two structures

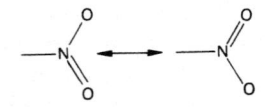

the two bonds between nitrogen and oxygen being equivalent to one another. The single N—O bond can alternatively be represented as N→O or $\overset{+}{N}$—$\overset{-}{O}$, the symbols having the same meanings as when used in formulae for metal complexes (see p. 56).

Examples

PICRIC ACID (2,4,6-trinitrophenol) (C.I. 10305; M.W. 299)

The phenolic hydroxyl group is ionized because of the electron-withdrawing effect of the three nitro groups. Picric acid is a yellow anionic dye of low molecular weight. It can also form "addition compounds" with a variety of substances. In these, the bonding is neither ionic nor covalent, its nature being incompletely understood (see March, 1977, for discussion). The addition compounds are charge-transfer complexes (see p. 54).

Picric acid is also used as a fixative (Chapter 2, where precautions attendant on its use are also mentioned). It is soluble in water (1.3%) and more soluble in alcohol and aromatic hydrocarbons, but xylene does not remove the dye from stained sections.

MARTIUS YELLOW (C.I. 10315; Acid yellow 24; Naphthol yellow; 2,4-dinitro-1-naphthol; M.W. 256 (sodium salt), 251 (ammonium salt), 273 (calcium salt), 234 (the unionized phenol))

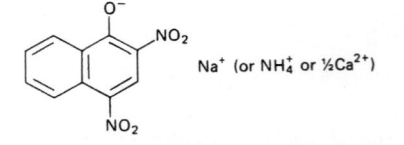

This dye is more deeply coloured (approaching orange) than picric acid and is somewhat more soluble in water and less soluble in alcohol. The ammonium and calcium salts (the latter less soluble in water) are also available. Like picric acid, Martius yellow is used as a cytoplasmic stain in conjunction with other anionic dyes.

A related compound, 2,4-dinitro-naphthol-7-sulphonic acid, is **flavianic acid**, well known to biochemists as a precipitant for the amino-acid arginine. The sodium salt of flavianic acid is **naphthol yellow S** (C.I. 10316), an anionic dye more soluble than Martius yellow.

5.9.3. Azo dyes

This very large group of dyes is of great importance in industry and to the biologist. It will be described in some detail; many points of theoretical and practical interest are conveniently illustrated by members of the azo series.

The chromophore is the azo group ($—N=N—$), which connects two aromatic ring systems. Dyes may contain one (monoazo), two (bisazo), three (trisazo), or, rarely, four (tetrakisazo) or more (polyazo) azo groups. The aromatic rings are usually benzene or naphthalene, and they may bear a variety of substituents.

5.9.3.1. Synthesis

Azo dyes are made by coupling diazonium salts with phenols or aromatic amines. The diazonium salts are made by reacting primary aromatic amines with nitrous acid, in acid solution, at temperatures at little above freezing:

$$Ar—NH_2 + HNO_2 + H^+ \rightarrow Ar—\overset{+}{N}\equiv N + 2H_2O$$

Most diazonium salts are unstable and cannot be isolated in pure form as solids, though some are available as stable complexes. Usually the solution containing the diazonium salt is used immediately. Excess nitrous acid can be decomposed by adding a little urea:

$$2HNO_2 + (NH_2)_2CO \rightarrow 3H_2O + 2N_2\uparrow + CO_2\uparrow$$

Coupling with phenols and amines occurs preferentially in the *para* position, but in the *ortho* position if another substituent or the site of fusion of rings is present *para* to the hydroxyl or amino group. With naphthols and naphthylamines, coupling is usually *ortho* to the hydroxyl or amine group. For example:

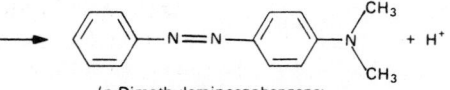

(Benzene diazonium salt)   (Dimethylaminobenzene)

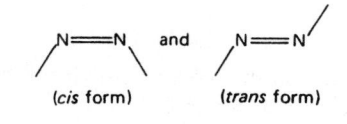

(*p*-Dimethylaminoazobenzene;
Oil yellow II; C.I. 11020;
Solvent yellow 2)

The product of this reaction is the well-known carcinogen "butter yellow" (which is neither a constituent of nor an additive to butter). It is insoluble in water and is not used in histology, but it is a dye on account of the auxochromic tertiary amine group. Solubility in water can be conferred by introducing a hydrophilic substituent. The sulphonic acid group is the one most often chosen and the sodium salt of the *para*-sulphonic acid derivative of butter yellow (formed from diazotized sulphanilic acid and *p*-dimethylaminobenzene)

is a water-soluble dye. It is rarely used as a stain but is familiar as an acid-base indicator, **methyl orange** (C.I. 13025; Acid orange 52).

In coupling with diazonium salts, phenols react as their anions whereas amines react as the unionized bases. Both these species exist in alkaline solutions. The diazonium ion, however, is changed in the presence of excess alkali into a diazotate anion, $Ar—N=N—O^-$, which does not participate in coupling reactions. Consequently, there is an optimum pH for every reaction between a diazonium salt and a phenol or amine. Usually a more strongly alkaline medium is required for coupling with phenols than with amines.

Azo dyes can also be produced by reactions of quinones with aryl hydrazines, but they are always manufactured by the coupling reactions of diazonium salts. For more complete accounts of the chemistry of azo compounds, the reader should consult a textbook of organic chemistry or the monograph of Zollinger (1961).

5.9.3.2. Structure

The azo linkage can exist in two isomeric forms:

$$\underset{\text{(cis form)}}{N\!=\!\!=\!\!N} \quad \text{and} \quad \underset{\text{(trans form)}}{N\!=\!\!=\!\!N}$$

In dyes, the *trans* configuration is usually present because of intramolecular hydrogen bonding between one of the azo nitrogen atoms and a substituent in the *ortho* position on at least one of the aromatic rings:

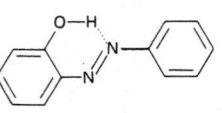

In some dyes, this arrangement provides a molecular shape conducive to the chelation of metal ions.

Phenolic azo dyes exist in tautomeric equilibrium with quinonoid forms:

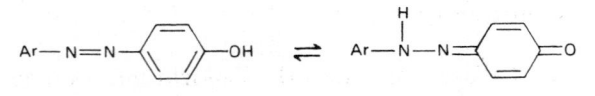

(Hydroxyazo tautomer)   (Hydrazone tautomer)

The dyes that are amines, however, usually occur only in the aminoazo form because of intramolecular hydrogen bonding.

5.9.3.3. Anionic azo dyes

The simpler anionic azo dyes are the "levelling" dyes. They are mostly monoazo compounds with sulphonic acid and phenolic substituents. An example much used in histology is orange G.

ORANGE G (C.I. 16230; Acid orange 10; M.W. 452)

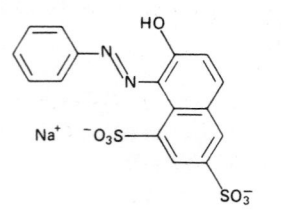

This dye is used as a cytoplasmic stain, usually in conjunction with other anionic dyes. It is very soluble in water but much less soluble in alcohol. Samples should contain at least 80% by weight of the anhydrous dye.

METANIL YELLOW (C.I. 13065; Acid yellow 36; M.W. 375)

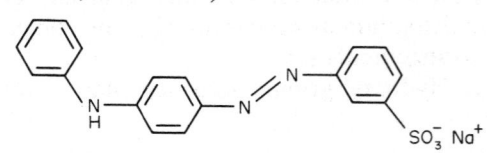

This dye, which is freely soluble in water and somewhat less so in alcohol, is used as a cytoplasmic stain in contrast to more deeply coloured dyes.

AMARANTH (C.I. 16185; Acid red 27; M.W. 604; also called azorubin S)

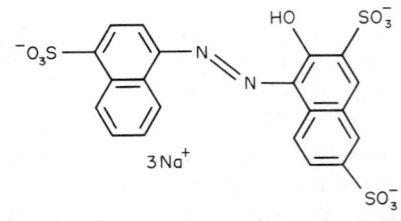

This red dye is used in combination with other anionic dyes, in methods for cytoplasm and connective tissue.

The "milling" dyes have larger molecules. They are bis- and tris-azo compounds. The "milling" dyes are used as anionic stains of large molecular size in methods for connective tissue (see Chapter 8). Some of the triphenylmethane dyes (see pp. 74–77) are used for the same purposes. Another group of anionic dyes with large molecules is that of the direct dyes, to be described below. Biebrich scarlet is a milling dye.

BIEBRICH SCARLET (C.I. 26905; Acid red 66; M.W. 557)

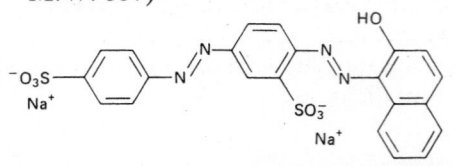

This dye is soluble in water but only slightly soluble in alcohol. It is used in several staining methods, as a counterstain or for collagen. Alkaline solutions of biebrich scarlet can be used alone for selective staining of strongly basic proteins (Spicer & Lillie, 1961).

5.9.3.4. Mordant azo dyes

These have functional groups capable of chelating chromium or similar metals. The commonest arrangements are:

(a) Hydroxyl groups *ortho* to both azo nitrogens:

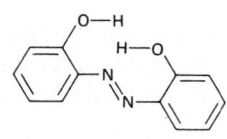

The chelate formed with an ion of a metal M has the structure

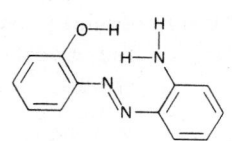

(Only three coordinate bonds to M are shown, but M will usually have coordination number 4 or 6)

It is possible for only one of the azo nitrogen atoms to be linked with one metal ion, the other being too far away as a consequence of the *trans* configuration of the azo group. However, it can be seen from the structural formula above that a metal with a coordination number higher than 3 could combine with another molecule of dye. In this way large aggregate molecules of the dye-mordant complex may be formed.

(b) Hydroxyl and amino groups *ortho* to both azo nitrogens:

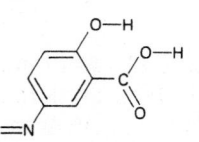

Chelates have structures analogous to that shown for (a) above.

(c) The salicylic acid arrangement on one of the aromatic rings:

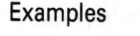

The formation of chelates by the salicylate ion is described on page 57.

Examples

SALICIN BLACK EAG (solochrome black A; C.I. 15710; Mordant black 1; M.W. 461)

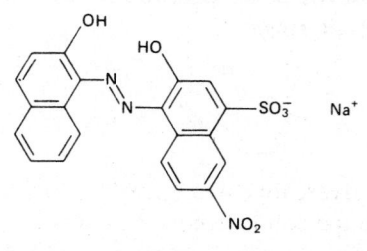

Soluble in water; slightly soluble in alcohol and acetone. The aluminium complex is fluorescent (orange emission) and is sometimes used as a fluorochrome.

CHROME ORANGE GR (Solochrome orange GRS) (C.I. 26520; Mordant orange 6; M.W. 470)

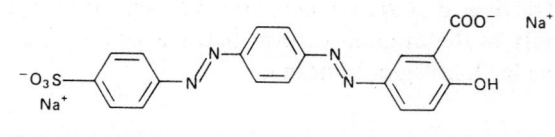

Soluble in water; slightly soluble in alcohol and acetone. Chrome orange GR is not used as a stain but it is an indicator, changing from yellow to red over the pH range 10.5–12.0.

5.9.3.5. Cationic azo dyes

Not many of these are used in histology. One that is encountered in a variety of techniques is Bismarck brown Y, which was the first commercially produced azo dye and is still used for colouring leather.

BISMARCK BROWN Y (Bismark brown G; vesuvin; leather brown; C.I. 21000; Basic brown 1; M.W. 419)

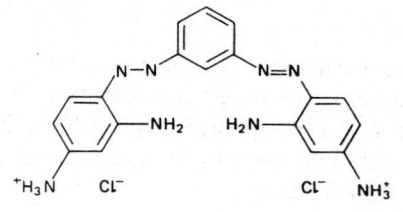

This dye is soluble in water and alcohol but insoluble in acetone and most other organic solvents. Samples should contain at least 45% of the anhydrous dyestuff.

5.9.3.6. Reactive azo dyes

Reactive dyes form covalent bonds with the dyed substrate and are used mainly on cotton (see p. 60), where the attachment is to the hydroxyl groups of cellulose. The commonest reactive auxochromes are the dichlorotriazinyl and monochlorotriazinyl groups. The following is a dichlorotriazinyl azo dye.

PROCION BRILLIANT RED M2B (C.I. 18158; Reactive red 1; M.W. 717)

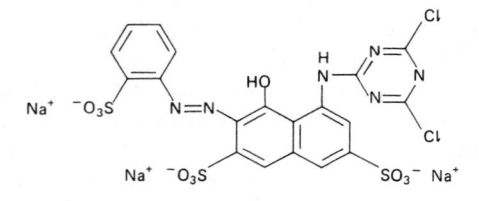

Reactive dyes are not used as stains for sections, but several applications in vital staining are cited by Lillie (1977). The constitutions of many reactive dyes, including PROCION YELLOW 4MR, which is used for injection into neurons through microelectrodes (see Kater & Nicholson, 1973), have not been revealed by the manufacturers.

5.9.3.7. Solvent azo dyes

These coloured substances are more soluble in lipids than in the solvents from which they are applied. Chemical dyeing does not occur. Solubilizing groups such as $-SO_3^-$ and $-NH_3^+$ are absent. Phenolic hydroxyl groups are usually present, but may be esterified without affecting the staining properties. Sudan black B is the most important histochemical reagent in this group (see Chapter 12).

Examples

SUDAN IV (C.I. 26105; Solvent red 24; M.W. 380)

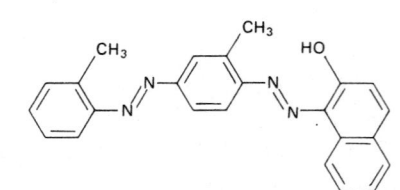

Sudan IV is suitable only for staining conspicuous accumulations of lipids, as in adipose tissue. The dye content should be at least 80% by weight. Sudan III (C.I. 26100; Solvent red 23; M.W. 352), which is less deeply coloured, lacks the two methyl groups of Sudan IV.

OIL RED O (C.I. 26125; Solvent red 27; M.W. 409)

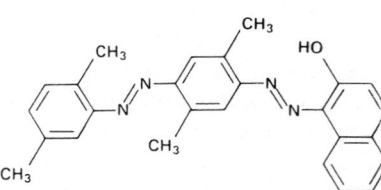

This dye is more intensely coloured than Sudan IV and therefore allows the resolution of smaller lipid-containing structures, such as intracellular droplets. Oil red O is often present as a contaminant in Sudan III and Sudan IV. All three dyes are mixtures of up to eight coloured components (Marshall, 1977).

SUDAN BLACK B (C.I. 26150; Solvent black 3; M.W. 457)

The dyestuff contains two major components (Pfüller et al., 1977), with the structures:

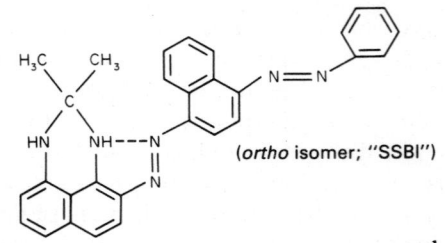

(ortho isomer; "SSBI")

and

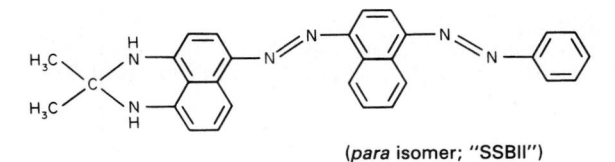

(*para* isomer; "SSBII")

The second formula (SSBII) corresponds to the Sudan black B (C.I. 26150; Solvent black 3; M.W. 457) of *Conn's Biological Stains* (Lillie, 1977). There is no satisfactory assay method for Sudan black B. Batches of the dye are evaluated by the Biological Stains Commission on the basis of their absorption spectra and their performance in methods for staining lipids.

Sudan black B can dissolve in hydrophobic lipids but, by virtue of its two potentially ionizable nitrogen atoms, may also behave like a cationic dye and bind to the hydrophilic phosphate-ester groups of phospholipids, in which it is more soluble than the other Sudan dyes. The *para* isomer (SSBII) is more strongly basic than the *ortho* (SSBI) because in the latter an internal hydrogen bond between one of the ionizable nitrogens and an azo nitrogen tends to inhibit ionization. A *bis-N*-acetylated derivative of Sudan black B can be prepared in the laboratory and some workers prefer this reagent to the commercially supplied dyestuff. Acetylated Sudan black B should not be able to form cations, but Marshall (1977) was unable to find in it any chromatographically distinct components that were not present in the parent dye.

5.9.3.8. Direct azo dyes

These, which are also called cellulose-substantive dyes, are capable of colouring cellulosic fibres (cotton and linen) without the help of mordants or special reactive auxochromic groups. Ionic auxochromes cannot be attracted by unionized hydroxyls, which are the only functional groups present in cellulose. The mode of action of the direct dyes has puzzled chemists for many years. The dyes are all anionic. Most are dis- or tris-azo dyes and all have long molecules with usually at least five aromatic rings (naphthalene counting as two). The rings must not be sterically hindered from assuming a coplanar configuration. In aqueous solution, direct dyes form molecular aggregates at room temperature but not at 90–100°C, the temperature at which they are used for dyeing textiles.

The entry of direct dyes into porous cotton or linen fibres is facilitated by interfacial effects (see p. 53). When the dye molecules are in the pores, they are probably close enough to the cellulose to be attracted by weak intermolecular forces. Hydrogen bonds are believed to form between the hydroxyl hydrogen atoms of the substrate and the extensive delocalized π-electron system of the dye (see also p. 55, and Gordon & Gregory, 1983). Additional dye molecules may be retained in the pores as a result of aggregation due to van der Waals forces between the hydrophobic parts of dye molecules (see Allen, 1971).

Some of the direct azo dyes are used in histological work, especially in techniques for the demonstration of amyloid, fibrin, and elastin (see Chapter 8). The various possible modes of dyeing of cotton may or may not be involved in these procedures, which are carried out at room temperature. The dyes can, of course, behave as ordinary anionic dyes towards cationic components of a tissue. In this respect the direct dyes resemble anionic azo dyes of the "milling" class described earlier.

Examples

CONGO RED (C.I. 22120: Direct red 28; M.W. 697)

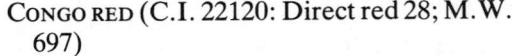

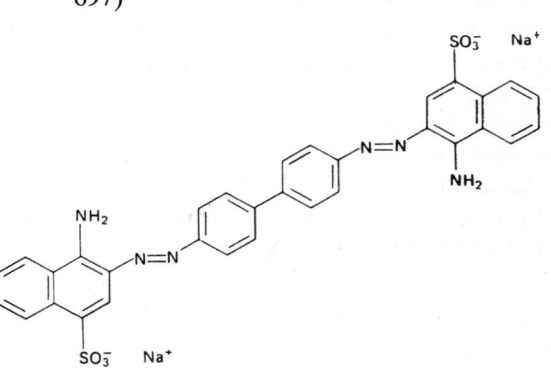

This dye is much more soluble in water than in alcohol. It is an indicator, changing from red to blue when the pH is less than about 3.0. It has several uses in histology, the best known being as a stain for amyloid. Samples should contain at least 75% by weight of the anhydrous dye.

CHLORAZOL BLACK E (C.I. 30235; Direct black 38; M.W. 782)

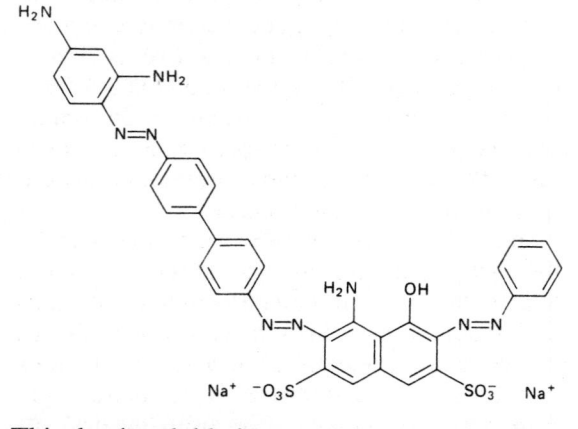

This dye is soluble in water and alcohol but not in most other organic solvents. It is used alone as a stain for sections, especially of plant tissues. The dye is amphoteric. In strongly acid solutions the molecule will bear a net positive charge.

5.9.3.9. Azoic dyes

Azoic dyes are insoluble azo compounds synthesized within or upon substrates by the combination of diazonium salts with suitable aromatic amines or phenols (known as "azoic coupling components"). In histological sections, attachment of the dye to protein, or solution in lipids, may occur as the dye is being formed. The importance of azoic dyes in microtechnique is due to their formation as the end-products of many histochemical methods. Some of these will be discussed in later chapters.

Diazonium salts and azoic coupling components are commonly used reagents in histochemistry. The diazonium salts are occasionally prepared immediately before use but more often they are bought ready-made as stabilized diazonium and tetrazonium salts. These are commonly zinc chloride double salts of the form

$$(R\overset{+}{-}N\equiv N)_2[ZnCl_4]^{2-}$$

Others are salts of sulphonic acids, such as the naphthalene-1,5-disulphonates,

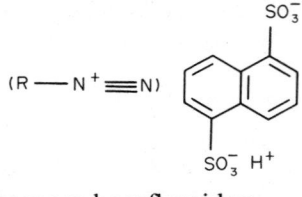

and yet others are borofluorides:

$$(R\overset{+}{-}N\equiv N)BF_4^-$$

A few are simple chlorides, $R\overset{+}{-}N\equiv N\ Cl^-$ or bisulphates, $R\overset{+}{-}N\equiv N\ HSO_4^-$. The diazonium salt is mixed with an inert diluent such as sodium sulphate, and other substances may also be present as stabilizers. The original amine usually accounts for less than half the weight of the commercially obtained powder. Stabilized diazonium salts are also called "azoic diazo components". They should **not** be referred to as "stable diazotates"; diazotate ions have the structure R—N=N—O⁻ and are formed when strong alkalis react with diazonium ions. The terms "tetrazonium" and "hexazonium" are used for salts with two and three diazotized amino groups per molecule.

(a) Examples of commonly used stabilized diazonium and tetrazonium salts

FAST RED B SALT (Diazotized 5-nitro-anisidine acid-1,5-naphthalene disulphonate; C.I. 37125; Azoic diazo 5; M.W. 467)

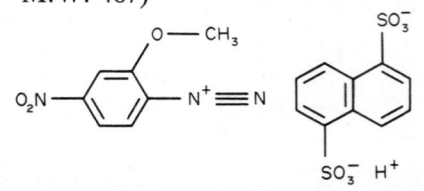

The primary amine represents 20% by weight of the commercial product (theoretical: 36% of formula weight). This diazonium salt is the one most

often used for staining argentaffin cells (see Chapter 17). It does not give satisfactory results in histochemical methods for hydrolytic enzymes.

FAST BLUE RR SALT (C.I. 37155; Azoic diazo 24; M.W. 776 including two molecules of diazo)

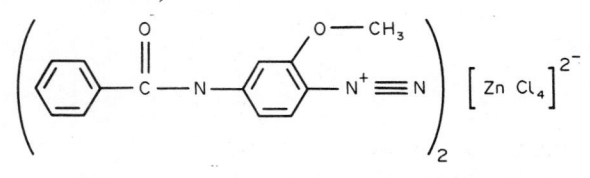

The primary amine accounts for 36% of the weight of the commercial product (theoretical: 71%). This diazonium salt is used in many methods for hydrolytic enzymes (see Chapter 15).

FAST BLUE B SALT (Tetraazotized o-dianisidine zinc chloride double salt; C.I. 37235; Azoic diazo 48; M.W. 476)

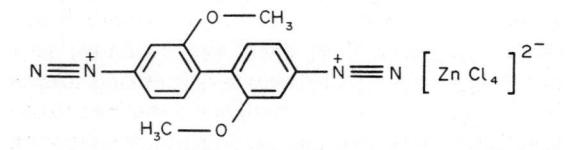

The theoretical content of the bifunctional primary amine (o-dianisidine) is 51% by weight but the commercial product contains about 20%. This is a "tetrazonium" salt with a diazotized amine group at each end of the molecule. It is used in the coupled tetrazonium reaction for proteins (Chapter 10).

FAST GARNET GBC SALT (C.I. 37210; Azoic diazo 4; M.W. 334)

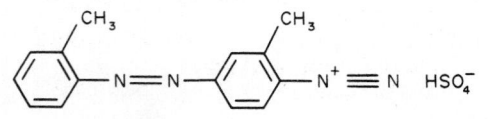

Contains 15–20% by weight of primary amine (theoretical: 67%). Used in some histochemical methods for enzymes. See also note under fast black K, below.

FAST BLACK K SALT (C.I. 37190; Azoic diazo 38; M.W. 836, including two molecules of diazo)

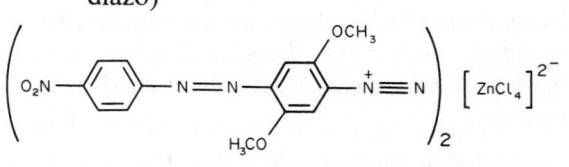

Lillie (1962) assumed that the primary amine represented 20% by weight of the commercial product (theoretical, from formula above, is 72%).

In fast garnet GBC salt and fast black K salt, azo linkages are already present in the diazonium ions. These salts are therefore coloured. Coupling with a phenol or amine produces a dis-azo dye, which has a darker colour.

Many dyes that are primary amines can be diazotized with nitrous acid to give coloured diazonium salts. Subsequent coupling with suitable aromatic molecules in sections of tissues results in the formation of new azo compounds with colours different from and often darker than those of the parent dyes. Pararosaniline and safranine O (Lillie *et al.*, 1968) are two dyes from which histochemically useful diazonium salts are easily prepared in the laboratory.

(b) Azoic coupling components

Most of the coupling components used in industry are derivatives of naphthalene, with hydroxyl, amine, and sulphonic acid groups substituted in various positions. An example is:

H-ACID   (8-amino-l-naphthol-3,6-disulphonic acid; M.W. 319)

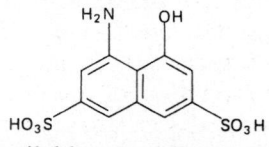

This is also available as a sodium salt (M.W. 341). Coupling occurs principally at position 2 (*ortho* to the hydroxyl group) in alkaline solution. This reagent is used in the coupled tetrazonium reaction (see Chapter 10) for the histochemical detection of proteins.

In enzyme histochemistry, azoic coupling components are released by enzymatic hydrolysis of certain of their derivatives, notably esters and amides. Such methods are discussed in Chapter 15. The coupling components most often liberated in histochemical methods for enzymes are naphthols:

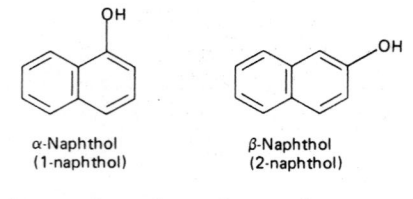

α-Naphthol
(1-naphthol)

β-Naphthol
(2-naphthol)

More intensely coloured azo dyes are formed from β-naphthol, but these are more soluble and form larger particles than the coupling products of α-naphthol. The advantages of both the simple naphthols are found in naphthol-AS (also known as benzosalicylanilide):

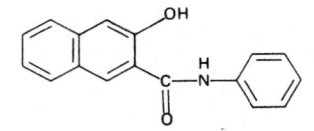

This compound couples with diazonium salts to give finely granular, brightly coloured insoluble products, but the rate of coupling is slower than with α- and β-naphthols.

5.9.4. Indamine and indophenol dyes

The chromophore is an indamine group (—N=) joining two rings—one aromatic and the other quinonoid. In the **indamine dyes** these rings bear, respectively, amine (—NRR′) and iminium (=N$^+$RR′) groups *para* to the bridge. In the **indophenol dyes** at least one ring carries an oxygen atom (phenol or quinone) *para* to the indamine linkage. In the general formulations, R and R′ may be hydrogen, methyl, ethyl, or similar radicals.

Dyes and pigments of this type are formed when the products of oxidation of amines react with other amines or with phenols. They are formed in the development of colour photographs and as end-products of some histochem-

ical reactions (see Chapter 16). The chemistry of the syntheses of indamine and related dyes is described by Weissberger (1966).

None of the synthetic indamine or indophenol dyes are important as histological stains, so only one example is given:

FAST BLUE Z (C.I. 49705; Solvent blue 22; M.W. 304)

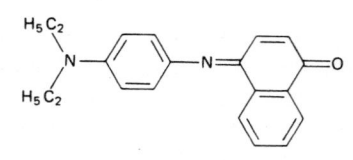

A blue indophenol dye, insoluble in water but soluble in alcohol and non-polar solvents. The tertiary amine function is potentially ionizable to give a cationic dye.

5.9.5. Polymethine dyes

Most of these dyes fade on exposure to light, so only a few are used in microscopy or for textiles. The relation of colour to chemical structure has been thoroughly studied in the cyanines (see Brooker, 1966), which are used as sensitizing agents in photographic emulsions.

The chromophore has the general structure:

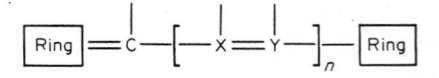

It forms a bridge between two charged aromatic rings, at least one of which is heterocyclic. If $n=0$, the bridge is a single methine (=CH—) group. If $n>0$ there are several possible structures, depending on the identity of X and Y. The **cyanine** dyes (X and Y both carbon; both rings heterocyclic) are occasionally used as stains. Related dyes include **hemicyanines** (X and Y both carbon; only one ring heterocyclic) and **azamethines** (X and/or Y is nitrogen). The structures of some of the polymethine dyes bear obvious resemblances to dyes of the azo, indamine and stilbene classes.

Example of a cyanine

CARBOCYANINE DBTC (No C.I. number; M.W.560)

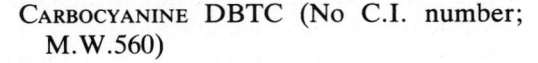

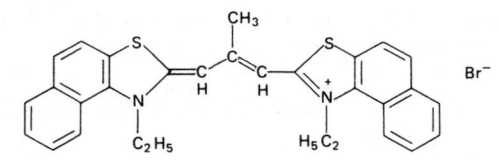

This dye, also known as "Stains-all", is used alone to stain sections in various shades of blue, purple, and red (Green & Pastewka, 1974a,b, 1979), but its usefulness is unfortunately rather limited because the stained sections fade rapidly on exposure to light.

5.9.6. Stilbene dyes and fluorescent brighteners

Stilbene is

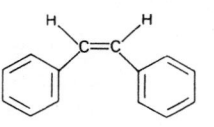

(the *cis*-isomer is shown). Several dyes include the stilbene configuration in conjunction with azo groups. Most of these dyes are mixtures of large molecules of uncertain structure and none are important as biological stains.

Fluorescent brighteners absorb ultraviolet radiation and emit blue or green light. A fabric containing one of these compounds appears more vividly white than it otherwise would. Thus fluorescent brighteners can offset the tendency of white materials to turn yellow with age. Most fluorescent brighteners are derivatives of 4,4'–diaminostilbene–2,2'-disulphonic acid,

$$HO_3S(NH_2)C_6H_3CH{=}CHC_6H_3(NH_2)SO_3H$$

Example

CALCOFLUOR WHITE M2R (C.I. 40622; Fluorescent brightening agent 28; M.W. 896)

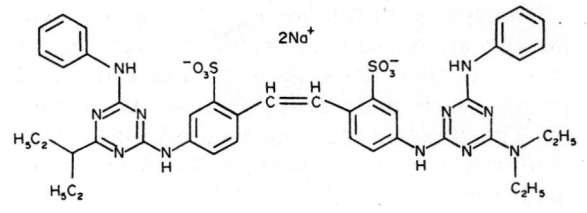

Fluorescent brighteners have been used in microscopy as fluorochromes, especially for plant tissues. Calcofluor white M2R binds to cellulose and does not enter living cells (Fischer *et al.*, 1985).

5.9.7. Quinoline dyes

In this small group of yellow dyes, the chromogen is quinophthalone:

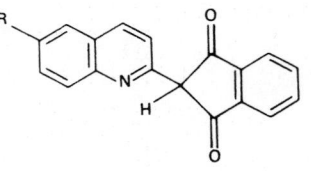

Sulphonic acid substituents, if present, confer water-solubility. The quinoline dyes are of no importance in microtechnique. The quinoline ring is present in other types of dyes, notably in some of the cyanines and their congeners.

5.9.8. Thiazole dyes

The benzothiazole ring

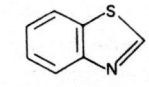

is the sole chromogen in dyes of this group. Dyes with other chromophoric systems (especially azo dyes) sometimes also contain benzothiazole rings.

Examples

PRIMULINE (C.I. 49000; Direct yellow 59) This greenish-yellow dye (amphoteric: here shown as anionic) is a mixture, the principal component of which is

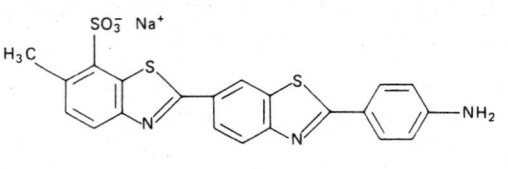

(M.W. 476)

It has been used as a vital stain and fluorochrome (e.g. Kuypers *et al.*, 1977). With textiles, primuline can function as a "direct" dye for cellulosic fibres. It can be diazotized on the fibre and then coupled with suitable phenols or amines to give more intense, faster colours.

THIOFLAVINE TCN (Thioflavine T) (C.I. 49005; Basic yellow 1; M.W. 319)

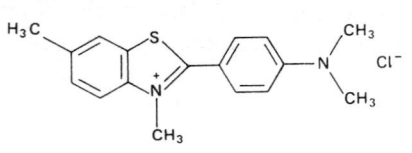

This yellow cationic dye can also serve as a fluorochrome. An insoluble yellow pigment is formed with phosphotungstomolybdic acid.

5.9.9. Hydroxyketone dyes

In these dyes, the chromogen is an array of conjugated double bonds including at least one carbonyl group. A hydroxyl group attached to an aromatic or quinonoid ring is also present. The group includes naturally occurring and synthetic dyes, of which only the former are important in histology. Brazilein and haematein are hydroxyketones formed by oxidation of the naturally occurring compounds brazilin and haematoxylin.

HAEMATOXYLIN (C.I. 75290; Natural black 1) Haematoxylin is extracted from the heart-wood of *Haematoxylon campechianum*, the logwood tree of South and Central America. Haematoxylin is very soluble in both water and alcohol but dissolves more rapidly in the latter. Haematein (see below) is only sparingly soluble in these sol-

vents but is much more soluble in ethylene glycol and glycerol, which are also solvents for haematoxylin.

Upon partial oxidation, haematoxylin is changed into **haematein**, which is the active ingredient of the "haematoxylin" stains used in histology. The oxidation may be brought about slowly by exposure to air (ripening) or rapidly by adding an oxidizing agent such as sodium iodate or ferric ions. One gram of haematoxylin crystals (with $3H_2O$ per dye molecule) is fully oxidized to haematein by 185 mg of $NaIO_3$. In practice, when the mordant is to be aluminium, smaller amounts of oxidizing agents are used because haematein itself is slowly oxidized by atmospheric oxygen to give useless products. A reservoir of initially unoxidized haematoxylin in a staining solution is believed to allow for the continued production of haematein by slow atmospheric oxidation and so prolong the useful life of the reagent.

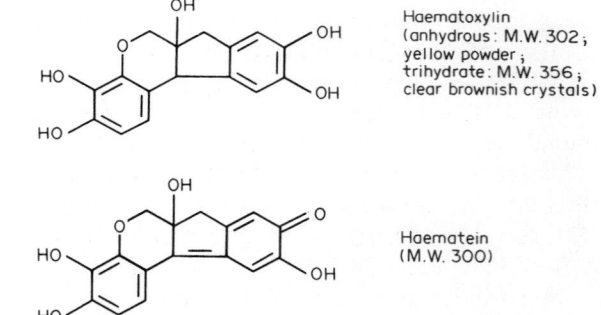

Haematoxylin
(anhydrous: M.W. 302;
yellow powder;
trihydrate: M.W. 356;
clear brownish crystals)

Haematein
(M.W. 300)

Ferric salts also serve to oxidize haematoxylin to haematein, so that iron-haematoxylin mixtures do not need to be "ripened". However, the concentration of ferric ions in such stains is usually greatly in excess of that needed to oxidize the dye and form the dye-metal complex, so the solutions are not indefinitely stable.

The main product of over-oxidation of haematein is an orange–yellow substance known as **oxyhaematein**. It is formed more quickly by oxidation with hydrogen peroxide, potassium permanganate, or sodium iodate than by exposure to oxygen. Marshall and Horobin (1972b) have provided strong (though not conclusive) evidence that oxyhaematein has the structure

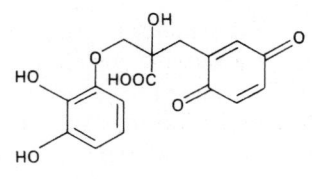

and have shown that it can behave as an ordinary anionic dye whose properties are unaffected by mixing with aluminium salts.

Staining with haematein is carried out in conjunction with a mordant, most often aluminium (for staining nuclei blue–purple) or iron (for staining nuclei, myelin or other elements blue-black, according to technique). Preformed dye–metal complexes are used for most purposes, though in the classical method of Heidenhain the mordant and the dye are applied sequentially to the sections. Many other mordants are used for special purposes. A list of mordants and structures demonstrated is given by Culling (1974, p. 160) and several recipes for mixtures containing haematoxylin are given by Bancroft & Cook (1984). Nuclear staining with metal–haematein complexes is described in Chapter 6. A method in which the dye is used for the localization of certain lipids is given in Chapter 12, and the staining of myelin is discussed in Chapter 18.

Spectrophotometric studies have shown that commercial samples of haematoxylin are usually quite pure (90–95% of anhydrous dyestuff). Materials sold as "haematein", however, are very variable, with dye contents ranging from 1.2 to 75.1% (Marshall & Horobin, 1974). When haematein is required, it is better to make it by oxidizing haematoxylin in the laboratory than to buy it.

BRAZILIN (C.I. 75280; Natural red 24)
Brazilin is extracted from brazil wood (*Caesalpina sappan* from Indonesia and *C. brasiliensis* from Central and South America). The country of Brazil was named after the discovery there of trees yielding the dyestuff, which had previously been imported into Europe from oriental sources. Brazilin differs from haematoxylin only in lacking one hydroxyl group. Upon oxidation, a red dye, **brazilein**, is produced.

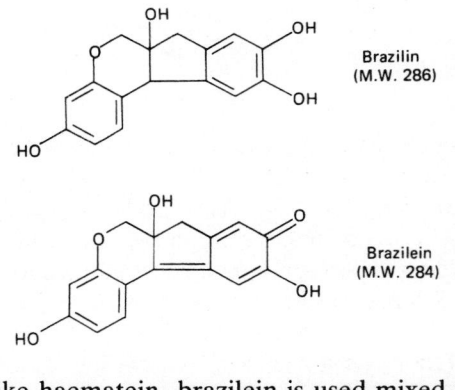

Like haematein, brazilein is used mixed with a metallic mordant. The use of the aluminium complex as a red nuclear stain is described in Chapter 6.

5.9.10. Arylmethane dyes

These are compounds in which three of the hydrogen atoms of methane are replaced by aromatic ring systems. The latter bear ionizable substituents and the rings resonate between quinonoid and truly aromatic configurations. The chromophoric system of conjugated bonds extends through all parts of the molecule, so that an unusually large number of canonical forms can be formulated for each dye. Diarylmethanes are not very important as biological stains, but the triarylmethane class includes many dyes of great value to the histologist and histochemist.

5.9.10.1. Diphenylmethane dyes

The only member of this group used in histology is auramine O.

AURAMINE O (C.I. 41000; Basic yellow 2; Anhydrous M.W. 304; commercial product has one molecule water of crystallization; M.W.322)

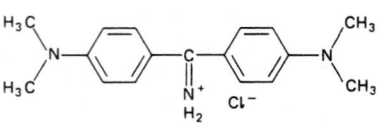

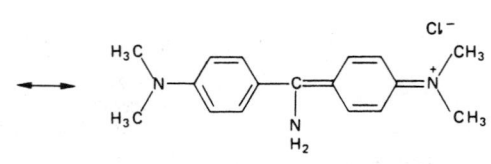

These formulae depict two of the possible resonance structures of the chromophoric system: an imine (above) and a quinonoid form. Auramine O is a yellow, fluorescent cationic dye used mainly in bacteriology (see Clark, 1981). Samples should contain at least 80% by weight of the anhydrous dye.

5.9.10.2. Triarylmethane dyes

The simplest members of this class are the triphenylmethanes. Of these, pararosaniline, a red cationic dye, has the most easily understood structure. This dye, which has many uses in histology and histochemistry, will therefore be described at some length. Other important triarylmethane dyes will be treated more briefly. Examples of cationic dyes will be followed by examples of anionic and mordant dyes.

PARAROSANILINE (C.I. 42500; Basic red 9; M.W. 324)

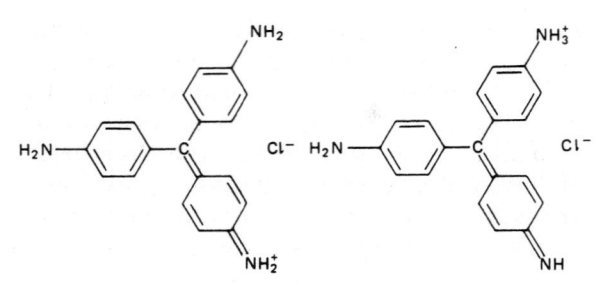

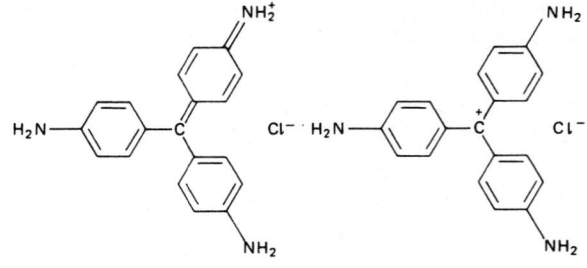

Four of the possible resonance structures are shown. It can be seen that any one of the three rings may transiently assume the aromatic or the quinonoid configuration and any may bear a positively charged nitrogen atom. Similar resonance occurs in all the triarylmethane dyes, anionic and cationic. By convention the iminium ion forms are usually shown in structural formulae. Acetate replaces chloride as the gegen-ion in some samples of pararosaniline. For notes on purity, see under basic fuchsine (p.75).

Colourless derivatives of three kinds are easily prepared from pararosaniline and related dyes. Reduction destroys the chromophoric structure and gives a **leuco compound**:

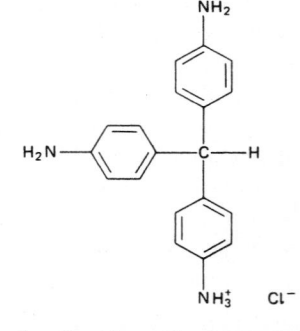

This is a salt of amine, the **leuco base**:

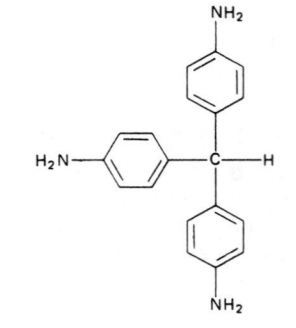

When pararosaniline is treated with strong alkali, an unstable **colour base** is first formed:

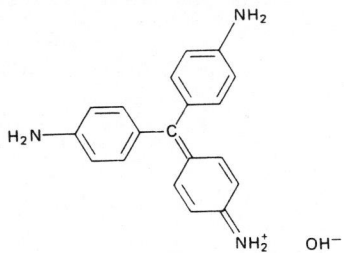

The unstable colour base rapidly isomerizes in solution to give the stable, colourless **carbinol base**:

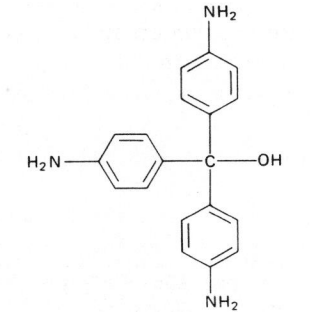

It can be seen that the carbinol base is a tertiary alcohol. It is named as a derivative of methanol (=carbinol; H_3COH).

The third way of decolorizing pararosaniline involves treatment with sulphurous acid, to give **Schiff's reagent**, the chemistry and uses of which are described in other chapters.

Other derivatives of pararosaniline are prepared in the laboratory for use in some special staining methods. These include a **hexazonium salt** (used in azo coupling procedures) and **aldehyde–fuchsine** (see Chapter 8).

BASIC FUCHSINE

This is a mixture of triphenylmethane dyes. Modern samples of basic fuchsine consist almost entirely of either **pararosaniline** or **rosaniline**. Older samples usually contained both dyes, together with smaller amounts of **new fuchsine** and **magenta II.** The gegen-ion is sometimes acetate rather than the chloride shown in the formulae below.

ROSANILINE (C.I. 42510; Basic violet 14; Anhydrous M.W. 338; crystals with $4H_2O$, M.W. 410)

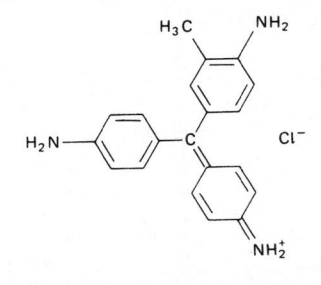

This dye is closely similar to pararosaniline and can replace it for all purposes except the preparation of aldehyde–fuchsine.

NEW FUCHSINE (Magenta III) (C.I. 42520; Basic violet 2; M.W. 366)

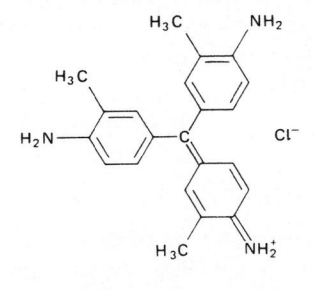

MAGENTA II (No C.I. number; M.W. 352)

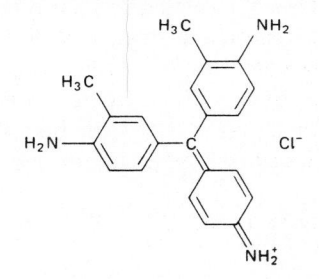

All the components of basic fuchsine are soluble in water and more soluble in alcohol. For critical work it is desirable to use pure pararosaniline. To be certified by the Biological Stains Commission, samples of basic fuchsine must contain at least 88% anhydrous dye (estimated as pararosaniline or rosaniline) and must perform satisfactorily in a variety of staining procedures.

CRYSTAL VIOLET (C.I. 42555; Basic violet 3; M.W. 408)

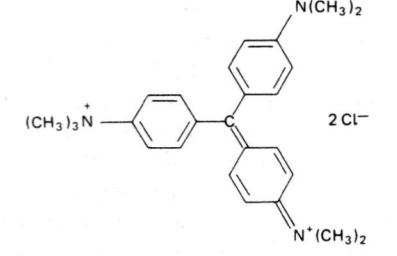

This *N*-hexamethylated derivative of pararosaniline is used in many bacteriological techniques as well as in histology. Samples should contain at least 88% by weight of the anhydrous dye.

METHYL GREEN (C.I. 42585; Basic blue 20; M.W. of the dichloride: 458.5. Lillie (1977) states that this dye is supplied as a zinc chloride double salt. This is [$C_{26}H_{33}H_3$]$^{2+}$ [$ZnCl_4$]$^{2-}$, M.W. 594)

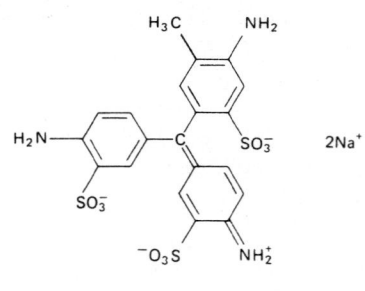

ETHYL GREEN (C.I. 42590; M.W. 517).
The quaternary methyl groups of methyl green are replaced by ethyl, and one of the chloride ions by bromide.

According to Lillie (1977) and Lyon *et al.* (1987), dyes sold as "methyl green" are nearly always ethyl green. These bluish-green dyes are used in several techniques, including the methyl green-pyronine method for nucleic acids. One of the alkyl groups on the quaternary nitrogen (left-hand side of formula above) detaches easily, so methyl green is always contaminated with crystal violet. The latter dye can be extracted from aqueous solutions of methyl green by shaking with chloroform in which the colour base of crystal violet is very soluble. Samples should contain at

least 65% of the anhydrous, zinc-free dye (M.W. 458.5).

ACID FUCHSINE (C.I. 42685; Acid violet 19)
This is a di- or trisodium salt of the trisulphonic acid derivative of basic fuchsine. The acid fuchsine derived from rosaniline is

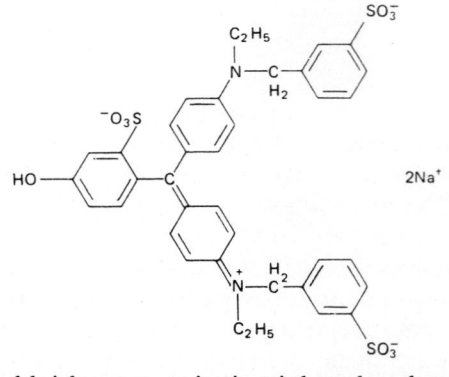

The one related to pararosaniline lacks the methyl group.

This red anionic dye has many uses, especially in conjunction with other anionic dyes, as will be explained in Chapter 8. It is very soluble in water but only sparingly so in alcohol.

For analytical purposes acid fuchsine is assumed to consist of equal parts of the anhydrous disodium salts of trisulphonated pararosaniline and rosaniline (average M.W. 578.5). Samples should contain at least 55% by weight of this hypothetical dyestuff.

FAST GREEN FCF (C.I. 42053; Food green 3; M.W. 809)

This bluish-green anionic triphenylmethane dye is used in methods for connective tissue and as a counterstain to methods that produce strong blues, purples, and reds. It is very soluble in water and considerably less so in alcohol.

A closely related dye is **light green SF** (light green SF yellowish; C.I. 42095; Acid green 5; M.W. 793), which lacks the phenolic hydroxyl group of fast green FCF. The two dyes can be used for the same purposes, but C.I. 42053 is preferred because it is less prone to fading. Samples of fast green FCF should contain at least 85% by weight of the anhydrous dye. Light green SF should be at last 65% pure.

ANILINE BLUE WS (water blue I; C.I. 42755; Acid blue 22; M.W. 738)

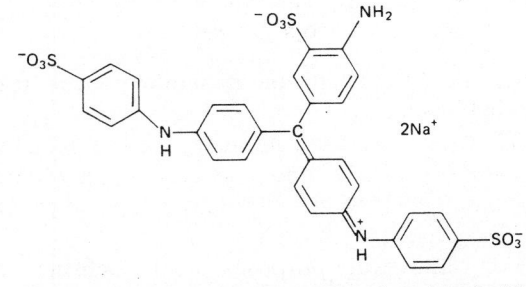

METHYL BLUE (soluble blue; ink blue) (C.I. 42780; Acid blue 93; M.W. 800)

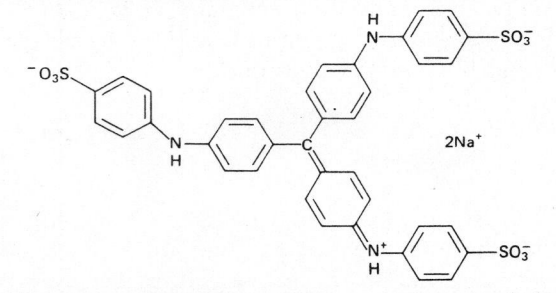

Both the above dyes and also a mixture of the two have been known in the past as "water blue", "soluble blue", and "aniline blue, water-soluble". The two dyes have identical staining properties. Both are very soluble in water and insoluble in alcohol. The colour changes to red with strong alkalis. They are used as anionic dyes of large molecular size, especially in methods for connective tissue. These dyes cannot be assayed very precisely. The criteria for certification by the Biological Stains Commission are given by Lillie (1977).

SPIRIT BLUE (C.I. 42775; Solvent blue 3) also known as "aniline blue, alcohol-soluble" is a mixture of diphenyl and triphenyl pararosanilines (i.e. aniline blue WS and methyl blue without the sulphonic acid groups). Spirit blue is a cationic dye but is insoluble in water. Its alcoholic solutions are occasionally used in histology.

CHROMOXANE CYANINE R (solochrome cyanine R; eriochrome cyanine R; C.I. 43820; Mordant blue 3; M.W. 492)

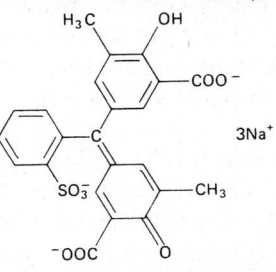

This anionic hydroxytriphenylmethane dye is employed in conjunction with a mordant. This is usually ferric iron in histological practice, though strongly coloured complexes are also formed with aluminium (red) and chromium (blue). Red and blue iron complexes are involved in staining (Kiernan, 1984a). The dye is used as a general-purpose blue and red single stain (Chapter 6), and in selective methods for nuclei (Chapter 6) and for myelin (Chapter 18). Chromoxane cyanine R is sold as a powder containing about 50% by weight of the dye, usually as its monosodium salt, not the trisodium salt shown above and in most catalogues and works of reference.

5.9.11. Xanthene dyes

The chromophoric system is

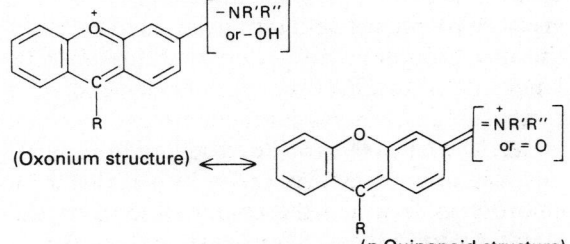

(Oxonium structure) ⟷

(*p*-Quinonoid structure)

R, R', and R" may be hydrogen or any of a variety of alkyl or aryl radicals. Other substituents may also be present. If R is a phenyl group, the dye will also be a triphenylmethane derivative in which two of the benzene rings are joined by an ether linkage. Many of the xanthene dyes are fluorescent and some are used primarily as fluorochromes. The *p*-quinonoid structure is (arbitrarily) used in the structural formulae for the examples of xanthene dyes given below. The dyes are classified by ionic charge rather than as hydroxy- and amino-xanthenes.

Examples

Cationic xanthene dyes

PYRONINE Y (pyronine G) (C.I. 45005; M.W. 303)

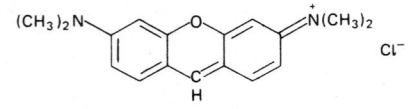

This red dye is used, together with ethyl or methyl green, in a method for nucleic acids (see Chapter 9). The dye is very soluble in water and alcohol. Samples should contain at least 45% by weight of anhydrous dye.

PYRONINE B (C.I. 45010) differs only in having ethyl instead of methyl groups on the nitrogen atoms, but it cannot be substituted for pyronine Y.

RHODAMINE B (C.I. 45170; Basic violet 10; M.W. 479)

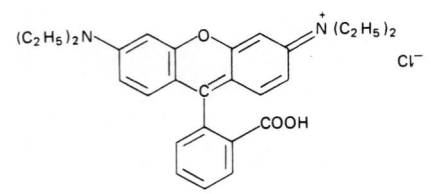

This red–violet dye emits a strong orange fluorescence when excited by either ultraviolet or green light. An isothiocyanate radical (S=C=N—) can be substituted in the carboxyphenyl group to give **rhodamine B isothiocyanate**, a reactive dye capable of conjugating with proteins to yield fluorescently labelled derivatives (see Chapter 19).

Anionic xanthene dyes

FLUORESCEIN SODIUM (uranin; C.I. 45350; Acid yellow 73; M.W. 376)

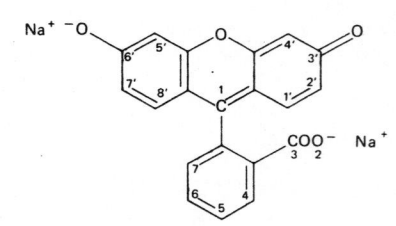

The numbering is as given by Lillie (1977).

This yellow anionic dye is used only as a fluorochrome. As with rhodamine B, an isothiocyanate can be prepared and this is used as a fluorescent label for proteins (see Chapter 19). Optimal excitation is by blue–violet or ultraviolet light and the fluorescent emission is green.

The halogenated derivatives of fluorescein are also fluorescent but are used principally as red dyes. Eosin is the most important.

EOSIN AND ITS CONGENERS

These red dyes are substituted fluoresceins with the structures indicated in Table 5.1. All the dyes in the table are red, in various shades. All are freely soluble in water, rather less soluble in alcohol, and are precipitated (as the colour acids) by mineral acids. All can be used for similar purposes, though certain dyes are traditionally associated with particular techniques. For certification by the Biological Stains Commission, the percentage by weight of anhydrous dyes in a sample must not be less than 80% (eosin), 85% (eosin B), 80% (phloxine B), or 80% (rose Bengal).

TABLE 5.1. *Eosin and related dyes*

Name and molecular weight	C.I. number	Substituent on fluorescein skeleton at carbon number									
		1'	2'	4'	5'	7'	8'	4	5	6	7
Eosin (eosin Y, yellowish; C.I. Acid red 87) M.W. 692	45380	H	Br	Br	Br	Br	H	H	H	H	H
Eosin B (C.I. Acid red 91) M.W. 624	45400	H	NO₂	Br	Br	NO₂	H	H	H	H	H
Phloxine (C.I. Acid red 98) M.W. 761	45405	H	Br	Br	Br	Br	H	Cl	H	H	Cl
Phloxine B (C.I. Acid red 92) M.W. 830	45410	H	Br	Br	Br	Br	H	Cl	Cl	Cl	Cl
Erythrosin (erythrosin Y; C.I. Acid red 95) M.W. 628	45425	H	H	I	I	H	H	H	H	H	H
Erythrosin B (C.I. Acid red 51) M.W. 880	45430	H	I	I	I	I	H	H	H	H	H
Rose Bengal (C.I. Acid red 94) M.W. 1018	45440	H	I	I	I	I	H	Cl	Cl	Cl	Cl

ETHYL EOSIN (eosin, alcohol soluble; C.I. 45386; Solvent red 45; M.W. 698)

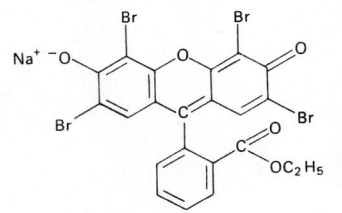

This, the ethyl ester of eosin, is only slightly soluble in cold water but dissolves freely in alcohol. It is used when staining with eosin is to be done from an alcoholic solution. Unlike the water-soluble red derivatives of fluorescein, ethyl eosin must be differentiated by alcohol rather than by water. Samples of ethyl eosin should contain at least 78% by weight of the anhydrous dye.

5.9.12. Acridine dyes

These dyes are similar to the xanthenes, but with a nitrogen rather than an oxygen atom linking the benzene rings. As with the xanthenes, the chromophoric system resonates between *o-* and *p*-quinonoid forms.

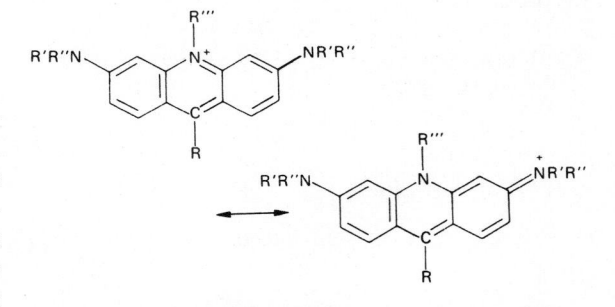

R, R', R" and R'''may be hydrogen atoms or alkyl or aryl radicals. If R is a phenyl group the dye will also be a triphenylmethane derivative with two of its benzene rings linked by a nitrogen atom. The acridine dyes used in histology are all strongly fluorescent.

Examples

ACRIFLAVINE (trypaflavine) (C.I. 46000; M.W. 260)

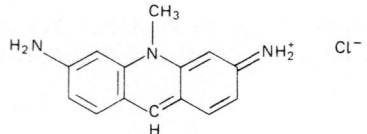

This may be used on the rare occasions when a yellow cationic dye is needed in ordinary light microscopy, and also as a fluorochrome. It is soluble in water and in alcohol. Acriflavine is better known as an antiseptic than as a biological stain.

ACRIDINE ORANGE (C.I. 46005; Basic orange 14; M.W. 302; Lillie (1977) states that this dye is usually supplied as a zinc chloride double salt. This would be $[C_{17}H_{20}N_3]_2^+$ $[ZnCl_4]^{2-}$, M.W. 438)

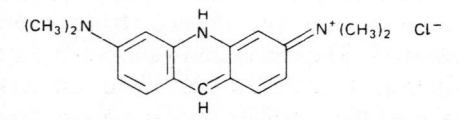

Used only as a fluorochrome, this dye is well known for its ability to impart fluorescent emissions of different colours to DNA and RNA (Bertalanffy & Bickis, 1956). It has also been used in carbohydrate histochemistry and as a vital stain.

5.9.13. Azine dyes

In these dyes the chromophore is a pyrazine ring

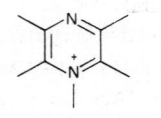

sandwiched between two aromatic systems, with o- and p-quinonoid configurations contributing to the resonance. These two structures are shown below for neutral red. For the other example only the p-quinonoid form is shown.

Examples

NEUTRAL RED (C.I. 50040; Basic red 5; M.W. 289)

Two of the canonical forms are shown. In another the positive charge is on the primary nitrogen atom.

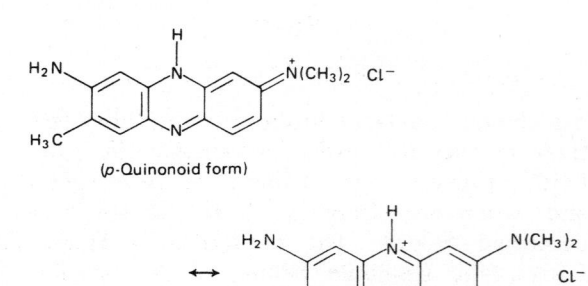

(p-Quinonoid form)

(o-Quinonoid form)

Neutral red is a cationic dye, freely soluble in water and alcohol. Aqueous solutions turn yellow on addition of alkali, the change occurring over the range pH 6.8–7.0. Very strongly acidic solutions are blue-green, and a brown precipitate is formed with an excess of alkali. The dye is valuable as a red cationic dye and is also used in some vital staining techniques. An acceptable sample of neutral red contains at least 50% by weight of the anhydrous dye.

SAFRANINE O (safranine, safranine T; C.I. 50240; Basic red 2; M.W. 351)

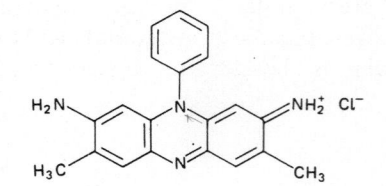

This is always a mixture of the compound formulated above with a variable amount of a related dye in which a methyl group is substituted on the phenyl radical, *ortho* to the site of attachment to nitrogen. It may have a yellowish or bluish cast, depending on the proportions of the two constituents. Safranine O is valuable as a red cationic dye. It dissolves in water and alcohol. Samples should contain at least 80% by weight of the anhydrous dye.

5.9.14. Oxazine dyes

The oxazine chromophore

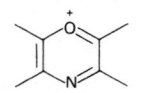

has obvious similarity to those of the azine and thiazine dyes and exists in *o*-quinonoid form (with positive charge attributed to the oxygen) and *p*-quinonoid form (as shown for the dyes described below). The natural dyes **orcein, litmus,** and **azolitmin** belong to the oxazine series. Orcein is used as a biological stain, litmus and azolitmin as pH indicators (see Lillie, 1977, for chemistry and properties). Only synthetic oxazines are considered below.

Examples

    CRESYL VIOLET ACETATE (No C.I. number; M.W. 321)

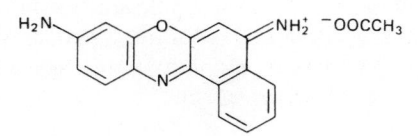

This dye is the modern equivalent of **cresyl fast violet (cresylecht violet)**. Both are often loosely termed "cresyl violet". They are useful as violet cationic dyes, soluble in water and alcohol. They are popular as "Nissl stains" for nervous tissue.

    GALLOCYANINE (C.I. 51030; Mordant blue 10; M.W. 337)

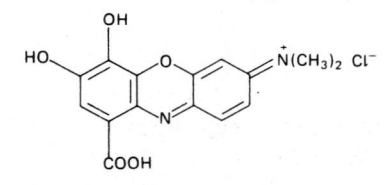

Gallocyanine is not used alone, but as a preformed dye-metal complex with the structure:

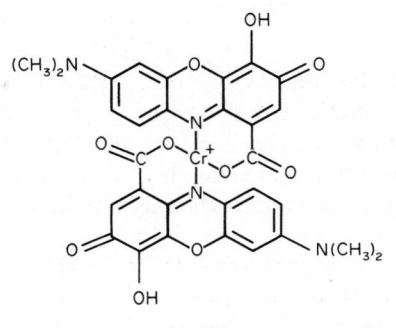

(Marshall & Horobin, 1972a). This complex is cationic, and is used principally to stain nucleic acids. The blue-grey colour is more resistant to extraction by water and alcohols than are simple cationic dyes. The chromium-gallocyanine staining solution is stable for only about one week, but it is possible to isolate the complex as a stable dry powder and dissolve it as needed (Berube *et al.*, 1966).

    NILE BLUE (C.I. 51180; Basic blue 12; M.W. 354)

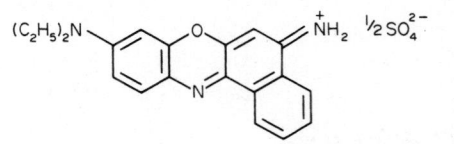

Nile blue is soluble in water, and can be used as an ordinary basic dye, which imparts blue and green colours to paraffin sections of fixed tissue. It has also been used in experimental embryology as a vital stain for transplanted tissues. For use in histochemistry, this dye is partly oxidized by boiling in dilute sulphuric acid. This generates an **oxazone**, known as **Nile red**:

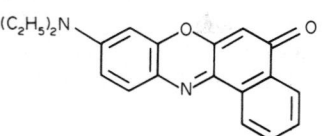

The oxazone, though insoluble in water, dissolves in aqueous solutions of Nile blue sulphate. The mixture can be used to stain lipids, in which the red compound dissolves. Furthermore, Nile red becomes fluorescent in a hydrophobic environment, so the oxidized Nile blue solution is also useful as a fluorochrome for lipids (see Chapter 12). A sample of Nile blue sulphate should be at least 70% dye for certification by the Biological Stains Commission.

5.9.15. Thiazine dyes

The chromophore of the thiazines is like that of the oxazines, with sulphur replacing oxygen. The thiazine dyes used in histology are all cationic and all except methylene green are blue or violet. All except methylene violet (Bernthsen) are soluble in water. In addition to their general usefulness as blue cationic dyes, some of the thiazines, especially the oxidation products of methylene blue, are used as eosinates in haematological staining. This will be discussed in Chapter 7. Most of the thiazine dyes can be used for metachromatic staining (see Chapter 11). Methylene blue is also used as a vital stain for nerve fibres (see Chapter 18).

The following examples include all the thiazine dyes that are important as biological stains. Structures are shown in the *p*-quinonoid configurations, with positive charges formally attributed to the most strongly basic nitrogen atoms.

THIONINE (C.I. 52000; M.W. 264)

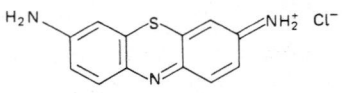

Thionine is a blue dye, soluble in water and in alcohol, but less so than most other thiazine dyes. An acceptable sample should contain at least 85% anhydrous dye. This dye must not be confused with thionine blue (see p. 83).

AZURE C (C.I. 52002; M.W. 278)

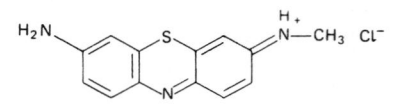

AZURE A (C.I. 52005; M.W. 292)

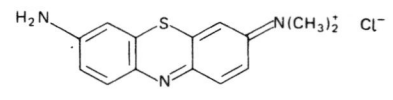

These are also blue cationic dyes, soluble in water and alcohol. Samples of azure A should contain at least 55% by weight of the anhydrous dye.

A symmetrical dimethylthionine (with one methyl group on each amine nitrogen atom) also exists and is said to have staining properties almost identical to those of azure A (see Lillie, 1977). Toluidine blue O (see p. 83) is also closely similar.

AZURE B (methylene azure; azure I; C.I. 52010; M.W. 306)

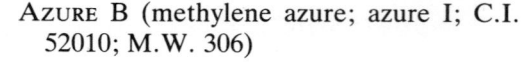

This blue dye, a product of the oxidation of methylene blue, is the most important coloured cationic component of the dye mixtures commonly used as haematological stains (see Chapter 7). It is soluble in water and in alcohol.

METHYLENE BLUE (C.I. 52015; Basic blue 9: M.W. 320)

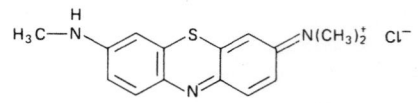

Many samples of methylene blue contain zinc chloride, which reduces the solubility in water and may render the dye too toxic for use as a vital stain (though this is disputed; see Richardson, 1969). If possible, a pharmacopoeial grade certified as zinc-free should be used. The dye is freely soluble in water and in alcohol. With time and exposure to air, methylene blue is oxidized to azure B, azure A, and other thiazine dyes. For use in bacteriology and haematology the dye is deliberately oxidized to give the mixture of dyes known as "polychrome methylene blue".

For certification by the Biological Stains Commission, a sample must contain at least 82% by weight of anhydrous methylene blue chloride and must be satisfactory as a nuclear and as a bacterial stain.

METHYLENE GREEN (C.I. 52020; Basic green 5; M.W. 365)

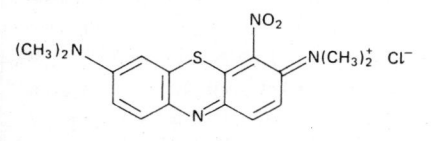

The exact position of the nitro group is uncertain. The dye is soluble in water and slightly soluble in alcohol.

THIONINE BLUE (C.I. 52025; Basic blue 25; M.W. 348)

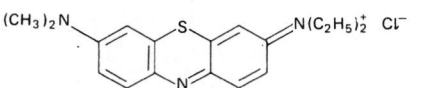

This is mentioned only to avoid confusion with thionine. Thionine blue, unlike thionine, does not give metachromatic effects.

NEW METHYLENE BLUE (C.I. 52030; Basic blue 24; M.W. 348)

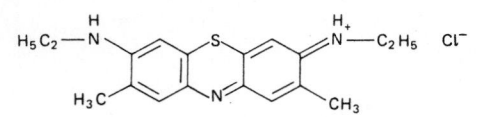

Metachromatic effects can be obtained with this dye, but it cannot replace methylene blue in vital staining methods. It is much more soluble in

water than is methylene blue, but the two dyes are about equally soluble in alcohol.

TOLUIDINE BLUE O (toluidine blue; C.I. 52040; Basic blue 17; M.W. 306)

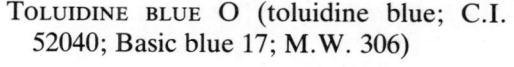
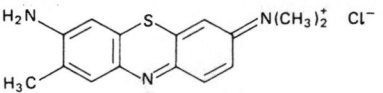

This is perhaps the most useful of all the blue cationic dyes, being similar to but cheaper than azure A. Metachromatic effects are easily obtained. Lillie (1977) points out that older samples of this dye are of the zinc chloride double salt (dye content about 60%) but that toluidine blue O is now sold in the zinc-free state (over 90% dye). An acceptable sample must contain at least 50% by weight of the anhydrous dye (M.W. 306). It is freely soluble in water and slightly soluble in alcohol.

METHYLENE VIOLET (BERNTHSEN) (C.I. 52041; M.W. 256)

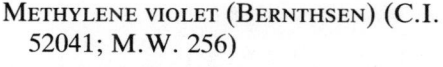
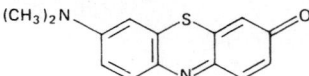

This dye is formed by oxidation of methylene blue and is a component of traditional staining mixtures for blood. The pure dye is insoluble in water but dissolves in the presence of other thiazine dyes. It is more soluble in chloroform. **Methyl thionoline** (with —NHCH₃ replacing —N(CH₃)₂ in formula above) and **thionoline** (with —NH₂ replacing —N(CH₃)₂ of methylene violet, Bernthsen) are related dyes also formed in the polychroming of methylene blue (Marshall, 1978).

Methylene violet (Bernthsen) should not be confused with methylene violet RR (C.I. 50205; Basic violet 5), which is an azine dye related to safranine O.

5.9.16. Indigoid and thioindigoid dyes and pigments

Natural **indigo** is the classical example of a vat dye (see p. 60): a soluble precursor, typically a colourless leuco compound, is applied to a fabric and then oxidized to an insoluble coloured substance. Indigo is manufactured by oxidizing indoxyl, which is now a synthetic product.

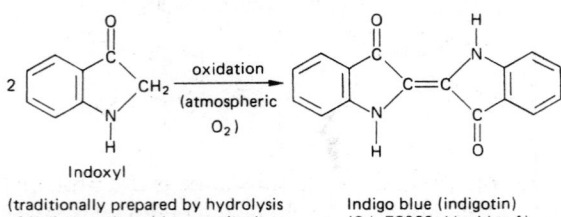

Indoxyl

(traditionally prepared by hydrolysis of *indican*, a glycoside occurring in *Indigofera* species)

Indigo blue (indigotin) (C.I. 73000; Vat blue 1)

For use in dyeing, the indigo blue is reduced, usually by sodium dithionite ($Na_2S_2O_4$), to its soluble leuco compound, **indigo white** (C.I. 73001):

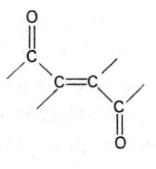

This, after application to the textile, is reoxidized by air to the insoluble indigo blue.

The chromophoric system of the indigoid dyes is the conjugated chain

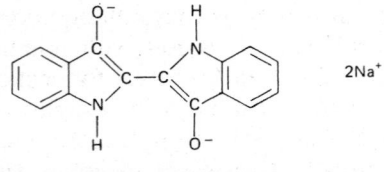

The colour is due to resonance among structures such as

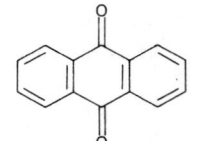

In the **thioindigoid** compounds the nitrogen of the indigo structure is replaced by sulphur.

Insoluble indigoid and thioindigoid compounds are produced as the coloured end-products of some histochemical reactions (see Chapter 15). In addition, one soluble indigoid dye is used in histology.

INDIGOCARMINE (C.I. 73015; Acid blue 74; M.W. 466)

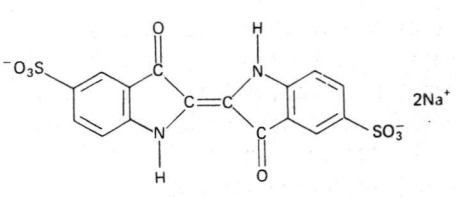

This blue anionic dye is soluble in water but almost insoluble in alcohol. It is used in a variety of procedures, the best known being a combination with picric acid (picroindigocarmine) used for staining connective tissue. Samples should contain at least 80% by weight of the anhydrous dye.

5.9.17. Anthraquinone dyes

Anthraquinone

with its *p*-quinonoid configuration, is the chromophore of all the members of this large class of dyes. Anionic, solvent, reactive, and mordant anthraquinones may all have potential value as biological stains, but only those of the last-named type have achieved any importance in microtechnique. Other anthraquinones, of great industrial importance, include vat dyes, disperse dyes, and pigments (Allen, 1971; Greenhalgh, 1976).

Examples

ALIZARIN RED S (alizarin red, water soluble; sodium alizarin sulphonate; C.I. 58005; Mordant red 3; M.W. 342)

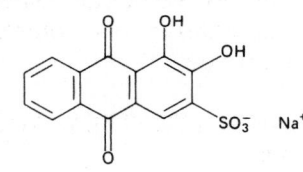

As an anionic dye this will stain tissues red. Its principal use, however, is as a histochemical reagent for calcium, with which an orange chelate is formed under appropriate conditions of use (see Chapter 13). The Biological Stains Commission tests alizarin red S as a stain for bone in whole mounts of small animals. Samples unsatisfactory for histochemical purposes are quite frequently encountered.

ALIZARIN (C.I. 58000; Mordant red 11) lacks the sulphonic acid radical. It was extracted from the roots of madder (*Rubia tinctoria*), but is now a synthetic product. The aluminium lake of alizarin is bright red and was once used for dyeing soldiers' uniforms. Alizarin has been used only rarely as a biological stain (Lillie, 1977).

KERNECHTROT (nuclear fast red; calcium red; C.I. 60760; M.W. 357)

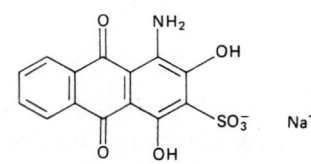

Like alizarin red S, this dye is sometimes used as a histochemical reagent for calcium. The aluminium complex is used as a red stain for nuclei. For the latter purpose kernechtrot is similar to carminic acid and brazilein.

CARMINIC ACID (C.I. 75470; Natural red 4; M.W. 492)

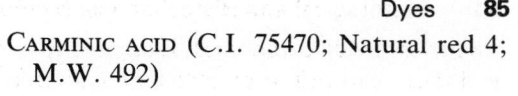
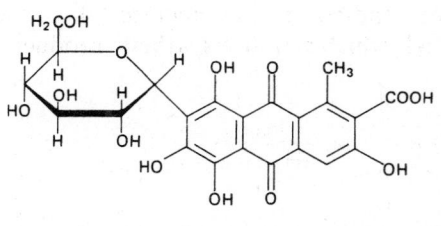

This red glycoside-like substance is the principal colouring matter of **cochineal**. Cochineal is made by grinding the dried bodies of female *Coccus* insects (from Central America). The cyclic side chain shown on the left side of the structural formula is derived from D-glucose, which is joined by a C-glycosyl linkage (without intervening oxygen) to the polyhydroxyanthraquinone carboxylic acid. The dye is used with aluminium as mordant as a red nuclear stain for sections and whole mounts. Carminic acid is slightly soluble in water and alcohol, but in the presence of an aluminium salt it dissolves in larger amounts, with the formation of the dye–metal complex. Solutions of the aluminium complex deteriorate after a few weeks of storage.

Carmine, which is also made from cochineal, is used in general histology and in a staining method for glycogen. It contains 20–25% of protein (some from the insects and some added during manufacture), together with calcium ions and an anionic dye–metal complex of carminic acid and aluminium. The structure of carmine is uncertain and probably varies according to the method of manufacture. Calcium may not always be present. A suggested formula is shown in Fig. 5.1. Since carmine is a material of uncertain composition it should not be used as a histological stain without preliminary testing, and supposedly histochemical methods involving its use must be viewed with suspicion.

5.9.18. Phthalocyanine dyes

Most of these are derivatives of **copper phthalocyanine**, a blue pigment of high physical and chemical stability.

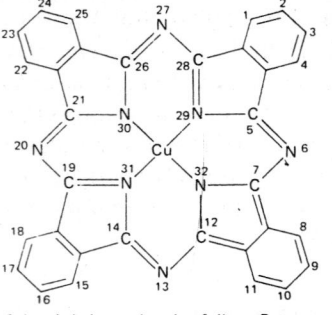

(Numbering of the phthalocyanine ring follows Patterson *et al.*, 1960)

The carbon and nitrogen atoms form a fully conjugated aromatic system, stabilized by resonance so that the *o*-quinonoid benzene ring (here shown at the lower right-hand corner) has no

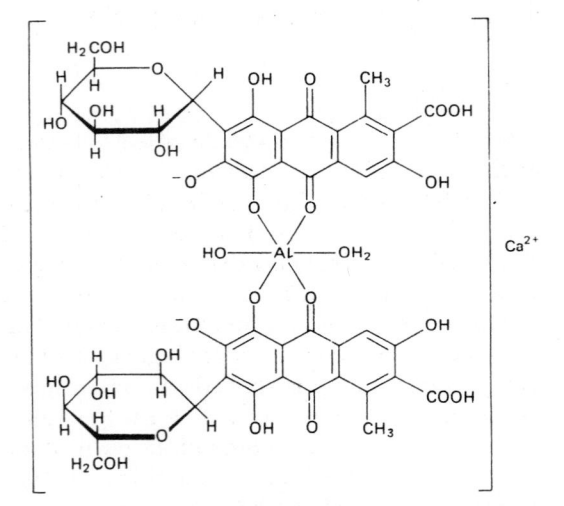

FIG. 5.1. Structure of carmine as suggested by Meloan *et al.* (1971). Their formula has been slightly modified to conform with the conventions used in this book. Ionization of the phenolic hydroxyl groups *ortho* to the sugar substituents is said to be necessary for chelation of Al^{3+} with the nearby oxygen atoms. Presumably the carboxyl groups are also at least partially ionized in solution, but it is not known whether carmine contains enough Ca^{2+} to balance these. Al^{3+} could also form complexes with the free quinonoid and α-phenolic oxygens shown on the sides of the carminic acid molecules facing away from the Al atom. This would give the dye–metal complex a net positive charge and might occur when Al^{3+} is present in excess, as in mixtures used for nuclear staining. The mode of attachment of the dye–metal complex to tissues has not been determined.

fixed location in the molecule. The bonds to the copper atom are shown as simple covalencies, though other representations are permissible (see p. 56). Some phthalocyanine pigments are synthesized on textile fibres by application of "phthalogen" dyestuff intermediates (Vollmann, 1971). In dyes, solubilizing groups are introduced as substituents on the phthalocyanine ring. Copper may be replaced by cobalt or other metals.

Examples

SIRIUS LIGHT TURQUOISE BLUE GL (C.I. 74180; Direct blue 86; M.W. 760)

Sulphonic acid radicals (as their sodium salts) are present on benzenoid rings at opposite corners of the molecule. This is the simplest phthalocyanine dye. It has been tried in histology but is not used in any important staining methods (see Lillie, 1977).

ALCIAN BLUE (alcian blue 8GS, 8GX or 8GN: C.I. 74240: Ingrain blue 1)

The exact chemical compositions of these dyes have not been revealed by Imperial Chemical Industries Ltd., the manufacturers, but Scott (1972a) analysed alcian blue 8GX and found that it contained four tetramethylisothiouronium groups,

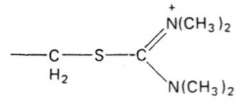

one on each of the benzenoid rings of the copper phthalocyanine structure, at positions 2 or 3, 9 or 10, 16 or 17, and 23 or 24. The dyestuff is a mixture of the four possible geometrical isomers. The other alcian blues are believed to bear either two or four thiouronium substituents. All these dyes are supplied mixed with boric acid, sodium sulphate, and dextrin. The solubilizing side-chains of alcian blue can be easily removed by base-catalysed hydrolysis, leaving insoluble copper phthalocyanine in the textile or tissue.

The dye **alcian' yellow** is not a phthalocyanine

but a monoazo dye which also contains the benzothiazole chromophore and a thiouronium auxochrome. **Alcian green** is a mixture of alcian blue with alcian yellow.

The most important application of alcian blue is in carbohydrate histochemistry (see Chapter 11). The other dyes mentioned above have been used for similar purposes. Unsatisfactory samples of alcian blue are often encountered, so it is necessary to test every new batch before applying the dye to important material.

LUXOL FAST BLUE MBS (C.I. Solvent blue 38; no C.I. number)

The "luxol" dyes are anionic chromogens which are balanced not by H^+ or Na^+ but by aryl-guanidinium cations. With gegen-ions of this type, of which diphenylguanidine is an example, the dyes are insoluble in water but soluble in alcohols. The structures of the dyes have not been disclosed by the manufacturers, but it is known that luxol fast blue MBS has a copper phthalocyanine chromogen, whereas LUXOL FAST BLUE ARN (C.I. Solvent blue 37) is an azo dye (Salthouse, 1962, 1963). Dyes with similar properties but of known chemical constitution can be produced in the laboratory from suitable anionic dyes and diphenylguanidine (Clasen et al., 1973).

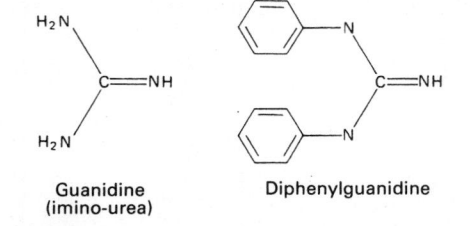

Guanidine (imino-urea)   Diphenylguanidine

The "luxol" dyes are used for staining myelin and will be further discussed in Chapter 18.

5.10. Structure-staining correlations

The objects or substances to which a dye will bind can be predicted, to some extent, from an examination of its structural formula. The estimation of physical and chemical properties of organic compounds, a field of great importance to many industries, is reviewed by Lyman et al. (1982). Horobin (1980, 1982) has discussed the subject critically from the viewpoint of biological staining. To evaluate a dye, one looks at the formula for information concerning electric charge, ability to form chelates or other covalent bonds, the shape and size of the dye molecule, and features favourable to non-ionic bonding (see also Section 5.5.)

5.10.1. Type of dye

The **overall charge** of the coloured ion is easily seen, and usually determines whether the staining will be principally of nuclei and acid carbohydrates (cationic dyes) or principally of cytoplasm and collagen (anionic dyes). Uncharged dyes may be reactive dyes (p. 66), lysochromes (p. 60) or mordant dyes (p. 57).

5.10.2. Size

The size of a dye molecule affects its **penetration** of different components of the tissue, and its **substantivity** (see p. 67). The molecular weight is a reasonable estimator of size. It should be remembered that many dyes consist of aggregated molecules in aqueous solution, but large molecules join to one another more readily than small ones, and form correspondingly larger particles. When a tissue is stained by two or more dyes with a charge of the same sign, the larger coloured ions enter the more porous parts of the tissue. For example, when eosin Y (M.W. 692) and methyl blue (M.W. 800) are used together, as in Mann's method (p. 94), collagen and chromatin become blue while nucleoli and most cytoplasms become red. Porosity is not the only factor in differential staining; some others are discussed in Chapters 6 and 8.

Except when covalent bonds are formed, the strength of the bonding between a tissue and any coloured substance depends largely on non-ionic forces of attraction. As explained in Sections 5.5.4–5.5.7, these forces are weak, and require intimate contact between dye and substrate. A large molecule that can be planar, such as a bisazo or a phthalocyanine dye will stick to a tissue more strongly than a small planar dye. There are some dyes that can never assume planar configurations because of steric hindrance. The triphenylmethanes, for example, have their rings arrayed like the blades of a propeller. Among such dyes, those with the largest molecules are the most difficult to remove from stained sections.

5.10.3. Hydrophobic-hydrophilic balance

The hydrophobic or hydrophilic nature of a dye is best indicated by its **partition coefficient** between a moderately hydrophobic solvent (such as octanol) and water. This figure can be estimated from the formula by attributing scores (see Horobin, 1980, 1982) to the different parts of the structure and summing the scores. The result (known as the Hansch π value) is most useful when it is compared with the values of dyes with known staining properties. Generally, the more hydrophobic dyes exhibit stronger binding in an aqueous environment. The most hydrophobic dyes are, of course, insoluble in water. Hydrophilic dyes are used in non-aqueous solution, for staining hydrophilic substrates. Hydrogen bonding probably contributes to the substantivity in these cases (see p. 55).

Another easily obtained value is the size (in number of bonds) of the **longest conjugated chain** in the molecule. This increases with the size of the chromophoric system, and high values are associated with strong non-ionic binding.

5.11. Exercises

Theoretical

1. Which of the following compounds are coloured and which could function as dyes?

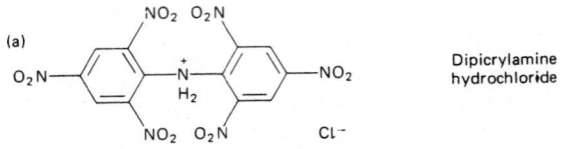

(a)   Dipicrylamine hydrochloride

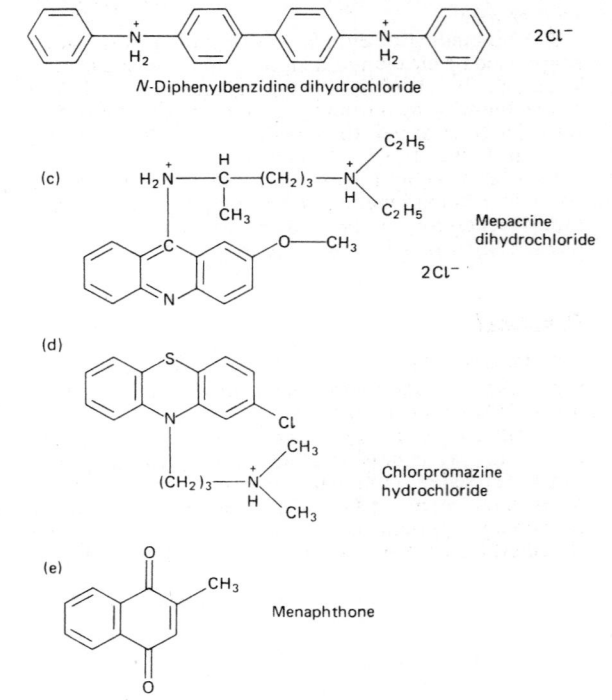

(b)   N-Diphenylbenzidine dihydrochloride   2Cl⁻

(c)   Mepacrine dihydrochloride   2Cl⁻

(d)   Chlorpromazine hydrochloride

(e)   Menaphthone

2. 3,4-benzpyrene

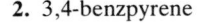

can be used as a fluorochrome. Is it a dye? What would you expect it to stain when applied to sections as a saturated solution in a saturated aqueous solution of caffeine? The caffeine serves only to enable the 3,4-benzpyrene to be dissolved in an aqueous medium.

3. Weigert's iron-haematoxylin contains 500 mg per ml of haematoxylin and 0.58 mg per ml of $FeCl_3.6H_2O$ (M.W. 270). This solution stains nuclei. Overstaining does not occur and the sections usually do not require differentiation.

If sections are mordanted in 5% iron alum, $NH_4Fe(SO_4)_2.12H_2O$ (M.W. 482) and then stained in 0.5% ripened haematoxylin, they go completely black (Heidenhain's method). Differentiation in 5% iron alum, if carefully controlled, permits the demonstration of many structures other than nuclei.

Explain the differences between the Weigert and Heidenhain methods of staining.

4. Urea readily forms hydrogen bonds with other substances. If aqueous solutions of some dyes are saturated with urea they lose their ability to stain sections. Which of the following dyes would you expect to be ineffective in a urea-containing solution: (a) Toluidine blue O, (b) Eosin, (c) Methyl blue, (d) Oil red O? Give reasons for your answers.

5. The binding of dyes by electrostatic forces can be suppressed by including a high concentration of a neutral salt such as NaCl or $MgCl_2$ in the staining solution. Which of the following dyes would you expect to be inhibited by salts: (a) Neutral red, (b) Procion brilliant red M2B, (c) Chlorazol black E, (d) Alcian blue?

6. What structural features of a dye molecule confer the ability to bind metal ions? Draw possible structures for dye–metal complexes of chromium(III) with (a) salicin black EAG, (b) gallocyanine, (c) brazilein.

Practical

7. Dissolve about 10 mg of basic fuchsine in 100 ml of water. Add 4% aqueous sodium hydroxide in aliquots of 0.5 ml. What happens and why?

8. Add single drops of ripened 5% alcoholic haematoxylin to: (a) slightly alkaline water, (b) slightly acid water, (c) water containing approximately 5 mg per ml potassium alum $(KAl(SO_4)_2.12H_2O)$, (d) water containing approximately 5 mg per ml iron alum $(NH_4Fe(SO_4)_2.12H_2O)$ or ferric chloride $(FeCl_3.6H_2O)$.

For (c) and (d) follow the addition of the haematoxylin with addition of two or three drops of 4% sodium hydroxide.

Observe the colours of the different mixtures. What do you deduce from these observations about the properties of haematein and of its complexes with Al^{3+} and Fe^{3+}?

9. Mix equal volumes of aqueous solutions of: (a) 1% neutral red with 1% fast green FCF; (b) 1% neutral red with 1% toluidine blue O; (c) 1% toluidine blue with 1% eosin; (d) 1% eosin with 1% fast green FCF.

Account for the observation that precipitates form with mixtures (a) and (c), but not with (b) and (d).

10. Dissolve 50 mg of fast blue B salt (tetraazotized *o*-dianisidine) in 30 ml of water at 4°C. Dissolve 30–40 mg of thymol in 2–3 ml of 95% ethanol. Pour the thymol solution into the solution of the tetrazonium salt and add three or four drops of 4% aqueous sodium hydroxide. Mix well and leave at 4°C for 20 min. Allow the reddish brown precipitate to settle and decant off as much as possible of the supernatant liquid. Add 20 ml of ethanol to the precipitate, which will dissolve.

What is the probable structural formula of the dye you have made? Thymol is 2-isopropyl-5-methyl phenol (M.W. 150). In view of its structure and the properties you have observed, how might this dye be used as a biological stain?

6

Histological Staining in One or Two Colours

For the study of microscopic anatomy and of pathological material it is usual to stain sections of tissue in such a way as to impart a dark colour to the nuclei of cells and a lighter, contrasting colour to the cytoplasm and extracellular structures. With simple methods of this type, of which the most widely practised is alum-haematoxylin and eosin ("H. & E."), it is not possible to make deductions concerning the chemical natures of any of the components of the tissue. Identifi-

cation of the different types of cells and of the fibres and matrices of the intervening connective tissue must be based solely on morphological criteria. The H. & E. method is perhaps the least revealing of the two-colour techniques because in most cell-types the alum-haematein component colours only the nuclei. Nevertheless, this procedure is used more than any other in the teaching of normal histology and by pathologists. Other simple staining methods provide more informative preparations than does H. & E., but familiarity with the latter has not yet led to contempt.

Nuclear stains and simple counterstaining procedures are considered in this chapter. A few methods for staining in various shades of one colour are also described. The histochemical rationales of the techniques are discussed, but methods of higher chemical specificity (e.g. for nucleic acids, carbohydrates, and functional groups of proteins) are not covered. The special methods for blood and for connective tissue are likewise described elsewhere (Chapters 7 & 8).

6.1. Nuclear stains

The nucleus of a eukaryotic cell contains the two **nucleic acids**—DNA (in the chromosomes) and RNA (in the nucleolus). Both are associated with **nucleoproteins**, which are rich in the basic amino acids arginine and lysine. The cations of these amino acids are bound electrovalently to phosphoric acid residues of the nucleic acids (see Chapter 9). The DNA and nucleoprotein of the chromosomes together constitute the material known as **chromatin**. In interphase cells the chromosomes are extended and cannot be seen individually. The chromatin seen in stained preparations of interphase nuclei may be evenly distributed through the nucleoplasm or aggregated in a pattern characteristic of the cell type. The dyes used as nuclear stains impart colour to the chromatin by binding either to the nucleic acids or to the nucleoprotein.

6.1.1. *Cationic dyes*

The modes of binding of cationic dyes to substrates were discussed in Chapter 5, where descriptions of several of the dyes of this type can be found. The dyes are applied from acidified solutions in order to ensure the ionization of their primary, secondary, or tertiary amine groups. Three types of anions capable of binding cationic dyes occur in tissues. These are the phosphates of nucleic acids, sulphate esters of certain macromolecular carbohydrates (see Chapter 11), and carboxylate ions of carbohydrates and proteins. As explained in Chapter 5, the ionization of carboxyl groups is suppressed if the concentration of hydrogen ions in the staining solution is too high (in practice, pH 3–4), while sulphate groups remain ionized even in strongly acidic media. Phosphoric acid is of intermediate strength. A basic protein, such as haemoglobin, contains an excess of free amino over carboxyl groups, so that the latter can be made available for binding cationic dyes only from alkaline solutions (e.g. at pH8). The staining properties of cationic dyes are profoundly influenced by the pH of their solutions. *If they are too acid, only the structures rich in sulphated carbohydrates will be coloured; if they are neutral or alkaline, everything will be stained.*

The interpretation of the results obtained with cationic dyes at various pH levels has been studied in some depth (see Gabe, 1976; Lillie & Fullmer, 1976, for reviews and discussion).

It is convenient to think of staining by cationic dyes as a process of ion-exchange, in which small cations such as H^+ are competitively displaced from the tissue by the larger cations of the dye (see Horobin, 1982). When the tissue and the dye solution are together for a long enough time, there is an equilibrium:

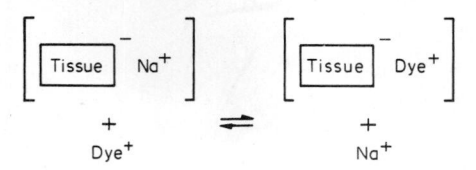

Binding of dye by non-ionic forces (see Chapter 5) favours staining (pulling the above reaction from left to right), whereas a high concentration of H^+ or Na^+ in the solution opposes staining (pushing the reaction from right to left). This simple explanation applies only to dilute solutions of dyes with small molecules. In most real staining techniques non-ionic forces cause aggregation of dye, so that more than one coloured molecule is bound at each anionic site in the tissue.

The following list will serve as a rough guide to the use of dilute (e.g. 0.01–0.2%) solutions of cationic dyes. It applies especially to those of the thiazine series.

pH 1.0   Only sulphated carbohydrate components are stained. Sulphonic acid groups or sulphuric acid esters can be artificially produced in the tissue (see Chapters 10 and 11) and these, too, will bind cationic dyes from strongly acid solutions.

pH2.5–3.0   Staining of the above, and also phosphate of nucleic acids (DNA and RNA), and of phospholipids (in frozen sections).

pH4.0–5.0   Staining of the above, and also carboxyl groups of carbohydrates and of the more acid proteins. Carboxyl groups of free fatty acids may be stained in frozen sections.

pH>5.0   Staining of the above with increasing staining of neutral and basic proteins with increasing pH.

With higher concentrations of dye, the rules set out above are not followed, and staining is generally less specific. The duration of exposure to the dye has only minor effects on the results, though longer times are needed to achieve the greatest possible depth of staining (equilibrium) if very dilute solutions are used. The end result is also affected by treatments after staining.

Most cationic dyes are rather readily extracted from sections by 70% alcohol. Absolute ethanol, *n*-butanol, and acetone are usually less active in this respect. In any critical work with cationic dyes the conditions of rinsing and dehydration must be carefully standardized. *The water used for rinsing the sections after staining should be buffered to the same pH as the dye solution*. Fixatives generally do not directly affect sulphate, phosphate, and carboxylate groups, but combination with or removal of amine groups has the effect of lowering the pH at which proteins are stained. Thus fixation by osmium tetroxide renders virtually all components of tissues stainable by cationic dyes at pH 4 or higher. The reason for this will be obvious from a consideration of one of the chemical actions of osmium tetroxide (see Chapter 2).

For general-purpose staining of nuclei, cationic dyes are used in dilute solution (0.1–0.5%) at pH 3–5.5. The aqueous solution of the dyes is usually acidified by addition of acetic acid or a suitable buffer. Objects that are coloured by cationic (basic) dyes are said to be **basophilic** (Baker, 1958, argues in favour of "basiphil"). When nuclei are stained by cationic dyes, basophilia will also be evident at sites of accumulation of cytoplasmic RNA such as the Nissl substance of neurons, and in structures containing sulphated carbohydrates such as mast cell granules, cartilage matrix, and many secretory products. In this propensity cationic dyes differ importantly from some of the mordant dyes used as nuclear stains.

As explained above, cationic dyes will not stain nuclei specifically. The following solutions are suitable for use as counterstains to other methods.

6.1.1.1. *Toluidine blue O* (C.I. 52040)

A 0.5% aqueous solution, acidified by adding approximately 1.0ml of glacial acetic acid to each 100ml of dye solution. Alternatively, make up in 0.1M acetate buffer, pH4.0. The optimum pH may vary between 3.0 and 5.5 according to the fixative and the tissue to be stained. Stain in this solution for 3min, rinse in water, dehydrate in 95% followed by two changes of absolute alco-

hol, clear in xylene, and mount in a resinous medium.

If too much colour is lost during washing or alcoholic dehydration, move the slides directly onto filter paper and blot firmly with two or three layers of filter paper. Transfer to a clean, dry staining rack or coplin jar and dehydrate in two changes (each 4–5min) of *n*-butanol. Clear and mount as above. If detachment of the sections seems imminent, the clearing in xylene may be omitted, but the resinous medium will then take longer to become fully transparent.

The dye may also be immobilized in stained structures by converting it to an insoluble molybdate. After staining, wash in water and immerse for 5min in 5% aqueous ammonium molybdate (=ammonium paramolybdate, $(NH_4)_6Mo_7O_{24}.4H_2O)$ solution (which may be used repeatedly until it becomes cloudy; usually stable for 1 or 2 weeks). Wash in running tap water for 2–3 min. The molybdate of toluidine blue is not extracted by water or alcohols and even resists counterstaining by acidic mixtures of dyes, such as that of van Gieson.

Nuclei and acid carbohydrate components are stained blue. Some carbohydrate-containing structures are metachromatically stained (red) as explained in Chapter 11. Cytoplasmic RNA (e.g. Nissl bodies of neurons) is also blue. Other thiazine dyes (e.g. Borrel's methylene blue, azure A) may be substituted for toluidine blue.

6.1.1.2. *Neutral red* (C.I. 50040)

Use a 0.5% aqueous solution for 1–5 min. Acidification with acetic acid (see above) is usually desirable. Wash, dehydrate, and clear as described for toluidine blue O.

Nuclei and other basophilic structures are stained red. Safranine O (C.I. 50240) gives a similar result, but longer staining times (often 30–60 min) are usually required.

6.1.1.3. *Methyl green* (C.I. 42585)

Instructions for preparing 2% aqueous methyl green and extracting chloroform-soluble contaminants are given by Pearse (1968 or 1985) and in the first edition of this book. Purified methyl (or ethyl) green is now commercially available. For staining, use a 0.2% solution at pH 4.0–4.5 (which is stable for a few months) and stain for 15 min. Rinse quickly in water, blot the slides dry, and dehydrate in two changes of acetone or of *n*-butanol (as described above for toluidine blue O). Clear in xylene and cover, using a resinous mounting medium.

Nuclei and other basophilic structures are stained bluish green.

6.1.1.4. Polychrome methylene blue

The staining solution is a mixture of thiazine dyes produced by oxidizing methylene blue. The oxidation (polychroming – see Chapters 5 and 7) may be accomplished in various ways. The following method, in which silver oxide is used as an oxidizing agent, is recommended for its simplicity and speed.

Borrel's methylene blue

Add 25 ml of 4% sodium hydroxide (NaOH) to 100 ml of 1% silver nitrate ($AgNO_3$). Allow the precipitate of silver oxide (Ag_2O) to settle, decant off the supernatant, and wash the precipitate by shaking and decantation, with two changes of 100 ml of water. Dissolve 1.0 g of methylene blue (C.I. 52015) in 100 ml of water, boil, and add to the washed Ag_2O. Boil for 5 min, leave to cool, then filter. This stock solution is diluted to five times its volume with water to give the working solution. Both the stock and the diluted solutions are stable for at least 3 years.

Method

Stain frozen or hydrated paraffin sections for 2 min in the diluted solution of Borrel's methylene blue. Rinse rapidly in running tap water, blot slides dry, and dehydrate in two changes (each 30 s, with agitation) of absolute ethanol. Clear in xylene and cover, using a resinous medium.

Result

All components of the tissue are stained in various shades of blue. Strongly acid carbohydrates stain metachromatically (red), especially if some differentiation is allowed to occur in the water wash. With deliberate overstaining, as described, the objects that are normally chromotropic will be dark blue or purple.

6.1.2. Anionic dyes

For reasons given in Chapter 5, anionic dyes are applied from acidic solutions. Used alone, an anionic dye will colour almost all the components of a tissue, but differential staining effects can sometimes be obtained when two or more dyes are applied simultaneously or in sequence. A method of this type is Mann's eosin-methyl blue. Nuclei, cartilage, and collagen acquire the blue colour, cytoplasms are stained pink or red by eosin. The reasons for these results are not fully known, but it is widely held that the large molecules of methyl blue are excluded by the smaller ones of eosin from the supposedly "dense" network of cytoplasmic protein molecules. Collagen and chromatin, with their presumed "looser" textures, would admit molecules of both dyes, but the eosin would diffuse out more rapidly than the methyl blue when the sections are washed following immersion in the mixture. The influence of molecular size on the staining properties of dyes will be discussed in more detail in Chapter 8. Nuclear staining by methyl blue was studied by McKay (1962) who found that the presence of nucleic acids was necessary. He was unable to obtain evidence for binding of the dye by ionic attraction and suggested that the staining of nuclei was analogous to the direct dyeing of cotton (see p. 67). McKay compared methyl blue with chlorazol black E, a direct azo dye. Chlorazol black E was found to behave like most other anionic dyes, including eosin, and was, therefore, thought to be bound to tissue principally by electrovalent forces. Useful staining of nuclei and cytoplasms can be obtained with chlorazol black E. This dye also stains elastin, but probably by a non-ionic mechanism (Goldstein, 1962).

6.1.2.1. *Mann's eosin-methyl blue*

The procedure given below is the "short method" of Mann (1902). The "long method" (Mann, 1902; also discussed at length by Gabe, 1976) is more controllable and is used for demonstrating intracellular objects such as secretory granules and viral inclusion bodies. Fixation is not critical, but superior contrast is obtained if mixtures containing mercuric chloride, potassium dichromate, or picric acid have been used.

Preparation of stain

> Mix: 1% aqueous eosin
> (C.I. 45380):   45 ml
> 1% aqueous methyl blue
> (C.I. 42780):   35 ml
> Water: 100 ml

The solution keeps for two years. Precipitates which may form in it do not matter but should be removed by filtration. Aniline blue WS (C.I. 42755) or any water-soluble "aniline blue" (see Chapter 5) may be used in place of methyl blue.

Procedure

1. De-wax and hydrate paraffin sections.
2. Stain in eosin-methyl blue for 10 min.
3. Wash in running tap water for a few seconds to remove excess dyes from slides.
4. Dehydrate in 95% and two changes of absolute alcohol.
5. Clear in xylene and cover, using a resinous mounting medium.

Result

Nuclei and collagen—blue; erythrocytes, cytoplasm, nucleoli—red. With this "short" method, the colours obtained are rather variable. Like Borrel's methylene blue, this is a "quick look" method for determining which of a series of sections or specimens are worthy of more critical examination.

6.1.2.2. *Chlorazol black E*

Staining solution

> Chlorazol black E (C.I. 30235):   3.0 g
> 70% aqueous ethanol: 300 ml

Dissolve the dye (magnetic stirrer; 30 min at room temperature), then filter the solution. It keeps for about 9 months.

Procedure

1. De-wax paraffin sections and take to 70% alcohol.
2. Stain in the chlorazol black E solution for about 10 min. The time is not critical.
3. Rinse in 95% ethanol, two changes, 1 min in each. (See *Note 2* below.)
4. Complete the dehydration in two changes of absolute alcohol.
5. Clear in xylene and cover, using a resinous mounting medium.

Result

Nuclei black. Elastin black. Other components of tissue in various shades of grey. Cytoplasm often has a greenish tinge and cartilage matrix is usually pinkish grey. Collagen is rather lightly stained. Cytoplasmic organelles such as mitochondria and secretory granules are well displayed in suitably fixed material. The colour does not fade. (See also *Note 3* below.)

Notes

1. The method will work after any fixation but, as with most other dye staining methods, greater contrast is obtained after fixation in a mixture containing picric acid, mercuric chloride or potassium dichromate than after fixation in formaldehyde alone.
2. Over-staining rarely occurs if the time at stage 2 is less than 20 min. Differentiation occurs slowly and controllably in 95% ethanol. Cannon (1937), the originator of the method, recommended terpineol as a differentiator. He also commented on the

similarity of the end result to that obtained with Heidenhain's iron-haematoxylin, a traditional but time-consuming technique involving a critical differentiation.

3. The most informative preparations are thin sections (5 μm or less) examined at high magnification. Other methods of staining with chlorazol black E are given by Clark (1981).

6.1.3. Mordant dyes

For nuclear staining, mixtures of mordant dyes with appropriate metal salts are applied to sections of tissue. The modes of action of mordants in binding dye molecules to their substrates have already been discussed in some detail in Chapter 5. There it was pointed out that some dye-metal complexes, such as that of chromium with gallocyanine, behave as if they were simple cationic dyes. Others, most notably haematein, are evidently true mordant dyes. These latter dyes are used in histology to obtain **selective staining of chromatin.** Nuclear staining by the commonly used aluminium-haematein mixtures is only partially inhibited by prior extraction of DNA from the tissue, and staining by iron-haematein is unaffected by such extraction. These and related dye-mordant complexes probably attach mainly to nucleoproteins, though staining is only partially prevented by chemical blocking of lysine and arginine residues (Lillie *et al.*, 1976a). Horobin (1988) considers it likely that the dye-metal complexes are bound to the chromatin by ionic and non-ionic forces, but this explanation may not account adequately for classical evidence for the interposition of metal ions between dye and substrate (see Chapter 5). Favourably conforming shapes of the nucleoprotein molecules and the dye-metal complex anions may result in reinforcement of the ionic attractions by the weaker, short-range van der Waals forces, and perhaps also by hydrogen bonds.

Haematein (the principal product of oxidation of haematoxylin) is used in solutions containing ferric, aluminium, or, more rarely, chromium ions. In other mixtures, used only for specialized purposes, the mordant may be a lead or copper salt, or phosphotungstic or phosphomolybdic acid. For nuclear staining, the solution must contain a considerably greater proportion of ferric or aluminium ions than of haematein and it must be acidified.

Staining with **alum-haematein** may be progressive or regressive. The former is usually preferred, but if overstaining has to be corrected the sections are differentiated in 70% alcohol containing a little hydrochloric or acetic acid. Aqueous acids may also be used, but their differentiating action is slower. Addition of more acid or more of the metal salt to the staining bath also suppresses the tendency to overstain, thereby making the mixture more selective as a progressive nuclear stain. Acids used for differentiation probably disrupt the bonding between metal and tissue rather than between metal and dye (Baker, 1960, 1962). The acid-lability of nuclei stained by alum-haematoxylin precludes counterstaining with solutions that are more than slightly acidic. Eosin is suitable but the van Gieson mixture (see Chapter 8) is not. The aluminium-haematein complex changes colour from reddish brown to bluish purple at about pH 7. The latter colour is the one desired, so stained sections are washed in tap water, to which a trace of alkali may have to be added, after staining. This process is called "blueing".

The **iron-haematein complex** has a deep blue-black colour and can be removed from stained sections only by strongly acid differentiating solutions. The progressive mode is the method of choice for nuclear staining, so the dye solutions are made with large excesses of ferric salt and acid (usually hydrochloric). **Weigert's haematoxylin** is a typical mixture of this type. The ferric ions eventually over-oxidize the haematein, so the stain deteriorates gradually, over a period of several days, until it is no longer usable. Some iron-haematoxylins have better keeping properties than Weigert's. The more stable iron-haematoxylin mixtures contain both ferric and ferrous ions. The reducing action of the latter is assumed to restrain the oxidation of the dye (see Lillie &

Fullmer, 1976). Almost any counterstain can be used after iron-haematoxylin.

It is also possible to use synthetic dyes that are cheaper than haematoxylin. A suitable substitute is **chromoxane cyanine R**, used with a ferric salt as mordant. Several other alternatives to haematoxylin have been described by Lillie *et al.* (1976b).

Often red is preferred to the colours imparted to nuclei by the metal-haematein complexes. This may be achieved by using an aluminium mordant with **carminic acid** ("carmalum"), **brazilein** ("brazalum"), or **kernechtrot** (nuclear fast red). Red nuclear stains are most often employed as counterstains to histochemical or other specialized methods that impart other colours to the objects in which the microscopist is primarily interested.

6.1.3.1. Haematoxylin and eosin

The method described here is typical of the many H. & E. procedures. The traditional recipe for Mayer's haemalum has been modified by reducing the amount of sodium iodate for the reasons given in Chapter 5. The citric acid in this solution probably serves only to maintain appropriate acidity. The function of the chloral hydrate is obscure. Do not expect a perfect result by following the numbered steps of this method uncritically: careful attention to the appended notes is essential.

Solutions required
A. *Mayer's haemalum*

Dissolve the following, in the order given, in 750 ml of water:

Aluminium potassium sulphate ($KAl(SO_4)_2.12H_2O$):	50 g
Haematoxylin (C.I. 75290):	1.0 g
Sodium iodate ($NaIO_3$):	0.1 g
Citric acid (monohydrate):	1.0 g
Chloral hydrate:	50 g

Make up with water to 1000 ml. Often keeps for a year but some batches lose their potency after only a few months. The solution may be reused many times. If the solution fails to stain nuclei properly, it may be over-oxidized. Try making another batch with only 50 mg of sodium iodate. See *Note 5* below for an alternative alum-haematein solution.

B. *Eosin*

Eosin (C.I.45380):	2.5 g
Water:	495 ml
Glacial acetic acid:	5 ml

This solution keeps for several months, and may be used repeatedly. Moulds often grow in it and need to be removed by filtration. Addition of a crystal of thymol to the solution helps to retard the growth of moulds.

Procedure

1. De-wax and hydrate paraffin sections. Frozen sections should be dried onto slides.
2. Stain in Mayer's haemalum (solution A) for 1–15 min (usually 2–5 min, but this should be tested before staining a large batch of slides). Overstained sections can easily be differentiated by agitating for a few seconds in 1% (v/v) concentrated hydrochloric acid in 95% alcohol, then washing thoroughly in tap water.
3. Wash in running tap water for 2 or 3 min or until the sections turn blue. If the tap water is not sufficiently alkaline to blue the sections, add a few drops of ammonium hydroxide (S.G. 0.9) **or** of saturated aqueous lithium carbonate **or** a small pinch of calcium hydroxide to about 500 ml of water and leave the washed sections in this for 30–60s, then rinse in tap water again. Examine the wet slide under a microscope to check that selective nuclear staining has been achieved. Any blue coloration of cytoplasm and connective tissue should be extremely faint. (See also *Note 3* below.)
4. Immerse slides in eosin (solution B) for 30s with agitation. (See also *Note 4* below.)
5. Wash (and differentiate) in running tap water for about 30s. (See *Note 1* below.)

6. Dehydrate in 70%, 95%, and two changes of absolute ethanol (with agitation, about 30s in each change; without agitation, 2–3min in each change). (See also *Note 2* below.)

7 Clear in xylene and cover, using a resinous medium.

Result

Nuclear chromatin—blue to purple; cytoplasm, collagen, keratin, erythrocytes—pink.

Notes

1. The ideal balance between the two components of the H. & E. stain is a matter of personal taste and is determined by the intensity of coloration due to the eosin. For a weaker counterstain, use 0.2% eosin or prolong the differentiation (stage 5). Differentiation also occurs in the 70% alcohol used for dehydration and, to a lesser extent, in the higher alcohols. In some objects stained by eosin, a yellow to orange cast can be discerned. This is most easily seen in erythrocytes. Staining by eosin should never be so strong that the nuclei are obscured.

2. For celloidin sections, avoid absolute alcohol and complete the dehydration in two changes (each 10min) of *n*-butanol. Frozen sections usually require shorter times in the dyes and longer times for differentiation, washing, dehydration, and clearing.

3. Poor nuclear staining may be due to prior excessive exposure of the tissue to acidic reagents (e.g. unneutralized formalin or decalcifying fluids). To restore the chromophilia, Luna (1968) recommends treatment of the hydrated sections with **either** 5% aqueous sodium bicarbonate ($NaHCO_3$) **or** 5% aqueous periodic acid, overnight, followed by a 5-min wash in water before staining.

4. A satisfactory alternative to eosin is **fast green FCF** (0.5% aqueous), which is differentiated by water more readily than by

alcohol. This dye stains acidophilic elements a bluish-green colour. It is valuable if some components of the section have already been stained pink or red with, for example, periodic acid-Schiff.

5. An alternative aluminium-haematein solution is Baker's (1962) "haematal-16". This contains 16 ions of Al^{3+} for every one molecule of haematein. Ethylene glycol is present to dissolve the haematein, which is only sparingly soluble in water. Haematal-16 is a slow, progressive nuclear stain and should be applied to the sections for 10–30min. Dissolve 0.94g of haematoxylin in a mixture of water (250ml) and ethylene glycol (250ml). Bubble air through the solution for about 4 weeks. Every few days make up again to 500ml with water to compensate for evaporation. Add to the above solution 500ml of ethylene glycol followed by 1000ml of an aqueous solution of aluminium sulphate ($Al_2(SO_4)_3.12H_2O$ 15.8g; water to 1000ml). Final volume is 2000ml. When the haematein is made by atmospheric oxidation, as described here, this haematal-16 solution is stable for approximately 2 years.

6.1.3.2. *Weigert's iron haematoxylin*

Stock solutions
Solution A

Haematoxylin (C.I. 75290):	5g
95% ethanol:	500ml
(Keeps for three years)	

Solution B

Ferric chloride ($FeCl_3.6H_2O$):	5.8g
Water:	495ml
Concentrated hydrochloric acid:	5ml
(Keeps indefinitely)	

Working solution
Mix equal volumes of A and B. Put A in the staining jar first for more rapid mixing. The mixture should be made just before using, but can be kept for about 2 weeks at 4°C.

Procedure

1. De-wax and hydrate paraffin sections.
2. Stain in the working solution of Weigert's haematoxylin for 5min (10min if the solution is more than a few days old).
3. Wash in running tap water.

Result

Nuclei blue-black (may be dark brown if the working solution has been stored for more than a few days).

Note

This should give selective nuclear staining without the need of differentiating. If overstaining occurs, the sections can be destained by immersion in acid alcohol (as for alum-haematein, see p. 96), though the process is slow. Overstaining can be prevented on future occasions by using a higher proportion by volume of solution B in the working mixture. Conversely, if staining is too slow or too weak, even with a freshly mixed working solution, the proportion of solution A should be increased. Slight grey coloration of the cytoplasm does not usually matter; it disappears when a counterstain is applied.

6.1.3.3. Lillie's modification of Weigert's iron haematoxylin

This stable solution is prepared as follows:

1. Make 1% alcoholic haematoxylin by dissolving 1.0g haematoxylin in 100ml of 95% ethanol.
2. Make the following solution:

Ferric chloride (FeCl$_3$.6H$_2$O):	2.5g
Ferrous sulphate (FeSO$_4$.7H$_2$O):	4.5g
Water:	298ml
Concentrated hydrochloric acid:	2.0ml

Stir until dissolved. (Filter if necessary, to remove undissolved impurities.)

3. Add the solution containing the iron salts to the alcoholic haematoxylin solution. The mixture goes black. It can be used immediately, in the same way as the original Weigert's iron haematoxylin. Lillie's solution is stable for 2–3 months at room temperature.

6.1.3.4. Brazilin

This dye (see p. 73) is closely related to haematoxylin. Its oxidation product (brazilein) forms with aluminium ions a lake which stains nuclei red. Since the colour (like that imparted by alum-haematein) is extracted by acids, alum-brazilein is best applied as a counterstain to follow a method that imparts green or blue colours to other components of the tissue.

A suitable mixture is **Mayer's brazalum**. This is made in exactly the same way as Mayer's haemalum (p. 96), but substituting brazilin (C.I. 75280) for haematoxylin. It is more stable than haemalum, probably because brazilin has only one pair of hydroxyl groups in the easily oxidized catechol configuration. The staining time is usually about 5min, with differentiation, if necessary, in acid alcohol. Nuclei are stained red.

6.1.3.5. Mayer's carmalum

Staining mixture (after Gatenby & Beams, 1950)

Carminic acid (C.I. 75470; not to be confused with carmine):	1.0g
Aluminium potassium sulphate, KAl(SO$_4$)$_2$.12H$_2$O:	10g
Water:	200ml

Heat until it boils, then allow to cool to room temperature. Filter. Add 1.0ml formalin (37–40% HCHO) as an antibacterial preservative. The solution loses most of its potency after 2–3 weeks.

Procedure

1. Stain hydrated sections for 10–30min. Overstaining does not occur.
2. Wash in running tap water for about 1min.
3. Dehydrate, clear, and cover.

Result

Nuclei crimson. Like alum-brazilein, carmalum is valuable as a counterstain. It gives pleasing appearances following indigogenic methods for esterases (see Chapter 15) or silver methods for nervous tissue (Chapter 18).

6.1.3.6. Alum-kernechtrot

Staining solution (after Humason, 1979)

Kernechtrot (nuclear fast red; C.I. 75470):	0.2g
Aluminium sulphate $(Al_2(SO_4)_3.18H_2O)$:	10g
Water:	200ml

Heat with stirring until it is nearly boiling, then leave overnight to cool. There is a substantial residue of insoluble material, which must be removed by decanting and filtering. Keeps for about a year; may need to be filtered before each use.

Procedure

1. Stain hydrated sections for 5min.
2. Wash in running tap water, about 1min.
3. Dehydrate, clear and cover.

Result

Similar to the appearance obtained with carmalum (see above).

6.1.3.7. Iron-chromoxane cyanine R

This method (Kiernan, 1984b) is derived from the methods of Llewellyn (1974, 1978), Hogg & Simpson (1975) and Clark (1979). It is for selective nuclear staining similar to that obtained with alum-haematein. A related method that stains in two colours is described on page 101. The dye is commonly sold under the names solochrome cyanine R, eriochrome cyanine R, and Mordant blue 3.

Solutions required

1. Staining solution

0.21M aqueous ferric chloride (5.6% w/v $FeCl_3.6H_2O$):	20.0ml
Chromoxane cyanine R (C.I. 43820):	1.0g
Concentrated (95–98% w/w) sulphuric acid:	2.5ml
Water:	to make 500ml

The ingredients take 2–5 min to dissolve, with stirring. Filtration should not be necessary. The solution can be kept and used repeatedly for at least 8 years. The same solution can be used for staining myelin (see Chapter 18).

2. Differentiating solution for nuclei

Ethanol:	500ml
Water:	500ml
Concentrated (12M) hydrochloric acid:	5ml

Keeps for several months, but use it only once.

Procedure

1. Stain hydrated sections in Solution 1 for 5min. (They may remain in the solution for 30min without harm.)
2. Wash in running tap water, 30s, or in 3 changes of distilled water. This is to remove unbound dye.
3. Immerse in Solution 2 with continuous agitation until only the nuclei are stained. This usually takes 5–30s. To check differentiation, rinse in water and examine under a microscope.
4. When differentiation is complete, wash in water.
5. Apply a counterstain, as desired, dehydrate, clear, and cover using a resinous mounting medium.

Result

Nuclei blue.

6.2. Anionic counterstains

In most of the general "oversight" methods used in histology a blue, purple, or black nuclear stain is followed by a paler, usually pink, counterstain, which colours all the other components of the tissue. **Eosin** is suitable for this purpose. It is an anionic dye, so it is bound principally by ionized cationic groups of protein molecules. The most numerous of these are the ε-amino group of lysine and the guanidino group of arginine. Nearly all proteins contain these two amino acids, so eosin and other anionic dyes are bound by almost all the structures present in any tissue. A few objects are, however, missed by anionic counterstains. Thus, the extracellular matrix surrounding the collagen fibres of connective tissue is composed largely of proteoglycans (see Chapter 11), which have negatively charged molecules and therefore cannot bind anionic dyes. The same is true of many glycoprotein secretory products and of the matrix of cartilage. Glycogen, a neutral polysacchride, is unable to bind anionic or cationic dyes. The granules of mast cells contain a basic protein, but this is already neutralized by heparin, a strongly acid proteoglycan also present in the granules. These are therefore stained by cationic dyes but not by mordant dyes or the usual anionic counterstains. Usually it is not possible to see very fine fibres (e.g. reticulin, nerve fibres) or cytoplasmic organelles in sections coloured by a single anionic counterstain. Although these delicate structures bind the dye, the degree of contrast is insufficient to permit their resolution in sections more than 0.5–1.0μm thick.

Anionic dyes other than eosin may also be used as counterstains, but those with strong tinctorial power (e.g. methyl blue, acid fuchsine) are usually avoided because they might obscure the primary staining of the nuclei. When a red nuclear stain has been used, **metanil yellow** or **fast green FCF** is suitable.

The composition of a solution of an anionic counterstain is rarely critical. The following usually work well if applied for about one minute. The lowest indicated concentration is sufficient for most specimens, but stronger solutions may be needed for very thin sections, for osmicated material, or for specimens that have spent months in formalin (see Chapter 5 for reasons for loss of acidophilia).

> **Eosin** (C.I. 45380):
> 0.1–0.5% in 1% acetic acid.
> **Fast green FCF** (C.I. 42053):
> 0.1–0.2% in 1% acetic acid.
> **Metanil yellow** (C.I. 13065):
> 0.25–0.5% in 0.25–0.5% acetic acid.

If the maximum amount of anionic dye is to be retained in the tissue, the wash after staining should be in 0.5% acetic acid, not water. There is usually some loss of dye into 70% ethanol. This is often desirable, but for maximum retention of colour, blot the slides after washing, and take directly to the first of 3 changes of absolute alcohol.

6.3. Single solution methods

It has already been pointed out that a cationic dye applied from a solution whose pH is higher than about 5 will stain nearly all the components of a tissue. Neutral or alkaline solutions of cationic dyes can therefore be used as single reagents for general purpose oversight staining. The contrast is less than when two dyes are used, but the speed and simplicity of a one-step method are sometimes advantageous. Although most anionic dyes are of no value when used alone, **chlorazol black E** (see p. 94) is exceptional, for reasons that are obscure. The blackness of this dye may possibly be more important than its physical and chemical properties in producing optical contrast in stained sections.

Mordant dyes can also be used by themselves: a moderate amount of detail can be seen in addition to the nuclei in a section slightly overstained by alum-haematein, alum-brazilein, or alum-carminic acid, but these dye-metal complexes are hardly ever used alone in this way. The **Heidenhain haematoxylin method** (see Exercise 3, p. 88) is a more complicated procedure, requiring critical differentiation but capable of revealing considerable cytoplasmic detail. Instructions for this technique are given by Gabe (1976) and Lillie & Fullmer (1976).

Chromoxane cyanine R used with an excess of a ferric salt has already been mentioned as a blue nuclear stain. If the ratio of iron to dye in the solution is low, nuclei are stained blue and cytoplasm and collagen are red (see below). Both colours are attributed to dye-iron complexes, because the metal-free dye imparts a different shade of red and its colour, unlike that from the mixture, is easily washed out of the sections by alkali. Experimental evidence indicates that the iron complexes of chromoxane cyanine R are bound to tissues by the dye moiety, without the interposition of iron atoms (Kiernan, 1984b).

Mann's eosin-methyl blue (see p. 94) is a mixture of anionic dyes used as a one-step oversight stain. Its possible mechanism of action has already been described. Mann's mixture is also used in pathology for the detection of viral inclusion bodies in diseased cells.

The **azure-eosin** technique (p. 101) is a valuable one-step staining method in which an anionic and cationic dye are applied simultaneously from a single solution at a carefully controlled pH. Results are similar to those obtained using the same dyes in sequence, but a greater variety of shades of colour is seen when the single solution is employed. This technique was developed from those used for staining blood. The properties of mixtures containing eosin and cationic thiazine dyes will be discussed in Chapter 7. The only disadvantage of the azure-eosin technique is the instability of the staining mixture. Instructions are given on p. 102, and a full account of the method and the results obtained with it is given by Lillie & Fullmer (1976).

6.3.1. Two colours with chromoxane cyanine R

This technique (Kiernan, 1984b) is a modification of the original method of Hyman & Poulding (1961) and of Chapman's (1977) variant.

Staining solution

0.21M aqueous ferric chloride (5.6% w/v $FeCl_3.6H_2O$):	2.5 ml
Chromoxane cyanine R (C.I. 43820):	1.0 g
Concentrated (95–98% w/w) sulphuric acid:	2.5 ml
Water:	to make 500 ml

The ingredients take 2–5 min to dissolve, with stirring. Filtration should not be necessary. *The pH must be 1.5*, or the correct colours will not be obtained. The solution can be kept and used repeatedly for at least 6 years, but the pH may change with storage, especially in the first few weeks after making the solution. It should be checked from time to time, and adjusted as necessary by adding a few drops of 1.0M hydrochloric or sulphuric acid, or 1.0M sodium hydroxide.

Procedure

1. Slides bearing hydrated sections are stained for 3 min in the above solution.
2. The stained slides are washed in 3 changes of water (must be distilled), each 20s, with agitation.
3. Dehydrate in 95% and 2 changes of absolute alcohol.
4. Clear in xylene or toluene, and cover, using a resinous medium.

Result

Nuclei blue-purple; cytoplasm pink to red; collagen mostly pink, but some fine fibres in loose connective tissue purple; myelin sheaths of nerve fibres blue-purple; red blood cells orange-red.

6.3.2. Azure-eosin method (Lillie's technique)

Staining solution

This is made from stock solutions, which all keep for several months.

Azure A (C.I. 52005)

(0.1% aqueous stock solution): 16ml

Eosin B (C.I. 45400)

(0.1% aqueous stock solution): 16ml

0.2M acetic acid: 6.8ml

0.2M sodium acetate: 1.2ml

Acetone: 20ml

Water: 100ml

This working solution should be mixed immediately before using. The quantity above suffices for one batch of 8–10 slides in a 100 or 125ml staining vessel. Its pH is 4.0, which is usually optimum for formaldehyde-fixed tissues (See also *Note 2*, below.)

Procedure

1. De-wax and hydrate paraffin sections.
2. Stain in the working solution for 1h.
3. Pour off the staining solution and replace it with acetone, three changes, each 45–60s, with agitation. (See also *Note 1*, below.)
4. Clear in two changes of xylene.
5. Apply coverslips, using a synthetic resinous mounting medium.

Result

Nuclei and cytoplasmic RNA blue. Cartilage matrix and other metachromatic materials red to purple. Muscle cells pink. Cytoplasm of other cells pale blue to pink according to type. Collagen and erythrocytes pink. (See *Notes 2* and *3* below.)

Notes

1. As an alternative to dehydration in acetone, the sections may be blotted dry and passed through two changes of *n*-butanol, each 3–5 min with occasional agitation.
2. If satisfactory results are not obtained, the pH of the staining solution should be

changed. The acetic acid and sodium acetate may be replaced by 8ml of 0.2M acetate buffer (see Chapter 20) of any pH from 3.5 to 5.5. Phosphate and other buffers may be used outside this range. Before applying this method to large numbers of slides, test it from pH 3.5 to 5.5 at intervals of 0.5 pH unit. Staining solutions buffered beyond this range will rarely be needed. Solutions of pH >4.0 are often needed after fixation in agents other than formaldehyde.

3. A full account of this technique, with descriptions of the tinctorial effects in normal and pathological tissues, is given by Lillie & Fullmer (1976).

4. Similar results can be obtained with Giemsa's stain, which is described in Chapter 7.

6.4. Exercises

1. What differences would you expect to see between two similar paraffin sections of an autonomic ganglion: one stained with H. & E., the other by the azure-eosin method?

2. Why is Mayer's haemalum stable for several weeks, whereas Weigert's iron-haematoxylin deteriorates after a few days?

3. Which of the following dyes would be suitable counterstains for paraffin sections in which the nuclei have been stained with chromoxane cyanine R? Give reasons: (a) neutral red; (b) erythrosin; (c) methyl green; (d) methyl blue; (e) orange G; (f) Bismarck brown Y.

4. When a specimen has stood for several months in an aqueous formaldehyde solution, it is often very difficult to obtain sufficiently intense staining by eosin when paraffin sections are subjected to the H. & E. procedure. Suggest a reason for this.

5. What result would you expect if you stained a paraffin section of a specimen fixed in Altmann's fluid (p. 29) with toluidine blue O at pH 4.0?

6. Making use of the dyes discussed in this chapter, devise a staining method which would colour nuclei but not cytoplasmic RNA red, and cytoplasm and collagen in some shade of green.

7. Discuss the chemical mechanisms involved in the staining of: (a) nuclei by mordant dyes; (b) nuclei by cationic dyes; (c) cytoplasm by anionic dyes.

7

Staining Blood and Other Cell Suspensions

It is frequently desirable to stain cells that have been smeared or otherwise deposited onto a glass slide. The cells may be naturally suspended in liquid, as in the case of blood or an inflammatory exudate, or they may have been artificially suspended by disaggregating a piece of tissue or a colony of cultured cells. Cells may be washed or scraped from an epithelial surface, as they are in the clinical diagnostic practice known as **exfoliative cytology**, or they may be obtained as "impression smears", by simply pressing the slide against a freshly cut surface of an organ or tumour. The present chapter is concerned with methods whereby whole cells are stained on the slide. Techniques in which cells are stained, sorted and counted while still suspended are

outside the scope of this book. Other excluded topics include blood grouping, methods for the immunological study of subpopulations of lymphocytes, and special methods for chromosomes.

Special methods of preparation and staining are used for identifying the types of cells in smears of blood and haemopoietic tissues. The same techniques can reveal abnormal cells and the presence of pathogenic protozoa such as malaria parasites. These methods are also valuable for the examination of cells derived from other fluids (e.g. urine, pleural exudates, cerebrospinal fluid), and in smears of exfoliated cells.

7.1. Preparation of films, smears and sediments

7.1.1. Preparation of blood films

Films are usually made with freshly drawn whole blood. Coagulation can be prevented by putting the blood in a specimen tube that contains an anticoagulant (e.g. sodium citrate: 25 mg of $Na_3C_6H_5O_7.2H_2O$; or EDTA: 0.5 mg of $Na_2EDTA.2H_2O$ for each 1 ml of blood). If no anticoagulant is used, the film must be prepared immediately (Fig. 7.1). The glass slides must be clean and have been degreased in alcohol or acetone. A small drop of blood is spread over the slide as shown. The film is then allowed to dry ("dry fixation"). For a preparation enriched in leukocytes, the **"buffy coat"** is used. In blood that has been centrifuged, this is the greyish layer seen on the upper surface of the deposit of packed erythrocytes. A suspension enriched in leukocytes can also be obtained by density gradient centrifugation of blood in a solution of a suitable inert polymer.

If it cannot be stained within a few hours, the film should be fixed (by immersing the slide in 95% alcohol for 5 min), rinsed in water, air-dried, and stored in a dust-free box. With the Romanowsky-Giemsa staining techniques (e.g. Leishman's, Wright's), the alcoholic solvent of the mixture of dyes serves as a fixative. **No fixative**

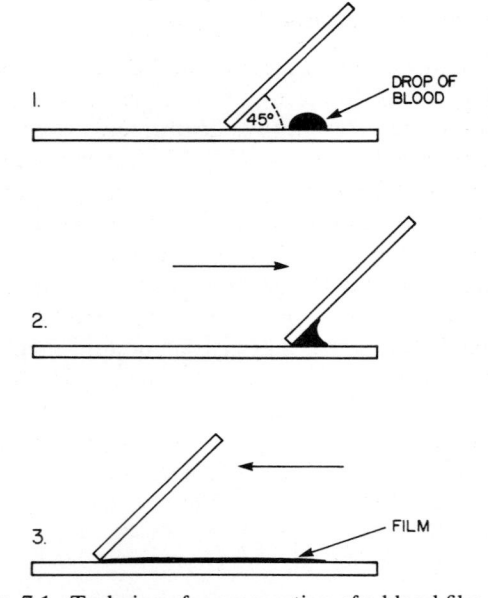

FIG. 7.1. Technique for preparation of a blood film.

Many methods are available for the disaggregation of lumps and for getting rid of mucus and other unwanted substances (see Mayall & Gledhill, 1977; Bancroft & Cook, 1984).

Unfixed cells will adhere to clean glass. The slide should then be fixed in ethanol or an acetic-ethanol mixture (Chapter 2) before the deposited cells have dried out. This is known as **wet fixation**, and it prevents much of the flattening and enlargement that occur in cells that have been allowed to dry. Even less distortion is seen if the cells are fixed by adding a fixative agent to the suspension, but some flattening is often advantageous, because the nuclear and cytoplasmic features are enlarged (see Boon & Drijver, 1986). For cells that have been fixed in the suspension, it is necessary to coat the slide with an adhesive. Chrome-gelatin (Chapter 4) is recommended.

7.1.3. Sedimentation methods

Cells suspended in saline or any other liquid less viscous than plasma will sink. Having sunk, they will adhere to a glass slide that forms the bottom of the container. An even layer can thus be obtained of cells that have not been squashed, torn, or otherwise distorted by smearing. A simple sedimentation method is illustrated in Fig. 7.2. This type of technique was originally introduced for use with cerebrospinal fluid (Sayk, 1954). The cells settle gently onto the slide as the liquid medium is slowly absorbed into the filter paper. Slides bearing sedimented cells should be wet-fixed and can be stored in 70% alcohol prior to staining.

Sayk's method is slow; the absorption of 2ml of fluid into the filter paper may take 2h. A Cytospin centrifuge (Shandon Elliott) or a Cytospin assembly usable with a conventional centrifuge deposits cells in the same way as gravity-powered sedimentation (Fig. 7.3). There is more flattening of the cells, but preparations can be made in 5 min from samples much smaller than those needed for making smears.

other than alcohol is permissible for blood films that are to be treated with neutral stains. Formaldehyde (even traces of the gas in the air, according to Boon & Drijver, 1986) and chromium compounds prevent the Romanowsky-Giemsa effect, presumably by combining with basic groups of proteins in nuclei, and altering the affinity for charged dye particles (Wittekind, 1983).

7.1.2. Preparation and fixation of smears

A smear of exfoliated cells may be prepared directly by spreading the material scraped from a surface over a glass slide, to make a thin layer. Alternatively, the material may be mixed with an excess of a physiological saline solution (see Chapter 20) and then spun for 5–10 min in a centrifuge (approximately $500 \times g$). The supernatant is discarded. The pellet is taken up in a small volume of serum or of 1% bovine albumin in saline, which is smeared onto a slide. The presence of protein in the suspending medium protects the cells from breakage during smearing.

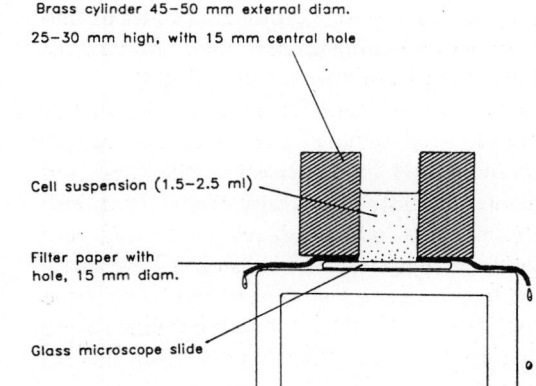

Brass cylinder 45–50 mm external diam.
25–30 mm high, with 15 mm central hole

Cell suspension (1.5–2.5 ml)

Filter paper with
hole, 15 mm diam.

Glass microscope slide

FIG. 7.2. Method recommended by Boon & Drijver (1986) for sedimentation of a cell suspension. The weight of the metal chamber (about 300 g) presses on the filter paper, slowing down the drainage of the suspending solution, thus allowing enough time for the cells to settle onto the slide.

7.2. Neutral stains

Useful staining reagents can be made by mixing aqueous solutions of suitable cationic and anionic dyes (e.g. methylene blue and eosin; neutral red and fast green FCF). The precipitate that forms contains ionic species of both dyes, and is almost insoluble in water but freely soluble in alcohol. It is also soluble in water containing an excess of either the anionic or the cationic dye. Mixed dyes of this type are often called "neutral stains". When the anionic component is eosin, the mixture is called an "eosinate". The most important applications of these solutions are in haematology and exfoliative cytology, but they may also be used for general staining of any tissue. Lillie's azure-eosin method (Chapter 6) is an example.

7.2.1. Theory of Romanowsky-Giemsa staining

The traditional stains for blood contain two or three thiazine dyes (methylene blue and its oxi-

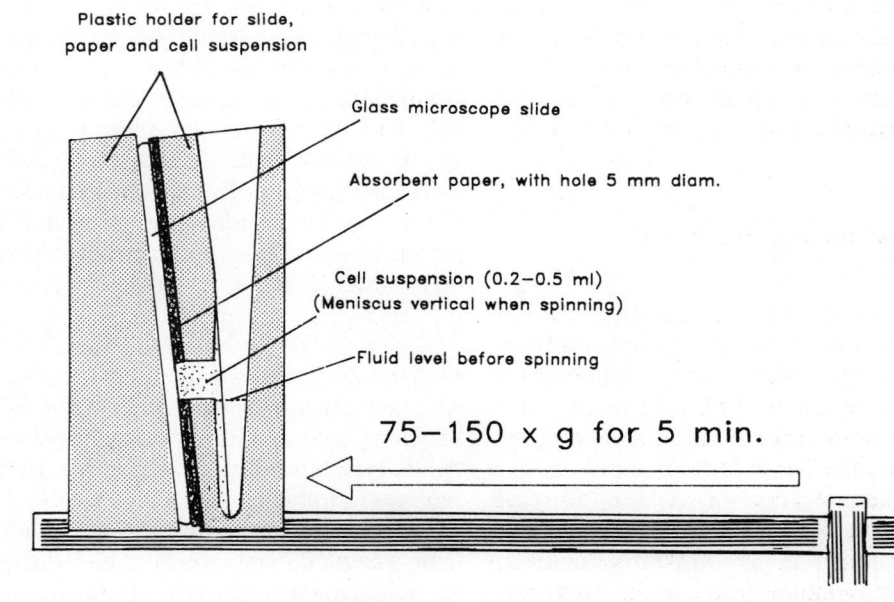

Plastic holder for slide,
paper and cell suspension

Glass microscope slide

Absorbent paper, with hole 5 mm diam.

Cell suspension (0.2–0.5 ml)
(Meniscus vertical when spinning)

Fluid level before spinning

75–150 × g for 5 min.

FIG. 7.3. Principle of cell sedimentation by centrifugation. The absorbent paper has about the same thickness as the slide. The ideal concentration of the cell suspension is determined by trial. 1–5×10^5 cells/mm^3 (or a 10- to 50-fold dilution of whole blood) is satisfactory for most purposes.

dation products) and eosin, dissolved usually in methanol and glycerol. The latter helps to stabilize the stock solution; it can be replaced by diethylamine hydrochloride, which does not increase the viscosity (Liao, Ponzo & Patel, 1981). These mixtures are known generically as "Romanowsky-Giemsa stains", after the Russian and German haematologists who first described their properties in 1891. The commercially available ready-mixed preparations are more reliable than those made in the laboratory from the separate ingredients. However, extempore mixtures are quite satisfactory for non-haematological purposes (see Lillie & Fullmer, 1976). The alcoholic solution of an eosinate must be diluted with water (or a buffer) immediately before use, to liberate the colorant ions in an active form. The pH of the diluted stain and of the water for rinsing is a critical factor, especially when using these mixtures of dyes to stain blood cells in sections of fixed tissues.

Haematological staining solutions have been developed empirically. The main source of variation among different batches is the rather unpredictable assortment of dyes produced by the oxidation of methylene blue (see Chapter 5). *It is now known that the properties of the Romanowsky-Giemsa stains are attributable to azure B and eosin* (Marshall, 1978; Wittekind, 1983). Other thiazine dyes, notably methylene violet (Bernthsen), are always present in the mixtures, but at low concentrations they do not have adverse effects. Although eosin is usually impure, its contamination with other xanthene dyes is unimportant (see Lillie, 1977). It is now possible to make standardized staining solutions that conform to internationally agreed standards. Eosin and purified azure B are the only dyes in such mixtures. Abnormalities in blood and marrow smears are seen only from time to time, so it is essential for the pathologist that the stains give consistent results, even in the hands of inexperienced technicians.

The different cells in a blood film are identified by virtue of the colours imparted to the nuclei, cytoplasm and cytoplasmic granules of the cells. Erythrocytes are coloured by the eosin anions, with intensity proportional to their intracellular concentrations of haemoglobin. Immature mammalian red cells (reticulocytes) also contain material stained by the cationic dyes, this representing the remnants of the nuclei of the cells. Leukocytes have purple (not blue) nuclei, and lilac-blue cytoplasm. Granules within the cytoplasm are red in eosinophils, blue in basophils and purple in neutrophils. Abnormal objects such as leukaemic white cells and malaria parasites also have characteristic tinctorial properties. The nuclei of protozoa should be coloured bright red by a satisfactory haematological staining solution.

The azure B in a diluted blood stain is present as dimers: two cations of dye are held together by van der Waals attractions between their planar aromatic phenothiazine rings, giving a particle with a charge of +2. The purple coloration of leukocyte nuclei, known as the Romanowsky-Giemsa effect, is due to (a) the binding of cationic dimers to nuclear DNA, and (b) the binding of eosin anions to the already bound azure B cations. Ionic and non-ionic forces (see Chapter 5) are involved in the binding of the dyes. The negatively charged phosphoric acid groups of DNA attract the azure B cations, and the attachment of the dye is reinforced by van der Waals and hydrophobic interaction between the aromatic rings of the dye and the purine and pyrimidine rings of the DNA (see also Chapter 9). Both types of force are also involved in the adherence of eosin to bound azure B (Horobin & Walter, 1987; Müller-Walz & Zimmermann, 1987).

7.2.2. Leishman's and Wright's stains

These are methods for blood films. The stock solutions can be bought ready-made or prepared by dissolving a powder, which is the precipitated and dried eosinate of a suitably polychromed methylene blue (see *Note 1* below).

A buffer solution, pH 6.4–6.8, is also required, but tap water, if not too alkaline, will often serve quite adequately in its place. Distilled water is usually rather too acid (pH around 5.5), owing to dissolved carbon dioxide derived from the atmosphere. A suitable solution for use with blood stains may be made by diluting a phosphate buffer (Chapter 20) with 5–10 times its volume of distilled water.

Procedure

1. Place slide with film on a horizontal staining rack over a sink.
2. Flood the slide with the stock solution of Leishman's or Wright's stain. Wait for 1 min. The solvent fixes the cells and plasma of the film. (See *Note 4* below.)
3. Add buffer solution in sufficient quantity to dilute the stain two or three times. (The excess liquid will spill over into the sink.) Leave for 3 min.
4. Wash off the stain with copious buffer solution (or tap water if satisfactory), making sure that the shiny skin of insoluble material on the surface of the stain does not deposit on the slide.
5. Stand the slide on end to drain and dry.
6. Clear and cover (see *Note 2* below) if permanent mount required.

Result

Erythrocytes—pink; nuclei of leukocytes—deep purple; cytoplasm of agranular leukocytes—pale blue or lilac; basophil granules—dark blue; neutrophil granules—purple; eosinophil granules—red to orange. Platelets—blue to purple. Nuclei of protozoa, including malaria parasites, are coloured bright red.

Notes

1. Both Leishman's and Wright's stains are made by dissolving in methanol the precipitated eosinates of polychromed methylene blue. They differ only in the methods of polychroming. If the stain is supplied as a powder, make a 0.15% solution in methanol. Shake well and leave to stand for 2 days or longer. Filter. The caps of the bottles in which the solution is kept should

not have metal liners; these have been shown to cause deterioration of the stains.

2. The preparations are commonly examined by placing immersion oil directly onto the dried films and examining with an oil-immersion objective, which is necessary for resolution of the cytological detail. A permanent preparation is made by clearing in xylene for about 2 min (**without** prior exposure to alcohols) and applying a resinous mounting medium and a coverslip.

3. Unsatisfactory staining can result from deterioration of the stain due to evaporation of the solvent or from using water at the wrong pH. It is occasionally necessary to vary the pH of the buffer between 6.0 and 7.0 in order to secure optimum coloration of the film.

4. If blood films cannot be stained within about 2 hr of preparation, they should be fixed by immersion for a few minutes in methanol. Stage 2 of the method described above is then unnecessary, and the films may be stained in freshly diluted Leishman's or Wright's solution.

7.2.3. Giemsa's stain

Giemsa's stain is sometimes used for blood films, but is more suitable for sections, with which more consistent results are obtained than with Leishman's or Wright's staining mixtures. Giemsa powder is a mixture of "Azure II" (21% w/w) with its eosinate (79% w/w). Azure II is believed to be a mixture of equal weights of methylene blue and azure B, originally marketed by the German firm of K. Hollborn, which bought Dr. G. Grübler's business (see Chapter 5, p. 59) in 1897 (Lillie, 1977). According to Lillie & Fullmer (1976), a Giemsa powder may be made by mixing methylene blue eosinate (4 g), azure B eosinate (5 g), azure A eosinate (1 g) and methylene blue (2 g if dye content is 85–88%, as the chloride). Giemsa powder obtained ready-made, from a reputable supply house and if possible certified by the Biological Stains Commission, is recommended for all purposes. The glycerol added to the methanol used as solvent has the effect of increasing the stability of the diluted Giemsa stain, so that it is possible to expose sections to the aqueous working solution for several hours if necessary.

Helly's fixative is recommended for haemo-poietic tissue. Decalcification may be necessary.

Preparation of stain

A. Giemsa stock solution

Giemsa powder:	2.0g
Glycerol:	132ml

Mix thoroughly in a 300ml bottle with an air-tight screw cap. Put the bottle containing the mixture in an oven at 60°C for 2h. Then add:

Absolute methanol:	132ml

Mix gently. When cool, tighten the screw cap and shake. Stable for about 5 years.

B. Buffer solutions

Reagents should be available for the preparation of acetate or phosphate buffers from pH 4.0 to pH 7 (see Chapter 20). Buffered water is made by diluting the buffer solutions 5–10 times with distilled water. See *Note 1* below.

Procedure

1. De-wax and hydrate paraffin sections.
2. Immerse slides for 5min in buffer solution at the optimum pH, as determined by experiment (see *Note 1* below). For Helly-fixed cancellous bone this is often pH 6.8. With formaldehyde- or Carnoy-fixed tissue, a pH as low as 4.0 is sometimes needed.
3. The stock Giemsa (solution A) is diluted with 50 volumes of buffered water at the same pH as that used in step 2. The slides are immersed in the diluted stain for 2h. Alternatively, preheat the diluted Giemsa solution in a closed coplin jar in the wax oven (60°C) for about 20min. Insert the slides and leave them for a further 15min. If slides are placed in cold, diluted Giemsa

and then put into the oven, allow about 30min for staining.
4. Rinse the slides rapidly in buffered water and differentiate if necessary as described in *Note 2* below.
5. Blot sections dry with filter paper.
6. Dehydrate in two changes (each 3–5min) of *n*-butanol. Alternatively, dehydrate rapidly in three changes of absolute ethanol.
7. Clear in xylene and cover.

Result

Nuclei—blue to purple; erythrocytes, collagen, and keratin—pink; leukocytes—as described above for blood films except that basophil granules (and also mast cell granules) are deep purple rather than dark blue; cartilage matrix—purple. Differentiation (see *Note 2*) accentuates the metachromasia (red-purple colour) of cartilage and of mast cell and basophil granules.

Notes

1. This and related methods are influenced greatly by the pH with which the section is equilibrated before staining, which should also be the pH of the staining solution itself. Distilled water is often satisfactory as a diluent for the stain and for washing. Check the pH of buffered water with a meter after dilution. Adjustment is sometimes necessary, especially with acetate buffers.
2. After rinsing with buffer, the section should be examined with a microscope. If the blue component of the stain is too strong, the slide should be differentiated by dipping it quickly into a 0.01% aqueous solution of glacial acetic acid and then returning it to the buffer. Several dips in the very weak acetic acid may be needed. If the pink component predominates, the pH of the stain is probably inappropriate.
3. Lillie & Fullmer (1976) strongly advocate this type of technique as a general oversight method, because more cytoplasmic details (granules, RNA, etc.) can be seen than with

haematoxylin and eosin. The azure-eosin method (Chapter 6) is similar in principle, but does not provide the same differential coloration of leukocytes as the Giemsa stain.

4. If blood-films are to be stained with Giemsa, fix them first in methanol. Dilute the stock Giemsa solution (A) with 40–50 volumes of distilled water and stain by immersing the slides for 15–40min. The time will vary with different batches of Giemsa powder. Rinse in water, dry, and, if desired, clear and mount as described for Leishman's and Wright's stains (Section 7.2.2).

7.2.4. Standardized Romanowsky-Giemsa method

This procedure, approved by the International Committee for Standardization in Haematology, makes use of pure dyes, which are commercially available but expensive. The procedure described below (Boon & Drijver, 1986) incorporates improvements recommended by Wittekind & Kretschmer (1987).

Stock solutions

1. Dye mixture

Combine the following two solutions:

Azure B thiocyanate:	1.0g
Dissolve in	
Dimethylsulphoxide (DMSO):	133ml
Eosin Y disodium salt:	0.33g

(This is the form in which eosin (C.I. 45380) is usually supplied. Its dye content is typically about 85%. Eosin for use in this technique is sometimes supplied as the dye acid, which may be assumed to be 100% pure, so 0.27g is equivalent to 0.33g of the disodium salt.)

Dissolve in
Methanol: 200ml

This stock solution is stored in a brown glass bottle at room temperature. It keeps for about 6 months and should not be used if it contains precipitate. The life of the solution can be prolonged by adding a few drops of 1.0M HCl, until the apparent pH (the reading on a pH meter with its electrode in the non-aqueous solution) is 4.

2. Buffer

Ideally, use 0.03M HEPES buffer, pH 6.5, but 0.033M phosphate buffer, pH 6.8, is satisfactory. See Chapter 20 for instructions for making buffer solutions.

3. Staining solution

Solution 1 (Dyes):	2ml	Mix when needed
Solution 2 (Buffer):	30ml	Use once and discard

Procedure

1. Make blood films or similar preparations and fix by immersion in either methanol or the stock dye solution (Solution 1, above), for 5min.
2. Immerse slides in the staining solution in a coplin jar or similar vessel:
   25min for blood films
   35min for bone marrow smears.
3. Wash in buffer (Solution 2), 1min.
4. Rinse briefly in water.
5. Drain and allow to dry. A coverslip may be applied, but is neither necessary nor desirable (See *Note 2* following the instructions for Leishman's and Wright's methods).

Result
As for Leishman's or Wright's stain.

7.3. Other methods using dyes

Any method for staining sections ought, in principle, to be equally applicable to fixed smears of whole cells. One of the main goals of clinical exfoliative cytology is the recognition of cells derived from malignant tumours. For this purpose, a staining method must show the shapes and sizes of the cells, provide crisp delineation of the nuclear chromatin, and demonstrate the presence in the cytoplasm of large amounts of ribonucleoprotein, or of deposits of keratin or mucus. The azure-eosin methods for blood cells (discussed earlier in this chapter) are often used for non-haematological purposes. Informative results can also be obtained with Mann's short eosin-methyl blue technique (Chapter 6, p. 94), and with bacteriological staining procedures (not covered in this book). The Papanicoloaou method, however, is the one most used by clinical pathologists.

7.3.1. Papanicoloaou staining of smears

The name of G. N. Papanicoloaou is associated with a family of techniques with which cytoplasms are coloured by eosin, mucus by light green, and keratin by orange G. Nuclear chromatin is stained blue with alum-haematein. After the nuclei have been stained, the small anions of orange G enter all the unstained components of the smear. This dye will bind to proteins only from a solution more acid than pH 3. The single ions and dimers of eosin and the large ions or aggregates of light green compete with one another for cytoplasmic binding sites, but they do not displace orange G from keratin. Single eosin ions enter erythrocytes, which become orange-pink (the colour of monomeric eosin, perhaps mixed with some residual orange G). Eosin dimers (red-pink) enter the cytoplasms of most cells, and are bound by proteins there. Light green is bound by mucus, which therefore acquires the bluish green colour of this dye. The optimum pH for competition between eosin and light green is 6.5. The pH of each dye solution is determined by the concentration of phosphotungstic acid (PTA) added to the solution. Hydrochloric acid may be substituted for PTA, but with some loss of distinctness in the selective staining. The large anions of PTA seem to assist in the exclusion of large dye particles (light green) from most cells.

These methods were probably derived empirically from procedures used for staining cells and connective tissue fibres in sections. The possible mechanisms of differential staining by mixtures of anionic dyes, with and without such additives as PTA, are discussed more critically in Chapter 8.

The following procedure (Boon & Drijver, 1986) is reliable because the pH of the dye mixtures is controlled, and all rinses and washings are clearly described.

Note that the term "pH" is used incorrectly in the instructions below. The reading obtained when the glass electrode of a pH meter is in an aqueous-alcoholic solution is not the correct value of the logarithm of the reciprocal of the hydrogen ion concentration. A more accurate figure, known as the pH*, can be calculated from the meter reading (see Perrin & Dempsey, 1974).

Solutions required

A. *Mayer's haemalum*
(Chapter 6, p. 96; any other alum-haematein that gives nuclear staining may be substituted.)

B. *Orange G with PTA*
Orange G (C.I. 16230):	1.0g
50% alcohol:	500ml
Phosphotungstic acid:	approx. 75mg

The pH, which must be between 2.0 and 2.8, is determined by the amount of phosphotungstic acid in the solution.

C. *Eosin and Light green with PTA*
Eosin (C.I. 45380):	900mg
Light green (C.I. 42095):	350mg
Phosphotungstic acid:	1.5g
50% alcohol:	500ml

The pH should be 4.6. Add solid lithium carbonate, a little at a time, with continuous stirring, until the pH is 6.5.

Solution A keeps for several months. Solutions B and C keep indefinitely. All may be used repeatedly.

Staining procedure

1. Wash the slides with the fixed smears in 50% alcohol. (At least one hour if the smears have been sprayed with a glutinous material such as polyethylene glycol to prevent drying; otherwise 1min.)
2. Rinse in tap water, 15 dips.
3. Immerse in Solution A (haemalum) for 2min.
4. Wash in tap water until no more colour comes away from the slides.
5. Rinse in 50% alcohol, 15 dips.
6. Immerse in orange G-PTA (Solution B), for 1min.
7. Two rinses, each 15 dips, in 50% alcohol.
8. Immerse in Solution C (eosin-light green-PTA), for 2min.
9. Rinse in 50% alcohol, 15 dips.
10. Dehydrate in two changes of absolute alcohol.
11. Clear in *t*-butanol.
12. Cover, using a resinous mounting medium.

Result

Nuclei blue-purple, with blue or red nucleoli. Cytoplasm pink or blue-green, according to cell-type. Keratinized cells orange. Erythrocytes pink to orange.

7.4. Enzyme methods for leukocytes

Some histochemical methods for enzymes are applied to blood and bone marrow smears, for the identification of normal and abnormal leukocytes, and for differential staining of B and T lymphocytes. The chemical principles of such techniques are explained in Chapters 14, 15 and 16. Practical instructions for two methods are given here. They differ in several practical details from comparable procedures used with sectioned tissues. For more information, and for other enzyme histochemical methods of haematological interest, see Elias (1982) and Hayhoe & Quaglino (1988).

7.4.1. Alkaline phosphatase in granulocytes

This enzyme occurs in the cytoplasm of granulocytes (neutrophils and basophils, but not eosinophils) and their immediate precursors, but not in myeloblasts. In myelogenous leukaemia, the abnormal circulating leukocytes contain little or no alkaline phosphatase.

The enzyme catalyses the hydrolysis of naphthol-AS phosphate at high pH. The released naphthol AS molecules couple immediately with a diazonium salt that is included in the incubation medium, to form an insoluble azoic dye. The following method (Rutenberg, Rosales & Bennett, 1965) is applied to air-dried films or smears.

Solutions required

Fixative

Methanol:	90ml	Keeps
Formalin (37–40% HCHO):	10ml	indefinitely at 0–4°C

Substrate stock solution

Naphthol-AS phosphate:	60mg	
N,N-dimethylformamide:	1.0ml	Keeps for several
Dissolve the ester in the solvent, then add:		months at 4°C
0.2M TRIS buffer, pH 9.1:	200ml	

Working incubation medium

Substrate stock solution:	20ml	Prepare immediately
Fast blue BBN salt:	20mg	before using
Dissolve, then filter.		

Neutral red (counterstain)

Neutral red (C.I. 50040):	0.2g
Water:	200ml

Keeps indefinitely. The pH may need to be adjusted. See Chapter 6, p. 92.

Procedure

1. Immerse slides with air-dried smears in the fixative, in a coplin jar, 30s at 0–4°C. For marrow smears, extract fat with cold acetone (see Chapter 12, p. 206).
2. Pour off fixative and replace with water (tap or distilled), 5 changes, then drain and leave until completely dry.
3. Mix the working incubation medium. Pour it into the coplin jar containing the slides, and leave for 15min.
4. Wash gently in 5 changes of tap or distilled water, and air-dry.
5. Counterstain nuclei with neutral red, for 3min.
6. Wash in running tap water for about 2s, to remove unbound dye.
7. Blot, and then allow to dry thoroughly in the air. A coverslip may be applied, using a resinous mounting medium, or immersion oil may be placed directly onto the dry film or smear. (Aqueous mounting media and alcohol extract the product of the histochemical reaction.)

Results

Sites of enzymatic activity appear as blue granules in the cytoplasm. Nuclei red-orange.

Notes

1. The alkaline incubation medium loosens cells from the slides. Losses are reduced by careful handling and the air-drying in steps 2 and 4.
2. Other substrates and diazonium salts can be used. Elias (1982) prefers naphthol AS-MX phosphate, and states that fast blue RR salt may used instead of fast blue BBN salt. (One gram of fast blue RR salt is equivalent to three of fast blue BBN salt, so only about 6mg would be needed in the method as described here.)

7.4.2. Esterases in monocytes and T cells

Various histochemically detectable esterases are present in the cytoplasm of monocytes. These enymes are remarkable in that their activities can survive fixation and heat, though they are inhibited by alcohol. An esterase that catalyses the hydrolysis of α-naphthyl acetate is present in monocytes and also in type T lymphocytes (Ferrari *et al.*, 1980).

The technique (Ranki *et al.*, 1980, incorporating some modifications from Elias, 1982) is similar to the one for alkaline phosphatase (Section 7.4.1). α-naphthol released by hydrolysis of the substrate couples with fully diazotized pararosaniline (known as "hexazonium pararosaniline", because the parent compound has three diazotizable amino groups). The final product of the reaction is therefore a large dye molecule derived from one molecule of pararosaniline and three of α-naphthol.

Solutions required

Formalin-acetone fixative

Dissolve 20mg anhydrous disodium hydrogen phosphate (Na_2HPO_4) and 100mg anhydrous potassium dihydrogen phosphate (KH_2PO_4) in 30ml water. Add 25ml of formalin (37–40% HCHO), and check that the pH is 6.6. Add 45ml acetone. Probably stable indefinitely in a tightly capped bottle. Use at 4°C.

Alternatively, use cold formal-calcium (see Chapter 2, p. 27), which keeps indefinitely and is useful for other purposes too.

Substrate stock solution

Dissolve 200mg of α-naphthyl acetate in 10ml of acetone. Keeps for several weeks in a tightly capped bottle.

Pararosaniline stock solution

Add 2.0g pararosaniline to 50ml of 2.0M hydrochloric acid (the concentrated acid diluted 6 times; see Chapter 20). Heat with stirring until it boils, leave to cool to room temperature, and filter. This stable solution is kept at 4°C.

Sodium nitrite solution

Dissolve 1.0g sodium nitrite ($NaNO_2$) in 25ml water. Keep at 4°C until needed. This solution should be made on the day it is to be used.

Hexazonium pararosaniline

This unstable reagent is prepared immediately before using. Mix equal volumes of the 4% aqueous sodium nitrite and the pararosaniline stock solution, stir and let stand for one minute.

Incubation medium

Immediately before using, mix:

Substrate stock solution:	0.5ml
0.07M phosphate buffer, pH5.0:	40ml
Hexazonium pararosaniline:	2.4ml

Add a few drops of 1.0M NaOH to bring the pH to 5.8.

Procedure

1. Air-dry the films or smears and fix by immersing the slides in fixative (formalin-acetone or formal-calcium) for 10 min at 4°C.
2. Wash in 3 changes of water and air-dry for at least 30 min.
3. Immerse for 60 min at room temperature in the freshly prepared incubation medium.
4. Wash in water, 3 changes. [Optionally, the slides may now be post-fixed in neutral buffered formaldehyde (Chapter 2, p. 25) which is said by Elias (1982) to improve the nuclear counterstain.]

5. Counterstain nuclei with either an alum-haematein or methyl green (see Chapter 6).
6. Rinse in 3 changes of water and air-dry. When completely dry, rinse in xylene and cover, using a resinous mounting medium.

Result

Sites of α-naphthyl acetate esterase activity red. Cytoplasm of monocytes is strongly positive. The cytoplasm of the T lymphocyte contains one or more red dots. Neutrophils and B lymphocytes are negative. All nuclei are coloured blue or blue-green by the counterstain.

Notes

1. Sodium fluoride (0.1–0.2M, in the incubation medium) inhibits the enzyme in monocytes and macrophages, but not in the granules of the T lymphocytes (Ranki *et al.*, 1980).
2. This method can be applied to paraffin sections. Acetone at 4°C (*not alcohol*) must be used for all dehydration and rehydration. Toluene or xylene may be used for clearing.

7.5. Exercises

Theoretical

1. How much information could you obtain about the cells contained in a blood-film stained with alum-haematoxylin and eosin?

2. Why are "neutral stains" dissolved in non-aqueous solvents and then considerably diluted with water immediately before using?

3. Why is it necessary to stain sections of formaldehyde-fixed tissues with an azure-eosin combination at a lower pH than that required for blood films?

4. What factors determine the shapes and sizes of cells as seen in films and smears? If suspended cells are fixed prior to deposition on the slide, how will their shapes and sizes differ from those of cells fixed after deposition?

Practical

5. Prepare some blood films and stain them with Leishman's or Wright's stain. Identify the various types of leukocyte. Stain a film with toluidine blue O at pH 4 (see Chapter 6, p. 92) after preliminary fixation in methanol. Does the result suggest any special purpose for which blood films might be stained with a thiazine dye alone?

6. Stain sections of decalcified bone in order to demonstrate the cell types in the marrow.

8

Methods for Connective Tissue

A connective tissue stain is employed when it is desired to identify and study the various extracellular fibrous elements of animal tissues. Smooth and striated muscle fibres, erythrocytes, and the cytoplasms of other cell types are also clearly shown by the same techniques. Nuclei are commonly stained in the same section in order to facilitate correlation of the multi-coloured histological picture with the more familiar appearance obtained with an oversight method such as haematoxylin and eosin. The amorphous ground substance of connective tissue, however, is most easily demonstrated by histochemical methods for proteoglycans (see Chapter 11).

The histological techniques for connective tissues fall into three categories: those based on the use of mixtures of anionic dyes to impart different colours to collagen and cytoplasm, the methods for reticulin (which are essentially histochemical in nature), and a variety of methods for elastin.

8.1. Collagen, reticulin and elastin

The two principal types of fibre found in connective tissue are **collagen** and **elastin**. Apart from their morphological characteristics (for which consult a textbook of histology), these fibres have certain physical and chemical properties which enable them to be stained selectively.

8.1.1. Collagen

Collagen is a family of basic glycoproteins containing high proportions of glycine and proline. Another amino acid, hydroxylysine, though present in smaller quantities, is found only in collagen. The conformation of the collagen molecule is maintained by hydrogen bonding between peptide groups (as in all proteins) and the peptide chains are joined to one another by various types of covalent linkage formed by condensation of hydroxylysine with lysine. This results in the formation of cables, each consisting of a triple helix of polypeptide strands. The carbohydrate components are mostly α-D-glucosyl and β-D-galactosyl residues, attached mainly to hydroxylysine. The triple helices are joined end to end and side to side by hydrogen bonds, electrostatic attractions, and covalent bridges to form structures visible in the electron microscope as collagen **fibrils** (see Kühn, 1987). The collagen **fibres** seen with the light microscope are bundles of these fibrils.

At least 11 types of collagen, identified by Roman numerals, are recognized on the basis of amino acid sequences, physical and immunological properties, and fibrillary ultrastructure. The common collagens of mammalian tissues are as follows:

Type I: Thick fibrils assembled in parallel bundles. This is the principal collagen of tendons.

Type II: Thin fibrils; in hyaline cartilage.

Type III: Thin fibrils. Occurs in skin and muscle, and also in basement membranes.

Type IV: The collagen molecules form a network rather than fibrils. This is the principal collagen of all basement membranes, where it is intimately associated with another protein, laminin, and a glycosaminoglycan, heparan sulphate.

For information on other types of collagen, see Weiss & Jayson (1982), Mayne & Burgeson (1987) and Kornblihtt & Gutman (1988).

As a glycoprotein, collagen is stained, though not strongly, by the periodic acid-Schiff method (see Chapter 11). It is possible to detect free aldehyde groups in collagen by means of reagents more sensitive than Schiff's (Davis & Janis, 1966). These may arise from the oxidative deamination of the side-chains of lysine, as in immature elastin (see below).

The term **reticulin** includes the basement membranes of epithelia and blood capillaries and very fine (including immature) collagen fibres. Both types of reticulin contain more carbohydrate than does ordinary collagen. In the electron microscope, the reticular fibril shows the characteristic banding pattern of collagen. The staining methods for reticulin probably demonstrate a carbohydrate-containing matrix in which the collagenous fibrils are embedded, rather than the fibrils themselves (Puchtler & Waldrop, 1978).

8.1.2. Elastin

About 90% of the volume of **elastic fibres** and **elastic laminae** consists of elastin, a protein different from collagen and reticulin. The remaining 10% is a glycoprotein that forms microfibrils within the largely amorphous elastin matrix (Franzblau & Faris, 1981). **Oxytalan fibres** are atypical elastin-containing structures in which the microfibrils predominate.

Elastin is a hydrophobic protein, rich in glycine, alanine, and valine. It contains remarkably few amino acids with ionizable side-chains. The peptide strands of elastin do not form fibrils but

are united by desmosine and isodesmosine linkages to form a material with a predominantly amorphous appearance in the electron microscope. A desmosine bridge is formed from four lysine side-chains. Three of these are oxidatively deaminated to give aldehydes, which then unite, together with one unoxidized lysine side-chain, to form an aromatic ring similar to that of pyridine:

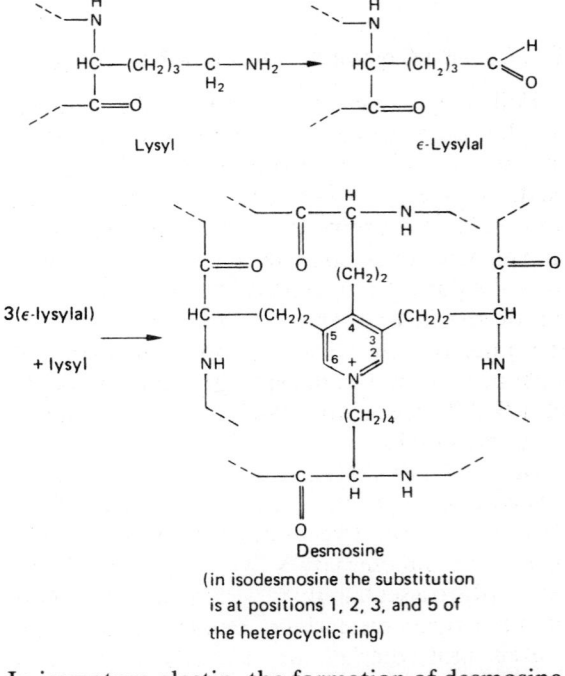

Desmosine
(in isodesmosine the substitution
is at positions 1, 2, 3, and 5 of
the heterocyclic ring)

In immature elastin, the formation of desmosine is incomplete, and free aldehydes can be detected histochemically (Nakao & Angrist, 1968; MacCullum, 1973). Elastin is resistant to digestion by most proteolytic enzymes. With ordinary staining methods it is acidophilic but contrasts poorly with other components of the tissue.

8.2. Methods using anionic dyes

Numerous techniques are available for staining with mixtures of dyes, but most are variants of one of two main groups. In the first group, typified by van Gieson's method, a mixture of two anionic dyes imparts one colour to collagen and another to cytoplasm, including that of muscle fibres and erythrocytes. The second group embraces the "trichrome" procedures in which two, three, or rarely four anionic dyes are used in conjunction with phosphotungstic or phosphomolybdic acid. Collagen, cytoplasm (of muscle and most other cells), and erythrocytes are all coloured differently, and other elements such as cartilage, fibrin, and secretory granules also acquire characteristic colours.

8.2.1. van Gieson's method—Principles

Following a nuclear stain with an iron-haematoxylin, the sections are immersed in a solution of acid fuchsine in near-saturated aqueous picric acid. Collagen is stained red, cytoplasm is yellow, and nuclei are brown or black. In related techniques other dyes (most frequently methyl blue) are substituted for acid fuchsine and occasionally a third dye is used, either by itself or mixed with the other two. In all the methods, the principal staining mixture is quite strongly acid, with a pH of 1.0–2.0 (see Clark, 1981; Lillie & Fullmer, 1976, for details).

Baker (1958) explained differential staining of this type on the basis of the different sizes of the molecules of the dyes used. He suggested that when two anionic dyes were applied simultaneously to a section there would be competition between them for cationic binding sites on the protein molecules of the tissue. The dye with smaller, more rapidly diffusing ions would penetrate quickly into a tightly woven proteinaceous matrix. There it would occupy the binding sites and exclude the dye whose particles were larger. This latter dye would, however, enter the loosely textured, more easily permeated parts of the tissue. It might not immediately attach to cationic groups there, however, because these would already have been occupied by the anions of the first, more diffusible dye. When the section is washed in water or alcohol, the small dye ions would be the first to leave the tissue and would come out of the more permeable regions first. The large dye ions could then attach themselves to the vacated cationic binding sites. Evidence

in support of this explanation was derived from experiments in which gelatin gels were stained by mixtures of eosin (fairly small molecules) and methyl blue (aggregates of large molecules). Concentrated gels were coloured red by eosin and dilute ones blue by methyl blue.

It is assumed in this hypothesis that fixed cytoplasm is in some way less permeable than fixed collagen, though there is no evidence other than that derived from the behaviour of dyes to support or refute such an assertion. In one variant of the van Gieson method (see Lillie & Fullmer, 1976) the sections are stained with fast green FCF, a dye of higher molecular weight than either picric acid or acid fuchsine, prior to immersion in the van Gieson mixture. Collagen is coloured red, erythrocytes green, and the cytoplasm of muscle yellowish green. From Baker's hypothesis, however, one would have expected green collagen, yellow erythrocytes, and probably red muscle. Another paradoxical observation is even more easily made. When sections are stained only with the van Gieson mixture (with no prior nuclear stain), nuclei are coloured yellow by the picric acid. However, nuclei are stained blue by Mann's eosin-methyl blue method (see Chapter 6). Thus, with one pair of anionic dyes the nuclei of cells show affinity for the smaller molecules, while with another pair they take up the larger molecules.

Lillie (1964) sought a chemical explanation for the differential dyeing of collagen and cytoplasm. He subjected sections to various pre-treatments before staining and found that:

(a) Treatment with nitrous acid abolished all staining by ordinary anionic dyes such as eosin and biebrich scarlet but did not prevent the red coloration of collagen by van Gieson's method. Nitrous acid removes amino groups: —NH_2 is probably usually replaced by —OH (see Chapter 10).

(b) Treatment with a mixture of acetic anhydride, acetic acid, and sulphuric acid abolished all acidophilia, including the affinity of collagen for acid fuchsine in the van Gieson mixture. The formerly acidophilic structures became basophilic, a property attributable to the formation of —NHSO₃⁻ groups with primary amines and of —OSO₃⁻ groups with hydroxyls. These changes were reversed by hydrolysis of the sulphate esters in a methanol-sulphuric acid mixture.

The reagents used for deamination and sulphation have small molecules and may reasonably be assumed to attack any parts of the tissue that are accessible to dye molecules. Binding of acid fuchsine to hydroxyl groups of serine, threonine, and hydroxylysine, presumably by hydrogen bonding to the nitrogen atoms of the dye, was tentatively suggested by Lillie (1964) as the mechanism of staining of collagen. His contention was supported by Puchtler & Sweat (1964a), who found that many direct cotton dyes were effective substitutes for acid fuchsine in mixtures of the van Gieson type. Such dyes (see p. 53) could account for their concen-mechanisms, though the occurrence of hydrogen bonding is questionable. Simple adsorption of the dyes (see p. 00) could account for their concentration in collagen, and closely apposed dye and protein molecules are likely to be held together by van der Waals forces and hydrophobic interactions (Puchtler, Meloan & Waldrop, 1988; and see Chapter 5). The staining of cytoplasm and erythrocytes by picric acid has not been studied critically. Molecular size may be significant, however, because other dyes with small molecules, such as Martius yellow and orange G, are used for the same purpose in other techniques. In all two-dye methods collagen is stained by the larger coloured particles while the smaller dye molecules stain cytoplasm (Horobin & Flemming, 1988).

8.2.2. van Gieson's method— Instructions

Solutions required

1. Weigert's iron-haematoxylin

Solution A.

| Haematoxylin (C.I. 75290): | 5g |
| 95% ethanol: | 500ml |

Keeps for 3 years

Solution B.

Ferric chloride (FeCl₃.6H₂O):	5.8g
Water:	495ml
Concentrated hydrochloric acid:	5ml

Keeps indefinitely

Working solution: Mix equal volumes of A and B. Put A in the staining jar or tank first for more rapid mixing. The mixture should be made just before using, but can be kept for 10–14 days at 4°C.

2. van Gieson's solution

| Acid fuchsine (C.I. 42685): | 0.5g | } Keeps |
| Saturated aqueous picric acid: | 500ml | } indefinitely |

See *Note 2* below.

3. Acidified water

Add 5ml acetic acid (glacial) to 1 litre of water (tap or distilled).

Procedure

1. De-wax and hydrate paraffin sections.
2. Stain in working solution of Weigert's haematoxylin for 5 min (10 min if the solution is more than 4–5 days old).
3 Wash in running tap water.
4. Stain in van Gieson's solution, 2–5 min. The time is not critical.
5. Wash in two changes of acidified water.
6. Dehydrate rapidly in three changes of 100% ethanol. This step also differentiates the picric acid.
7. Clear in xylene and mount in a resinous medium.

Result

Nuclei—black or brown; collagen—red; cytoplasm (especially smooth and striated muscle), keratin and erythrocytes—yellow.

Notes

1. An alternative nuclear stain is iron–alum–chromoxane cyanine R (see p.99). If this is used, the nuclei will be blue or greenish blue. Stage 2 may also be replaced by a 5-min staining in toluidine blue O (p.92), followed by washing in water and insolubilization of the dye by immersion of the slides for 5 min in 5% aqueous ammonium molybdate. The toluidine blue O colours nuclei and cytoplasmic RNA blue, cartilage, mast cell granules, and other metachromatic materials red–purple. A green colour is formed in the cytoplasms of many epithelial and secretory cells, presumably as a consequence of staining by both the blue and the yellow dyes.

2. Owing to the variable dye contents of samples of acid fuchsine, a van Gieson mixture may perform unsatisfactorily. If stained sections are too red (e.g. orange muscle fibres), add 2.5ml of concentrated hydrochloric acid to the van Gieson staining solution (No.2 above). This nearly always corrects the colours, but if the red is still excessive throw away about one quarter of the mixture and top up the remaining three quarters to the original volume with saturated aqueous picric acid.

   If red is too weak (inadequate staining of collagen), add another 0.1g of acid fuchsine. Once the mixture has been adjusted to give correct staining, it will retain its properties for several years.

8.2.3. Other mixtures of two anionic dyes

The following mixtures are used in exactly the same way as the van Gieson's solution in the preceding method, except where otherwise indicated.

8.2.3.1. Picro-indigocarmine

Indigocarmine (C.I. 73015):	0.50g
Saturated aqueous solution of picric acid:	200ml
(Keeps indefinitely)	

This gives yellow cytoplasm and blue or blue-green collagen. The most pleasing colours are obtained if the nuclei have first been stained red with basic fuchsine (e.g. 0.2% in water or 1% acetic acid). The method is then known as "**Cajal's trichrome**", but it should not be confused with the trichrome methods discussed later in this chapter.

8.2.3.2. Picro-aniline blue

Aniline blue WS *or* methyl blue (see Chapter 5):	0.1g
Saturated aqueous solution of picric acid:	200ml

This gives yellow cytoplasm and blue collagen, including basement membranes and reticulin, which are often missed by indigocarmine or by the acid fuchsine of van Gieson's solution. If the periodic acid-Schiff technique (see Chapter 11) is interposed between the nuclear staining with iron-haematoxylin and picro-aniline blue, basement membranes and many secretory products are coloured bright pink, and the whole procedure is then known as **Lillie's allochrome method**.

8.2.4. Trichrome methods—Principles

The name "trichrome" identifies staining techniques in which two or more anionic dyes are used in conjunction with either phosphomolybdic or phosphotungstic acid. These acids may be mixed with the dyes, or solutions of the individual reagents may be applied sequentially to the sections in various ways. Whatever technique is employed, the result is a selective colouring of collagen by one of the dyes. Cartilage and some mucous secretions acquire the same colour as collagen, but their intensity of staining is usually less. If one other dye is applied, it stains cytoplasm and erythrocytes. If two other dyes are used, they impart their different colours to erythrocytes and to the cytoplasms of other types of cell. Secretory granules are variously stained: sometimes the same colour as collagen, sometimes the same colour as erythrocytes. Nuclear staining can also be obtained with some of these methods, but it is more usual to stain the nuclei black with an iron–haematoxylin before carrying out the trichrome procedure. The trichrome techniques reveal fine collagenous and reticular fibres and basement membranes.

8.2.4.1. The heteropolyacids

Phosphomolybdic acid (PMA) and phosphotungstic acid (PTA) are known as heteropolyacids. They are formed by coordination of molybdate or tungstate ions with phosphoric acid. They are sold as hydrated crystals, which are freely soluble in water to give strongly acid solutions:

$$H_3PO_4.12MoO_3.24H_2O \longrightarrow 3H^+ + [PMo_{12}O_{40}]^{3-} + 24 H_2O$$

Phosphomolybdic acid   Phosphomolybdate
(=*dodeca*molybdophosphoric   anion
          acid)

$$H_3PO_4.12WO_3.24H_2O \longrightarrow 3H^+ + [PW_{12}O_{40}]^{3-} + 24 H_2O$$

Phosphotungstic acid   Phosphotungstate
(= *dodeca*tungstophosphoric   anion
          acid)

The oxidation number of Mo and of W in these compounds is +6. The heteropolyacids are decomposed by alkalis to give molybdate (MoO_4^{2-}) or tungstate (WO_4^{2-}) and dibasic phosphate ions (see Cotton & Wilkinson, 1980). In aqueous solutions, PTA forms complex anions, $[PO_4(WO_3)_{12}]^{3-}$ as shown above, but these decompose if the pH rises above 2.0, to give $[(PO_5)_2(WO_3)_{17}]^{10-}$ and $[PO_6(WO_3)_{11}]^{7-}$ (Rieck, 1967).

The heteropolyacids are able to bind to tissues from aqueous or alcoholic solutions. Baker (1958) called them "colourless anionic dyes". Sites of attachment of PMA are easily demonstrated by subsequent treatment of the sections with either ultraviolet radiation or a chemical reducing agent such as stannous chloride. A blue mixture of insoluble oxides of Mo(V) and Mo(VI), with compositions such as $MoO_2(OH)$ and $MoO_{2.5}(OH)_{0.5}$, is formed. It is known as molybdenum blue. A corresponding but less intensely coloured tungsten blue, formulated as $WO_{2.7}$, can also be produced, but sites of binding of PTA to tissues have been more thoroughly studied under the electron microscope, with which the electron-dense tungsten-containing deposits can be accurately localized.

The various studies of the binding of heteropolyacids to tissues (Baker, 1958; Puchtler & Isler, 1958; Bulmer, 1962; Puchtler & Sweat, 1964b; Everett & Miller, 1974; Hayat, 1975) are not all in agreement with one another, but the following facts appear to be undisputed:

1. Chemical studies indicate that PTA binds to proteins and amino acids but not to carbohydrates. Both the heteropolyacids are used as precipitants for proteins, amino acids, and alkaloids.
2. Applied at pH<1.5, PTA imparts electron density to carbohydrate-containing structures, but at pH>1.5 it binds to proteins. PTA can oxidase carbohydrate hydroxyl groups to aldehydes, so the electron-dense deposits resulting from staining at pH<1.5 may be of insoluble compounds in which the oxidation state of the tungsten is less than +6.
3. Collagen fibres bind large amounts of PMA. Cytoplasm binds smaller amounts. Nuclei of cells have very little affinity for PMA. PTA behaves similarly.
4. Affinity for PMA and PTA is depressed or abolished if amino groups in the tissue are first removed by treatment with nitrous acid or esterified by reaction with acetic anhydride or benzoyl chloride. In light microscopy there is no evidence for binding of PMA or PTA to carbohydrates or to hydroxyl groups of amino acids.
5. Methylation of sections results in increased attachment of PMA to all parts of the tissue,

including erythrocytes. Methylating agents add methyl groups to amine nitrogen atoms (increasing their basicity) and to hydroxyl oxygen atoms (forming ethers or glycosides).

6. Structures that have bound PMA become stainable by cationic dyes.

7. Treatment with PMA or PTA affects stainability by anionic dyes. The effects are variable:

   (a) There is considerable suppression of the staining of all parts of the tissue by some anionic dyes, including picric acid, Martius yellow, eosin, orange G, and biebrich scarlet. The amount of suppression is greater in collagen than in cytoplasm.

   (b) There is similar suppression of cytoplasmic staining by other dyes, including aniline blue WS, light green SF, fast green FCF, and acid fuchsine, but collagen is stained with only slightly reduced intensity by these dyes after treatment of the sections with a heteropolyacid.

   (c) Treatment of sections with PMA or PTA either before or at the same time as staining with aniline blue WS, light green SF, or fast green FCF has the effect of preventing the attachment of these dyes to materials other than collagen, cartilage matrix, and certain carbohydrate-containing secretory products.

   (d) If sections are treated with PTA, stained with aniline blue WS, and then exposed to 6M urea, a reagent which disrupts hydrogen bonds, the dye is removed, but the PTA remains attached to the collagen in the tissue. The treatment with urea may not, however, be a specific test for hydrogen bonding.

8.2.4.2. *How do trichrome methods work?*

Three hypotheses have been advanced to account for the differential colouring of tissues by anionic dyes used in association with heteropolyacids.

In the **first theory**, championed by Baker (1958), it is held that the anions of the dyes and of the heteropolyacids compete with one another for cationic binding sites and that the textures of the various structural components of a tissue determine their penetration by molecules of different size. PMA and PTA form anions intermediate in size between those of the dyes generally used as cytoplasmic stains and those which stain collagen. The smallest molecules would therefore enter and bind to the supposedly dense network of proteins forming the stroma of the erythrocyte, which would not be penetrated by the heteropolyacid. The much less "dense" matrix of the collagen fibre would accommodate both the ions of PMA or PTA and those of a dye with large molecules such as aniline blue WS or light green SF. Small dye molecules could also penetrate the collagen fibres but, since they diffuse rapidly, they would be able to enter and leave freely. The larger, more slowly diffusing molecules of the heteropolyacids and of such dyes as aniline blue WS would remain in the collagen and become attached to cationic sites there. Dyes with molecules of intermediate size (e.g. acid fuchsine) would compete with PMA or PTA for binding sites in the cytoplasm of muscle fibres and other cells. These dye molecules would be too big to enter erythrocytes in the presence of dyes with small molecules, but they would be small enough to escape from collagen fibres more quickly than the largest dye molecules. Bulmer (1962) demonstrated that heteropolyacids were bound to tissues only when the latter contained ionizable amino groups, but agreed with Baker (1958) in attributing the trichrome staining effects to differential permeability of cytoplasm and collagen to large and small molecules.

A similar postulated mechanism was discussed earlier in relation to the differential staining of cytoplasm and collagen by the van Gieson method. The objections raised there also apply to the application of this hypothesis to the trichrome techniques. Furthermore, it is difficult, in terms of this theory, to account for the fact that treatment with heteropolyacids induces basophilia, as well as selective though depressed affinity for certain anionic dyes, in collagen.

The **second hypothesis** accounts adequately for the basophilia produced by treatment of collagen with PMA or PTA. Puchtler & Isler (1958) proposed that cationic dyes were attracted by the free negatively charged groups of the collagen-bound ions of heteropolyacid. For example:

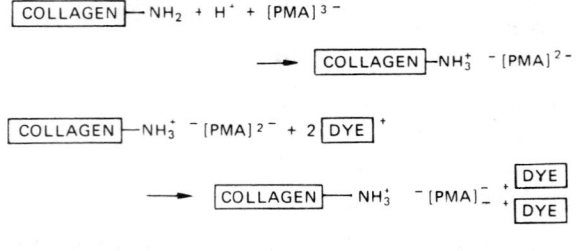

Thus a function similar to that of a mordant is attributed to the heteropolyacid. Puchtler & Isler noted that the dyes used to stain collagen in the trichrome techniques are all amphoteric. They suggested that these dyes were bound by ionic forces to the PMA or PTA, which was itself attached electrovalently to the tissue. The cytoplasmic stains used in trichrome techniques are wholly anionic dyes, so they would not attach to the free negatively charged sites of the bound heteropolyacid molecules.

Several objections to this hypothesis can be made. The staining of collagen is attributed to its content of amino acids with basic side-chains: lysine, arginine, and histidine. Haemoglobin, the principal protein of erythrocytes, is more basic than collagen but it is not similarly stained in the trichrome procedures. If collagen did contain more cationic binding sites than cytoplasm, mordanting by PMA or PTA would be expected to result in intensification of staining by amphoteric dyes, since two molecules of dye could attach to each of the bound trivalent anions of the heteropolyacid. However, all investigators agree that pre-treatment with PMA or PTA reduces the intensity of staining by aniline blue WS and similar dyes, even in collagen. The matrix of cartilage and the granules of mast cells, although they are composed of strongly acid proteoglycans (see Chapter 11), are not ordinarily stained strongly by amphoteric dyes such as acid fuchsine and aniline blue WS, so these dyes do not behave as if they were cationic. Finally, the effect of 6M urea on trichrome-stained sections (see above, p. 121) indicates that the heteropolyacid and the dye are bound to collagen by different mechanisms.

A **third explanation** for the actions of PMA and PTA in trichrome procedures was offered by Everett & Miller (1974), who provided evidence of two different modes of binding of these acids to tissues. Attachment by electrovalent forces was thought to occur in cytoplasm and to inhibit there the binding of anionic dyes. The binding of the heteropolyacids to collagen was believed to be non-ionic, so that staining by anionic dyes was not prevented. The principal objection to this hypothesis is that it does not account for the fact that the staining of collagen after treatment with PMA or PTA can be effected by some anionic dyes but not by others.

From the foregoing discussion it will be seen that there is, as yet, no satisfactory explanation of the effects of heteropolyacids on the stainability of tissues by different anionic dyes. The relationships between the applications of PTA in light and electron microscopy also require elucidation.

8.2.5. Two trichrome methods— Instructions

The original trichrome techniques of Mallory, Heidenhain ("AZAN") and Masson are done in stages, with control over the differentiation of each colour. (See Luna, 1968; Gabe, 1976; Clark, 1981 for detailed accounts of these methods.) The two procedures described here are one-step methods. They are technically simpler than the classical trichromes, but do not always work as well. The method of Cason (1950) with aniline blue as the collagen stain is for thin (less than 7 µm) sections. That of Gabe (1976) can be applied to thicker sections, because fast green FCF colours the collagen less intensely than aniline blue.

The **fixative** for tissues to be stained by any trichrome method should not be a simple formaldehyde solution with no other active ingredients, and it should not contain glutaraldehyde. Mercury-containing mixtures such as SUSA give excellent results, and Bouin is also satisfactory. If you have used a mercury-containing fixative, don't forget to treat the sections with iodine and thiosulphate before staining (see Chapter 2, where there is also some more information on choice of fixatives).

The nuclei are stained by the dye with molecules of intermediate size (red in both the methods that follow), but it is often desirable to stain them black instead, with an iron-haematoxylin (Chapter 6). Prior nuclear staining reduces the brilliance of the colours in cytoplasm and collagen, but usually contributes to the overall morphological clarity of the stained preparation.

8.2.5.1. Cason's trichrome

Solutions required

A. Weigert's iron–haematoxylin (working solution) (see p. 118).

B. Cason's trichrome solution

Water:	200 ml

Dissolve in order stated:

Phosphotungstic acid:	1 g	
Orange G (C.I. 16230):	2 g	Keeps for years
Aniline blue WS (C.I. 42755):	1 g	
Acid fuchsine (C.I. 42685):	3 g	

Procedure

1. De-wax and hydrate sections.
2. Stain in Weigert's haematoxylin, 5 min.
3. Wash in running tap water, 2 min.
4. Immerse in Cason's trichrome solution, 5 min.

5. Wash in running water, 3–5 s.
6. Blot slides dry with filter paper.
7. Dehydrate rapidly in three changes of 100% ethanol.
8. Clear in xylene and cover, using a resinous medium.

Results

Collagen—blue; cytoplasm, muscle—red; keratin, erythrocytes—orange; nuclei—brown (but sometimes blue). The pre-staining with iron-haematoxylin makes the trichrome colours a little unpredictable with some material. If the nuclear stain with iron–haematoxylin is omitted, most nuclei are coloured red: others may be blue or unstained.

8.2.5.2. Gabe's trichrome

Solutions required

A. An iron–haematoxylin such as that of Weigert or Lillie (see Chapter 6).

B. One-step trichrome solution

Amaranth (C.I. 16185):	2.5 g
Phosphomolybdic acid:	2.5 g
Fast green FCF (C.I. 42053):	1.0 g
Water:	500 ml
Glacial acetic acid:	5.0 ml

Stir until all dissolved, then add

Martius yellow (C.I. 10315):	0.5 g

Stir for 1 h, then filter to remove undissolved Martius yellow. The solution is ready for use immediately, may be used repeatedly, and is stable for at least 5 years.

Naphthol yellow S (C.I. 10316), 0.05 g, may be used instead of Martius yellow. This dye dissolves completely and filtration is not needed.

C. Acidified water

Add 5 ml acetic acid (glacial) to 1 litre of water (tap or distilled).

Procedure

1. De-wax and hydrate sections.
2. (Optional) Stain nuclei with an iron haematoxylin (*Solution A*).
3. Wash in running tap water, 2 min.
4. Immerse in Gabe's trichrome mixture (*Solution B*) for 10–15 min.
5. Wash in two changes of acidified water.
6. Shake off most of the water and dehydrate directly in 100% alcohol, 3 changes. (There is no need for speed in the washing or dehydration, because the colours are not extracted.)
7. Clear in xylene and cover, using a resinous mounting medium.

Results

Nuclei black if stained with iron-haematoxylin; otherwise red. Cytoplasm pink, red or greyish-purple. Erythrocytes yellow (ideally, but sometimes pink). Collagen fibres bluish green. Mucus, matrix of cartilage, and some secretory granules are also green.

8.3. Methods for reticulin

8.3.1. Chemical principles

Reticulin is stained by virtue of its content of hexose sugars. The simplest method, therefore, is the periodic acid-Schiff (PAS) procedure, which will be described, along with other techniques of carbohydrate histochemistry, in Chapter 11. There are, however, older methods for reticulin in which the fibres and basement membranes are rendered visible by deposition upon them of finely divided metallic silver. These older methods show reticulin in more striking contrast to the background than does PAS, so they will be discussed briefly alongside other methods for connective tissue.

The histochemical basis of the methods for reticulin has been shown by Lhotka (1956) and by Velican & Velican (1970) to be closely similar to that of the PAS procedure. Adjacent hydroxyl groups of the hexose sugars of the glycoprotein are oxidized to aldehydes by potassium permanganate or periodic acid. The oxidation is sometimes followed by an empirically discovered "sensitizing" treatment with a ferric or uranyl salt. The chemical rationale of the "sensitization" is unknown, but the result is enhancement of contrast in the final preparation. The aldehydes then reduce the silver diammine ion, $[Ag(NH_3)_2]^+$, to the metal. Silver diammine is produced by adding ammonium hydroxide to an aqueous solution of silver nitrate. A precipitate of silver oxide is formed, but dissolves as the addition is continued:

$$2NH_4OH + 2AgNO_3 \longrightarrow 2NH_4NO_3 + Ag_2O \downarrow + H_2O$$

$$Ag_2O + 4NH_4OH \longrightarrow 2[Ag(NH_3)_2]^+ + 2OH^- + 3H_2O$$

The sequence of reactions involved in the silver staining of reticulin is as follows:

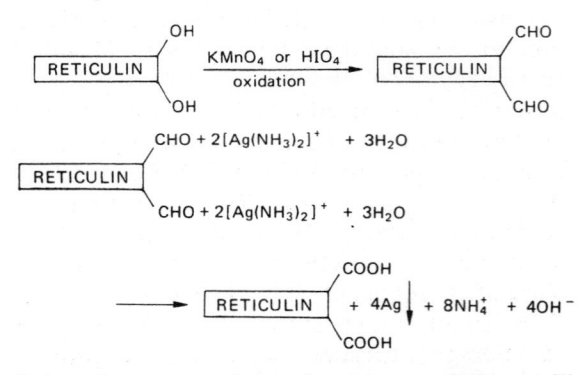

It is to be expected that four atoms of silver will be deposited at the site of each glucosyl, galactosyl, or other reactive sugar residue. This is not enough to provide adequate visibility, but fortunately it is possible to precipitate more silver upon that initially deposited by a process similar to photographic development. This occurs when the incompletely washed sections are transferred to an aqueous formaldehyde solution. Residual silver diammine ions are reduced to silver by the formaldehyde. Metallic silver catalyses the reaction, so precipitation of the metal due to the action of formaldehyde occurs mainly at the original sites of the sugar molecules of the reticulin. Yet further contrast may be obtained by the process known as gold toning. The chemistry of development and toning will be described in connection with neurological staining (Chapter 18) in which the two processes are used extensively.

In some techniques, the aldehyde-detecting reagent is not silver diammine but a complex of silver with hexamethylene tetramine. This substance, $(CH_2)_6N_4$, is a solid cyclic condensation product of ammonia and formaldehyde. It forms with silver a complex cation similar to the one formed with ammonia. Hexamethylene tetramine (also called hexamine or methenamine) can be weighed accurately, whereas an aqueous solution of ammonia loses potency with time after a new bottle has been opened.

8.3.2. A silver technique for reticulin

This method (Gordon & Sweets, 1936) is one of many available silver methods for reticulin. The oxidizing agent is $KMnO_4$ rather than HIO_4, probably because the latter was not used in histochemistry when methods of this type were invented. The function of the iron alum is obscure, but it has been shown to improve the histological picture (Velican & Velican, 1972). Any fixative may be used. The sections must be firmly attached to the slides with coagulated albumen or chrome-gelatin if they are to remain in place after treatment with alkaline reagents.

Nearly all fixatives are suitable. After mixtures containing potassium dichromate or chromic acid, it is sometimes necessary to omit the oxidation with potassium permanganate.

Solutions required

A. *Acid permanganate*

Potassium permanganate
  ($KMnO_4$):   1.0g
Water: 95ml
3% aqueous H_2SO_4:   5ml
  Prepare just before use from a stock 6% aqueous $KMnO_4$. The addition of H_2SO_4 is not necessary.

B. *1% oxalic acid*

Oxalic acid
  ($HOOC.COOH.2H_2O$):   5g
Water: to 500ml

  Keeps indefinitely.

C. *Iron alum*

Iron alum ($NH_4Fe(SO_4)_2.12H_2O$):   10g
Water: 500ml

  Prepare on the day it is to be used. Alternatively, use 4% ferric chloride ($FeCl_3.6H_2O$), which keeps indefinitely.

D. *Ammoniacal silver solution*

Stock solutions:
  1. 10% aqueous silver nitrate.
  2. Ammonium hydroxide (28% NH_3).
  3. 4% aqueous sodium hydroxide.

Working solution: Add ammonium hydroxide drop by drop to 10ml of 10% $AgNO_3$ until the brown precipitate of Ag_2O is almost (not quite) re-dissolved. Add 7.5ml of 4% NaOH, followed by a few more drops of ammonium hydroxide, until the newly formed precipitate is just dissolved. Be careful not to add too much ammonia. Swirl the solution for a few seconds after adding each drop, because dissolution of the precipitate is not quite instantaneous. Make up to 100ml with water. **This solution should be made just before use and discarded afterwards by washing it down the sink with plenty of water.** Ammoniacal silver solutions decompose on evaporation to form explosive "fulminating silver", a mixture of silver amide and silver nitride.

E. *Reducer*

Neutralized formalin
  (40% HCHO which
  has stood over marble):   10ml   } Prepare just before using
Water: 90ml

F. Yellow gold chloride

Sodium tetrachloroaurate ($NaAuCl_4.2H_2O$):	1 g	Keeps for several months
Water:	500 ml	

This solution may be re-used repeatedly.

G. Sodium thiosulphate

Sodium thiosulphate
($Na_2S_2O_3.5H_2O$):   25 g
Water: to 500 ml

Keeps indefinitely, but use only once.

Procedure

1. De-wax and hydrate paraffin sections.
2. Oxidize for 1 min in acid permanganate (solution A).
3. Wash in water.
4. Immerse in 1% oxalic acid (solution B) until the sections are white. Usually about 30s.
5. Wash in water (three changes).
6. Treat with iron alum (solution C), 10 min.
7. Wash in water (three changes).
8. Immerse slides in the ammoniacal silver solution (solution D) 5–10s.
9. Rinse in water (once only). See *Note 1* below.
10. Place in formaldehyde reducer (solution E), 30s.
11. Wash in water (three changes). See *Note 2* below.
12. Tone in 0.2% yellow gold chloride (solution F), 2 min. (See also *Note 3* below.)
13. Wash in water (two changes).
14. Immerse in sodium thiosulphate (solution G), 3 min.
15. Wash in water (three changes).
16. Dehydrate through graded alcohols, clear in xylene and cover.

Result

Reticulin—black. Other elements in shade of grey-purple.

Notes

1. This rinse is a critical step in the method. If it is excessive, staining of reticulin will be inadequate. If it is insufficient there will be non-specific precipitation of silver on the sections. If working with slides in a coplin jar, pour out the ammoniacal silver solution, fill up with water, agitate for about 5s, pour out the water, and immediately pour in the reducer.
2. The sections should be examined after stage 11. If staining is excessive, the method should be repeated from stage 6. This will produce an effect equivalent to differentiation. Probably some of the finely divided metallic silver in the specimen is oxidized to silver ions, which are then precipitated as silver chloride:

$$Ag + Fe^{3+} \rightarrow Ag^+ + Fe^{2+}$$
$$Ag^+ + Cl^- \rightarrow AgCl \downarrow$$

This salt would dissolve in the slight excess of ammonia present in the ammoniacal silver solution:

$$\underset{\text{(solid)}}{AgCl} + 2NH_3 \rightarrow [Ag(NH_3)_2]^+ + Cl^-$$

Any undissolved silver chloride would be removed at stage 14 of the method:

$$\underset{\text{(solid)}}{AgCl} + 2S_2O_3^{2-} \rightarrow [Ag(S_2O_3)_2]^{3-} + Cl^-$$

3. Palladium toning (Krikelis & Smith, 1988) is a substitute for gold toning: At stage 12, immerse the slides for 10 min in 0.05% potassium hexachloropalladate (K_2PdCl_6) in 4M HCl. The silver deposits become uniformly black, without the purplish background associated with gold toning. Palladium toning is therefore preferable when the preparations are to be photographed.

8.4. Methods for elastin

The methods for staining elastin are many and varied, and there is no completely satisfactory explanation for their modes of action.

8.4.1. Binding of dyes to elastic fibres

Dyes are certainly not attracted to elastin by electrostatic forces (see Baker, 1958 and Horobin, 1982 for review of evidence). Thermodynamic studies, based on measurement of half-staining times with orcein at different temperatures, indicate that elastic fibres have low permeability to this dye (which stains them selectively), but that there are large numbers of weak (non-ionic) dye-binding sites per unit volume of substrate (Friedberg & Goldstein 1969). It was once generally supposed that hydrogen bonding was primarily involved (see Pearse, 1968). However, Horobin and James (1970) showed that elastic fibres could be coloured by dyes incapable of forming hydrogen bonds. These investigators found that the only feature held in common by all of a large number of dyes that stained elastin was the presence of at least five aromatic rings in the molecule. Chemical blocking methods for many reactive groups did not appreciably reduce the staining of elastin by such dyes, and the prevention of hydrogen bonding caused only slight inhibition. Horobin and James proposed that large dye molecules were bound to elastic tissue in the same way that they are thought to be bound to some textile fibres, namely by van der Waals forces. The nature of this type of intermolecular attraction has been explained in Chapter 5. It should be remembered that direct dyes are anionic and also colour all acidophilic components of a tissue, though not always as intensely as elastin, when applied from aqueous solutions.

Ordinary anionic dyes, such as eosin, stain elastin lightly. Further evidence of hydrophobic bonding of dyes to elastin was obtained by Horobin and Flemming (1980), who correlated several structural features of dye molecules with the ability to bind to elastic fibres. The most effective dyes were those whose molecules had the longest chains of conjugated bonds (alternating single and double bonds; see Chapter 5), and the greatest numbers of hydrophobic groups.

Unfortunately, the traditional methods for elastin are not based on synthetic dyes of known structure, such as those investigated by Horobin and colleagues. The staining agents still most frequently employed are the following:

(a) Weigert's resorcin-fuchsine, a compound made by boiling basic fuchsine with resorcinol and ferric chloride.
(b) Orcein, a mixture of oxazine dyes, derived from a lichen or made synthetically.
(c) Verhoeff's stain, a mixture of haematoxylin, ferric chloride, iodine and potassium iodide. This also stains nuclei and the myelin sheaths of nerve fibres.
(d) Aldehyde-fuchsine, made by reaction of pararosaniline with acetaldehyde. The resulting product colours elastic fibres and laminae and also binds to strong acid (sulphonic or sulphate-ester) groups, and aldehyde groups in the tissue.

The composition of resorcin-fuchsine is not known, but the reagent is likely to contain several compounds in which aromatic rings have been added to the triphenylmethane structure, providing large molecules with long conjugated chains and a relative deficiency of hydrophilic amino groups. The components of orcein contain 4 or 6 conjugated six-membered rings, with methyl and phenolic hydroxyl side-chains (see Lillie, 1977). It is probable, therefore, that these dyes also stain elastin by hydrophobic bonding at multiple sites. Verhoeff's stain and aldehyde-fuchsine are considered in the next two sections of the chapter.

8.4.2. Verhoeff's stain

Experiments with mixtures of varying composition were carried out by Puchtler and Waldrop (1979), who found that increasing the concentration of iodine increased the stability of the solution, but reduced the intensity of staining of elastin. The iodine in an aqueous solution containing iodide is present as triodide (I_3^+) ions. In some variants of Verhoeff's method (e.g. Clark, 1981) the iodine is omitted, though some must be formed from oxidation of iodide ions by ferric chloride. Puchtler and Waldrop suggest that the iodide ions might serve as ligands that bind together iron atoms that are complexed to haematein molecules. The resulting complexes would be large, coloured molecules and might be expected to bind to elastin by van der Waals forces.

The following method is that of Musto (1981).

Solutions required

The stock solutions (A, B, C) are all stable for at least a year. The working solution (D) is made before using and used only once.

A. Haematoxylin: 2% in 95% ethanol.

B. Acidified ferric chloride:

$FeCl_3.6H_2O$:	12.4g
Water:	495ml
Concentrated HCl:	5ml

C. Iodine: 2% in 4% aqueous potassium iodide (KI)

D. Working solution:

Solution A:	30ml
Solution B:	20ml
Solution C:	10ml

Procedure
1. Any fixative may be used. Paraffin sections are dewaxed and hydrated. Musto (1981) recommends a treatment with Bouin's solution for 10min at 55–60°C (or overnight at room temperature) for sections of formalin-fixed material. If this is done, the slides must be washed in running water until the sections are no longer yellow.
2. Stain in the working solution (D) for about 45min. The staining is progressive, and should be checked at intervals.
3. Apply a suitable counterstain. Van Gieson's picro-fuchsine (Chapter 6) is suitable.

Result

Elastic fibres, nuclei, and myelin sheaths (if present) black. Cytoplasm and collagen coloured according to the counterstain used.

8.4.3. Aldehyde-fuchsine

One of the simplest, though not the most specific, of techniques is Gomori's aldehyde-fuchsine method. The reagent is made by treating pararosaniline with acetaldehyde (formed by acid-catalysed depolymerization of paraldehyde). The reaction occurs slowly, so that the solution matures and then deteriorates over the course of several days. Alternatively, the product of the reaction may be precipitated, collected and dried, and then dissolved when needed. Despite much investigation, the active compounds in aldehyde-fuchsine have not been certainly identified. The properties vary according to the way in which the stain is prepared (Mowry, 1978), but all the variants are effective in colouring elastin. Redissolved precipitated aldehyde-fuchsine stains less brightly than an optimally matured solution, and absorption spectra indicate that the solutions have different compositions (Nettleton, 1982).

Aldehyde-fuchsine is widely presumed to contain condensation products of acetaldehyde with the amino groups of pararosaniline, with such structures as

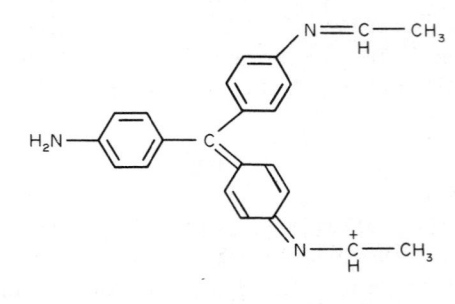

(Bangle, 1954; Buehner *et al.*, 1979), and

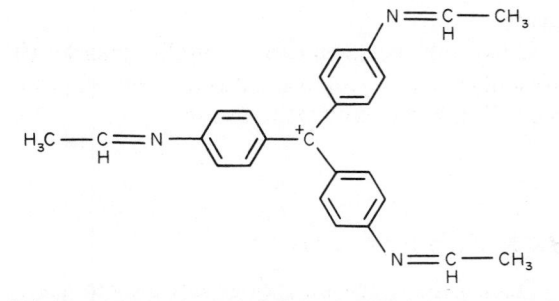

(Lichtenstein & Nettleton, 1980). Horobin and James (1970) suggested that large heterocyclic molecules, formed by combination of structures like those illustrated above, are also present. The binding to elastin is non-ionic (Goldstein, 1962), but extremely resistant to extraction. Such properties are expected of a dye with large molecules (see Chapter 5), but could also be attributable to the formation of covalent bonds between the dye and its substrate.

In addition to elastic fibres and laminae, aldehyde-fuchsine stains the sites of half-sulphate esters (see Chapter 11), and sulphonic acid or aldehyde groups artificially introduced into the tissue (Sumner, 1965; Gabe, 1976). If sections are treated with an oxidizing agent prior to staining, many carbohydrate components will be coloured (hydroxyls oxidized to aldehydes), as will sites of high concentration of disulphide groups (cystine oxidized to cysteic acid; see Chapter 10). The B cells of the pancreatic islets contain insulin, which is rich in cystine, but are stained by fresh (not by reconstituted) solutions of aldehyde-fuchsine even without prior oxidation. The reason for this is unknown. In unoxidized sections, any sample of aldehyde-fuchsine stains elastin, the granules of mast cells and cartilage matrix.

Preparation of stain

There are many ways of preparing aldehyde-fuchsine. The following (from Gabe, 1976) has the advantage of yielding a solid product which is stable for at least 6 years. It is important to use pararosaniline (C.I. 42500) or a sample of basic fuchsine consisting only of this dye (Mowry & Emmel, 1977).

Dissolve 1.0g of pararosaniline in 200ml of water (heat to boiling, then allow to cool to room temperature). Add 2.0ml of concentrated hydrochloric acid and 1.0ml of paraldehyde. Leave for 24h, or longer if a pink ring appears when the deep purple mixture is spotted onto filter paper. Filter. Discard the filtrate. Wash the residue with 50ml of water, then dry the filter paper and its contents in an oven at 60°C. Collect and keep the aldehyde-fuchsine powder.

Working solution:

Aldehyde-fuchsine powder:	0.25g
70% ethanol:	200ml
Acetic acid (glacial):	2.0ml

Stable for at least 2 years
Leave overnight to dissolve. Filtration is not needed.

Procedure
1. De-wax and hydrate paraffin sections.
2. Stain in aldehyde-fuchsine for 5min.
3. Rinse in running tap water to remove most of excess dye.
4. Rinse in 95% ethanol containing 0.5% (v/v) concentrated hydrochloric acid until any remaining excess of dye is removed (usually 20s. Longer times do no harm.)
5. Apply counterstain if desired (alum-haematoxylin and fast green FCF is a suitable combination).
6. Wash, dehydrate in graded alcohols, clear, and cover.

Result
Elastic fibres and laminae—red to purple.

Note
Aldehyde-fuchsine is also used to stain the endocrine pancreas, the adenohypophysis, mast cells, and neurosecretory material (see Chapter 18). The different techniques vary somewhat and are described by Luna (1968), Pearse (1968), and Mowry (1978).

8.4.4. Orcein

This is one of the oldest, and probably the easiest method for elastin, but the colour is less intense than that imparted by Verhoeff's method or by aldehyde-fuchsine. Bound orcein is not easily extracted, so it may be followed by almost any counterstain for nuclei, cytoplasm or collagen. If the nuclei are to be stained with an iron-haematoxylin, however, this should be done before staining the elastic fibres with orcein. Clark (1981) gives instructions for several multi-dye techniques that include orcein. As with the other methods for elastin, almost any fixation may be used.

Staining solution

| Orcein: | 1g |
| 70% alcohol: | 100ml |

Heat to about 60°C, with stirring. Cool, add 1ml concentrated hydrochloric acid, and filter to remove any undissolved material. Keeps for several months. It is used at 37°C.

Procedure

1. Take the sections to 70% alcohol.
2. Stain in the orcein solution for 30–60 min at 37°C.
3. Rinse in 70% alcohol and examine. If there is a brown background (in collagen and nuclei), differentiate in acid-alcohol (1ml concentrated hydrochloric acid in 99ml 70% alcohol). This usually takes only a few seconds.
4. Wash in running tap water, at least 30s.
5. Counterstain as required, dehydrate, clear and cover.

Result

Elastic fibres and laminae brown.

8.5. Exercises

Theoretical

1. What are the chemical differences between collagen and elastin? Do these differences account for the stainability of the two substances by different techniques?
2. What is reticulin? Suggest a reason why the silver methods for reticulin stain this substance more intensely than collagen.
3. Discuss the role of phosphotungstic or phosphomolybdic acid in histological techniques for the differential staining of cytoplasm and collagen.
4. Devise a staining procedure whereby nuclei, muscle, collagen, and elastin would all be differently coloured.
5. Lison (1955) stained sections of rat's liver with a trichrome mixture containing orange G and aniline blue. The percentage of yellow nuclei varied with the thicknesses of the sections as follows:

4μm	0%
6μm	2.5%
8μm	16.2%
10μm	30.1%
12μm	37.7%
14μm	43.5%

The nuclei that were not yellow were blue. How can these observations be explained? [*Hint*: The diameter of a nucleus in a fixed, dehydrated and embedded piece of liver is approximately 6μm.]

Practical

6. De-wax and hydrate some paraffin sections of any convenient tissue. Immerse in 2% aqueous phosphomolybdic acid for 15min. Wash in water and then immerse in a freshly prepared 0.5% aqueous solution of stannous chloride ($SnCl_2.2H_2O$) for 5min. Wash, dehydrate, and mount in a resinous medium. What structures in the tissue are stained blue? What chemical reactions are responsible for the production of the colour?

9

Methods for Nucleic Acids

9.1. Chemistry and distribution of nucleic acids

The nucleic acids are macromolecular compounds with the general structure:

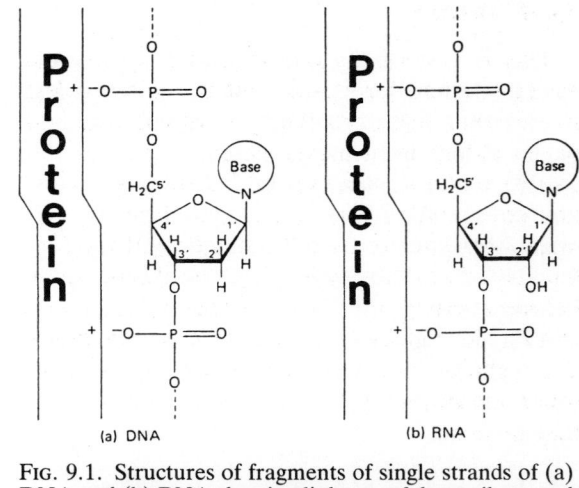

FIG. 9.1. Structures of fragments of single strands of (a) DNA and (b) RNA showing linkages of deoxyribose and ribose with phosphoric acid and with purine or pyrimidine bases. The negative charges on the phosphoric acid groups are neutralized by nucleoprotein cations. The numbers 1′ to 5′ identify the carbon atoms of the pentose sugars. Thickened lines indicate that the furanose rings lie perpendicular to the plane of the paper, with carbons 2′ and 3′ nearest to the leader. (See also Chapter 11 for structural formulae of sugars.)

The aromatic rings of the bases are stacked in the centre of the double helix. The sugars, phosphoric acids and nucleoproteins are on the outside.

guanine and the pyrimidine base cytosine are present in DNA and RNA. The fourth base, a pyrimidine, is thymine in DNA, and uracil in RNA. The structural formulae of these bases can be found in textbooks of biochemistry. The constituent units of a nucleic acid molecule are **nucleotides,** each consisting of a base, a sugar and a phosphate group. A **nucleoside** is formed from a base and a sugar only. In deoxyribonucleic acid (DNA) the sugar is deoxyribose (the full name is

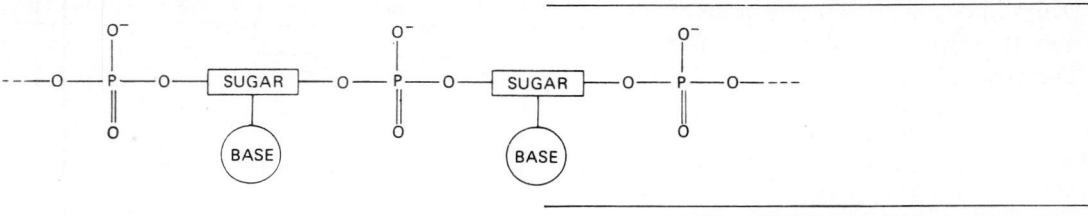

The phosphate and sugar moieties are shown in detail in Fig. 9.1. The bases are purines and pyrimidines. These are heterocyclic aromatic compounds. The two purine bases adenine and

α-2-deoxyribofuranose), whereas it is ribose (α-ribofuranose) in ribonucleic acid (RNA). For a detailed account of the chemistry of nucleic acids, see Mainwaring *et al*. (1982).

131

Most of the DNA in a eukaryotic cell is in the chromosomes of the nucleus. There the molecules of DNA form paired strands, held in the famous double helical configuration discovered by Watson & Crick (1953). The chains of nucleotide units are held together by weak forces (hydrogen bonds and hydrophobic interactions) between **complementary bases:** guanine pairs with cytosine, and adenine pairs with thymine. The backbone of the DNA helix consists of alternating phosphate and deoxyribose subunits, which form the outside of the spiral. The bases, which are hydrophobic, are directed into the centre of the helix. There they form a stack, with their planar aromatic rings perpendicular to the long axis of the macromolecule, and separated from one another by 0.36nm. The external diameter of a DNA double helix is 1.8nm. In eukaryotic cells the negative charges on the ionized free hydroxyl groups of the phosphoric acid residues are neutralized by the basic side-chains of the accompanying molecules of the **nucleoproteins**. The latter include strongly basic proteins of the type known as **histones**, which are rich in lysine and arginine and have molecular weights in the range 12,000–20,000. The nucleoproteins also include species with very large molecules, of molecular weight 10^7–10^9.

RNA has a more open structure than DNA. Most of it is in the ribosomes (ribosomal RNA or rRNA) and in the nucleolus, and most is in single-stranded form. Most of the messenger RNA (mRNA) in a cell is, at any one time, attached to the ribosomes; with the much smaller transfer RNA (tRNA) molecules move between the ribosomes and the cytosol. Strands of RNA are folded in quite a complicated manner, owing to interactions between pairs of bases: guanine with cytosine, and adenine with uracil. The spacing between bases is 0.36nm. During the process of transcription, similar base-pairs are formed with chromosomal DNA and mRNA in the nucleus. The ribosomal and nucleolar RNA, like the chromosomal DNA, is associated with nucleoprotein. The nucleic acids of prokaryotic cells (organisms without nuclei in their protoplasm) have their phosphoric acid groups balanced by inorganic ions (Na^+, Mg^{2+}) and by certain amines.

The histochemical procedures discussed in this chapter demonstrate DNA only in the nucleus and RNA only in the nucleolus and ribosomes, although it is known that the nucleic acids are by no means confined to these situations. The nucleic acids of prokaryotes (such as bacteria) can also be stained, but their localizations cannot be accurately resolved by light microscopy. Chapter 19 contains a brief introduction to *hybridization histochemistry*, in which a particular species of nucleic acid (such as the mRNA for a known protein) can be demonstrated by virtue of its affinity for suitably labelled complementary RNA or DNA molecules.

Both the nucleic acids can be stained with cationic dyes, and DNA can be demonstrated selectively by the Feulgen reaction. The specificity of a histochemical method for DNA or RNA can be confirmed by specific removal of either of these substances by enzymatic hydrolysis or chemical extraction.

9.2. Demonstration of nucleic acids with cationic dyes and fluorochromes

The chemistry of the staining of nucleic acids by cationic dyes has already been discussed in Chapter 6. The process may be considered to be histochemical when conditions are carefully standardized and suitable control procedures are carried out. Any basic dye may be used for the purpose, but the specificity of staining of nucleic acids in any structure must be confirmed by chemical or enzymatic extraction, because nucleic acids are not the only substances capable of binding coloured cations. Controls are particularly important in the case of cytoplasmic RNA, which cannot easily be distinguished from the basophilic secretory products of some cells. The dye-metal complex formed from gallocyanine and chromic ions (see Chapter 5) is a cationic dye and is also used to stain DNA and RNA. The chrome alum-gallocyanine method of Einarson (1951) is used in quantitative micro-photometric studies of the nucleic acid contents of cells because the bound dye-metal complex is more stable towards light, water, and organic solvents than are ordinary cationic dyes. There are some simple fluorescent or coloured compounds, however, whose molecules can fit tightly into the hydrophobic domains of nucleic acids, and provide quite specific staining (see Gurr *et al.*, 1974). Some of these are now considered.

9.2.1. Intercalating fluorochromes

A planar aromatic ring of suitable dimensions that bears at least one polar group is able to fit, like the filling of a sandwich, between the stacked bases of a nucleic acid molecule. Some common cationic dyes, including the phenothiazines and oxazines, fulfil these structural requirements, though these can bind to macromolecular anions by other mechanisms too. The mechanism of intercalation into DNA requires transient opening of the helix to admit the coloured or fluorescent molecule. This opening does not involve unwinding of the helix, because hinge-like bending can occur at certain sites (Sobell *et al.*, 1977; see Fig. 9.2).

The most widely used fluorescent intercalating chromophores are acridines, phenanthridines, and benzimidazoles. These are taken up from extremely dilute solutions by living or fixed cells. The fluorescence efficiency is increased by binding to nucleic acid, and there is sometimes a change in the colour of the emitted light (Morthland *et al.*, 1954; Burns, 1972; Hilwig & Gropp,

1972). With some of these reagents, it is possible to distinguish between the two nucleic acids by the colour of the fluorescence (Bertalanffy & Bickis, 1956) or by adjusting the pH of the solution to favour selective staining (Hilwig & Gropp, 1975).

Ethidium bromide
(M.W. 394.3)

Hoechst 33258 (Bisbenzimide trihydrochloride, pentahydrate) (M.W. 624.0)

Another cationic fluorochrome is **quinacrine** (mepacrine) (see p. 88). This exhibits enhanced fluorescence when it is intercalated into an adenine-thymine pair, but the fluorescence is suppressed by guanine (Weisblum & de Haseth, 1972). Consequently, regions of DNA rich in adenine-thymine pairs fluoresce brightly, and regions in which guanine-cytosine pairs predominate appear darker. This is the reason why mitotic chromosomes stained with quinacrine exhibit patterns of bright and dark transverse bands, which are useful for identification, description, and the recognition of abnormalities.

9.2.2. Fluorescent demonstration of nucleic acids

The procedures are simple, and vary only in details of the composition of the staining solutions. The concentrations of fluorochromes are not critical, and those recommended by different authors often vary by 3 orders of magnitude. After staining, the preparations are washed in a solution (water, buffer, culture medium, etc.)

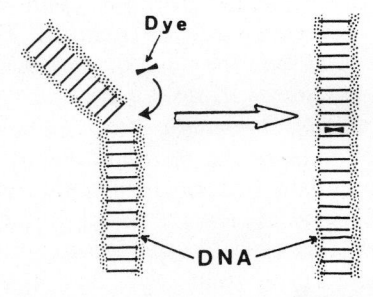

FIG. 9.2. Transient bending of a DNA helix to admit a planar dye or fluorochrome ion, which intercalates between base pairs in the hydrophobic centre of the macromolecule.

identical to that in which the fluorochrome was dissolved.

Fluorochrome solutions

Use one of the following. See below for the results that can be expected. The stock solutions are stable for months if kept in dark bottles, or in a dark cupboard. Refrigeration (4°C) retards microbial contamination.

Acridine orange. A stock solution of acridine orange (C.I. 46005), 1.0mg per ml in water, is diluted 10-fold, immediately before use, with 0.1 M phosphate buffer, pH 6.0 (see Chapter 20).

Ethidium bromide. 1.0 µg per ml, in 0.1 M phosphate buffer, pH 7.2. This solution (Franklin & Locker, 1981) will stain both nucleic acids. See also *Note 1* below.

Hoechst 33258 (also known as **bisbenzimide** though the latter name has also been applied to Hoechst 33342, a closely related fluorescent bisbenzimidazole dye): 0.5 µg per ml in a balanced salt solution (see Chapter 20), pH 7.0–7.4. At this pH, the fluorochrome is selective for DNA. To stain both nucleic acids, add HCl to lower the pH to 2. *If staining is done at pH 2, the preparation must be washed afterwards at pH 7,* to avoid loss of the fluorescent reagent (Hilwig & Gropp, 1975).

Procedure

1. Fix specimens, suspended cells, or cultured monolayers in any fluid that preserves nucleic acids (see Chapter 2). Acetic-alcohol mixtures are recommended.
2. Rinse the preparation in the buffer or salt solution used as solvent for the fluorescent reagent.
3. Stain for 15 min at room temperature.
4. Wash in three changes of the buffer or salt solution.
5. Apply a coverslip, using a mixture of equal volumes of glycerol and buffer (or balanced salt solution) as the mounting medium. Examine immediately.

Instructions for fluorescence microscopy are summarized in Table 9.1. These fluorochromes are easily seen even if the equipment cannot provide illumination at the optimum wavelength.

TABLE 9.1 *Absorption and emission maxima of fluorochromes that intercalate into nucleic acids.*

Fluorochrome	Absorption	Emission
Acridine orange	500nm (blue)	530nm (green–DNA); >600nm (red-brown–RNA)
Ethidium bromide	520nm (green)	610nm (orange)
Hoechst 33258	365nm (near UV)	480nm (blue)

Results

Fluorescence due to DNA is seen in nuclear chromatin. Fluorescence due to RNA is seen in nucleoli and cytoplasm.

Notes

1. If ethidium bromide is applied from a strongly acid solution (0.25 M HCl, pH 0.6), RNA is destroyed by hydrolysis and the method becomes specific for DNA. The staining of DNA is prevented by prior methylation (see Chapter 10) for 3 h at 55°C in 0.1 M HCl in absolute methanol. After this treatment, the reagent binds only to RNA (Cox *et al.*, 1982). If ethidium bromide is added to a culture medium for 2 h, the fluorescence of the living cells is due almost entirely to RNA (Burns, 1972).
2. The preparations are not permanent, so they should be examined and photographed immediately. If the washed slides are thoroughly air-dried, rinsed in xylene, and then coverslipped with a non-fluorescent resinous mounting medium, some fluorescence is preserved, though with reduced brightness.

9.2.3. Methyl green-pyronine and related methods

With these techniques two cationic dyes are applied simultaneously under carefully controlled conditions of concentration and pH. This results in differential colouring of DNA (bluish-green with methyl green) and RNA (pink to red with pyronine Y). There has been much speculation and disagreement on the mechanism of this staining (Baker & Williams, 1965; Scott, 1967), but all agree that the shapes and sizes of the dye molecules (see Chapter 5) determine which colour will be imparted to each nucleic acid (see Horobin, 1982).

It seems probable that the large propeller-shaped cation of methyl green is attracted electrostatically to the phosphate groups on the outside of the DNA helix. Non-ionic bonds to the sugar units of DNA and to the associated nucleoprotein may help to hold the dye in position. Pyronine Y has a planar structure and could be expected to intercalate between the base pairs of a nucleic acid. It does not enter the DNA molecule, however, because its ions are excluded by the methyl green on the outside of the helix. RNA, with its more open structure, easily admits pyronine cations, which become firmly intercalated between the base pairs, neutralize the phosphate anions, and exclude methyl green, which cannot intercalate.

Support for the mechanism summarized above comes from the observation that pyronine Y can be replaced by thionine (Roque *et al.*, 1965), whose cations are much the same size and are comparably hydrophobic. It is also possible to replace the methyl green with a cationic surfactant, so that RNA is the only nucleic acid stained, by a dye with small planar molecules (Bennion *et al.*, 1975)

Methyl green-pyronine and related methods are valuable general histological stains, especially for nervous and lymphoid tissues. Cationic fluorochromes are useful as counterstains, in conjunction with fluorescent immunohistochemical procedures (Jones & Kniss, 1987). In critical histochemical work, however, it is always necessary to use extracted or enzyme-treated control sections to check the specificity of staining of DNA or RNA. In methods using two dyes, basophilic materials other than nucleic acids are stained mainly by the pyronine or other planar cationic chromogen.

9.2.4. A methyl green-pyronine procedure

The simple method given below (Høyer *et al.*, 1986) is for use with pure samples of both dyes, which have become available in recent years (see Chapter 5; also Lyon *et al.*, 1987). Older samples of the dyes, especially methyl green, will not work with the quantities given below, even if they are from batches certified by the Biological Stains Commission. Detailed instructions for getting the method to work with impure dyes were given in the first edition of this book. Techniques can also be found in Pearse (1968 or 1985) and Bancroft & Cook (1984).

Solutions required

A. 0.01 M phthalate buffer, pH 4.0

Potassium hydrogen phthalate (KHC$_8$H$_4$O$_4$):	1.02g
Water	490ml
0.01 M HCl: approximately	1.0ml
Check the pH, then add water to make volume up to :	500ml
(See *Note 1* below)	

B. Dye mixture

Pyronine Y (C.I. 45005; dye content >90%):	0.03g
Methyl or ethyl green (C.I. 42585 or 42590; must be close to 100% dye content, and free of violet impurities):	0.15g
Buffer (solution A):	100ml

Procedure

1. Paraffin sections of material fixed in Carnoy or aqueous formaldehyde are de-waxed and hydrated.
2. Stain in the dye mixture (Solution B) for 5 min.
3. Rinse in two changes of water, 5 s in each. Shake to drain off most of the water.
4. Dehydrate by agitating vigorously in each of 3 changes of *n*-butanol, about 1 min in each. The volume of butanol must be adequate (about 300 ml for a rack of 10-12 slides, or 50 ml for 1 or 2 slides in a coplin jar).
5. (Optional) Clear in xylene.
6. Apply coverslip, using a resinous mounting medium.

Results

DNA green or bluish-green. RNA bright pink or red. Sulphated carbohydrates (mast cell granules, cartilage matrix, some types of mucus) are coloured metachromatically (orange) by the pyronine.

Notes

1. The pH of the staining solution may be varied within the range 4.0-4.6. pH 4.0 has been found optimal for most material (see Høyer *et al.*, 1986).
2. The histochemical specificity of staining, especially of RNA, should be checked by selective enzymatic or chemical extraction as described later in this chapter.

9.3. The Feulgen method

9.3.1. Chemical principles

When sections of tissue are treated with 1.0M hydrochloric acid at 60°C for 5–20 min (according to the fixative used, see below), most of the RNA is broken down to soluble substances and lost from the section, but DNA is only partly hydrolysed. The purine and pyrimidine bases of the DNA are removed from the deoxyribose resi-dues, which remain in their original positions and are capable of reacting as aldehydes (Overend & Stacey, 1949). The Feulgen hydrolysis does not break the ester linkages between the phosphoric acid and sugar units of the DNA.

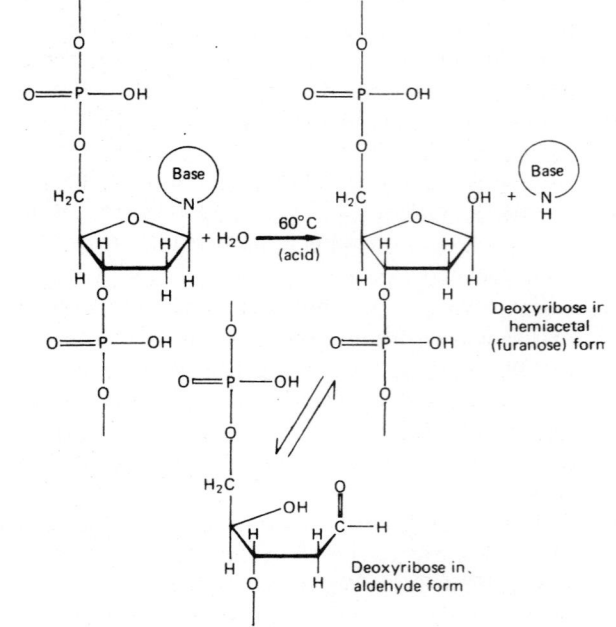

The exact reasons for these changes and for the fact that RNA does not behave similarly are imperfectly understood. Several theories are discussed at some length by Pearse (1968, 1985). The ribosyl residues of any RNA that is not removed during the hydrolysis do not react as aldehydes. This property is probably due to the presence of a hydroxyl group at position 2' of ribose.

The aldehyde groups liberated by the Feulgen hydrolysis are stained with Schiff's reagent, and the chemistry of this process will be considered in Chapter 10. Other reagents, notably silver diammine and naphthoic acid hydrazide, can also be used to detect aldehydes formed from partially hydrolysed DNA.

The Feulgen reaction is, in practice, a highly

specific method for DNA, though its specificity may be checked by incubating control sections with deoxyribonuclease. Any free aldehydes initially present in the tisse (i.e. introduced by fixative or present in immature elastic tissue and some lipids) will give false positive staining, but this will be obtained when the acid hydrolysis is omitted (except, rarely, in the case of some lipids), so is unlikely to cause confusion.

Pre-existing aldehydes are easily destroyed before staining by treating the sections with sodium borohydride. This manoeuvre (see Chapter 10) is always necessary with tissue fixed in glutaraldehyde (Kasten & Lala, 1975). It should be noted that Bouin's fluid and acid decalcifying agents can attack DNA and may hydrolyse it excessively with consequent loss of stainability by the Feulgen method. For the histochemistry of nucleic acids it is desirable to avoid the use of fixatives containing picric acid, and chelating agents are preferred to acids when decalcification is necessary.

The intensity of staining by the Feulgen method is proportional to the concentration of DNA. Thus it is possible to make micro-photometric and morphometric observations and determine the quantities of DNA in the nuclei of cells. Important studies of the cell cycle have been carried out in this way.

9.3.2. Feulgen staining in practice

The optimum time of hydrolysis by 1.0N HCl at 60°C varies with the fixative:

Carnoy:	8min
Formaldehyde:	8min
Helly:	5min
SUSA:	18min

(Data summarized from Pearse, 1968.) See also *Note 4* below.

Preparation of Schiff's reagent

This is not the only way to make Schiff's reagent. For other methods, see Pearse (1968) and Lillie & Fullmer (1976).

It is important to use either basic fuchsine designated "special for DNA" or pure pararosaniline.

Boil 400 ml of water, add 1.0 g of basic fuchsine, cool to room temperature, and filter. Add 1.0ml of thionyl chloride (**cautiously:** the reaction $SOCl_2 + H_2O \rightarrow SO_2 + 2HCl$ is very rapid. Do not pipette thionyl chloride by mouth. Use fume hood.) Allow to stand in a stoppered bottle overnight. Shake the solution with about 2g of activated charcoal, which should remove all residual colour. A slight yellowish-brown tinge does not matter, but the solution must not be pink. Filter to remove charcoal and store in a tightly closed bottle at 4°C.

Schiff's reagent will keep for about 6 months, but should be discarded if a pink colour develops or if it no longer smells strongly of sulphur dioxide. See *Note 3* below for an alternative to Schiff's reagent.

Other solutions required

A. 1.0M hydrochloric acid

B. Bisulphite water

Approximately 5g of sodium or potassium metabisulphite (anhydrous) dissolved in 1 litre of water and 3–5ml of concentrated hydrochloric acid added. This solution must be freshly made and put into three staining tanks.

C. A counterstain, if desired

Fast green FCF (0.5%, aqueous) is suitable.

Procedure

1. Pre-heat the 1.0M hydrochloric acid to 60°C in a covered container (see *Note 1*) in the 60°C oven. Take Schiff's reagent out of refrigerator and allow it to warm to room temperature.
2. De-wax and hydrate paraffin sections.
3. Immerse in 1.0M HCl at 60°C for the optimum time. (See also *Note 4c*).
4. Rinse in water (two changes).
5. Immerse in Schiff's reagent, for 15–30min

(until nuclei are stained; the sections should be pale pink).

6. Transfer slides directly to bisulphite water; three changes, each 10–15s with agitation (see *Note 2*).
7. Wash in running tap water for 2min.
8. Counterstain (if desired). Wash briefly in water (which also differentiates fast green FCF).
9. Dehydrate, clear, and mount in a resinous medium.

Result

DNA (nuclear chromatin; chromosomes) deep pink to purple.

Notes

1. Do not allow water vapour to contaminate any melted wax that is also in the oven.
2. The purpose of the bisulphite rinse is to prevent the re-colorization of Schiff's reagent which occurs on dilution with water alone, and can cause artifactual background staining. Demalsy & Callebaut (1967) recommend omission of this rinse and prescribe washing in **copious** running tap water instead.
3. Horobin & Kevill-Davies (1971a) have shown that an acidified alcoholic solution of basic fuchsine (not decolorized with SO_2) will stain aldehyde groups. This solution is made as follows:

Basic fuchsine:	1g
Ethanol:	160ml
Water:	40ml
Concentrated hydrochloric acid:	2ml

Keeps for a few weeks. Without the hydrochloric acid, it is indefinitely stable. The HCl can be added before use. Store in a tightly closed bottle

After staining for 20min in this solution, the slides are washed in absolute or 95% alcohol to remove all excess dye, cleared, and covered. (See Chapter 10 for discussion of histochemical detection of aldehydes.)

4. Controls.
(a) Omit the acid hydrolysis (step 3). Nothing should be stained. A positive result may indicate pre-existing aldehyde groups derived from the fixative (especially glutaraldehyde) or hydrolysis of DNA during fixation (picric acid does this) or decalcification. Elastin of immature animals normally has free aldehyde groups (see Chapter 8), but is unlikely to be mistaken for DNA. The pseudo-plasmal reaction of some lipids (see Chapter 12) can be a cause of direct reactivity with Schiff's reagent in frozen sections, especially after fixation and storage in formaldehyde solutions.
(b) The times for hydrolysis given above are only approximate. In critical work the optimum time must be determined experimentally for each batch of sections. The optimum hydrolysis is that which yields the most darkly stained nuclei. If an unexpected negative result is obtained, stain a hydrolysed section with a basic dye: the DNA may have been completely depolymerized and extracted.
(c) From the results of a chemical study of the acid-catalysed hydrolysis of DNA (Kjellstrand, 1977), it has been suggested that the highest yields of aldehydes might be attained by using concentrations of HCl considerably higher than 1.0M, at or only slightly above room temperature.
5. If staining seems to be too weak, even with an optimum hydrolysis, the Schiff's reagent is probably unsatisfactory.

9.4. Enzymatic extraction of nucleic acids

9.4.1. Enzymes as reagents in histochemistry

Sections of tissue can be treated with the pancreatic enzymes ribonuclease (RNase) and deoxyribonuclease (DNase) in order to remove, selectively, the two nucleic acids. Other enzymes are also used in histochemistry to remove (or prevent the staining of) such substances as collagen, glycogen, and various carbohydrate constituents of mucosubstances. Some caution is required in interpreting the results of enzymatic treatments, and the following conditions must always be satisfied in order to draw valid conclusions:

1. The substrate must be present in a suitable state for being attacked by the enzyme. Fixatives containing picric acid, mercury, or chromium should be avoided since these substances are inhibitory to many enzymes. Fixation in formaldehyde prevents the subsequent digestion of collagen and reticulin by collagenase, but for most other enzymes formaldehyde (aqueous or alcoholic) and mixtures containing alcohol and acetic acid as the main active ingredients are the least objectionable fixatives.

2. The preparation of the enzyme used must be free from contamination by other enzymes which might produce confusing effects. The presence of proteolytic enzymes in a specimen of RNase, for example, would be undesirable. Similarly, DNase must be free of RNase.

3. The conditions of incubation must be adequate for complete action of the enzyme upon all of the substrate present in the section. The use of an identically fixed and processed control section known to contain the substrate is recommended.

4. Non-specific extraction of the substrate should not occur and should be sought in control sections incubated only in the buffer or other solvent used for the enzyme. Buffer solutions alone sometimes extract appreciable quantities of RNA and glycogen from sections. In some circumstances it is preferable to dissolve the enzyme in water, even if this does not provide the optimum pH for the reaction.

5. Any necessary co-factors or co-enzymes must be provided. Neuraminidase, for example, will function only in the presence of calcium ions, and DNase requires magnesium.

9.4.2. Ribonuclease and deoxyribonuclease

Enzymes from various animal and plant sources catalyse the decomposition of nucleic acids. In histochemical practice, enzymes extracted from the bovine or porcine pancreas are usually used. The enzymes should be of the highest available purity, regardless of expense, for the reasons given in the preceding section. Cruder preparations are cheaper and will work, but when they are used there can be no certainty that the observed alterations in stainability of the tissue are due solely to removal of RNA or DNA. The reason for using an enzyme is that it acts with high specificity upon its substrate. With a mixture of enzymes, this specificity will be lost.

Ribonuclease (systematic name: polyribonucleotide 2-oligonucleotido-transferase (cyclizing); E.C. 2.7.7.16*) acts on the phosphate group attached to position 3' of those ribose units of RNA that bear pyrimidine bases at position 1'. The phosphate-ester linkage to position 5' of the adjacent sugar residue is transferred to position 2' of the first ribose unit—the one bearing the pyrimidine base.

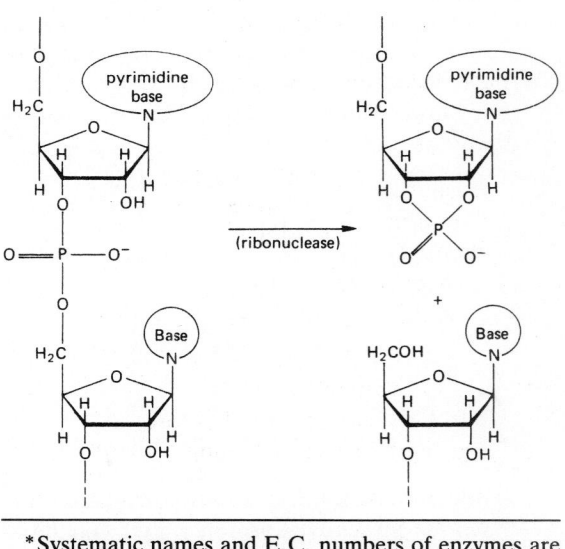

*Systematic names and E.C. numbers of enzymes are explained in Chapter 14.

The long molecule of RNA is thus broken into small, soluble oligonucleotides. These products are lost from the tissue by diffusion into the RNase solution and the water used for subsequent washing.

Deoxyribonuclease (systematic name: deoxyribonucleate oligonucleotido-hydrolase; E.C. 3.1.4.5) of pancreatic origin catalyses the attack of water upon multiple phosphoric acid ester linkages to position 5′ of deoxyribose. The DNA is thus hydrolysed to yield soluble oligonucleotides. A related enzyme is deoxyribonuclease II (systematic name: deoxyribonucleate 3′-nucleotidohydrolase; E.C. 3.1.4.6), extracted from spleen and from various micro-organisms, which brings about hydrolysis at position 3′ of deoxyribose. The latter enzyme has been used less often in histochemistry than pancreatic DNase.

9.4.3. Techniques of application

DNase and RNase are expensive, so they can be used only in small quantities, applied as drops to individual sections. This is conveniently done in a petri dish containing damp gauze or filter paper to saturate the atmosphere inside with water vapour and prevent evaporation of the drop of enzyme solution. The dish containing the slide is placed in an incubator at 37°C. One section on a slide can be treated with enzyme while another is treated only with the solvent. Any other sections on the slide remain dry. In this way the effects of both the enzyme and its solvent can be examined on adjacent serial sections.

Enzyme solutions

The enzymes should be free of other interfering enzymes. Very small quantities may be weighed with a torsion balance. The solutions of the enzymes can be recovered after use (by sucking off the slide with a syringe and needle) and stored frozen at −20°C, but the potency declines with such storage.

Ribonuclease is used as a solution containing 0.2–0.5mg of enzyme per ml of water. Incubate for 1h at 37°C.

Deoxyribonuclease is used as a solution containing 0.05mg of enzyme per ml in TRIS buffer (pH 7.5) containing $0.2_M MgSO_4$. The usual 0.2M TRIS buffer should be diluted five times with water. Incubate for 3–6h at 37°C.

Sections of central nervous tissue are valuable for determining the adequacy of extraction of nucleic acids. The Nissl substance of neurons (stainable by cationic dyes) should be completely removed by an adequate RNase treatment. The nuclear chromatin (but not the nucleoli) of all cells should fail to stain with cationic dyes or by the Feulgen method after adequate treatment with DNase. The control sections of central nervous tissue should have been fixed and processed in parallel with other specimens being studied.

9.5. Chemical extraction of nucleic acids

It is also possible to extract nucleic acids non-enzymatically from sections of tissue. Trichloroacetic acid (4% aqueous, 80°C, 15min) removes both DNA and RNA. Perchloric acid (10% aqueous $HClO_4$, 4°C, overnight) removes RNA but not DNA, but a more drastic treatment with perchloric acid (5% aqueous $HClO_4$, 30min, 60°C) extracts both nucleic acids. These chemical procedures are based on the catalysis by acids of the hydrolysis of the bonds between the sugars and the bases and between the sugars and phosphoric acid. The phosphate-ester bonds of RNA are more labile to hydrolysis than those of DNA. It is also possible to remove RNA by treatment with hot 1.0M hydrochloric acid, as in the Feulgen hydrolysis. The extraction of RNA by perchloric acid also removes the purine and pyrimidine bases from DNA, thus engendering stainability of this nucleic acid by Schiff's reagent. The reagents used for chemical extraction are less expensive than RNase and DNase, but they are also less specific for the nucleic acids and their actions are more difficult to control.

Perchloric and trichloroacetic acids are highly caustic and must be handled carefully. They should be discarded by flushing down the sink with copious tap water.

9.6. Exercises

Theoretical

1. How would you proceed to carry out the Feulgen reaction on sections of a tissue fixed in glutaraldehyde?

2. Using sections of Carnoy-fixed central nervous tissue as test objects, how would you determine (a) whether a sample of DNase is substantially free of RNase, (b) whether a sample of RNase is substantially free of DNase?

3. What types of dyes are suitable for staining the nucleoprotein of the nuclei of mammalian cells? What would be the effect of prior extraction of DNA upon the results of such staining?

4. Why do enzymes such as RNase and DNase work more rapidly and more completely on sections of tissue after fixation in alcohol–acetic acid mixtures than after fixation in aqueous formaldehyde or glutaraldehyde?

Practical

5. Stain paraffin sections of lymphoid and nervous tissues by the methyl green-pyronine method. Apply control procedures to identify cytoplasmic RNA in the presence of other basophilic materials.

6. Stain sections of the tissues by the Feulgen method or with a basic fluorochrome selective for DNA. Why do some nuclei become more intensely coloured than others when nearly all contain the same amount of DNA?

7. Bearing in mind the limitations of enzymes as histochemical reagents, attempt to prove that:

(a) The Nissl substance of the neuron is stained by pyronine Y on account of its content of RNA.

(b) Mast cell granules (skin and tongue are suitable tissues) must owe their stainability with pyronine Y to a substance other than RNA.

(c) The Feulgen reaction in the nuclei of cells of the adrenal cortex is due to the presence of DNA.

(d) The stainability of nuclei (in any tissue) by alum-haematoxylin is not dependent upon the presence of either type of nucleic acid.

10

Organic Functional Groups and Protein Histochemistry

This chapter is devoted to the histochemistry of the side-chains of some amino acids and the reactions of a few functional groups artificially produced by the action of reagents upon tissues. The histochemical reactions of other organic compounds such as carbohydrates, lipids, nucleic acids, and amines are described in other chapters.

As elsewhere in this book, no attempt is made to describe more than a small selection of the methods available for the various organic functional groups. **For rarely used techniques, the theory is explained briefly, but practical instructions are not given.** Functional group histochemistry is treated at length by Lillie & Fullmer (1976), and much information can also be found in the works of Gabe (1976), James & Tas (1984) and Pearse (1985).

Blocking reactions common to many methods are given at the end of the chapter, in Section 10.11.

10.1. General considerations and methodology

Functional groups are detected histochemically by making use of their characteristic chemical reactions to produce coloured compounds. The techniques can be properly understood and intelligently used only when the underlying organic chemistry is constantly kept in mind. For more information the reader should consult a textbook of organic chemistry. Those of Noller (1965), Morrison & Boyd (1973), and March (1977) are recommended. Some histochemically valuable reactions are not included in general texts, but many of them are described by Feigl (1960) and Glazer (1976).

The methods discussed in this chapter are for functional groups that form part of the macromolecular structure of the tissue. Diffusion and extraction by solvents prior to staining cannot, therefore, give rise to false localizations. The reactions can all be carried out on frozen or paraffin sections of suitably fixed material. It is necessary, however, to take into account the chemical reactions of fixation. The non-additive coagulant fixatives do not cause chemical changes in tissues, so alcohol-based mixtures such as Carnoy's fluid are ideal for the histochemical study of proteins (except for a few secretory products not fixed by alcohol). Formaldehyde is also acceptable, because most of its reactions with the side-chains of amino acids are reversed when the fixative is washed out (see Chapter 2). The same is true of picric acid and, except when sulphydryl groups are to be demonstrated, of mercuric chloride. Chromium trioxide, potassium dichromate, osmium tetroxide, and glutaraldehyde are best avoided: they all produce imperfectly understood irreversible chemical changes in proteins.

Often a chromogenic reaction for a functional group is not as specific as we would like it to be. It is then necessary to examine control sections in which the group has been chemically altered (i.e. "blocked") so that it can no longer take part in the reaction that yields a visible deposit. Many of the blocking reactions are also of low specificity, however, so it is frequently necessary to use more than one of them in order to discover, by elimination, the sites at which a single functional group is present in the tissue. Sometimes it is possible to undo the effect of blockade of a functional group by judicious application of a second reagent. This strategy is valuable when the blocking of one functional group is reversible but that of another is not.

Despite the intrinsic interest and great instructional value of functional group histochemistry, the methods derived from this branch of the science are used only on rare occasions in routine histological and pathological practice and in the investigation of biological problems. This is unfortunate because the techniques are often valuable for displaying structural features of tissues and have the added advantage of providing some chemical information.

The major histochemically demonstrable organic functional groups present in animal tissues are listed in Table 10.1 (pp. 144–145).

10.2. Alcohols

Aliphatic hydroxyl groups occur in carbohydrates, and also in the widely distributed amino acids serine, a primary alcohol, and threonine, a secondary alcohol. Hydroxylysine and hydroxyproline, which occur principally in collagen, are also secondary alcohols. Some lipids are alcohols of high molecular weight, but in tissues they are usually esterified.

10.2.1. Histochemical detection

Hydroxyl groups in tissues can be detected by converting them to sulphate esters and then staining with a cationic dye at pH 1.0 or lower (see p. 91). This procedure demonstrates all hydroxyl groups. Those of proteins cannot be distinguished from those of carbohydrates, which are generally more abundant. The dye will also bind to preexisting sulphate-ester groups in the tissue, such as those of some proteoglycans and of sulphatide lipids. The native sulphate esters are recognized by staining control sections that have not been subjected to the sulphation treatment. Both natural and artificially produced sulphate esters are hydrolysed, with loss of the sulphate group, by exposure to hot acidified methanol.

Various reagents are available for sulphation of hydroxyl groups in sections. Concentrated sulphuric acid is the simplest, but is physically injurious. It works well, however, with semi-thin sections (0.5–1.0 μm) of plastic-embedded material. A mixture of sulphuric acid and diethyl ether is milder. An even more gentle reagent is 0.25% sulphuric acid in a mixture of acetic acid and acetic anhydride (Lillie, 1964), but this also brings about N-sulphation of amino groups.

TABLE 10.1. *Histochemically demonstrable functional groups*

Functional group	Structure	Occurrence
Hydroxyl	—OH	Proteins: serine, threonine hydroxyproline, hydroxylysine All carbohydrates (see Chapter 11) Some lipids (see Chapter 12)
Phenol	Ar—OH	Proteins: tyrosine In plants: tannins Small molecules: serotonin, catecholamines (see Chapter 17)
Carboxyl	—COOH	Proteins: C-terminus of all peptide chains; side-chains of glutamic and aspartic acids Carbohydrates: most glycoproteins and proteoglycans (see Chapter 11) Lipids: free fatty acids (see Chapter 12)
Amino	—NH$_2$	Proteins: N-terminus of all peptide chains; side-chain of lysine Some carbohydrates (see Chapter 11) Some lipids (see Chapter 12) Small molecules (see Chapter 17)
Guanidino		Proteins: side-chains of arginine
Indolyl		Proteins: side-chains of tryptophan Small molecule: serotonin (see Chapter 17)
Sulphydryl (= thiol)	—SH	Proteins: side-chain of cysteine
Disulphide	—S—S—	Proteins: cystine bridges between peptide chains
Sulphate ester	—OSO$_3$H (—OSO$_3^-$ + H$^+$)	Carbohydrates: some proteoglycans (see Chapter 11) Lipids: sulphatides (see Chapter 12) Also introduced artificially in histochemical sulphation procedures
Sulphonic acid	—SO$_3$H (—SO$_3^-$ + H$^+$)	Does not occur naturally Produced or introduced artificially in several histochemical methods
Phosphate esters	$$-OP(OH)_2;$$ $$-OP(OH)O-$$	Nucleic acids (see Chapter 9) Some lipids (see Chapter 12) Some proteins, notably casein (in mammary gland and in milk), with —OH of serine esterified by H$_3$PO$_4$

Functional group	Structure	Occurrence
Aldehyde	—CHO	Proteins: elastin of immature animals (see Chapter 8) Lipids, when oxidized by air (see Chapter 12) Produced artificially in many histochemical methods
Ketone	$\diagdown$C=O$\diagup$	Lipids: some steroids (see Chapter 12) Produced artificially in a few histochemical procedures
Olefinic double bond	—C=C—	Lipids (see Chapter 12)

10.2.2. Blocking procedures

The reactivity of hydroxyl groups is blocked by base-catalysed acylation with either acetic anhydride or benzoyl chloride:

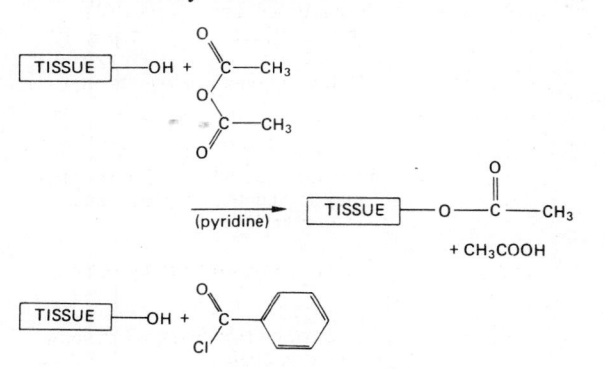

These reactions generate esters at the sites of hydroxyl groups. Acetic anhydride also produces N-acetylation of some amino groups. Benzoyl chloride causes benzoylation of almost all amino groups. The acylation reactions also block the reactivity of tyrosine, tryptophan, and histidine residues. The histochemical reactivity of arginine, however, is only slightly depressed by acylation. Esters formed with hydroxyl groups and amides formed by N-acylation of amines are hydrolysed, with reversal of the blockade, by exposure of the sections to alkaline ethanol, a procedure known in histochemical parlance as "saponification".

Diisopropylfluorophosphate (DFP) and related organophosphorus compounds combine irreversibly with the hydroxyl group of serine, but not in all proteins. The value of these reagents in histochemistry is therefore limited to their uses as inhibitors of certain serine-containing enzymes (see Chapter 15).

10.3. Carboxylic acids

Free carboxyl groups occur in glycoproteins and proteoglycans (Chapter 11), in fatty acids (Chapter 12), and in proteins. Every protein molecule has a terminal carboxyl group, but this is usually outnumbered by the side-chain carboxyls of glutamic and aspartic acids.

10.3.1. Histochemical detection

Cationic dyes will bind to carboxylic acids when the latter are ionized. For this to be so, the pH of the dye solution must usually be 4.0 or higher, so sulphate esters of proteoglycans and phosphoric acid groups of nucleic acids are also stained (see Chapter 6). Dyes are therefore of limited use for the identification of sites of carboxyl groups in tissues.

The **acid anhydride method** supposedly demonstrates only the carboxyl groups of proteins. Sections are first treated with a solution of acetic anhydride in pyridine at 60°C. This reagent is thought to react in different ways with C-terminal and with side-chain carboxyl groups:

Both reactions result in the covalent binding of a naphthol group at the site of the carboxyl of the protein molecule. The naphthol can easily be made to couple with a diazonium salt to yield a coloured product (see Chapter 5).

The carboxyl groups of carbohydrates do not react with acetic anhydride to give mixed anhydrides (Karnovsky & Mann, 1961), though there is no obvious reason why they should not behave in the same way as those of protein side-chains. A chemical study by Stoward & Burns (1971) indicated that the principal effect of hot acetic anhydride in pyridine was the formation of ketones at the sites of C-terminal carboxyl groups and that the side-chains of glutamic and aspartic acids formed mixed anhydrides only to a very slight extent. The histochemistry of carbohydrates with free carboxyl groups is discussed in Chapter 11.

10.3.2. Blocking procedures

Carboxyl groups are easily blocked by converting them to their methyl esters. This may be accomplished with a variety of reagents. The simplest is 0.1M HCl in methanol:

$$\boxed{\text{TISSUE}}\text{—COOH} + \text{HOHC}_3 \xrightarrow[\text{(acid)}]{} \boxed{\text{TISSUE}}\text{—COOCH}_3 + \text{H}_2\text{O}$$

(The active component of the reagent may be methyl chloride or methylene chloride rather than methanol itself.)

Other methylating agents occasionally used by histochemists are methyl iodide, diazomethane, and a solution of thionyl chloride in methanol. Methylation also blocks amino groups, sulphydryl groups, and the phosphoric acid moieties of nucleic acids. The reagent brings about hydrolysis of sulphate esters. It is a blocking reaction of greater importance in carbohydrate histochemistry than in the examination of carboxyl groups of proteins. Methyl esters of carboxylic acids are hydrolysed by saponification with alkaline ethanol.

The product of the first reaction (Barnett & Seligman, 1958) is a ketone and combines with the next reagent to be applied to the sections, 2-hydroxy-3-naphthoic acid hydrazide (HNAH) as described in Section 10.10.2. The reaction of acetic anhydride with side-chain carboxyls yields a mixed anhydride (Karnovsky & Fasman, 1960). This also combines with HNAH. The reaction is analogous to the acylation of an amine:

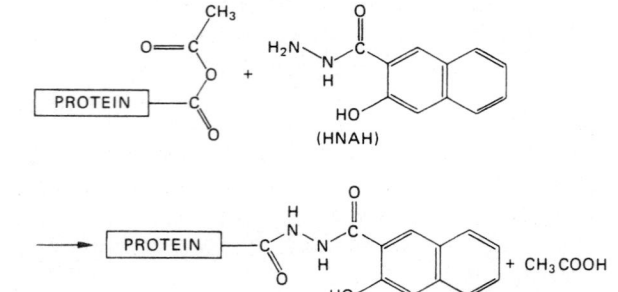

10.4. Amino groups

This section is concerned with primary amino ($-NH_2$) groups of proteins. Some important soluble amines are also detectable histochemically (see Chapter 17), but these are no longer present in sections of tissues prepared by ordinary techniques. The most numerous amino groups of proteins are those in the ε-position on the side-chain of lysine. N-terminal amino groups are less abundant because there is only one for each peptide chain.

Fixatives that combine covalently with amines, such as aldehydes, should be avoided if possible. They always impair staining of amino groups and sometimes prevent it completely. Strong oxidizing agents are also unsuitable for fixation, since they can cause oxidative deamination of proteins.

10.4.1. Histochemical detection

There are two main groups of methods for the demonstration of the amino groups of proteins. The first group consists simply of staining the tissue with a suitable anionic dye. This must be a dye that binds to its substrate exclusively by ionic attraction: usually one with fairly small molecules. Anionic dyes of high molecular weight, which are bound to tissues by forces other than electrostatic ones, are unsuitable. The mechanisms of staining by anionic dyes have already been discussed in Chapters 5, 6, and 8. Staining is to be expected at the sites of all tissue-bound cations, including amino and guanidino groups of proteins and organic bases such as choline that occur in some lipids. Blocking reactions (see Section 10.4.2) and extraction of lipids (see Chapter 12) must be used in order to establish that the acidophilia of a structure is due to the presence of amino groups.

The methods of the second group depend upon the formation of coloured deposits as the result of covalent combination of reagents with either intact or chemically modified amino groups. Many techniques are available. A typical one is the **hydroxynaphthaldehyde method**. The reagent, 2-hydroxy-3-naphthaldehyde, con-

denses with the protein-bound amino group to form an imine (also known as an azomethine or Schiff's base):

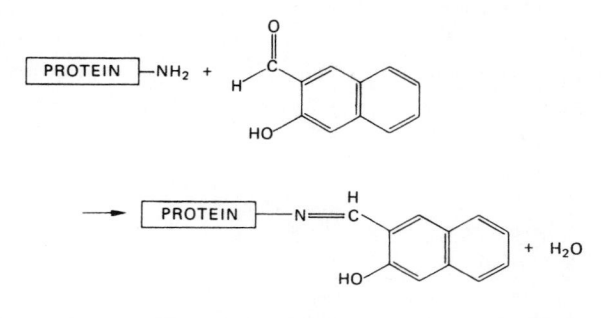

Thus, a naphthol becomes covalently bound at the site of the original NH_2 group. In the second stage of the method, the naphthol is made to couple with a diazonium salt to form an azo dye. The diazonium salt will, of course, also couple with the aromatic ring of tyrosine and with some other amino acids (see Section 10.9.2). However, the colours produced with these amino acids are different from and paler than that of the naphtholic azo dye. In any case they are easily allowed for in control sections that have not been treated with hydroxynaphthaldehyde.

A more recent method for the amino group is the **dithiocarbamylation reaction** (Tandler, 1980), in which small pieces of fixed tissue are immersed in a mixture of carbon disulphide and a strong base, triethylamine. Dithiocarbamyl derivatives of primary amino groups are formed:

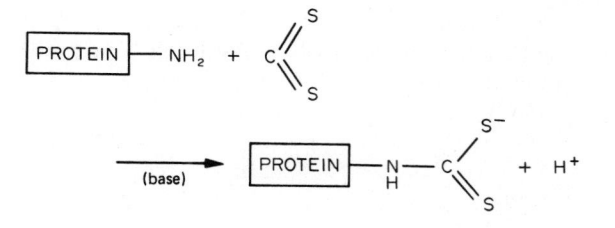

After washing out excess carbon disulphide, the tissue is exposed to lead acetate. A lead dithiocarbamate complex is first formed, but it decomposes and yields lead sulphide, which is insoluble, brown-black in colour, and also dense enough to impart contrast in the electron microscope.

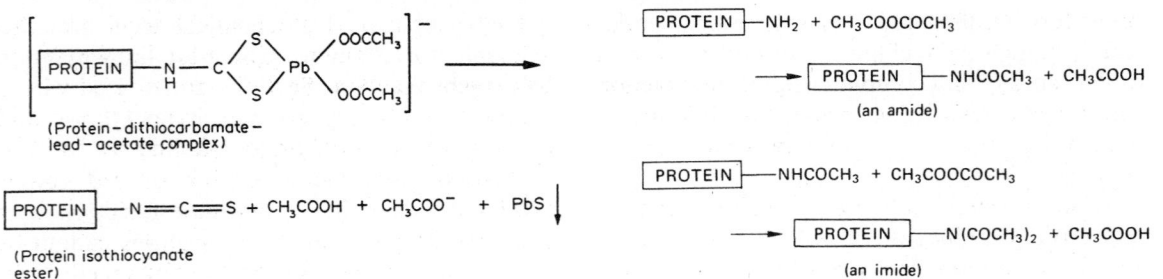

(Protein–dithiocarbamate–
lead–acetate complex)

(Protein isothiocyanate
ester)

The guanidino group of arginine does not react with carbon disulphide, which is a reagent of high specificity for the —NH_2 group. The sensitivity of the method is not high, however, so dithio-carbamylation can be used for histochemical localization of proteins that owe their basicity to lysine.

10.4.2. Blocking reactions

Primary amines may be blocked either by removing them or by converting them to amides or sulphoamino compounds.

Deamination is most easily accomplished by treating the sections with nitrous acid, which is usually applied as a solution of sodium nitrite in aqueous acetic acid. The reaction is most simply expressed as:

but is much more complicated in reality (White & Woodcock, 1968). Nitrous acid also reacts with tyrosine (see Section 10.7.1) and with sulphur-containing amino acids, but these reactions are unlikely to cause confusion in the testing of histochemical methods for amino groups. The guanidino group of arginine is not removed by treatment with nitrous acid (Lillie *et al.*, 1971). Strong oxidizing agents such as osmium tetroxide and potassium permanganate also bring about deamination of lysine side-chains, but their actions are rather unpredictable.

Acylation of amines is brought about by reaction with either acetic anhydride or benzoyl chloride:

Equivalent reactions occur with benzoyl chloride. The reaction of acylation is catalysed either by a weak base such as pyridine or by a strong acid such as perchloric acid. *Longer times and higher temperatures are needed for the base-catalysed acetylation and benzoylation of amines than for the esterification of hydroxyl groups by the same reagents.* The amides and imides produced by these reactions are hydrolysed, with regeneration of the amines, by "saponification" with an alcoholic solution of potassium hydroxide. For example:

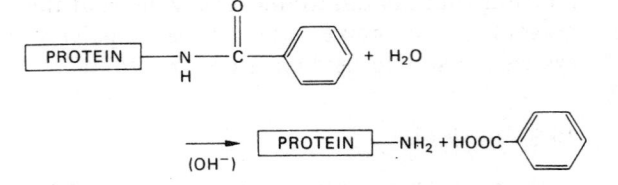

When sections are treated with acetic anhydride in the presence of a little sulphuric acid, amino groups are blocked in a few minutes and their sites become strongly basophilic. Although some acetylation may occur, the principal reaction is probably **N-sulphation**:

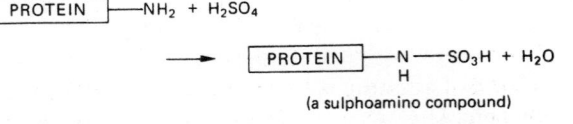

(a sulphoamino compound)

The product ionizes as a strong acid, thus accounting for the induced basophilia. This blockade is not reversible by saponification, but it is reversed by treatment with hot acidified methanol, which presumably causes breaking of the nitrogen-sulphur bond. It should be noted that the chemistry of this sulphation-acetylation procedure (Lillie, 1964) is still poorly understood, though the technique is useful on account of the rapidity and completeness of the reaction. The reagent also brings about O-sulphation of hydroxyl groups, as has already been described (p. 143).

Primary amines are converted to secondary and tertiary amines by **alkylation** with reagents such as methanolic hydrogen chloride:

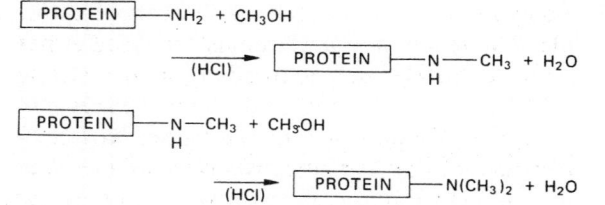

This reaction, which is slower than the methylation of carboxyl groups, does not prevent the binding of anionic dyes because the products are even more strongly basic than primary amines. **Arylation** of amino groups can be achieved by reaction with 2,4-dinitrofluorobenzene, but this reagent also combines with the side-chains of cysteine, histidine, and tyrosine.

10.5. Arginine

Most proteins contain some arginine, but the highest concentrations are found in the histones associated with DNA and in certain cytoplasmic granules. The side-chain of this amino acid is strongly basic, owing to its guanidino group

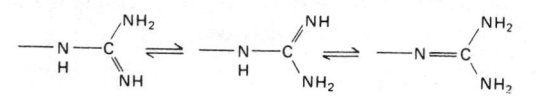

The easiest way to stain sites of high concentration of arginine is to use an **alkaline solution of an anionic dye** (e.g. biebrich scarlet (Spicer & Lillie, 1961) or fast green FCF (see James & Tas, 1984) at pH 8–9). Amino groups are not ionized at high pH. To stain the histone proteins of chromatin it is necessary first to remove the DNA by chemical or enzymatic extraction (see Chapter 9), because the phosphate groups of the nucleic acid compete with and exclude dye anions.

Highly specific histochemical tests are also available for arginine. The best known is the **Sakaguchi reaction.** Sections are treated with a solution containing sodium hypochlorite and α-naphthol, made strongly alkaline by addition of sodium hydroxide. A pink or red colour develops at the sites of arginine residues. It is unstable, so permanent preparations cannot be made. Baker (1947) tested the Sakaguchi reaction with several pure compounds of known structure and concluded that coloured products were formed by reaction of the reagent with substances containing the arrangement

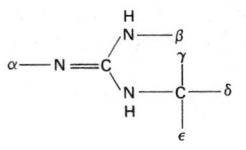

in which α and β were −H or −CH₃ and γ, σ and ε could be any radicals. Arginine is the only known component of fixed tissues to conform to this structural requirement.

The chemistry of the Sakaguchi reaction is not fully understood, but a plausible mechanism was proposed by Lillie *et al.* (1971). They discovered that an identical pattern of staining could be obtained with an alkaline solution of the sodium salt of 1,2-naphthoquinone-4-sulphonic acid (NQS):

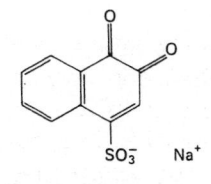

The alkalinity was produced by barium or strontium hydroxide. (The hydroxides of sodium, calcium, and magnesium did not permit full development of the colour.) The product of the reaction with NQS, unlike that formed in the Sakaguchi reaction, resisted extraction by organic solvents, and was permanent in resinous mounting media. Studies with blocking reactions indicated that NQS stained arginine specifically. Lillie *et al.* (1971) speculated that the coloured product of the reaction of the guanidino group with NQS was

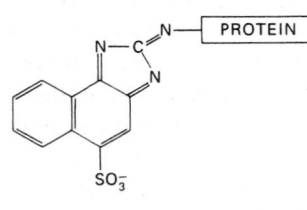

These authors also suggested that in the Sakaguchi reaction α-naphthol was oxidized by hypochlorite to a mixture of 1,2- and 1,4-naphthoquinones, and that the former condensed with guanidino groups to give a coloured product. It was later suggested (Lillie & Fullmer, 1976) that cleavage of the chromogenic part of the structure from the remainder of the arginine residue might occur during the course of staining to give

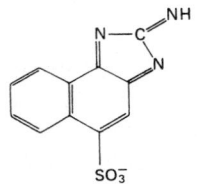

The sodium salt of this anion is soluble, but the barium salt is insoluble. Formation of such a product would account for the requirement for barium ions in the NQS reagent and for the observation that only weak histochemical reactions could be obtained when 1,2-naphthoquinone was substituted for its sulphonic acid derivative.

The histochemical specificity of the Sakaguchi and related reactions is so high that blocking reactions are rarely considered necessary, though several are available. Lillie et al. (1971) used an alkaline alcoholic solution of **benzil** to convert the guanidino group to an unreactive imidazole:

Another reagent, used by biochemists, is the trimer of 2,3-butanedione (also called diacetyl trimer), which reacts with arginine in neutral solution.

10.6. Tryptophan

Tryptophan is a ubiquitous component of proteins, so histochemical methods are employed only for the study of sites of high concentration of this amino acid, such as the Paneth cells of the intestine, the zymogen granules of the exocrine pancreas and several other secretory products. Amyloid (see Chapter 11, p. 175) also contains much tryptophan.

Histochemical techniques are based on the reactions of the indole ring, which remains free to react when the carboxyl and amino groups of tryptophan are incorporated into peptide linkages. Other naturally occurring indole-containing compounds are soluble and are therefore unlikely to be confused with tryptophan in paraffin sections. A possible exception is the serotonin of argentaffin cells and of mast cells of rodents (see Chapter 17). Reactions for tryptophan can be obtained after almost any fixation but are strongest following formaldehyde or an alcoholic fixative. Picric acid, mercuric chloride, and potassium dichromate partly inhibit staining.

The most popular histochemical method for tryptophan is based on the addition of *p*-dimethylaminobenzaldehyde (DMAB) to the indole ring, which occurs at the position adjacent to the nitrogen atom. Sections are first treated with a solution of DMAB in acetic acid to which a strong acid (perchloric, hydrochloric, or a mixture of the two) has been added. The DMAB

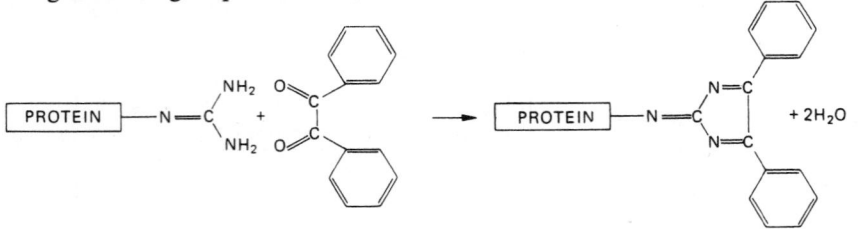

combines with the indole ring in an acid-catalysed hydroxyalkylation reaction:

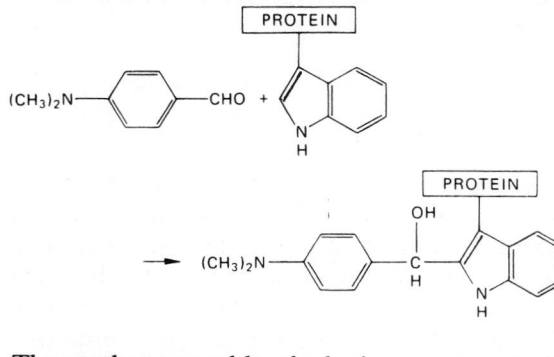

The product resembles the leuko-compound of a diarylmethane dye. In the second stage of the method an oxidizing agent, usually a strongly acidified solution of sodium nitrite, is applied. A blue colour develops due to the formation of a product that may have a structure such as

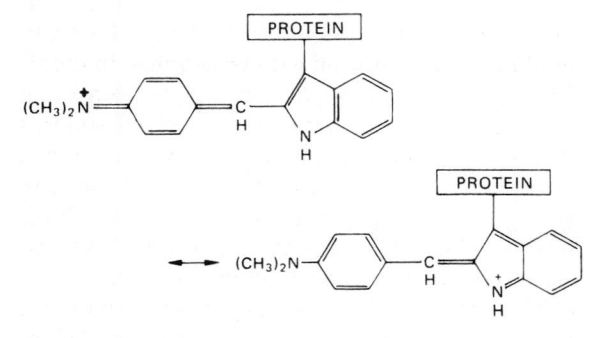

The resonance structure of such a compound shows similarity to those of diphenylmethane and cyanine dyes (see Chapter 5).

Glenner (1957) suggested the product of the reaction was a "rosindole" triarylmethane dye:

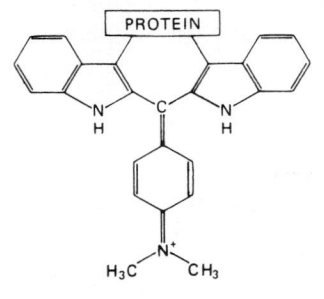

Such a compound could only be formed with fixed proteins if the two indole rings were exactly the right distance apart to be able to combine with one molecule of DMAB. A triarylmethane dye is a likely product of the reaction of DMAB with solutions of indole-containing compounds but an unlikely one under the conditions of the histochemical procedure.

There is no reaction of high specificity for preventing the characteristic reactions of the indole group of tryptophan. Strong oxidizing agents cause opening of the heterocyclic ring: an alkaline solution of potassium persulphate is suitable for this purpose. The products of the reaction have not been identified and the blockade is irreversible. Performic and peracetic acids (see pp. 155–156) may also be used.

10.7. Tyrosine

Tyrosine is another ubiquitous amino acid. Its histochemical demonstration may be considered indicative of the presence of proteins. Several techniques for the detection of the simple phenolic side-chain are available. Other phenols such as serotonin, the catecholamines, and tannins will give the same reactions, but protein-bound tyrosine and the serotonin of argentaffin cells are the only reactive compounds likely to be present in fixed animal tissues.

Millon's classical test for protein was adapted for histochemical use by Baker (1956). A red colour develops when tissues are treated with an acid solution containing mercuric and nitrite ions. All investigators agree that the reaction is specific for phenolic compounds.

In the first stage of the reaction, nitrous acid reacts with the phenolic ring causing C-nitrosation, principally *ortho* to the hydroxyl group

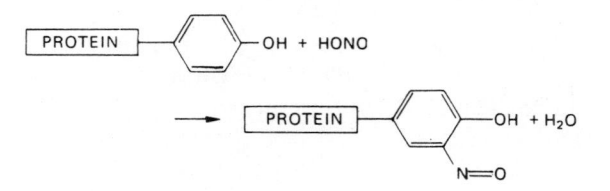

The nitroso compound forms with mercuric ions a stable chelate, which is red and may have a structure such as

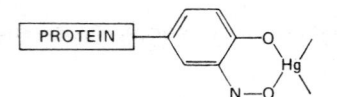

The other two coordination positions of the mercury atom may be satisfied by another nitrosophenol molecule (if one is suitably positioned in the fixed tissue) or by an inorganic ligand such as water.

In the absence of a metal such as mercury, further reaction of tyrosine with nitrous acid results in the formation of a diazonium salt:

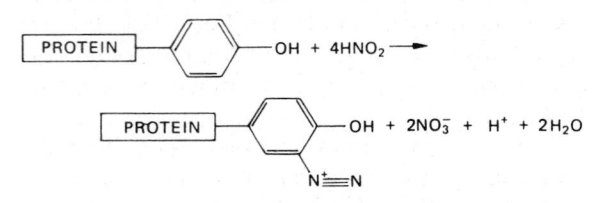

This product is rapidly decomposed by light, but in darkness it is stable enough to be coupled with a naphthol to form an azoic dye. These reactions form the basis of another histochemical technique for tyrosine.

As a phenol, tyrosine can couple with exogenous diazonium salts to form azo dyes, but a similar reaction occurs with the imidazole ring of histidine and probably also with the indole ring of tryptophan. Diazonium salts detect proteins but lack specificity for any individual amino-acid side-chains.

The phenolic hydroxyl group of tyrosine can be blocked by **acylation** with acetic anhydride or benzoyl chloride. The aromatic rings of the resultant esters no longer possess the reactivity of phenols towards nitrous acid and other reagents. Aliphatic hydroxyls and amino groups are also acylated (see pp. 145, 148).

The hydrogen atoms of phenolic rings are quite easily replaced by iodine atoms, so adequate treatment with a solution of **iodine** will prevent the histochemical reactions of tyrosine. Iodination can also cause oxidation of free sulphydryl

groups of cysteine and partial inhibition of the reactions of tryptophan.

Blockade of tyrosine can also be achieved with **tetranitromethane**, which introduces a nitro group *ortho* to the hydroxyl group. However, this reagent also decomposes tryptophan and the sulphur-containing amino acids.

10.8. Cysteine and cystine

10.8.1. Some properties of cysteine and cystine

The free sulphydryl (also known as a mercaptan or thiol) group of **cysteine** is very reactive, mainly because it is a strong reducing agent. In ordinary paraffin sections, sulphydryl groups are demonstrable only if an unreactive fixative, such as a mixture based on alcohol and acetic acid, has been used and if exposure of the tissue to atmospheric oxygen has been kept to a minimum at all stages of processing. Fixatives that react with the —SH group, such as mercuric chloride, oxidizing agents, and, to a lesser extent, aldehydes, must be avoided.

Suitably positioned sulphydryl groups may be oxidized in pairs to yield disulphides:

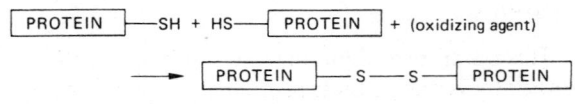

This is the normal mode of formation of **cystine**, the predominant sulphur-containing amino acid in both living and fixed tissues. Strong oxidizing agents, such as the permanganate ion and performic and peracetic acids, will cleave the disulphide bridge of cystine to yield two molecules of cysteic acid, which is an aliphatic sulphonic acid:

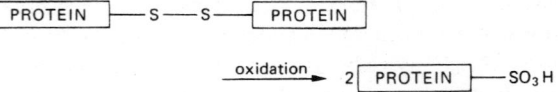

Cystine is easily reduced to cysteine. The reducing agents used in histochemistry include the alkali metal cyanides

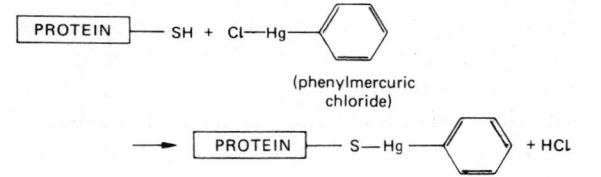

```
PROTEIN —S—S— PROTEIN + CN⁻ + H₂O  ──────→  2 PROTEIN —SH + CNO⁻
                                        (OH⁻)
```

and a variety of compounds containing the sulphydryl group. For example:

```
PROTEIN —S—S— PROTEIN + 2HSCH₂COO⁻ + 2Na⁺  ──────────→  2 PROTEIN —SH + ⁻OOCCH₂—S—S—CH₂COO⁻ + 2Na⁺
                     (sodium thioglycollate)   (pH 9 to 11)
```

(dithiothreitol)

Dithiothreitol (Cleland, 1964) can be used in a less strongly alkaline solution than sodium thioglycollate, but is more expensive.

The other histochemically important property of the sulphydryl group is its ability to combine with both inorganic and organic compounds of mercury. For example:

```
PROTEIN —SH + Cl—Hg—⟨phenyl⟩
                    (phenylmercuric
                       chloride)

──→  PROTEIN —S—Hg—⟨phenyl⟩ + HCl
```

10.8.2. Histochemical methods for cysteine

It is possible to demonstrate sulphydryl groups either by virtue of their reducing properties or by the binding of chromogenic mercury-containing reagents. These methods may be applied directly to sections, to detect any cysteine that has survived fixation, or they may be applied to sections previously treated for the reduction of —S—S— to —SH. In the latter case, cysteine and cystine will both be demonstrated.

The simplest and probably the most sensitive technique based on the reducing properties of the sulphydryl group is the **ferric ferricyanide method**. A solution containing the ions Fe^{3+} and $[Fe(CN)_6]^{3-}$ yields an insoluble blue pigment when it is acted upon by a reducing agent. According to Lillie & Donaldson (1974) it is the ferric ion rather than the ferricyanide which is reduced by the —SH groups, so the pigment

would be expected to be ferrous ferricyanide, also known as Turnbull's blue. However, it is known that this substance is identical to Prussian blue, the pigment precipitated when ferric and ferrocyanide ions meet one another. Prussian blue is ferric ferrocyanide, $Fe_4[Fe(CN)_6]_3$, but its crystals also include ions of sodium or potassium and molecules of water (see Cotton & Wilkinson, 1980).

The ferric ferricyanide reaction is positive with reducing agents other than cysteine, including serotonin, catecholamines, and some of the metabolic precursors of melanin. The reagent is also reduced by lipofuscin pigment and (sometimes) by elastin, for reasons that are incompletely understood. In the histochemical study of cysteine and cystine, blocking reactions must be used to confirm the specificity of staining.

Several organic compounds of mercury are used as histochemical reagents for the sulphydryl group (see Cowden & Curtis, 1970; Lillie, 1977). Examples are:

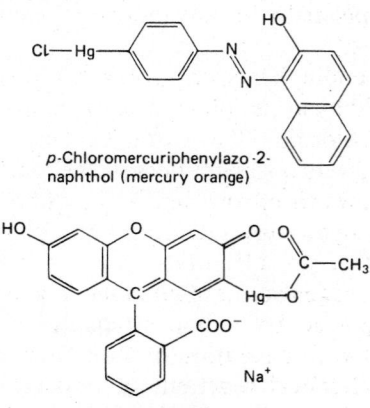

p-Chloromercuriphenylazo-2-naphthol (mercury orange)

Fluorescein mercuric acetate

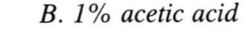

Mercurochrome (merbromin) (M.W. of disodium salt with $3H_2O = 805$)

These and other mercurials react rather slowly with sulphydryl groups to give covalent coloured adducts. Reaction is favoured by an aprotic polar solvent such as N,N-dimethylformamide, in which the active anion of the reagent is not hindered by dipole-dipole attractions to the solvent molecules as would occur in water.

10.8.2.1. Ferric ferricyanide reaction

This is a rather general reaction for reducing groups, as explained on p. 153. If the sections are treated for conversion of disulphide to sulphydryl groups (see p. 168) between stages 2 and 3, —SH generated from —S—S— will be demonstrated. Controls with sulphydryl blockade are necessary if any demonstrated site of reduction is to be attributed to cystine or cysteine. Free sulphydryl groups naturally present in tissues are seen to best advantage after fixation in Carnoy or a similar non-reactive coagulant fixative.

Solutions required

A. *Ferric ferricyanide reagent*

This is made from two stock solutions, both of which are indefinitely stable:

(i) Potassium ferricyanide
   $K_3Fe(CN)_6$:   24mg
   Water: 60ml
(ii) Ferric chloride ($FeCl_3.6H_2O$):   5.0g
   Water: 500ml

The working solution, prepared immediately before use and used only once, is made by mixing 10ml of (i) with 30ml of (ii).

B. *1% acetic acid*

Glacial acetic acid: 1.0ml } Keeps
Water: to 100ml } indefinitely

Procedure

1. De-wax and hydrate paraffin sections. If desired reduce —S—S—to—SH (see Section 10.11.9). Blocking reactions for —SH (Section 10.11.10) may be carried out after or instead of the reduction, with control sections.
2. Immerse in solution A for 10 min.
3. Rinse in 1% acetic acid (solution B). See *Notes 1* and *2*.
4. Dehydrate through graded alcohols, clear in xylene, and mount in a resinous medium.

Results

Sites of strong reduction blue. Sites of weak reduction (most of the "background") yellow-green. See Lillie & Burtner (1953) for a long list of positively reacting substances.

Notes

1. The wash in dilute acetic acid is probably unnecessary. The idea is to preserve as much as possible of the Prussian blue, which is soluble in alkalis and might be slightly extracted by a neutral liquid such as tap water.
2. A counterstain may be applied after stage 3 if considered necessary. A pink or red nuclear stain (see Chapter 6) is convenient, but usually the background coloration is adequate for recognition of the architecture of the tissue.

10.8.2.2. Mercurochrome method for cysteine

Solutions required

A. Mercurochrome reagent

Mercurochrome
  (= merbromin):   20 mg   ⎫
Water: 0.5 ml   ⎬ Stable for several weeks
Dissolve, then add: ⎪
N,N-dimethylformamide: 100 ml   ⎭

B. N,N-dimethylformamide
Required for washes

Procedure

1. De-wax and hydrate paraffin sections. If desired, apply a reaction for the conversion of cystine to cysteine. Wash in water, drain, and proceed to stage 2.
2. Immerse in mercurochrome reagent (solution A) for **either** 1h **or** 48h (for fluorescence or conventional microscopy, respectively).
3. Wash in two changes of N,N-dimethylformamide, each 3 min.
4. Dehydrate in absolute ethanol (two changes), clear in xylene, and cover, using a non-fluorescent resinous mounting medium.

Result

Bright green fluorescence (excitation by blue or ultraviolet) at sites of sulphydryl groups, after 1h at stage 2.

Red to pink staining of sulphydryl sites after 48h at stage 2.

Notes

1. The specificity of staining may be confirmed in control sections by applying a sulphydryl-blocking procedure before stage 2 of the technique. If non-specific background staining is troublesome (mercurochrome is an anionic dye, but has little staining power in the solvent used), try washing it out with water or 70% alcohol.
2. Some batches of mercurochrome are unsatisfactory for this method (Lillie, 1977) for unknown reasons.

10.8.3. Cysteic acid methods for cystine

Cystine may be demonstrated without prior reduction to cysteine by oxidizing it to cysteic acid (see p. 152). This sulphonic acid is ionized even at very low pH and can therefore bind cationic dyes from strongly acid solutions. Alcian blue (at pH 0.2), thiazine dyes (at pH 1.0), and aldehyde-fuchsine are often used for the staining of cysteic acid. The other properties of these dyes, especially their attachment to sulphated carbohydrates, must, of course, be allowed for in suitable control sections.

Several other methods for cysteine and cystine are described in the larger textbooks of histochemistry. No histochemical method is yet available for methionine, the third sulphur-containing amino acid.

10.8.3.1. Cysteic acid method for cystine

This is a slightly modified version of the technique of Adams & Sloper (1955). Fixation is not critical, but neurosecretory material is not preserved by alcoholic fixatives.

Solutions required

A. Performic acid

This reagent should be prepared in a 150–200 ml conical flask. A magnetic stirrer should be used, but if one is not available, mix the ingredients in a 200 ml beaker and stir with a glass rod, taking great care to avoid splashing and spillage. See *Note 4* below.

Formic acid (the 98% acid is recommended, but the 88% acid is suitable):   80 ml

Hydrogen peroxide (30%; "100
  volumes available oxygen"):   8ml
Concentrated sulphuric acid (96%):   1ml

Stir vigorously at intervals for 1h. This solution is highly corrosive. It should be free of bubbles when used. It keeps for 24h. Dilute with at least 1 litre of tap water before discarding. See *Note 5* below for alternatives to performic acid.

B. Alcian blue

Alcian blue (see p. 86):   6g ⎫
Water: 200ml ⎬ Keeps for several months
Concentrated sulphuric   ⎪
  acid (96%):   5.5ml ⎭

Heat to 70°C, cool to room temperature and filter. With some batches of alcian blue it is impossible to obtain a 3% solution, but this does not matter.

Procedure

1. De-wax and hydrate paraffin sections. Coating with a film of nitrocellulose (see Chapter 4) is advisable, though not always necessary. Blot dry with filter paper.
2. Immerse in performic acid (solution A) or alternative (see *Note 5* below) for 5min.
3. Rinse in four changes of tap water, without agitation, for a total of 10min.
4. Transfer to 70% and then to absolute alcohol. Blot the sections to flatten creases and return to water. Place slides on a hotplate until just dry.
5. Stain in alcian blue (solution B) for 1h.
6. Wash in three or four changes of tap water until all excess dye is removed.
7. Counterstain if desired (see *Note 1* below).
8. Wash in water, blot dry, dehydrate in two changes (each 5min) of *n*-butanol, clear in xylene, and cover. See *Note 2* below.

Result

Sites of high cystine concentration (>4% of total amino-acid content of protein) blue. Sulphated carbohydrates are also blue (see *Note 3*).

Notes

1. Suitable counterstains are: 0.5% aqueous neutral red or safranine (for nuclei, etc.); 0.5% aqueous eosin (for general pink background). The alcian blue resists extraction by most of the commonly used histological reagents.
2. If the sections look as if they will come off the slides, mount into a resinous medium directly from the second change of *n*-butanol.
3. To control for the specificity for cystine, omit the oxidation (stage 2). Sulphate esters (see Chapter 11) are stained in oxidized or unoxidized sections. Their stainability can be prevented by prior methylation (see p. 166).
4. Perfomic and peracetic acids can react explosively with copper and its salts. Use clean glassware and remember to dilute the reagent before pouring it down the sink.
5. An alternative oxidizing agent is peracetic acid, $H_3C-C\overset{\displaystyle O}{\underset{\displaystyle O-OH}{\Big\langle}}$. This is much more stable than performic acid and is available commercially as a 40% solution. Peracetic acid can also be made in the laboratory: thoroughly mix 40ml of acetic anhydride with 10ml of 30% hydrogen peroxide; leave to stand for 24h and then add an equal volume of water. A less noxious but less specific oxidizing agent is an acidified permanganate solution (Pasteels & Herlant, 1962; see Adams, 1965, and Gabe, 1976, for discussion): sections are immersed for 5min in a freshly prepared 0.5% solution of potassium permanganate ($KMnO_4$) in 2% sulphuric acid and then rinsed in 1% aqueous oxalic acid ($H_2C_2O_4.2H_2O$) to remove the brown stain of manganese dioxide from the sections. A further wash in water precedes staining with alcian blue. Oxidation with permanganate is acceptable when the cysteic acid method is being used for purely histological purposes such as the study of cell types in the hypophysis or

endocrine pancreas. Performic or peracetic acid must be used for critical histochemical identification of cystine-containing proteins.

6. Other cationic dyes in acid solution (pH<1.0) may be substituted for alcian blue.

10.8.4. Blocking procedures

Two convenient blocking agents for the sulphydryl group are N-ethylmaleimide and iodoacetic acid:

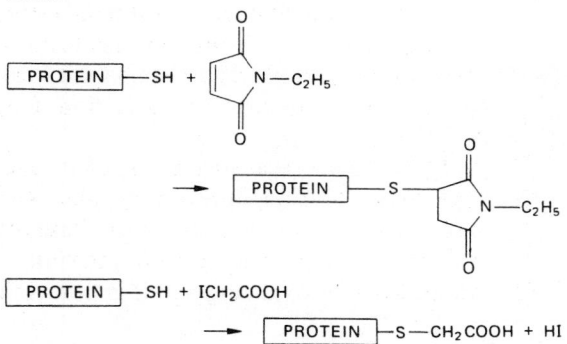

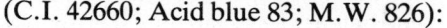

Both these reagents completely block the reactivity of SH groups. See Section 10.11.8 for technical details. Other blocking agents, including iodine and various compounds of mercury, are likely to give incomplete or easily reversible blockades.

10.9 Methods for proteins in general

The histologist or pathologist often wants to know whether or not an intracellular inclusion or an extracellular material is composed largely of protein. For the simple staining of protein, techniques are available that do not depend upon the presence of individual amino acids. Two will be considered here.

10.9.1. Demonstration of proteins with dyes

Staining by anionic dyes, especially those with small molecules, is due largely to electrostatic attachment to the basic side-chains of lysine and arginine. Such staining has already been discussed in Chapters 5 and 6 and in Section 10.4.1 of this chapter. Dyes with large molecules, especially those used industrially for the direct dyeing of cotton, are probably bound to tissues by hydrogen bonds, van der Waals forces, or charge-transfer bonds as well as by ionic attraction (see Chapter 5). Consequently such dyes are likely to stain nearly everything in a section.

An anionic dye is useful for the demonstration of protein only if the intensity of the colour produced at any site in the tissue is proportional to the local concentration of protein there. A triphenylmethane dye, which has been shown to bind stoichiometrically to proteins under histochemical conditions, is BRILLIANT INDOCYANINE 6B (C.I. 42660; Acid blue 83; M.W. 826):

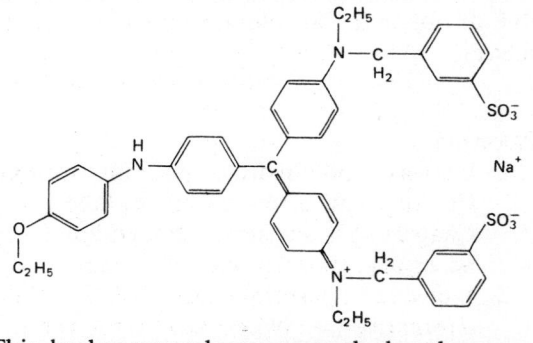

This dye has several synonyms, the best known of which are brilliant blue R and coomassie brilliant blue R-250. It is used by biochemists for staining proteins in acrylamide gels. Brilliant indocyanine 6B is almost insoluble in water, so sections are stained from a solution of the dye in a mixture of ethanol and acetic acid. The chemical specificity of the technique and the mode of binding of the dye are in need of investigation, but it has been shown that staining is prevented by prior treatment of plastic-embedded sections with proteolytic enzymes (Cawood *et al.*, 1978).

10.9.1.1. Brilliant indocyanine 6B for proteins

This technique is suitable for paraffin sections of tissues fixed in formaldehyde, Carnoy or Bouin. See *Note* below.

Solutions required

A. Acetic-ethanol

Glacial acetic acid: 100ml ⎱ Keeps
Absolute ethanol: 300ml ⎰ indefinitely

B. Dye solution

Brilliant indocyanine 6B (see p. 157
for synonyms):   40mg
Acetic-ethanol (solution A above): 200ml

The dye dissolves completely and filtration is not needed. The solution is stable for at least 5 years.

Procedure
1. De-wax and hydrate paraffin sections and take to absolute ethanol.
2. Stain in the dye solution (B) for 30 min.
3. Drain slides and immerse in acetic–ethanol (solution A) for 5min with occasional agitation.
4. Rinse in 95% ethanol, dehydrate in absolute ethanol, clear in xylene, and cover, using a resinous mounting medium.

Result
Proteins are stained a bright royal blue. The depth of colour is proportional to the local concentration of protein.

Note
In the original description of this technique (Cawood *et al.*, 1978), plastic-embedded sections were stained for 24h and stage 3 lasted 20min. When paraffin sections are used, there is no increase in intensity of staining after 15–20min and subsequent washing in acetic–ethanol removes hardly any dye from the sections. The specificity for proteins is not fully established. Carbohydrate-rich components of tissues, such as the matrix of cartilage, are very weakly stained by comparison with cytoplasm and collagen. Nuclei are recognizable in the stained sections but are no more strongly coloured than the surrounding cytoplasm.

10.9.2. *The coupled tetrazonium reaction and related methods*

Alkaline solutions of diazonium salts combine with proteins in at least three different ways. Thus the expected coupling reaction occurs with the phenolic side-chains of tyrosine:

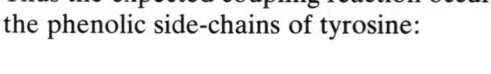
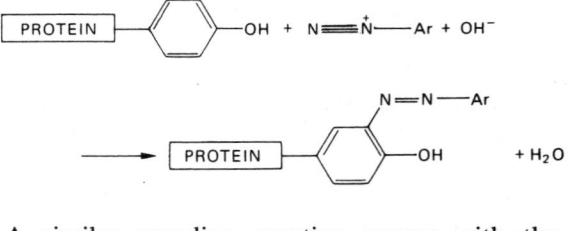

A similar coupling reaction occurs with the imidazole ring of histidine. The nitrogen atoms of this aromatic heterocycle confer some of the properties of an aromatic amine to the ring:

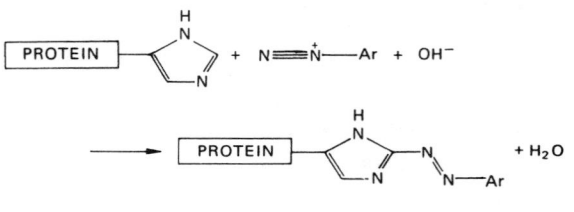

Coupling probably also occurs *ortho* to the nitrogen atom in the indole ring of tryptophan. A different reaction takes place with the ε-amino group of lysine:

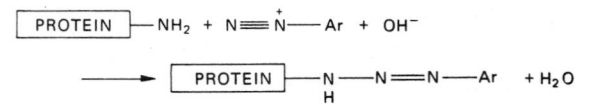

The product is a **triazene**. Similar reactions may occur with the secondary amine nitrogen atoms of proline and hydroxyproline and with the guanidino and sulphydryl groups of arginine and cysteine respectively. The nitrogen atoms of the purine and pyrimidine bases of the nucleic acids almost certainly do not couple with diazonium salts (see Lillie & Fullmer, 1976). Thus a diazonium salt applied to a section of tissue will be covalently bound to several organic functional groups, all of which are found only in proteins.†

In the coupled tetrazonium reaction (Danielli, 1947), the diazonium salt used is tetraazotized benzidine, which has to be freshly prepared in the laboratory from benzidine and nitrous acid:

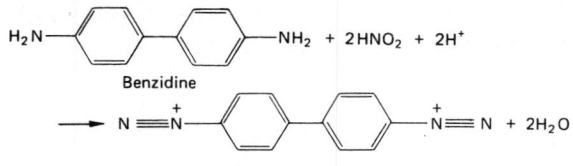

An alternative, more convenient reagent is the stabilized tetrazonium salt derived from *o*-dianisidine, known as fast blue B salt (see p. 69 for formula). It will be noticed that each of these compounds bears two diazonium groups per molecule. Usually only one of these will couple with a reactive site in a section, because not often will two coupling sites in the fixed tissue be exactly the right distance apart to react with the same molecule of reagent.

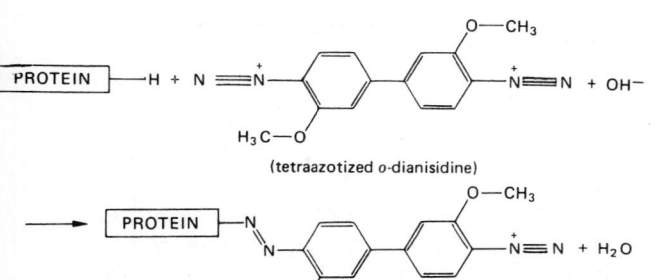

Azo compounds formed by coupling of simple diazonium salts with proteins are only feebly coloured. However, the free —$\overset{+}{N}\equiv N$ group of a bound tetrazonium salt is able to couple with any phenol or aromatic amine subsequently applied to the section, to give a strongly coloured dye containing two azo linkages. H-acid (see p. 69) is a suitable azoic coupling component for this purpose:

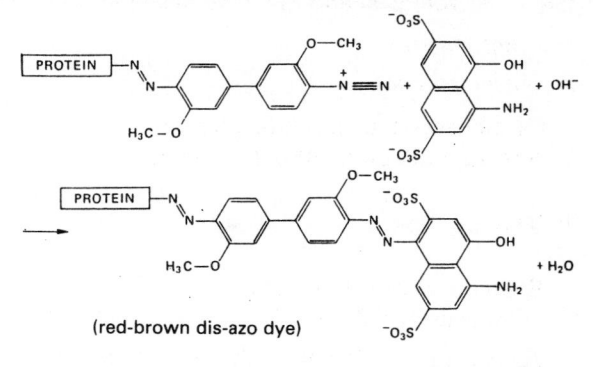

(red-brown dis-azo dye)

As an alternative to the coupled tetrazonium reaction, one may use a diazonium salt (with only one $N^+\equiv N$ group) that is itself an azo dye. Such a salt couples with proteins to give dis-azo dyes in a single-step reaction. A coloured stabilized diazonium salt suitable for staining proteins is fast black K salt (see p. 69). Freshly diazotized safranine can also be used (Lillie *et al.*, 1968).

In attempts to make the coupled tetrazonium reaction specific for aromatic amino acids, sections are commonly treated with dilute hydrochloric acid after the first coupling, to destroy triazenes, but the efficacy of this manoeuvre is questionable.

10.10. Aldehydes and ketones

Free aldehyde groups are largely absent from tissues‡ but they are often produced in the course of histochemical manipulations. The Feulgen reaction (Chapter 9) and the PAS reaction (Chapter 11) are the most commonly used methods of this type. Ketones occur naturally as ketosteroids, for which histochemical techniques have been described, and as intermediate products in some staining methods (e.g. see p. 146), but are generally less important to the histochemist than are aldehydes.

---

†Exceptions are certain phenols: the catecholamines of chromaffin cells and serotonin in argentaffin cells (see Chapter 17), but these substances are absent from most tissues and their presence can be detected by other methods. Phenolic compounds of plants, such as tannins, may also be expected to couple with diazonium salts.

†They occur in immature elastin (see Chapter 8), in lignin, in lipids that have been oxidized by air (see Chapter 12), and in specimens fixed in glutaraldehyde (see Chapter 2).

10.10.1. *Schiff's reagent*

Before reading this section the reader may wish to review the chemistry of pararosaniline and related triphenylmethane dyes (Chapter 5, pp. 74–75). In this account it is assumed that Schiff's reagent is prepared from pararosaniline, but equivalent chemical reactions occur with the other components of basic fuchsine.

Schiff's reagent is made by treating a solution of basic fuchsine with sulphurous acid. This weak acid is formed when sulphur dioxide combines with water:

$$H_2O + SO_2 \rightleftharpoons H_2SO_3 \rightleftharpoons H^+ + HSO_3^- \rightleftharpoons 2H^+ + SO_3^{2-}$$

These equilibria are displaced to the left when $[H^+]$ rises. Sulphurous acid, containing free sulphur dioxide, is prepared in the course of making Schiff's reagent, either from the reaction of a bisulphate (sodium metabisulphite, $Na_2S_2O_5$, is usually used) with a mineral acid, or of thionyl chloride with water (see p. 137), or by bubbling gaseous sulphur dioxide through an aqueous solution of the dye. Whatever method is used, a sulphonic acid group is added to the central carbon of the triphenylmethane structure

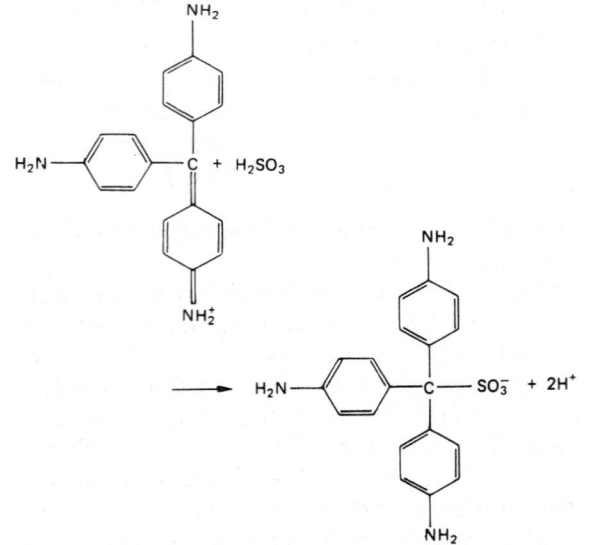

Schiff's reagent therefore consists of a solution of this colourless derivative of the dye in water, which contains an excess of sulphurous acid.

Schiff's reagent reacts with aldehydes, but not with ketones, to form brightly coloured products. This is not simply a matter of recolorizing the dye. New compounds are formed with colours somewhat different from that of basic fuchsine. The chemistry of the histochemical reaction is still not fully understood, but the results of three investigations (Dujindam & van Duijn, 1975; Gill & Jotz, 1976; Nettleton & Carpenter, 1977) concur in attributing the structure

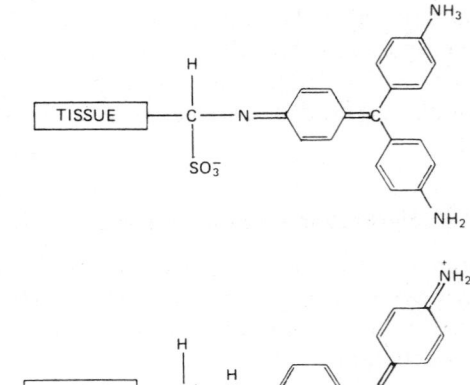

or

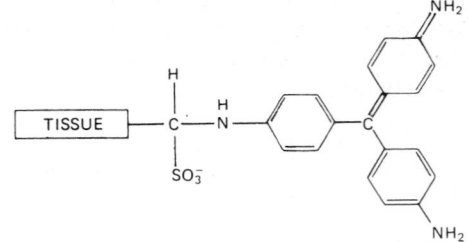

to the principal coloured compound formed when Schiff's reagent reacts with tissue-bound aldehyde groups. Both these compounds are alkylsulphonic acid derivatives of pararosaniline, and the two structures would be expected to exist in tautomeric equilibrium. There is still some uncertainty as to whether the aldehyde group first reacts with sulphurous acid to form a bisulphite compound, which then condenses with an amino group of Schiff's reagent, or whether the amino group of the reagent first reacts with the aldehyde to form an imine (= Schiff's base, anil, or azomethine), which is subsequently sulphonated. The second stage of either scheme must involve removal of the sulphonic acid group from the dye molecule and re-establishment of the triphenylmethane chromophore.

Several dyes and fluorochromes other than basic fuchsine will serve as substitutes for Schiff's reagent if their solutions are treated with sulphur dioxide, but usually these "pseudo-Schiff" reagents are coloured. The most widely used ones are derived from thionine with bisulphite (van Duijn, 1956) and from acriflavine with thionyl chloride (Ornstein *et al.*, 1957). The latter is useful for the demonstration of aldehyde-containing components of tissues in fluorescence microscopy. The traditional Schiff's reagent also gives a fluorescent product with aldehydes, but its orange-brown emission is rather weak, and contrasts poorly with a dark background.

The **preparation of Schiff's reagent** is described on page 137, in conjunction with the Feulgen technique. It is used as described on pp. 137, 186 and 214.

10.10.2. Hydrazines and hydrazides

Organic hydrazines are compounds in which one hydrogen atom of hydrazine, H_2N-NH_2, is replaced by an organic radical. They have the general structure

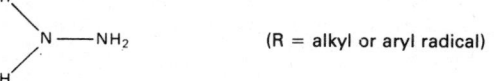

(R = alkyl or aryl radical)

Hydrazides are similar but with an acyl in place of R in the above general formula. Hydrazines and hydrazides condense readily with both aldehydes and ketones:

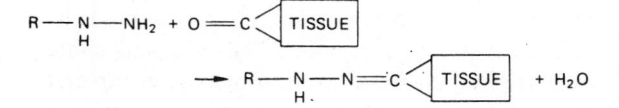

The reagent $RNHNH_2$ in this equation will be potentially useful for the demonstration of tissue-bound carbonyl groups if R is either coloured or able to form a coloured compound by reaction with another reagent. Several hydrazines and hydrazides serve as histochemical reagents in this way. They include the following:

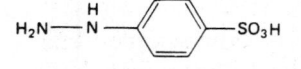

Phenylhydrazine-4-sulphonic acid (Shackleford, 1963): confers affinity for cationic dyes at low pH on sites of aldehydes

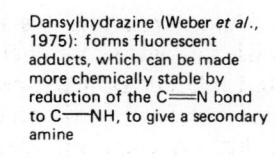

Dansylhydrazine (Weber *et al.*, 1975): forms fluorescent adducts, which can be made more chemically stable by reduction of the C=N bond to C—NH, to give a secondary amine

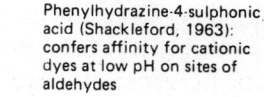

Salicylhydrazide (Stoward, 1968b): forms fluorescent adducts. Chelation with certain metals (Zn^{2+}, Al^{3+}) can be employed to distinguish between products of condensation with aldehydes or ketones

2-Hydroxy-3-naphthoic acid hydrazide (HNAH): this is the most popular reagent of its type. Sites of attachment are made visible by coupling with a diazonium salt

Phenylhydrazine ($C_6H_5NHNH_2$) is a useful reagent for blocking the reactivity of carbonyl compounds.

10.10.2.1. Hydroxynaphthoic acid hydrazide

The HNAH reagent is made up as follows:

2-hydroxy-3-naphthoic acid hydrazide:	200 mg	Keeps for several weeks
Absolute ethanol:	100 ml	
Glacial acetic acid:	10 ml	
Water:	90 ml	

Dissolve with aid of gentle heat (up to 70°C) and vigorous stirring.

Sections of tissue in which aldehyde or ketone groups have been produced are immersed in the HNAH reagent for 1h at 60°C or for 3h at room temperature. They are then washed in three changes (each 10min) of 50% ethanol, taken to water, and reacted at 0–4°C with fast blue B salt (1mg per ml, pH 7.5, 3min). The sections are then washed, dehydrated, cleared, and mounted in a resinous medium. Sites of aldehydes and ketones are shown in blue and purple. There is also a pink to light brown background staining, which has no histochemical significance.

10.10.3. Aromatic amines

The simplest reaction of a primary amine with an aldehyde is

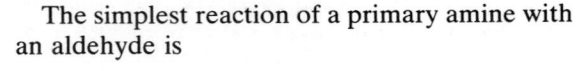

The product, containing the C=N linkage, is variously known as an **imine**, an azomethine, a Schiff's base, or an anil. It is stable only when R or R' is an aromatic ring. If the amine is present in excess, as it must be when it is a reagent applied to a section of a tissue, two of its molecules may react with one aldehyde group:

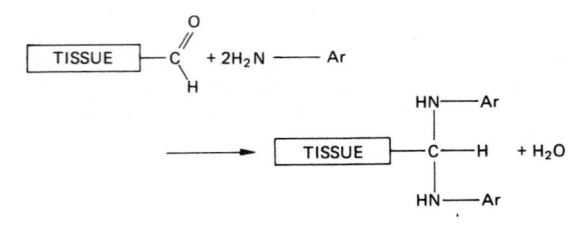

The product is known as an **aminal.** Its formation is favoured when the group Ar in the above equation carries an electron-withdrawing substituent (see Sollenberger & Martin, 1968). In an elegant study of the reaction of *m*-aminophenol with tissue-bound aldehydes, Lillie (1962) provided convincing evidence that aminals† were the principal products of the reaction, which was acid-catalysed and occurred in the absence of water. Sites of attachment of *m*-aminophenol could be demonstrated by coupling with a diazonium salt to give coloured products with structures such as

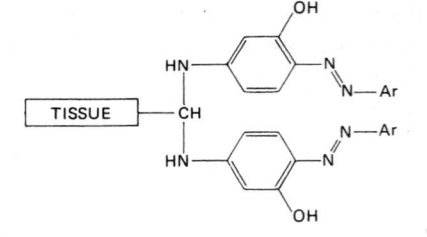

Aromatic amines are also used as blocking agents for aldehyde groups. Blockade is achieved more rapidly with *m*-aminophenol than with aniline. Both reagents are used as solutions in glacial acetic acid.

Many dyes are primary aromatic amines. An acidified alcoholic solution of basic fuchsine (Horobin & Kevill-Davies, 1971a) is a useful substitute for Schiff's reagent (see p. 138). The dye in this solution almost certainly condenses with aldehydes to form aminals in the same way as does *m*-aminophenol (Horobin & Kevill-Davies, 1971b). A solution of acriflavine in ethanol, used similarly, serves as a fluorescent reagent for aldehydes and has some advantages over the "pseudo-Schiff" reagent prepared by treatment of the same dye with sulphurous acid (Levinson *et al.*, 1977).

10.10.4. Ammoniacal silver nitrate

An aqueous solution of ammonia contains the dissolved gas in equilibrium with ammonium hydroxide:

$$NH_3 + H_2O \rightleftharpoons NH_4OH$$

Ammonium hydroxide ionizes as a weak base, so the solution is alkaline:

$$NH_4OH \rightleftharpoons NH_4^+ + OH^-$$

When an ammonia solution is added in small aliquots to aqueous silver nitrate, a black or brown precipitate of silver oxide is first formed:

$$2Ag^+ + 2OH^- \longrightarrow Ag_2O \downarrow + H_2O$$

With further addition of ammonia the precipitate dissolves. The solution contains the silver diammine ion, $[Ag(NH_3)_2]^+$:

$$Ag_2O + 4NH_3 + H_2O \longrightarrow 2[Ag(NH_3)_2]^+ + 2OH^-$$

---

†Lillie (1962) and Lillie & Fullmer (1976) use the old term "diphenamine bases" for aminals. These compounds have also been called secondary amines in the histochemical literature, but "aminal" is the correct name for a compound containing the grouping

$$\begin{array}{c} \quad\; H \;\;|\;\; H \\ C-N-C-N-C \\ \quad\quad\; | \\ \quad\quad\, H \end{array}$$ just as the better-known acetals

include the structure $\quad C-O-\overset{\displaystyle |}{\underset{\displaystyle H}{C}}-O-C.$

The silver diammine ion rapidly oxidizes aldehydes (but not ketones) and is itself reduced to metallic silver:

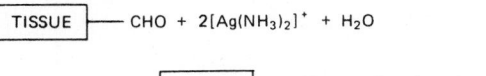

The precipitated silver is black, and the deposits can be further intensified by gold toning or by physical development (see Chapter 18), so the test is a very sensitive one. The silver is electron opaque, so ammoniacal silver nitrate solutions are sometimes used as substitutes for Schiff's reagent in ultrastructural histochemical studies. Other complexes of silver, notably that formed with hexamethylenetetramine (also known as methenamine or hexamine), may be used instead of silver diammine.

Unfortunately the histochemical specificity of this method is low. The silver complexes are also reduced by sulphydryl groups and by o-diphenols (see Chapter 17). The staining method for reticulin described in Chapter 8 is an example of a useful histological technique based on the detection of artificially produced aldehyde groups.

10.10.5. Other methods for aldehydes

Three other methods for the localization of aldehydes are worth mentioning, but none are much used.

(a) Aldehydes have a peroxidase-like property (see Feigl, 1960) in that they catalyse the oxidation of p-phenylenediamine by hydrogen peroxide. This reaction is also used as a spot-test in analytical chemistry.

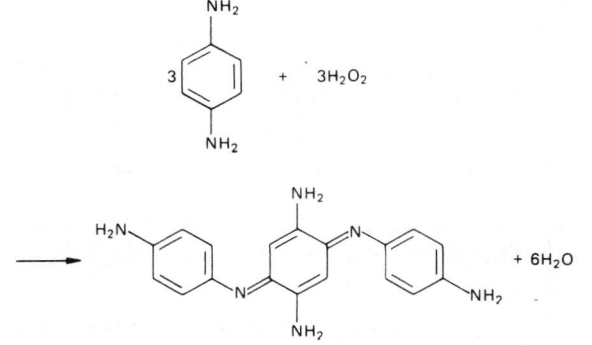

The product, which is known as Bandrowski's base, is darkly coloured and unstable, but is said to be stabilized by treatment with a solution of gold chloride (Scarselli, 1961). The structure of Bandrowski's base and the reactions involved in its formation are discussed by Corbett (1971).

(b) Another reagent used in spot-tests is 3-methyl-2-benzothiazolone hydrazone (MBTH) (Sanwicki et al., 1961). It combines with aldehydes to form blue and green dyes and is a sensitive and specific histochemical reagent. The coloured product can be stabilized for a few days by treatment of the stained sections with ferric chloride or phosphomolybdic acid (Davis & Janis, 1966; Nakao & Angrist, 1968), but permanent preparations cannot be made.

(c) The aldehyde may be converted to its bisulphite addition compound:

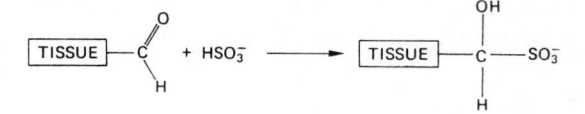

The addition compound is a sulphonic acid, so it will bind cationic dyes such as toluidine blue (Romanyi et al., 1975) or alcian blue (Klessen, 1974) from solutions at pH 1.0 or lower.

10.10.6. Blocking procedures

Three blocking procedures for the carbonyl group are histochemically valuable. Most of the other available methods are unduly time-consuming, are damaging to the sections, or are too easily reversible. The three useful blocking agents are:

(a) **Phenylhydrazine:**

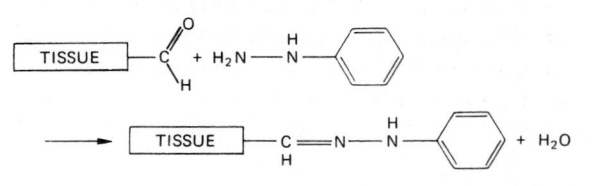

This is used as an aqueous solution of its hydrochloride or acetate.

(b) A solution of **m-aminophenol** in glacial acetic acid. The reaction is discussed in Section 10.10.3.

(c) **Sodium borohydride:**

$$4RCHO + NaBH_4 + 4H_2O \rightarrow$$
$$4RCH_2OH + B(OH)_3 + Na^+ + OH^-$$

This reducing agent is used as a mildly alkaline aqueous solution. It reduces aldehydes to primary alcohols and ketones to secondary alcohols. No other organic functional groups likely to be present in tissues are reduced, but it is worth noting that the C=N linkages of imines and hydrazones are reduced to C—N by this reagent (Billman & Diesing, 1957). Sodium borohydride is the reagent of first choice for irreversible histochemical blocking of the carbonyl group. It is important that the pH of the solution be greater than 8.2. Neutral or less alkaline solutions will not reduce all the aldehyde groups present in a section (Bayliss & Adams, 1979).

10.10.6.1. Borohydride reduction procedure

Take sections to water and transfer to a freshly prepared solution:

Sodium phosphate, dibasic
  (Na_2HPO_4):   0.5 g
Water: 50 ml   } Use at once
  Dissolve, then add:
Sodium borohydride
  ($NaBH_4$):   25 mg

Leave slides in this for 10 min, with occasional agitation to release bubbles of hydrogen from surfaces of slides. Wash in four changes of water.

Caution. Sodium borohydride releases hydrogen on contact with acids. Do not use this reagent near any naked flame. Do not acidify its solutions. Discard the alkaline solution by flushing down the drain with several litres of tap water.

After reduction, aldehyde groups such as those formed by Feulgen hydrolysis (Chapter 9) or periodate oxidation (Chapter 11) can no longer be demonstrated with Schiff's reagent or by other methods.

10.11. Chemical blocking procedures

The reactions described in this section bring about modifications of the organic functional groups of macromolecules. They are used as control procedures in protein and carbohydrate histochemistry.

10.11.1. Sulphation

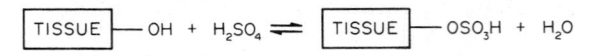

TABLE 10.2 *Some blocking and unblocking reactions.*

Functional group	Blocking reaction	Product of blocking reaction	Unblocking reaction
—OH	Acetylation (3h)	—OOCCH₃	Saponification
	Benzoylation	—OOCC₆H₅	Saponification
	Sulphation	—OSO₃H	Methylation (2h)
—COOH	Methylation (2h)	—COOCH₃	Saponification
$\begin{array}{c} O \\ \parallel \\ -OP(OH)_2 \\ -OP(OH)O- \\ \parallel \\ O \end{array}$	Methylation (24h)	$\begin{array}{c} O \\ \parallel \\ -OP(OCH_3)_2 \\ -OP(OCH_3)O- \\ \parallel \\ O \end{array}$	Saponification
—OSO₃H (of carbohydrates)	Methylation (24h)	—OH	Irreversible
—NH₂	Acetylation (48h)	—NHCOCH₃	Saponification
	Benzoylation	—NHCOC₆H₅	Saponification
	Sulphation	—NHSO₃H	Methylation (2h)
	Deamination (HNO₂)	—OH, etc	Irreversible

TISSUE —NH$_2$ + H$_2$SO$_4$ ⇌ TISSUE —NHSO$_3$H + H$_2$O

Solutions required

A. Sulphation reagent

Pack ice around a 100 ml conical flask containing 25 ml of diethyl ether. Slowly add 25 ml of concentrated sulphuric acid (which should have been pre-cooled by leaving at 4°C for 30–60 min). The mixture becomes hot. Use it when it has cooled to room temperature. After use, pour the mixture carefully into about 500 ml of tap water and then flush down the sink.

B. A cationic dye

A dilute solution of a thiazine dye (see Chapter 6, p. 92) at pH 1.0 is suitable. Alcian blue at pH 1.0 (see Chapter 11, p. 177) may also be used.

Procedure

1. Take sections to absolute ethanol and then into diethyl ether.
2. Place the slides in the sulphation reagent (A) for 5–10 min.
3. Rinse in 95% alcohol and then immerse in water, two changes, each 30 s.
4. Stain with a suitable cationic dye (solution B).
5. Wash in water, dehydrate appropriately for the dye used, clear in xylene, and mount in a resinous medium.

Notes

1. A control section which has not been sulphated is necessary to demonstrate the distribution of sulphated mucosubstances.
2. Sulphation-induced basophilia is most prominent in carbohydrate-containing structures. Comparison with a section stained by the PAS method (see Chapter 11) is useful for distinguishing between carbohydrates and proteins.
3. Sites of sulphate esters are stained. When thiazine dyes are used, metachromatic effects are often seen. If the staining (Step 4) is omitted, sulphation is a blocking reaction of low specificity for hydroxyl groups. Some amino groups are *N*-sulphated, and also become basophilic.

10.11.2. Acetylation and benzoylation

Acetylation reagent

Pyridine (anhydrous):	24 ml	Use once only. Stable for a few days
Acetic anhydride:	16 ml	

Benzoylation reagent

Benzoyl chloride:	2 ml	Use once only. Stable for a few days
Pyridine (anhydrous):	38 ml	

Procedure

De-wax paraffin sections and take into absolute ethanol, and then into pyridine. Transfer to either the acetylation or the benzoylation reagent in a tightly capped container. Conditions are as follows:

Mild acetylation: 3h at 37°C.
Complete acetylation: 48h at 37°C.
Benzoylation: 24h at room temperature.

After treatment, wash the slides in absolute ethanol, and take to water.

Effects

Mild acetylation blocks hydroxyl groups. Complete acetylation blocks hydroxyl and amino groups. Benzoylation has the same effect as complete acetylation. These blockades are largely reversed by saponification (see Section 10.11.5).

10.11.3. Sulphation-acetylation

This is the fastest way to block amino groups (Lillie, 1964). Sulphation of hydroxyl groups also occurs.

Reagent

Acetic anhydride:	10ml	Stable for
Glacial acetic acid:	30ml	about 1
Concentrated		week. Use
sulphuric acid:	0.1ml	once only

Procedure

De-wax paraffin sections and take to absolute ethanol and then into glacial acetic acid. Transfer to the reagent for 10 min at room temperature. Wash in running tap water for 5 min.

Effects

Amino and hydroxyl groups are blocked, mainly by sulphation. This effect is reversed by treatment for 3h at 60°C with the reagent used for methylation (see Section 10.11.4, below). Some acetylation may also occur. Histochemical reactions of arginine are unimpaired. It is possible however that N-sulphation of guanidino groups occurs (preventing the attachment of anionic dyes and inducing basophilia) but is reversed by hydrolysis in the strongly alkaline reagents used in specific methods for arginine.

10.11.4. Methylation and desulphation

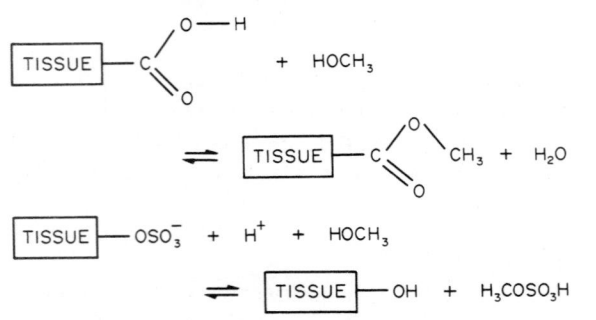

Reagent

Methanol (absolute):	60ml	Use
Concentrated		immediately
hydrochloric acid:	0.5ml	

Procedure

De-wax paraffin sections, take into absolute ethanol, and then into absolute methanol. Transfer to the reagent in a tightly screw-capped coplin jar, which should be no more than two-thirds full, at 60°C for the required time (see below). Rinse in 95% ethanol and bring to water.

Effects

The chemical changes produced depend on the length of time for which the reaction is allowed to proceed:

2h: Carboxyl groups are esterified.
Sulphate groups introduced experimentally are released with re-establishment of original $-NH_2$ or $-OH$.
There is partial hydrolysis of naturally occurring sulphate esters (of carbohydrates) and partial esterification of phosphate-ester groups (of nucleic acids).

24h: Carboxyl and phosphate groups are esterified.
All sulphate esters are hydrolysed. Primary amines are largely converted to tertiary amines. This increases their acidophilia but prevents chemical reactions characteristic of the $-NH_2$ group.

The effects of methylation, **except for removal of sulphate groups,** are reversible by saponification.

10.11.5. Saponification

This procedure is used for reversal of effects of acylation and methylation. The reagent is alkaline and may remove sections from slides. Coating with a film of celloidin is sometimes helpful in preventing losses of sections.

Reagent

Water:	25 ml	Use once only. Stable for a few days
Potassium hydroxide (KOH):	1.0 g	
Absolute ethanol:	75 ml	

Procedure

Take sections to 70% ethanol, then immerse in the reagent for 30 min at room temperature. Rinse with minimum agitation in 70% ethanol, blot to flatten sections if necessary. Rinse in water. Allow the sections to dry in air if they look as if they are only loosely adherent to their slides. Proceed with the staining method.

10.11.6. Deamination with nitrous acid

Reagent

Sodium nitrite (NaNO$_2$):	7.0 g	Use immediately
Water:	94 ml	
Dissolve, then add		
Glacial acetic acid:	6.0 ml	

Procedure

De-wax and hydrate paraffin sections. Immerse in the reagent for 24 h. Wash in running tap water for 5 min.

Effects

Primary amine groups are removed. The guanidino group of arginine is only slightly affected. Most of the deamination occurs in the first 4 h, but it is occasionally necessary to apply the reagent for 48 h.

10.11.7. Benzil blockade of arginine

Reagent

Benzil:	1.6 g	Prepare just before using. Use only once
Ethanol (absolute):	32 ml	
Water:	8 ml	
Sodium hydroxide (NaOH):	0.8 g	

This is a saturated solution. Some benzil may remain undissolved.

Procedure

De-wax paraffin sections and take to 95% ethanol. Transfer to the reagent and leave for 1 h at room temperature. Rinse in 70% ethanol and then in three changes of water.

Effects

Histochemical reactions of arginine are prevented. Acidophilia generally is depressed, indicating that there may also be partial blockage of amino groups.

10.11.8. Sulphydryl blocking reagents

Either of the following will block sulphydryl groups effectively. Sections are hydrated and then treated as described below. Both reagents should be freshly dissolved.

N-ethylmaleimide

Dissolve 625 mg of this compound in 50 ml of phosphate buffer, pH 7.4. Treat sections for 4 h at 37°C, rinse in 1% acetic acid, wash thoroughly in water.

Iodoacetate

Dissolve 0.9 g of iodoacetic acid in 40 ml of water. Add 1.0 M (4%) NaOH until the pH is 8.0. Make up to 50 ml with water. Treat sections for 18–24 h at 37°C. Wash thoroughly in water.

10.11.9. *Reduction of cystine to cysteine*

Many reagents are available for the conversion of —S—S— to —SH. Two are given below. The first of these, sodium thioglycollate, is the most widely used. The second, dithiothreitol, is just as effective and can be used in a less strongly alkaline solution, but is more expensive.

Thioglycollate reduction

Take sections to water and immerse for 15 min in:

Thioglycollic acid
  (HSCH$_2$COOH):   5.0ml
Water: 80ml   ⎫ Use at
4% (1.0N) sodium   ⎬ once
  hydroxide (NaOH):   ⎭
Add dropwise until the pH is 9.5
Then add water: to 100ml

Wash carefully in five changes of water.

Dithiothreitol reduction

Take sections to water and immerse for 30 min in:

Dithiothreitol   ⎫ Stable for
  (Cleland's reagent): 1.5g ⎬ about 3 days
0.03M phosphate   ⎬ in tightly
  buffer, pH 8.0:   50ml ⎬ closed
                         ⎭ container

Wash carefully in five changes of water.

Effects

Disulphides are reduced to thiols, for which histochemical tests become positive.

10.11.10. *Blocking aldehydes and ketones*

Borohydride reduction, explained with practical instructions in Section 10.10.6 (p. 163), removes the carbonyl group of aldehydes and ketones. The following two procedures add aromatic substituents to the carbonyl group. **Phenylhydrazine** blocks aldehydes and ketones, but

m-aminophenol probably combines only with aldehydes (see Section 10.10.3).

Phenylhydrazine

Take sections to water and transfer to the following solution in a screw-capped coplin jar:

Phenylhydrazine: 5ml   ⎫ Prepare
  (**Caution:** poisonous: ⎬ just
  avoid contact with skin)   ⎬ before
Glacial acetic acid: 10ml   ⎬ using
Water: to 50ml   ⎭

Leave the slides in this reagent in the wax oven (about 60°C) for 3h. Rinse in three changes of 70% alcohol and then in water. *Alternatively*, use a fresh 0.5% aqueous solution of phenylhydrazine hydrochloride, for 2h at room temperature, and then wash in water.

In some circumstances, blockade by phenylhydrazine is reversible by acids (including the H$_2$SO$_3$ of Schiff's reagent), and this is exploited in carbohydrate histochemistry (see Chapter 11).

Amine-aldehyde condensation

Take sections to absolute ethanol and then into glacial acetic acid. Transfer to the following solution:

m-Aminophenol: 5.5g
Glacial acetic acid: 50ml

Keeps for about 2 weeks. May be used three or four times.

Leave in this, at room temperature, for 1h. Rinse the slides in 95% alcohol and take to water.

This procedure can be extended to provide a staining method for aldehydes. After reaction with *m*-aminophenol and washing as described above, the slides are immersed for 2min at 0–4°C in a freshly prepared solution containing 150 mg fast black K salt (see p. 69) in 50ml of 0.1M barbitone–HCl buffer, pH 8.0. The slides are then washed in three changes of 0.1M hydrochloric acid (each 5min) to remove excess diazonium salt (and perhaps also to decompose triazenes, see

p. 159), rinsed in water, dehydrated, cleared, and mounted in a resinous medium. Sites of aldehydes are stained black to deep purple. Background staining due to azo-coupling with proteins (see p. 159) also occurs, in various shades of pink and brown. The theory and practice of this technique are discussed in great detail by Lillie (1962).

10.12. Exercises

Theoretical

1. The cytoplasmic granules of the Paneth cells of the intestine are stained strongly by eosin in the H. & E. and the azure–eosin techniques. How would you set about determining which cationic groups are responsible for the acidophilia of the granules?

2. What would you expect to happen when a section of tissue is treated with a proteolytic enzyme of broad specificity such as trypsin? Could such an enzyme be used to prove that an object seen in a routinely stained section is composed of protein?

3. What would be stained in a tissue containing cartilage (whose matrix is rich in sulphated proteoglycans) subjected to the following treatments:

(a) methylation for 24h at 60°C; (b) sulphation in ether–sulphuric acid reagent; (c) staining with 0.1% azure A at pH 1.0?

Is it possible, using this or any other procedure, to demonstrate selectively the hydroxyl groups of proteins?

4. How may azo dyes be formed at the sites of (a) aromatic amino acids, (b) amino groups, (c) ketones?

11

Carbohydrate Histochemistry

Although "carbohydrate" applies to all the sugars and their derivatives, the only such substances available in sections of fixed tissue are those in which the sugars form parts of lipids, nucleic acids and mucosubstances. The glycolipids are considered in Chapter 12 and the nucleic acids in Chapter 9. **Mucosubstance** is a collective term that includes **polysaccharides, glycoproteins** and **proteoglycans.**

11.1. Constituent sugars of mucosubstances

Mucosubstances are identified histochemically by virtue of the properties of their constituent sugars. The monosaccharide residues most commonly encountered in polysaccharides, glycoproteins, and proteoglycans will now be described. The chemistry is explained at greater length by Brimacombe & Webber (1964), Gottschalk (1972), and Cook & Stoddart (1973). Textbooks of biochemistry also contain accounts of carbohydrate chemistry, and an introductory review intended for histochemists is given by Barrett (1971).

11.1.1. Structural formulae

The structural formulae are shown as rings that lie in a plane perpendicular to the paper. The thickened lines represent bonds in the side of the ring that is nearer to the reader. All these sugars exist as six-membered pyranose rings. The numbering system is shown below for β-D-glucose. In the α-anomers of sugars of the D-series, the hydroxyl group at position C1 is directed downwards. The formulae for the L-enantiomers are obtained by envisaging the images in a mirror held **in the plane of the ring**. In α-L-sugars, therefore, the hydroxyl at position C1 is directed upwards and C6 lies below rather than above C5. **Anomers** differ only in the configuration at position C1. **Epimers** differ in the direction of the hydroxyl group at one carbon atom other than C1. Thus, α-D-glucose and β-D-glucose are anomers. Galactose and mannose are epimers of glucose, but not of one another.

In the complex carbohydrates, monosaccharide units are joined together by **glycosidic** linkages. Each unit or residue may be called a glycosyl group. Carbon atom C1 of one sugar is connected via an oxygen atom to one of the carbon atoms (most commonly C3, C4, or C6) of another sugar. In the names of glycosides, the linkage is indicated in an abbreviated form such as α-1→4, which shows which carbon atoms are joined and what the configuration is at C1. It should be noted that the formulae as they are printed here do not accurately represent the shapes of either the sugar units or the larger molecules. Many of the bonds have to be bent or distorted in order to make the structures fit tidily onto the page.

11.1.2. *Some monosaccharide units*

All but xylose and NANA are hexoses and their carbon atoms are numbered as shown for β-D-glucose.

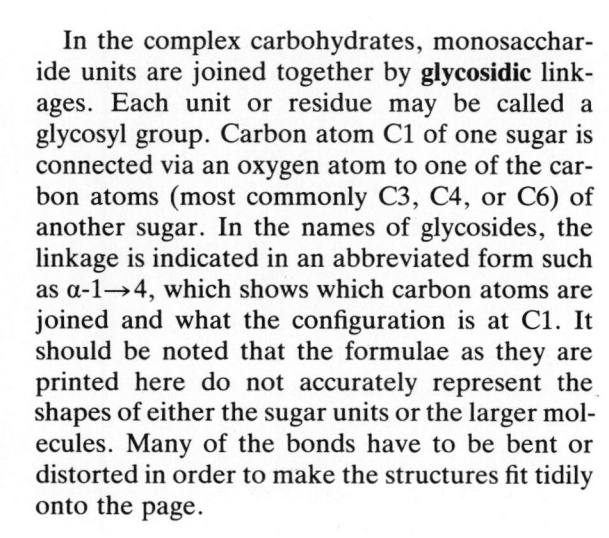

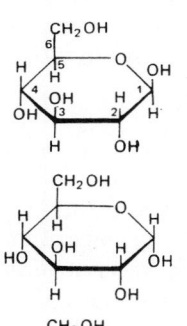

β-D-glucose. β-D-Glc. This is the easiest structure to remember since the hydroxyl groups are placed alternately above and below the ring

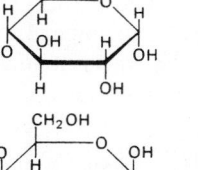

α-D-glucose. α-D-Glc. Differs from the β-anomer only in that the hydroxyl at C1 is directed downwards

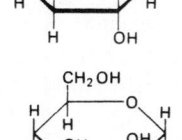

β-D-galactose. β-D-Gal. The C4 epimer of β-D-glucose. A common component of glycoproteins

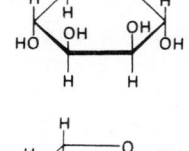

α-D-mannose. α-D-Man. The C2 epimer of α-D-glucose. Abundant in glycoproteins

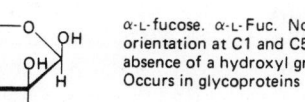

α-L-fucose. α-L-Fuc. Note the orientation at C1 and C5 and the absence of a hydroxyl group on C6. Occurs in glycoproteins

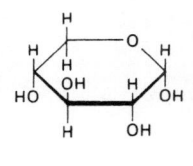

α-D-xylose. α-D-Xyl. This sugar commonly joins long polysaccharide chains to the side-chain hydroxyl group of serine or threonine of the protein core of glycoprotein and proteoglycan molecules. It is never present in large quantities in mucosubstances

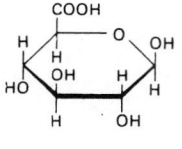

β-D-glucuronic acid. β-D-GlcUA. This is a β-D-glucose in which the primary alcohol group at C6 has been oxidized to carboxyl group, which ionizes as a weak acid: $-COOH \rightleftharpoons -COO^- + H^+$. Occurs in some proteoglycans

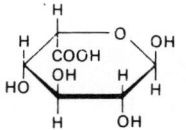

α-L-iduronic acid. α-L-IdUA. Closely similar to β-D-glucuronic acid, differing only in the configuration at C5. Note that the D-enantiomer would have a very different structure from that of any of the four glucuronic acids

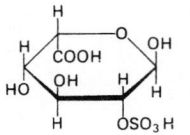

α-L-iduronic acid-2-sulphate. α-L-IdUA-2-OSO$_3$H. The hydroxyl group at position C2 of α L-iduronic acid is esterified by sulphuric acid

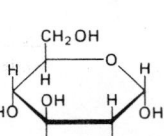

α-N-acetylglucosamine. α-D-GlcNAc. Also known as 2-acetamido-2-deoxy-α-D-glucose. The group $-NHCOCH_3$ replaces $-OH$ at position C2 of α-D-glucose

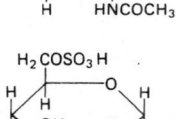

β-N-acetyl glucosamine-6-sulphate. β-D-GlcNAc-6-OSO$_3$H. Esterified by sulphuric acid at C6

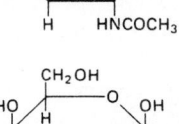

β-N-acetyl galactosamine. β-D-GalNAc. Also known as 2-acetamido-2-deoxy-β-D-galactose. The group $-NHCOCH_3$ replaces $-OH$ at position C2 of β-D-galactose

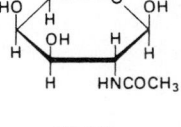

β-N-acetylgalactosamine-4-sulphate. β-D-GalNAc-4-OSO$_3$H. The hydroxyl group at C4 of the preceding sugar is esterified by sulphuric acid. This sulphate ester group ionizes as a strong acid: $-OSO_3H \rightarrow -OSO_3^- + H^+$

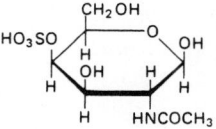

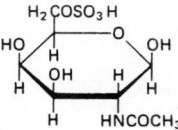

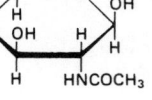

β-N-acetylgalactosamine-6-sulphate. β-D-GalNAc-6-OSO$_3$H. Another strongly acid sulphated sugar

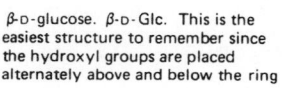
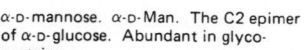

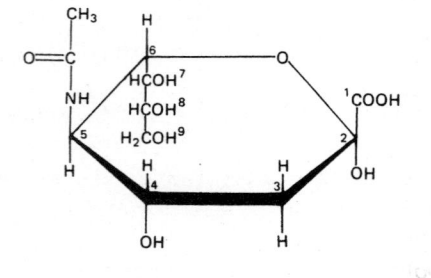

2-Sulphoamino-2-deoxy-α-D-glucose-6-sulphate. α-D-GlcNSO$_3$H-6-OSO$_3$H. This carries two strong acid groups, the sulphate ester on C6 and the —NHSO$_3^-$ group on C2. It occurs in heparin

α-N-acetylneuraminic acid (NANA). This is the most abundant of the **sialic acids**, which are nonoses (numbering system shown). Other sialic acids differ in the substituents at C5. In mucus, any or all of the hydroxyls at C4, 7, 8 and 9 may be esterified by acetic acid. Like the uronic acids, sialic acids are weak acids. The hydroxyl at C2 is the only one involved in glycosidic linkages.

11.2. Classification and composition of mucosubstances

The macromolecular carbohydrates may be classified on the basis of their chemistry, their behaviour in selected histochemical tests, or their distribution in nature. Pearse (1985) presents a system of classification in which an attempt is made to correlate chemical compositions with histochemically discernible properties, but with the advent of techniques for the demonstration of individual monosaccharides residues (see Section 11.3.3) this scheme is now somewhat outdated. Classifications based solely on staining properties, such as those of Cook (1974) and Culling (1974), are valuable in diagnostic pathology, but they lead to identifications which do not correspond very closely with known chemical entities. Indeed, the mucosubstances known to biochemists cannot all be distinguished by means of histochemical procedures. The differing approaches of the two disciplines have also led to a wealth of confusing terminology. In this account, the recommendations of Reid & Clamp

(1978) are followed as far as possible. Ambiguous terms such as "mucin", "mucoid", "mucopolysaccharide", "mucoprotein", "sialomucin", "sulphomucin", and several others that abound in the literature are avoided.†

The following scheme is not so much a classification as a descriptive list. It includes the major mucosubstances of vertebrate animals and a few others. For each mucosubstance the constituent sugars and the forms of glycosidic linkage are given, together with the principal organic functional groups that are present and the histochemically relevant physical properties.

MUCOSUBSTANCES
(Macromolecular compounds composed in whole or in part of carbohydrate)

POLYSACCHARIDES
Composed entirely of carbohydrate (polyglycosides)

PROTEOGLYCANS
Long polysaccharide chains, covalently attached to a relatively small protein core

GLYCOPROTEINS
Proteins bearing numerous covalently bound oligosaccharide chains

11.2.1. Polysaccharides

The following polysaccharides are all **homopolysaccharides**, composed of single monosaccharide units. The chains may be unbranched, typically when the same glycosidic linkage occurs throughout the molecule, or branched when some of the units are connected through more than one hydroxyl group.

Glycogen. ---D-Glc(α-1→4)D-Glc---. The chain is branched because there are some ---D-Glc(α-1→6)D-Glc---linkages in each molecule of glycogen. This polysaccharide is fairly soluble in water, and is best preserved by alcoholic fixatives.

The only important functional group in glycogen is the hydroxyl group. Adjacent hydroxyls,

---

†The term "mucosubstance", although not recommended by Reid & Clamp (1978), is retained here to embrace all carbohydrate-containing macromolecular compounds.

occur in every monosaccharide unit.

This is known as the **glycol** formation (also called vicinal diol, *vic*-glycol or 1,2-diol) and it is present in many other mucosubstances.

Starch. This plant polysaccharide has the same general chemical structure as glycogen, differing only in the size and conformation of its molecules.

The unbranched form is **amylose**, and the branched type, which is similar to glycogen, is **amylopectin.** (Pectin, the principal polysaccharide of the middle lamellae of plant cell walls, consists largely of galacturonic acid units.)

Cellulose. - - -D-Glc(β-1$\rightarrow$4)D-Glc - - -. This, the principal component of plant cell walls, is absent from animal tissues, with the exception of the exoskeleton in tunicates. It differs from starch and glycogen in being a β- rather than an α-polyglucoside. Cellulose is insoluble in all the reagents commonly used in microtechnique.

Chitin. - - - D-GlcNAc(β-1$\rightarrow$4)D-GlcNAc - - -. This is the principal component of the exoskeleton in insects, crustacea, and various other invertebrates. It also occurs in some fungi and algae. Chitin never occurs in a "pure" state. It constitutes about half the dry weight of exoskeletal material, the balance consisting largely of protein and of insoluble salts of calcium.

11.2.2. Proteoglycans

The carbohydrate components of proteoglycans are **heteropolysaccharides,** each long chain being composed of repeating units of two or more different monosaccharide units. The polysaccharide chains of proteoglycans, considered in isolation, are often called "glycosaminoglycuronans" or **glycosaminoglycans.** The term "mucopolysaccharide" is often used as synonym for either "proteoglycan" or the heteropolysaccharide component. The repeating units always include a nitrogen-containing sugar and an acid sugar. The latter, which may be a uronic acid or a sulphate-ester of a hexose, confers affinity for cationic dyes. The repeating units given below

for some individual proteoglycans are not completely constant: smaller quantities of other monosaccharides are also present in the molecule, and the distribution of sulphate-ester groups is somewhat variable. For longer accounts of these compounds, see Brimacombe & Webber (1964) and Jaques (1978). The latter author proposed a new system of nomenclature, but it is not widely used.

Hyaluronic acid. The repeating unit is: - - - D-GlcUA(β-1$\rightarrow$3)D-GlcNAc(β-1$\rightarrow$4)- - -. This proteoglycan exists in a variable state of polymerization. The polymers of low molecular weight are soluble in water and have low viscosity. The high polymers are more viscous and are less easily extracted by water. Glycol and carboxyl groups are present in the β-D-glucuronic acid component of each repeating unit.

Chondroitin-4-sulphate (synonym: chondroitin sulphate A). The repeating unit is: - - - D-GlcUA(β-1$\rightarrow$3)D-GalNAc-6-OSO$_3$H(β-1$\rightarrow$4) - - -. Insoluble in all commonly used histological reagents and has glycol, carboxyl, and sulphate-ester groups.

Chondroitin-6-sulphate (synonym: chondroitin sulphate C). The repeating unit is: - - - D-GlcUA(β-1$\rightarrow$3)D-GalNAc-6-OSO$_3$H(β-1$\rightarrow$4) - - -. Chemically and physically this is closely similar to chondroitin-4-sulphate, differing only in the position of the sulphate-ester group.

Dermatan sulphate (formerly "chondroitin sulphate B"). The repeating unit is: - - -L-IdUA (α-1$\rightarrow$3)D-GalNAc-4-OSO$_3$(β-1$\rightarrow$4) - - -. This is another sulphated proteoglycan, differing from chondroitin-4-sulphate only in the configuration at one carbon atom of the repeating disaccharide unit.

Keratan sulphate I. The repeating unit is: - - - D-Gal(β-1$\rightarrow$4)D-GlcNAc-6-OSO$_3$H(β-1$\rightarrow$3) - - -.

Keratan sulphate II. The repeating unit is: - - - D-Gal(β-1$\rightarrow$4)D-GalNAc-6-OSO$_3$H(β-1$\rightarrow$3) - - -. These closely related proteoglycans are insoluble in all commonly used histological reagents. Sulphate-ester groups are present but glycol formations are absent, since position C3 of the D-galactose is involved in the glycosidic linkage.

Heparin. The composition of heparin is still not

fully known. The protein core of the molecule bears several long heteropolysaccharide chains, the principal repeating unit of which is

protein-carbohydrate linkages are *O*-glycosidic bonds to the side chain of serine, but there are also some *N*-glycoside linkages to basic amino

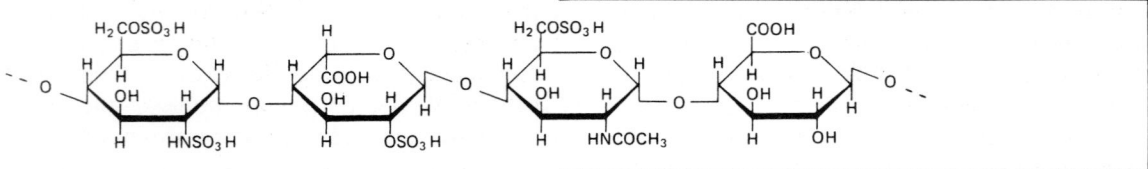

All the glycosidic linkages are α-1→4. The monosaccharide units are variously sulphated derivatives of L-iduronic and D-glucuronic acids and *N*-acetyl-D-glucosamine. Small amounts of D-galactose and D-xylose are also present (see Jeanloz, 1975; Jansson *et al.*, 1975). All the organic functional groups that occur in mucosubstances are present in heparin, but sulphate groups predominate.

Heparin occurs principally in the cytoplasmic granules of mast cells, but it is possible that lower concentrations are also present in other cells and in connective tissues (Jaques *et al.*, 1977).

Heparitin sulphate. This substance may occur both within and on the surface of animal cells, but it has been studied mainly by chemical and pharmacological methods, since it is a by-product of the manufacture of heparin from animal tissues. At least four different compounds are embraced by the name "heparitin sulphate". They are similar to heparin but contain fewer sulphate and more *N*-acetyl groups. Heparitin sulphate was once thought to be a metabolic precursor of heparin, but this is now thought to be unlikely (Silbert *et al.*, 1975; Backstrom *et al.*, 1975). Heparitin sulphate has also been called **heparan sulphate** and heparin monosulphuric acid.

A **proteoglycan molecule** consists of chondroitin, dermatan or keratan sulphate molecules, which are typically 50–200nm long, joined to a core protein that is some 300nm long. Oligosaccharide chains similar to those of glycoproteins (5–15 sugar units) are also attached to the core protein. In an extracellular matrix, the proteoglycan molecules are joined, through a smaller linking protein, to strands of hyaluronic acid of indefinite (>2µm) length (Fig. 11.1). Most of the

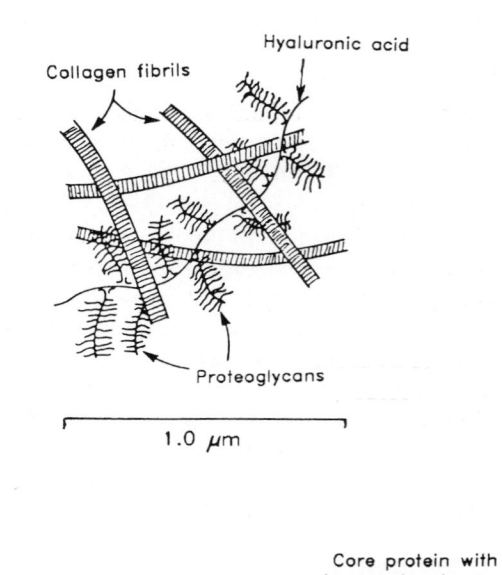

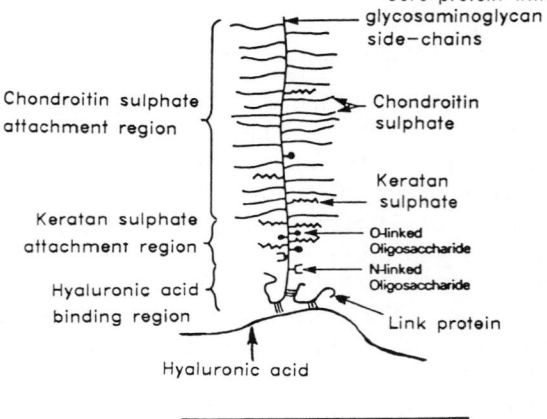

FIG. 11.1. Structural relations of collagen, hyaluronic acid and proteoglycans in a tissue (cartilage matrix). (After Hascall & Hascall, 1981; copied with permission.)

acids. Proteoglycans and hyaluronic acid form a three-dimensional macromolecular mesh that occupies the spaces between collagen fibrils. The extracellular matrices of tissues vary in their proportions of collagen, proteoglycan and hyaluronic acid, and in the relative abundances of the different glycosaminoglycans.

Proteoglycans are preserved chemically intact by all non-oxidizing fixatives, though there may be extraction of some heteropolysaccharide material, such as the lower polymers of hyaluronic acid. It is advisable, when investigating a previously unstudied tissue, to try more than one fixative mixture. A proteoglycan in the rat's brain, for example, has been shown to be histochemically detectable after fixation in formalin-acetic acid or Bouin, but not after fixation in pure formaldehyde solutions (Lai *et al.*, 1975). Cetyl-pyridinium chloride (Williams & Jackson, 1956; see also Exercise 6 at the end of Chapter 2) has been proposed as a special fixative for the precipitation of proteoglycans, though it can depress staining by competition with cationic dyes. Cyanuric chloride (Goland *et al.*, 1967), which combines covalently with hydroxyl groups, has also been used as a fixative for mucosubstances, though it would be expected to prevent many histochemical reactions.

11.2.3. Glycoproteins

These mucosubstances are more numerous and more varied than the polysaccharides and proteoglycans. The oligosaccharide chains may consist of from 2 to 12 monosaccharide units. The commonest sugars in glycoproteins are β-D-galactose, α-D-mannose, α- and β-N-acetylglucosamine, α- and β-N-acetylgalactosamine, α-L-fucose and sialic acids. The two last-named always occupy terminal positions (farthest from the protein). In a branched side-chain, however, these terminal sugars can be nearer to the polypeptide core than some of the residues in the interior of the chain (see Podolsky, 1985). Examples of glycoproteins are:

Serum proteins (including immunoglobulins).

Blood-group specific substances. These occur on the surfaces of erythrocytes.

Secretory products. Both exocrine and endocrine glands secrete glycoproteins. Some of these, especially in the alimentary canal, contain sulphated sugars and sialic acids. Histochemical methods cannot determine whether such substances are single glycoproteins or mixtures.

Constituents of the glycocalyx. Glycoproteins form the "cell coat" on the outside surface of the plasmalemma of every cell. The glycocalyx is an integral part of the cell membrane.

Collagen. This is unusual in that α-D-glucosyl units form a major part of the carbohydrate component. See also Chapter 8.

Amyloid. Deposits of amyloid accumulate in various organs as a consequence of chronic inflammatory diseases. A rare condition known as primary amyloidosis is also encountered occasionally by pathologists. The deposits are of a glycoprotein, which probably contains sialic acids, sulphated sugars, and neutral monosaccharide residues (see Glenner *et al.*, 1980).

Histochemical demonstration of glycoproteins is based on the presence of carboxyl, hydroxyl and sulphate-ester groups. They are preserved by most fixatives, but mixtures containing oxidizing agents (which would attack the glycol formation) should not be used. The carbohydrate components of glycolipids are similar to those of glycoproteins, and the two groups of substances can be confused with one another when frozen sections are used. It is probable that in many sites, especially in mucus that lubricates epithelial surfaces, several different glycoproteins are present. The histochemist can identify and localize various functional groups and some individual monosaccharide residues, but cannot determine whether these occur in the same or different molecules.

The techniques of carbohydrate histochemistry fall into three categories: those in which **dyes** and related reagents are used, **chemical methods**, and methods involving use of **lectins**. All types of method can also be used in conjunction with chemical blocking procedures for reactive groups and with enzymatic degradation of specific mucosubstances. Several techniques exist other than the ones described in this chapter, and it is also possible to demonstrate many mucosubstances by immunohistochemical methods.

11.3. Histochemical methods using dyes

Four techniques will be discussed: the uses of the alcian blue, the observation of metachromasia, staining with a cationic dye in conjunction with a metal salt, and the use of anionic dyes with large molecules.

11.3.1. Alcian blue

This dye (see Chapter 5) binds to carboxyl and sulphate-ester groups at pH 2.5, but only to the latter at pH 1.0. With the more strongly acid solution, carboxyl groups are not ionized and therefore cannot electrostatically attract the cations of the dye. It is therefore possible to identify with some degree of certainty mucosubstances that owe their acidity wholly to carboxyl groups, but it is not possible to tell whether a substance stained at both pH levels owes its acidity only to sulphate groups or to both sulphate and carboxyl groups. If, however, a section is subjected before staining to an adequate treatment with hot, acidified methanol (see Chapter 10, p. 166; also this chapter p. 194), the sulphate-ester groups and probably also most of the sialic acids will be removed, and the carboxyl groups will be esterified. Nothing will be stained by alcian blue. The section can next be subjected to a "saponification" procedure (see p. 195), which will cause hydrolysis of the methyl esters and restore the carboxyl groups. These will regain their stainability by alcian blue at pH 2.5. The sulphate esters, however, will have been irreversibly removed and will no longer be detectable. Thus, provided that at least six sections of the same tissue are available, it is possible to determine whether affinity for alcian blue at pH 1.0 and 2.5 is due solely to sulphate esters or to both sulphate ester and carboxyl groups coexisting at the same site.

Alcian blue has also been used mixed with an inorganic salt (usually magnesium chloride) at various ionic strengths. The cations of the salt compete with those of the dye for binding sites in the tissue. The dye is used at a high pH (5–6) so that the results will not be complicated by incomplete ionization of carboxyl groups. The highly dissociated acids ($-OSO_3H$ in this case) can bind the dye in the presence of high concentrations of the sale, whereas the weaker acids ($-COOH$) cannot. However, the truth of this assertion, the basis of the "critical electrolyte concentration" method, has been questioned (Horobin & Goldstein, 1974), and Horobin (1982, 1988) considers it more likely that dye solutions of increasing ionic strength are bound by substrates of increasing porosity.

It is quite feasible to study acid mucosubstances by staining with cationic dyes other than alcian blue. Some effects of pH on the affinities of thiazine dyes for different anions were briefly reviewed in Chapter 6. The main reason for using alcian blue in carbohydrate histochemistry is the fact that this dye does not usually stain nuclei or cytoplasmic deposits of RNA in sections, though it is known to be able to bind to nucleic acids in solution (Scott et al., 1964).

Studies with molecular models (Scott, 1972b) have revealed that the four tetramethylisothiouronium groups attached to the phthalocyanine ring of alcian blue 8GX make the dye molecule too big to fit between the coils of the DNA helix. Electrostatic attraction between the phosphate groups and the auxochromes is weakened by distance and there can be no close-range interaction (van der Waals forces, etc., see Chapter 5, Section 5.5) between the aromatic rings of the dye and the purine and pyrimidine rings of the DNA. Possibly the access of the large molecules of alcian blue to the phosphate groups of nucleic acids in a fixed tissue is also hindered by the presence of the associated nucleoproteins. Steric hindrance may also explain the failure of this dye to stain RNA. Sometimes, especially in the tissues of very young animals, nuclei are stained by alcian blue at pH 2.5. Such staining may be due at least in part to acid mucosubstances present in the chromosomes of dividing cells (Ohnishi et al.,

1973). Another advantage of alcian blue over most other cationic dyes is that it is not extracted from stained sections by water, alcohol, weak acids, or solutions of other dyes used for counterstaining. The PAS procedure (see below) is frequently applied to tissues already stained with alcian blue.

Solutions required

A. Alcian blue, pH 1.0

> Alcian blue: 1.0g
> 0.1M hydrochloric acid: 100ml
>> Stability variable. Sometimes the dye precipitates after 2 or 3 weeks. Other batches are stable for at least 1 year. Filter before using.

B. Alcian blue, pH 2.5

> Alcian blue: 1.0g
> 3% aqueous acetic acid: 100ml
>> Keeps for several years.

Procedure

1. De-wax and hydrate paraffin sections.
2. Stain in solution A **or** solution B for 30 min.
3. Rinse in 0.1 M HCl or 3% acetic acid (the solvent for the dye), then wash in running tap water for 3 min.
4. (Optional.) Apply a pink or red counterstain if desired.
5. Dehydrate in graded alcohols. Alcian blue is not removed by alcohol, but the counterstain may be differentiated.
6. Clear in xylene and mount in a resinous medium.

Result

All acid mucosubstances are stained at pH 2.5. Only sulphated mucosubstances are stained at pH 1.0.

Notes

1. Not all batches of dye are satisfactory. Use a sample designated 8GX, 8GS or 8GN. A solution of alcian blue that is ineffective at room temperature will sometimes perform adequately at 58–60°C.
2. Mucosubstances with carboxyl or sulphate-ester groups may be distinguished from one another if control sections are methylated and saponified (See Section 11.7). Neuraminidase (Section 11.6) and mild acid hydrolysis (Section 11.7) may be used to identify glycoproteins that owe their acidity to sialic acids.
3. Ordinary red cationic dyes, especially safranine (Spicer, 1960), may be used in conjunction with alcian blue. Differential staining effects have been described in carbohydrate-containing structures such as mast cell granules (Jasmin & Bois, 1961; Combs *et al.*, 1965), but the tinctorial variations have been shown by Tas (1977) to have no histochemical significance. For further information concerning the histochemical properties of alcian blue, see Scott *et al.* (1964) and Quintarelli *et al.* (1964a,b).

11.3.2. Metachromasia

When a cationic dye imparts its own colour to an object, the staining is said to be **orthochromatic**. In some circumstances the dye ions bind to a substrate in such a way as to alter the wavelength of the absorbed light such that the observed colour of the stained object is different from that of the dye. This phenomenon is known as **metachromasia** and substrates which are stained metachromatically are said to be **chromotropic**. In most cases the metachromatic colour of the dye is of a longer wavelength than the orthochromatic colour. Blue dyes, such as thionine and toluidine blue, stain chromotropic materials in shades of red and purple. The red and purple colours produced by such dyes have been called γ- and β-metachromasia, respectively, but this distinction is probably of little importance. Only γ-metachromasia is of interest in the histochemistry of mucosubstances. The change in colour is due to a shift to shorter wavelengths (a hypsochromic shift) in the absorption spectrum of the dye and to an associated reduction in the intensity of the colour.

The metachromatic effect is produced when the coloured ions of a dye are brought in close proximity to one another. This occurs when anionic radicals of the substrate are close together, as in some proteoglycans. The dye-substrate complex of a metachromatically stained object has the form

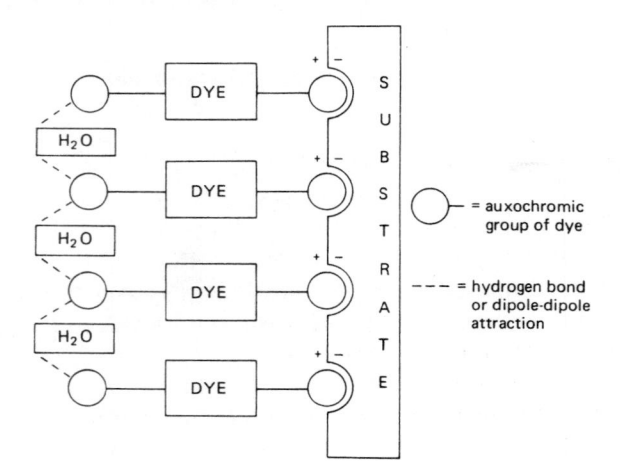

The water molecules interposed between the stacked dye ions are believed to be necessary for modifying the distribution of electrons in the chromophoric system in such a way as to reduce the wavelength at which light is maximally absorbed (Bergeron & Singer, 1958). The dyes with which metachromatic effects can be obtained are thiazines, oxazines, azines, and xanthenes (see Chapter 5). These have planar molecules and can be formulated with their positively charged auxochromic groups on either side of the systems of fused rings. When the auxochromes are more bulky than the $—\overset{+}{N}(CH_3)_3$ group, metachromatic staining does not occur (Taylor, 1961), probably because there is not room for the interposition of water molecules. The bound water may resist extraction by dehydrating agents: indeed, it is generally agreed that metachromasia must persist in dehydrated, cleared preparations if it is to have any significance in relation to macromolecular carbohydrates (Kramer & Windrum, 1955).

Another cause of metachromasia may be the formation of dimers of dye molecules in the staining solution and the subsequent attachment of these dimers to anionic sites in the tissue (see Schubert & Hamerman, 1956; Wollin & Jaques, 1973 for discussion). This does not seem a very probable mechanism, however, since dye solutions display metachromasia only when they are concentrated, but metachromatic histological staining is easily obtained from very dilute solutions.

Acid mucosubstances are not the only sources of metachromasia found in animal tissues. Nucleic acids are also chromotropic in some circumstances, though their metachromasia usually reverts to orthochromasia after dehydration. Despite the low histochemical specificity of metachromatic staining with cationic dyes, the method is useful for the morphological study of structures known to contain proteoglycans, such as mast-cell granules and the intercellular matrices of cartilage and other connective tissues.

Solution required

A 0.05–0.5% (w/v) solution of toluidine blue O (C.I. 52040) or azure A (C.I. 52005), or thionine (C.I. 52000) in 1% (v/v) aqueous acetic acid. Keeps indefinitely, but filter before each use.

Procedure

1. De-wax and hydrate paraffin sections.
2. Stain in toluidine blue for 1–5 min (see *Note 1*).
3. Wash in water.
4. Dehydrate in 70%, 95%, and two changes of absolute ethanol (see *Note 2*).
5. Clear in xylene and mount in a resinous medium.

Result

Orthochromatic colour (nuclei, cytoplasm of some cells, Nissl substance of neurons)—blue. Metachromatic colour—red.

Notes

1. If staining is excessive, add more acetic acid to the toluidine blue. The pH of the solution should be approximately 4.0.
2. The dye is differentiated by the 70% ethanol. The sections should be **pale** blue before being cleared. If the staining is initially weak, blot the sections after washing and dehydrate in two changes (each 3–5 min) of *n*-butanol.

11.3.3. Heath's aluminium-basic dye method

A dilute (approximately 10^{-3}M) solution of a cationic dye in 0.1–0.5M aluminium sulphate gives selective staining of structures that contain sulphated mucosubstances. Heath (1961) investigated this phenomenon thoroughly. The most effective dyes were found to be *N*-methyl azines and thiazines, including neutral red and toluidine blue. Heath also found that the same result could be obtained by treating sections with the aluminium salt before the dye, and that aluminium could be replaced by other metals that readily formed coordination compounds. Specificity for sulphate esters was confirmed by ascertaining that the stainability was prevented by methylation of the tissue, and not restored by saponification (see also Chapter 10 and later in this chapter). Strong acid groups artificially introduced into tissues (by sulphation of hydroxyls or oxidation of cystine; see Chapter 10)) were also found to be stained by aluminium-basic dye combinations.

Aluminium ions are bound, by coordination or ionic attraction, to all parts of a section immersed in a solution of $Al_2(SO_4)_3$. It seems probable that the metal forms complexes with carboxyl groups of proteins and carbohydrates, and perhaps also with the phosphate groups of nucleic acids, leaving sulphates as the only anionic groups free to combine with a cationic dye.

Solutions required

Aluminium-dye solution

Neutral red (C.I. 50040) or	
Toluidine blue O (C.I. 52040):	0.1g
Water:	100ml
Aluminium sulphate $(Al_2(SO_4)_3 . 18H_2O)$:	5.0g

Boil, allow to cool, and filter to remove insoluble material; then add:

5% (w/v) aqueous $Al_2(SO_4)_3 . 18H_2O$: to make 300ml.

70% ethanol, for differentiation of stained sections.

Procedure

1. De-wax and hydrate sections. Remember to remove mercury deposits, if necessary.
2. Stain for 5–30 min.
3. Rinse in water
4. Differentiate in 70% ethanol for 20–30s, or until the background (nuclei, cytoplasm, collagen) is colourless.
5. If desired, apply a suitably contrasting counterstain, such as eosin or fast green FCF (see Chapter 6).
6. Complete the dehydration, in 95% and 2 changes of 100% alcohol, clear in xylene and cover, using a resinous mounting medium.

Result

Sites of sulphated mucosubstances (mast cells, cartilage matrix, some goblet cells, etc.) red-purple with toluidine blue or red with neutral red.

Note

Heath (1962) strongly recommends a dye he called "nuclear fast red" for this method. Kernechtrot (C.I. 60760), the anthraquinone dye usually known as nuclear fast red, is not suitable for this method because its aluminium complex stains nuclei. Confusion about the identity of a suitable dye may account for the lack of popularity of this method, which is simple and reliable and deserves to be used more widely.

11.3.4. Dye methods for glycogen and amyloid

Anionic dyes that exist in solution as large molecules or aggregates of molecules are used for staining collagen, as explained in Chapters 5 and 7. Some of these dyes can also be used for the demonstration of polysaccharides. Neutral polysaccharides like **glycogen** have no charged groups, so the binding and retention of a dye necessarily occur by non-ionic mechanisms. The classical stain of this kind for glycogen is Best's Carmine technique, in which the dye is dissolved in a mixture of water and alcohol. The structural formula of carmine (Chapter 5, p. 86) shows a large molecule with numerous hydrophilic substituents (carboxyl, phenolic hydroxyl, and the hydroxyls of the sugar moieties). Glycogen, a macromolecule, also bears great numbers of hydroxyl groups. Horobin (1982) has suggested that this may be one of the few instances in which hydrogen bonding plays an important part in a staining mechanism. Evidence in support of this idea comes from the observation that the intensity of staining of glycogen by carmine is reduced if the proportion of water in the solvent of the dye is increased. Water competes for hydrogen bonding sites on the molecules of dyes and substrates, as explained in Chapter 5 (p. 55).

The staining of **amyloid** by Congo red and other direct cotton dyes is non-ionic, and has been attributed to hydrogen bonding to the carbohydrate component of the substrate (Puchtler et al., 1964). However, the staining of amyloid by these dyes is retarded by addition of alcohol, and under certain conditions of application Congo red can simultaneously stain amyloid and hydrophobic objects such as elastic fibres (Mera & Davies, 1984). Hydrophobic and van der Waals forces are therefore probably the most important factors involved in the binding of the dye (see also Chapter 5, Section 5.5). Amyloid stained by Congo red acquires a conspicuous **dichroism:** the deposits show as bright green areas on a black background when viewed in a polarizing microscope with crossed polarizer and analyser. (For information on polarizing microscopy, see James, 1976.) Other materials stained by Congo red, such as collagen, do not have this property. The dichroism indicates crystal-like alignment of dye molecules in a highly ordered structure, in this case the β-pleated sheet structure of the protein component of amyloid (Wolman & Bubis, 1965; see also Glenner et al., 1980; Pearse, 1985).

Different proteins are present in the amyloids associated with different diseases. These can be distinguished on the basis of the preservation or destruction of their Congo red dichroism after such treatments as autoclaving or treatment with alkaline guanidine. Immunohistochemical methods specific for the proteins are preferred, however, when antisera are available (Elghetany & Saleem, 1988).

Amyloid can also be stained with basic dyes. These often give unusual metachromatic colours, presumably due to aligned glycosaminoglycan chains within the proteinaceous matrix. Thus, although amyloid contains carbohydrate, its staining properties are determined largely by the physical structure and orientation of the protein molecules.

11.3.4.1. Best's carmine for glycogen

An alcoholic fixative should be used to preserve glycogen (see Chapter 5). Diffusion artifacts occur with aqueous fixatives.

Solutions required

A. *Nuclear counterstain*

An iron- or alum-haematoxylin (see Chapter 6).

B. Stock solution of carmine

Water:	60 ml
Carmine:	2.0 g
Potassium carbonate (K_2CO_3):	1.0 g
Potassium chloride:	5.0 g

Boil for 5 min, in a 500 ml flask (large volume because of effervescence). Cool, filter, and add the filtrate to 20 ml of strong ammonium hydroxide (S.G. 0.880).

Store in a dark place. Keeps for 2–3 months

C. Working staining solution

Stock solution (above):	15 ml
Ammonium hydroxide (S.G. 0.880):	12.5 ml
Methyl alcohol:	12.5 ml

Make up as required. Keeps for about one week, and can be re-used, but its potency declines with time.

D. Best's differentiator

Absolute methyl alcohol:	40 ml
Absolute ethyl alcohol:	80 ml
Water:	100 ml

Stable, but usually mixed before using

Procedure

1. De-wax the slides and coat with celloidin as described in Chapter 4, p. 44. Take to water.
2. Stain the nuclei with solution A (see Chapter 6). Wash in water.
3. Stain in the working solution of carmine (C) for 15 min (more or less, according to intensity obtained).
4. Transfer (without rinsing) directly to Best's differentiator (solution D). Agitate gently for about 10 s.
5. Wash in 2 changes of 95% alchohol and examine. If necessary, repeat the differentiation (step 4 above).
6. Complete the dehydration in 2 changes of 100% alcohol, clear in xylene and mount, using a resinous medium.

Result

Glycogen bright crimson. Fibrin, mast cell granules and some mucus pink. Nuclei black or blue, according to counterstain.

Notes

1. Control sections may be incubated in amylase, which removes glycogen (see p. 192 for instructions.
2. The periodic acid-Schiff method (pages 182–188) also stains glycogen, and is frequently used for the purpose.

11.3.4.2. Congo red method for amyloid

This method (Highman, 1946) is suitable for paraffin sections of formaldehyde-fixed material. The differentiation in alkali removes most of the dye that is initially bound to collagen and cytoplasm.

Solutions required

A. Congo red solution

Congo red:	1.0 g
50% (v/v) ethanol:	200 ml

Keeps indefinitely

B. Differentiating solution

Ethanol:	160 ml
Water:	40 ml
Potassium hydroxide:	0.4 g

Dissolve, then add water if necessary to bring the final volume to 200 ml. (Keeps for months, but replace if it goes brown.)

C. Nuclear stain

An iron-haematoxylin or an alum-haematoxylin (see Chapter 6).

Procedure

1. De-wax and hydrate the sections.
2. Stain with Congo red (Solution A) for 1 to 5 min, until sections are deep red.
3. Differentiate in Solution B, 2–10 s, then rinse in copious distilled water and exam-

ine. The only red colour remaining in the section should be in elastic fibres and laminae and in strongly acidophilic structures such as the cytoplasmic granules of eosinophils and Paneth cells. Amyloid deposits are also stained, but may be quite inconspicuous in the wet section. The differentiation may be repeated if necessary, and the staining may be repeated if differentiation has been excessive.

4. Counterstain nuclei with an iron- or alum-haematoxylin (see Chapter 6).
5. Wash in water, dehydrate in acetone, clear in xylene and cover, using a resinous mounting medium.

Result

Amyloid deposits red. With polarizing microscopy, green dichroism in amyloid. This may reveal deposits that are hardly visible with ordinary illumination. Nuclei blue or black, according to counterstain.

11.4. Chemical methods—the periodic acid-Schiff method

The chemical technique most extensively used in carbohydrate histochemistry is the periodic acid-Schiff (PAS) reaction which, as will be seen below, is positive with structures containing neutral† hexose sugars and/or sialic acids. Chemical methods are also available for sulphate-ester groups, for sialic acids, and for some amino sugars, though the methods are often not of high specificity. These latter techniques will not be discussed here.

In the PAS method sections are treated with periodic acid, which oxidizes glycols to aldehydes. The aldehydes are then rendered visible by reaction with Schiff's reagent.

---

†The term "neutral sugar" though not strictly accurate in the chemical sense, is used by histochemists for monosaccharide residues that do not have sulphate-ester, carboxylic acid, or nitrogen-containing functional groups. Glucose, galactose, mannose, and fucose are the principal neutral sugars present in mucosubstances.

11.4.1. Periodate oxidation

Periodic acid, $HIO_4.2H_2O$, is used in histochemistry as a 1% aqueous solution (0.044M) and is normally allowed to act upon sections of tissue for 5–10min at room temperature, though oxidation for 30min is necessary for some specimens (Culling & Reid, 1977). The sodium salt of the acid may also be used. The effect of this treatment is a selective oxidation by the periodate ion of hydroxyl groups attached to adjacent carbon atoms (i.e. glycols), with fission of the intervening carbon-to-carbon bond and production of two aldehydes:

$$\begin{matrix} H-\overset{|}{C}-OH \\ | \\ H-\overset{|}{C}-OH \\ | \end{matrix} + IO_4^- \rightarrow \begin{matrix} H-\overset{|}{C}=O \\ \\ H-\overset{|}{C}=O \\ | \end{matrix} + IO_3^- + H_2O$$

Where there are three neighbouring carbon atoms bearing hydroxyl groups a similar reaction occurs but with the elimination of a molecule of formic acid:

$$\begin{matrix} H-\overset{|}{C}-OH \\ | \\ H-\overset{|}{C}-OH \\ | \\ H-\overset{|}{C}-OH \\ | \end{matrix} + 2IO_4^- \rightarrow \begin{matrix} H-\overset{|}{C}=O \\ \\ \\ \\ H-\overset{|}{C}=O \\ | \end{matrix} + HCOOH + 2IO_3^- + H_2O$$

Glycol groupings are present in the neutral sugars and in the uronic and sialic acids and some of the N-acetylamino sugars.

Periodic acid can also oxidize the α-amino-alcohol formation

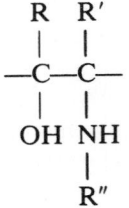

where R, R', and R'' may be hydrogen or alkyl radicals (Nicolet & Shinn, 1939). This configuration occurs in the amino acids serine and threonine (but not when these are incorporated into peptide linkages), in the side-chain of hydroxylsine, and also in sphingosine, an amino alcohol present in certain lipids. The only common sugars that are α-amino alcohols are glucosamine and galactosamine, but these do not occur in significant quantities in tissues. The nitrogen-containing sugars of proteoglycans and glycoproteins, *N*-acetylglucosamine and *N*-acetylgalactosamine, contain the structural arrangement

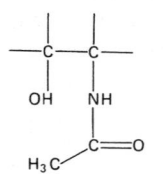

They are α-*N*-acetylamino alcohols and, as such, are almost completely resistant to oxidation by periodate (Nicolet & Shinn, 1939; Carter *et al.*, 1947). The *N*-acetyl hexosamines could be oxidized by periodate if they formed glycosidic linkages at position C6, leaving C3 and C4 with free hydroxyl groups. In mucosubstances these sugars are connected to others through glycosidic linkages involving position C3 or C4, so they are never responsible for stainability by the PAS method.

Lead tetraacetate (Glegg *et al.*, 1952) and sodium bismuthate (Lhotka, 1952) have been used for the same purposes as periodic acid. They act in the same way, but as histochemical reagents they have not been investigated as thoroughly as periodic acid.

If the histochemical situation were as simple as the foregoing account might lead one to believe, all carbohydrate-containing structures would be expected to yield aldehydes with periodic acid. This is not the case, however. It has been shown that proteoglycans (consisting of uronic acids and acetylhexosamines, sulphated or not) are **not** PAS-positive (Hooghwinkel & Smits, 1957). In sections of fixed tissue, periodic acid produces aldehydes from glucosyl, galactosyl, mannosyl, and fucosyl residues (Leblond *et al.*, 1957) and sialic acids. These are components of glycogen and of glycoproteins.

Proteoglycans are PAS-negative because the usual treatment with periodic acid fails to oxidize the glycol formations at C2–C3 in glucuronic and iduronic acids. Scott & Harbinson (1969) have shown that uronic acids are not attacked by periodic acid on account of repulsion of the periodate ion by the carboxylate anions and perhaps also by the nearby sulphate-ester anions, when these are present. This electrostatic effect can be overcome by using periodate for a much longer time and at a higher temperature than usual. Although the uronic acid components of mucopolysaccharides yield aldehydes under these more rigorous conditions, the *N*-acetyl hexosamines are still unaffected. Scott & Dorling (1969) have developed a modified PAS technique, based on the principle outlined above, for the selective demonstration of uronic acid-containing mucosubstances (i.e. proteoglycans). A preliminary treatment with periodate is followed by reduction with sodium borohydride:

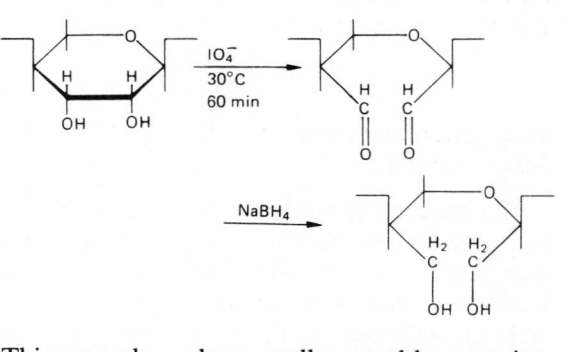

This procedure changes all neutral hexoses into compounds that can no longer yield aldehydes by reaction with periodate. In the next stage of the technique, the sections are again exposed to periodate, but for a time sufficient to cause oxidation of the glycol groups of uronic acids:

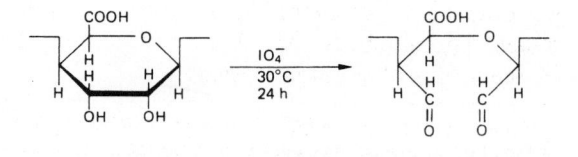

The aldehyde groups produced by the second oxidation are demonstrated with Schiff's reagent. Scott & Dorling's modified PAS method therefore demonstrates mucosubstances whose uronic acid residues have free hydroxyl groups at positions C2 and C3. From Section 11.2.2 of this chapter it can be seen that these are hyaluronic acid, chondroitin-4-sulphate, chondroitin-6-sulphate, dermatan sulphate, and heparin. The polysaccharides, the glycoproteins, and the keratan sulphates will not be stained.

The versatility of periodic acid as a histochemical reagent is further exemplified in its ability, under appropriate conditions, to produce aldehydes selectively from sialic acid residues. The only potentially periodate-reactive part of a sialic acid glycosidically linked at position C2 (see formula for NANA, p. 172) is the side-chain (C7, C8, and C9) attached at C6, which bears three adjacent hydroxyl groups. The side-chain reacts with periodate much more rapidly than do the glycols of hexoses (see Hughes, 1976). By using very dilute periodic acid for a short time it is possible to oxidize only the sialic acid residues of glycoproteins:

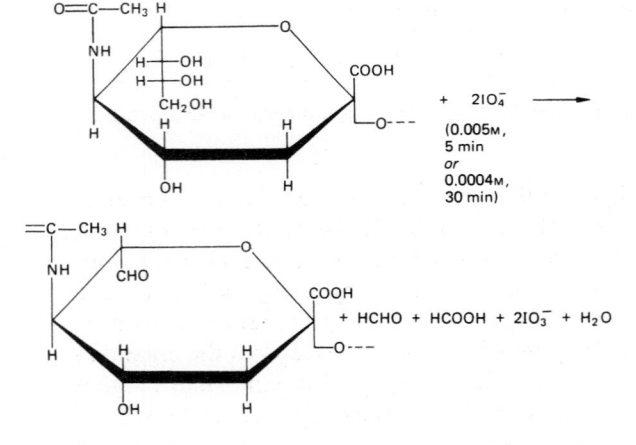

The resultant aldehyde can be demonstrated either with Schiff's reagent (Roberts, 1977) or with a fluorescent hydrazine (Weber *et al.*, 1975). Volz *et al.* (1986, 1987) have determined that a strongly acidified very dilute solution of periodic acid used at 40°C provides optimum selective oxidation of sialic acid side-chains.

In some intestinal glycoproteins, *O*-acetyl groups replace the hydroxyls at position C7, C8,

or C9 of NANA. These acetylated sialic acids can be recognized histochemically by means of a sequence of treatments.

(1) Oxidation by periodic acid (glycol→aldehyde). It is important to convert all available glycol groups to aldehydes; thus, sections are oxidized for 30–120min in 0.044M periodic acid.

(2) Reduction by borohydride (aldehyde→primary alcohol). (PAS staining carried out after this reduction should give a negative result if the initial oxidation has been adequate.)

(3) Saponification with alcoholic potassium hydroxide (acetyl ester→hydroxyl).

(4) Second oxidation with periodate (glycol→aldehyde).

The production of aldehydes is sought, with Schiff's reagent, after stages 1 and 4.

The side-chains of the sialic acids react as shown in Fig. 11.2.

Unsubstituted and C9-acetylated sialic acids are Schiff-positive only after the first periodate oxidation and cannot be distinguished from one another. The C7-acetylated sialic acid is Schiff-positive after both oxidations and the C8-acetylated compound is Schiff-positive only after the second oxidation. If acetyl groups are present on any two or all three of the positions C7, C8, and C9, positive Schiff reactions will be obtained only after the second oxidation with periodate. It must be emphasized that this histochemical analysis (Reid *et al.*, 1978) involves oxidation of sialic acids and neutral hexoses, so it is applicable only when it has already been proved that the sialic acids are the sole substances responsible for stainability by the PAS method at the sites being investigated.

Another method, related to the procedure described above, is that of Culling *et al.* (1976). In this, the first oxidation with periodic acid is followed by staining with a "pseudo-Schiff" reagent (see Chapter 10, p. 161) made from thionine. The sections are then saponified, oxidized for a second time with periodic acid, and stained with conventional Schiff's reagent. Sialic acids which bear no *O*-acetyl groups or are acetylated only at C9 are stained blue. Those acetylated at C8 are red and those acetylated at C7 or at more

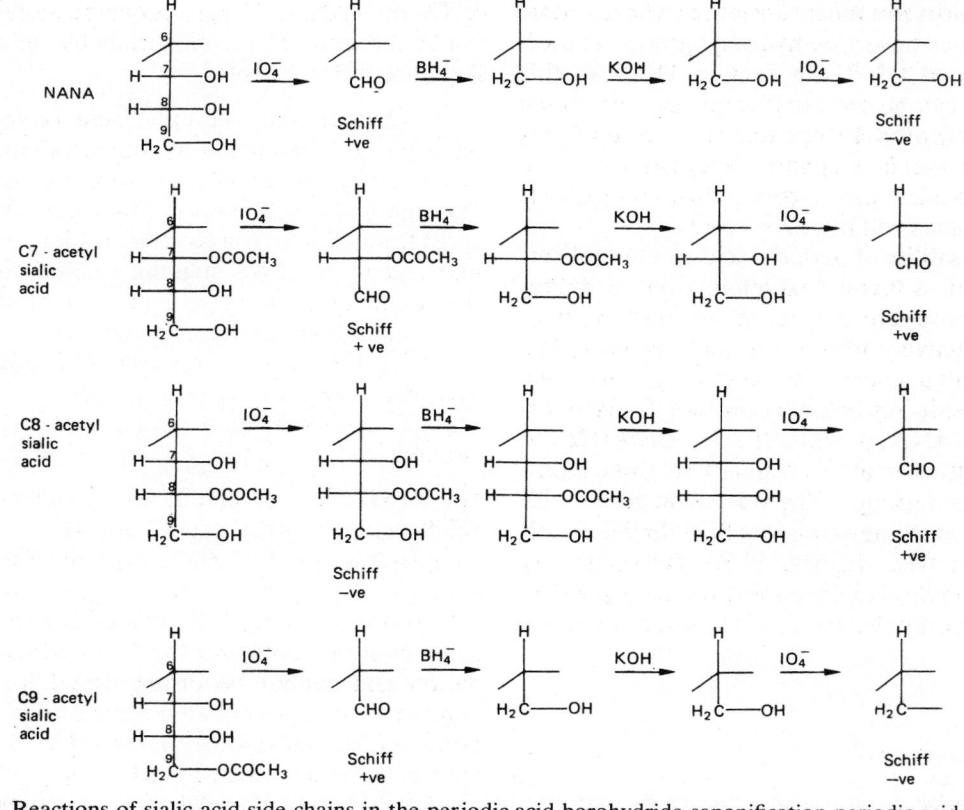

Fig. 11.2. Reactions of sialic acid side-chains in the periodic acid-borohydride-saponification-periodic acid-Schiff staining method.

than one of the positions C7, C8, and C9 are purple.

Two strategies are available to enhance the selectivity for sialic acids of these PAS procedures. The simpler is the use of selective (mild) oxidation, which leaves the neutral hexoses unchanged. The other procedure makes use of the fact that the products of periodate oxidation of sialic acids and neutral hexoses react differently with phenylhydrazine. The single aldehyde group resulting from oxidation of a sialic acid side-chain gives an ordinary phenylhydrazone (see Chapter 10), which is unstable in an aqueous acidic medium such as Schiff's reagent. The aldehyde is regenerated, and promptly reacts to give a coloured product. In contrast, the two joined aldehydes produced by oxidation of a hexose (see p. 183) react with one molecule of phenyl-

hydrazine to form a cyclic compound. The latter is not affected by Schiff's reagent. Consequently, treatment of a section of an animal tissue with periodic acid, followed by phenylhydrazine, and then by Schiff's reagent, gives selective staining at the sites of sialic acids (except for the unoxidizable C8-acylated ones). If a Schiff-like reagent made from thionine is used to colour the sites of sialic acids blue, it is then possible to stain the section again by the ordinary PAS method, to show the distribution of neutral hexoses in red. If neutral hexoses are absent, or if selective (mild) periodate oxidation is used, a saponification can be interposed between the treatment with thionine-Schiff and the second periodate oxidation. A second PAS routine will then stain the C8-acylated sialic acids (Park et al., 1987; Reid at al., 1987).

11.4.2. PAS procedures

11.4.2.1. Standard PAS method

Solutions required

A. *Periodic acid solution*

> Periodic acid (HIO$_4$.2H$_2$O):   2.0g
> Water: 200ml

Keeps for several weeks and may be used repeatedly, but should be discarded if it goes brown.

B. *Schiff's reagent*

See under the Feulgen technique (p. 137).

C. *Solutions for counterstaining*

A suitable sequence: alum-haematoxylin for nuclei, with fast green FCF as a counterstain. For critical evaluation of the PAS reaction it is preferable to use no counterstain.

Procedure
1. Allow the Schiff's reagent to warm to room temperature.
2. De-wax and hydrate paraffin sections. (See *Note 1*).
3. Oxidize for 10–30 min in periodic acid (solution A).
4. Wash in running tap water for 3min.
5. Immerse in Schiff's reagent (solution B) for 20min.
6. Transfer to copiously running tap water and leave to wash for 10min (see *Note 2*).
7. Apply counterstains, as desired.
8. Dehydrate in graded alcohols, clear in xylene, and cover, using a resinous mounting medium.

Result

Hexose-containing and sialic acid-containing mucosubstances pink to bright purplish red. In an adequately stained preparation reticulin, basement membranes, and mucous glands should stand out sharply against the background of faintly coloured or unstained components of the tissue.

Notes
1. Remove mercurial deposits if introduced by the fixative. A control slide should be treated with Schiff's reagent **without** prior oxidation by HIO$_4$. This will indicate whether any reactive aldehyde groups (from the fixative, for example) were previously present in the tissue. If found, such aldehydes must be chemically blocked **before** the treatment with HIO$_4$. If pink staining occurs in the absence of aldehydes, the Schiff's reagent has deteriorated.
2. A bisulphite rinse is often recommended between stages 5 and 6, but Demalsy & Callebaut (1967) have shown that this causes some fading of the stain and is not necessary provided that the Schiff's reagent is washed away very rapidly.
3. Glycogen can be identified if the PAS reaction is applied in conjunction with amylase-digested control sections (see p. 192). Neuraminidase and/or mild acid hydrolysis will assist in the identification of sialic acids. The PAS reaction is blocked by acetylation, though this procedure has little analytical value.
4. If frozen sections are used, some glycolipids are stained. A pseudoplasmal reaction is also usually obtained with frozen sections (see Chapter 12), which are therefore unsuitable for the study of glycoproteins unless lipids are extracted before staining.

11.4.2.2. Specialized PAS methods

The theoretical basis of these procedures, which should be used in conjunction with appropriate enzymatic digestions and other controls, is explained and discussed in Section 11.4.1 of this chapter.

(a) Sialic acids

To generate aldehydes from sialic acids (NANA and possibly others), oxidize in a freshly prepared solution of **either** sodium metaperiodate (NaIO$_4$; 0.0086% aqueous, for 30min) **or** periodic acid (0.114% aqueous, for 5min). Transfer the slides directly into 1.3% sodium sulphite (Na_2SO_3) for 5min to arrest the oxidation. Wash in running tap water for 5min and then proceed with stages 5–8 of the standard technique.

(b) O-acetylated sialic acids

Treat hydrated sections as follows:

1. Oxidize in 1% periodic acid ($HIO_4.2H_2O$), 2 h.
2. Wash in running tap water for 10min.
3. Stain some of the slides with Schiff's reagent (stages 5–8 of standard technique).
4. Immerse the remaining slides in 0.05% sodium borohydride ($NaBH_4$) in 1% Na_2HPO_4 for 10min. Use this solution within 30min of dissolving the $NaBH_4$. **Caution:** see p.164.
5. Wash in four changes of tap water.
6. Immerse in 0.5% potassium hydroxide (KOH) in 70% ethanol for 15min.
7. Wash gently in four changes of tap water.
8. Oxidize for a second time with 1% periodic acid for 10min.
9. Stain with Schiff's reagent (stages 5–8 of standard technique).

If positive reactions due to hexoses have been excluded, staining at stage 3 above is due to NANA or to sialic acids acetylated at position C7.†

Staining obtained at stage 9 but not at stage 3 is due to sialic acids acetylated at C8 or at more than one of the positions C7, C8, and C9. A positive result at stages 3 **and** 9 indicates a sialic acid acetylated at C7. The results may be difficult to interpret if mucosubstances containing different sialic acids occur at the same site.

(c) Uronic acids

Treat hydrated paraffin sections as follows:

1. Oxidize for 1h at 30°C in 2% aqueous sodium metaperiodate ($NaIO_4$).
2. Wash in running tap water for 10min.
3. Treat with freshly dissolved 0.1% sodium borohydride in 1% Na_2HPO_4 for 10min.
4. Wash in running tap water for 10min.
5. Oxidize for a second time in 2% $NaIO_4$ for 24h at 30°C.
6. Wash in running tap water for 10min.
7. Stain with Schiff's reagent (stages 5–8 of the standard technique).

A positive result is seen at the sites of proteoglycans with free hydroxyl groups at positions C2 and C3 of their uronic acid residues. Hexoses and sialic acids are unstained.

11.4.3. Artifacts associated with the PAS method

A structure coloured pink or purple by the PAS technique cannot be assumed to contain periodate-reactive sugars unless the following causes of false-positive staining have been excluded.

1. Aldehydes may be initially present in the tissue. These will be stained by Schiff's reagent without prior oxidation by periodic acid and may be derived from the fixative (especially glutaraldehyde) or from atmospheric oxidation of olefinic linkages in unsaturated lipids (the **pseudoplasmal** reaction). Fixatives containing mercuric chloride may produce aldehydes from plasmalogen phospholipids (the **plasmal** reaction). Lipids (see Chapter 12) rarely interfere with the interpretation of PAS staining in paraffin sections.

Blocking procedures for pre-existing aldehyde groups have been described in Chapter 10. Sodium borohydride is the aldehyde-blocking agent of first choice for use in carbohydrate histochemistry.

2. Periodic acid can produce aldehydes from olefinic bonds of lipids as well as from glycols. Consequently, a genuinely positive PAS reaction in a frozen section may not be due to glycoprotein or glycogen, or even to glycolipid, especially if

---

† From the scheme on page 185, C9-acylated sialic acids should also stain, but in practice it is found that substitution at C9 prevents oxidation of the C7–8 glycol formation (P. E. Reid, personal communication 1988).

an aqueous mounting medium is used. Adams (1965) describes a modified PAS method in which the reactivity of unsaturated linkages is chemically suppressed before the oxidation of carbohydrates by periodic acid. Sphingosine, an amino alcohol present in several lipids, is also oxidized by periodic acid (see p. 199).

Glycolipids (which are commonly PAS-positive on account of their content of galactose and sialic acids) are distinguished from mucosubstances by virtue of the solubility of the former in a hot mixture of methanol and chloroform. Some glycolipids are largely extracted by water.

3. Hydroxylysine, an amino acid which occurs only in collagen (see Chapter 8), might be expected to be PAS-positive, but Bangle & Alford (1954) showed that the stainability of collagen by the PAS method was due almost entirely to its carbohydrate content. The hydroxylysine may not be present in sufficient quantity to be histochemically detectable, or its amino groups may be involved in an amide linkage with some part of the collagen molecule.

4. Some batches of Schiff's reagent prove unsatisfactory and may give either false positive or false negative results. An untried sample of the reagent should be tested on sections in which the correct localizations of PAS-positive structures are already known. An aldehyde-blocking reaction applied after oxidation with periodate will serve as a control for false positive coloration by an unsatisfactory sample of Schiff's reagent.

11.5. Lectins as histochemical reagents

The word "lectin" was originally applied to proteins that were extracted from plants and that had the property of agglutinating mammalian erythrocytes (see Boyd, 1970). Such proteins are also called **phytohaemagglutinins.** One lectin molecule can bind to two or more glycoprotein molecules on the outside surfaces of cells, and thereby join the cells together. Comparable proteins are now known from fungal and animal sources, and any "carbohydrate-binding protein of non-immune origin that agglutinates cells or precipitates polysaccharides or glycoconjugates" is considered to be a lectin (Liener et al., 1986).

11.5.1. Properties of lectins

It has been known for many years that whereas some lectins agglutinate all types of red cells, others are selective for particular blood groups in much the same ways as are circulating antibodies. The blood group determinant substances are glycoproteins located on the surfaces of the erythrocytes and it has been shown that lectins bind to specific carbohydrate moieties of these glycoproteins.

The lectins can bind to appropriate carbohydrates whether these be located on the surfaces of erythrocytes or elsewhere. Måkelå (1957) showed that the specific agglutination of red cells of particular blood groups was due to affinity of the lectins for the terminal monosaccharide residues of the determinant glycoproteins. The blood group O glycoprotein, for example, has terminal α-L-fucosyl residues, and the lectins that agglutinate group O erythrocytes bind to L-fucose and its α-glycosides. Lectins that agglutinate erythrocytes of all blood groups are those that bind to sugars such as mannose and N-acetylglucosamine, which are universally distributed on the surfaces of cells. Irrespective of specificity for human blood groups, all the lectins are remarkably selective with respect to the types of sugar to which they bind. Lectins are classified into 5 groups according to their affinities for different sugars (see Table 11.1). Binding is principally to the terminal sugar of a polysaccharide or oligosaccharide, and can be competitively inhibited by adding the free sugar or an appropriate glycoside to a solution of the lectin. The affinity of a lectin is often influenced, however, by the next one to three units in the chain (see Goldstein & Poretz, 1986). Consequently a particular cell-type or glycoprotein may be able to bind different amounts of different lectins of the same group.

The binding of a lectin molecule to a carbohydrate does not involve the formation of covalent bonds. It is similar in nature to the attachment of an antigen to its specific antibody (see Chapter

19). The molecules of some lectins incorporate ions of calcium and of a transition metal (usually manganese), and the presence of these metal ions is essential for the carbohydrate-binding activity.

Lectins are proteins with large molecules (M.W. 20,000–300,000), so it is possible to attach dye molecules covalently to some of their free amino groups without interfering with the carbohydrate-binding properties. The most frequently used labels are fluorochromes (especially fluorescein and rhodamine derivatives), histochemically demonstrable enzymes (horseradish peroxidase, alkaline phosphatase), ferritin (an electron dense protein), and [³H]acetyl groups (for subsequent autoradiography). Bound lectins are sometimes detected by immunohistochemical methods (see Chapter 19), using anti-lectin antisera.

11.5.2. Techniques for use of lectins

Histochemical methods involving the use of lectins have much in common with immunohistochemical techniques (see Chapter 19). In the simplest type of procedure, a solution of the fluorescently labelled lectin, usually at a concentration of about 0.1 mg per ml in water or saline, usually buffered to pH 7.0–7.6, is applied to the sections of tissue for 15–60 min. Excess reagent is washed off and the preparation is either mounted in an aqueous medium or dehydrated, cleared, and mounted in a non-fluorescent resinous medium. Fluorescence is observed at sites of binding of the lectin.

Materials needed

Fluorescent conjugates of lectins can be prepared in the laboratory (Roth, Binder & Ger-

TABLE 11.1. *Some lectins used as histochemical reagents*

Source of lectin (Name, where available)	Common abbreviation[a]	Specific affinity[b]
GROUP 1. **Affinity for glucose and mannose**		
Canavalia ensiformis (Concanavalin A)	Con A	α-Man > α-Glc > GlcNAc
Lens culinaris (Lentil lectin)	LCA	α-Man > α-Glc > GlcNAc
Pisum sativum (Pea lectin)	PSA	α-Man > α-Glc = GlcNAc
GROUP 2. **Affinity for *N*-acetylglucosamine**		
Phytolacca americana (Pokeweed mitogen)	PAA *or* PWM	GlcNAc-β-1,4-GlcNAc = Gal-β-1,4-GlcNAc
Solanum tuberosum (Potato lectin)	STA	GlcNAc-β-1,4-GlcNAc
Triticum vulgare (Wheat germ lectin)	WGA	GlcNAc-β-1,4-GlcNAc > β-GlcNAc > Sialic acids
GROUP 3. **Affinity for galactose and *N*-acetylgalactosamine**		
Arachis hypogaea (Peanut lectin)	PNA	Gal-β-1,3-GalNAc > α- and β-Gal
Dolichos biflorus (Horse gram lectin)	DBA	GalNAc-α-1,3-GalNAc >> α-GalNAc
Glycine max (Soybean lectin)	SBA	α- and β-GalNAc > α- and β-Gal
Maclura pomifera (Osage orange lectin)	MPA	α-GalNAc > α-Gal
Phaseolus vulgaris (Kidney bean lectin)	PHA	Gal-β-1,4-GlcNAc-β-1,2-Man
Ricinus communis (Castor bean agglutinin I)	RCA₁ *or* RCA₁₂₀	β-Gal > α-Gal >> GalNac

(*continued*)

TABLE 11.1. *Some lectins used as histochemical reagents (continued)*

Source of lectin (Name, where available)	Common abbreviation[a]	Specific affinity[b]
GROUP 4. **Affinity for L-fucose**		
Anguilla anguilla (Eel lectin)	AAA	α-L-Fuc
Lotus tetragonobolus (*Tetragonobolus purpureus; Asparagus pea lectin*)	LTA	α-L-Fuc
Ulex europaeus (Gorse lectin I)		α-L-Fuc
GROUP 5. **Affinity for sialic and uronic acids**		
Limax flavus (Slug lectin)	LFA	N-Acetylneuraminic acid >N-Glycolylneuraminic acid
Limulus polyphemus (Limulin; Horeshoe crab lectin)	LPA	N-acetyl(*or* -glycolyl)neuraminic acid-α-2,6-GalNAc >N-acetylneuraminic acid
Sambucus nigra (Elderberry bark lectin)	SNA	N-Acetylneuraminic acid-α-2,6(-Gal *or*-GalNAc)
Aplysia depilans (Aplysia gonad lectin)	AGL	Galacturonic acid>>D-Gal
Bovine or porcine lung, pancreas, salivary glands (Aprotinin; Bovine trypsin inhibitor)		Uronic and sialic acids[c]

[a]The terminal A in most of the abbreviations stands for "agglutinin". A subscript number indicates one of a number of lectins extracted from the same source (e.g. UEA_1), or sometimes the $M.W. \times 10^{-3}$ (e.g. $RCA_1 = RCA_{120}$, whose M.W. is 120,000).

[b]See Section 11.1.2 of this chapter for the structures of the sugars corresponding to the abbreviations in this column of the table. The information on specificity is derived largely from Toms & Western (1971), Cook & Stoddart (1973), Nicolson (1974), Goldstein & Poretz (1986), Benhamou *et al.* (1988) and Taatjes *et al.* (1988). For lectins that bind to more than one terminal sugar, ligands are listed in order of decreasing affinity (>>indicates much greater affinity; >greater, and = equal affinity for the lectin.)

[c]Aprotinin is not a lectin, but the fluorescently labelled polypeptide can be used to stain glycoproteins and glycosaminoglycans that owe their acidity to carboxyl rather than to sulphate groups (Kiernan & Stoddart, 1973; Stoddart & Kiernan, 1973). Fluorescent aprotinin also binds to the proteolytic enzymes it inhibits, but this latter affinity does not permit the use of the reagent for histochemical demonstration of the enzymes in sectioned tissues.

hard, 1978), but the procedure is not as easy as the labelling of most other types of protein. Most histochemists now use labelled lectins from commercial suppliers, and these are entirely satisfactory. **It is important to buy the correct lectin**, especially when more than one is available from the same plant (e.g. *Ulex europaeus* or *Ricinus communis*). A few lectins are exceedingly toxic, but most are harmless. Warnings about toxicity are given in catalogues and on the outsides of containers. The supplier should also provide information about the purity and specificity of each product.

It is usual to buy 0.5 or 1mg of a fluorescently labelled lectin, and the almost invisible amount of freeze-dried powder will be in a vial that can hold 1–10 ml. The most commonly encountered labels are fluorescein and rhodamine derivatives. There is little to choose between the two fluorochromes; I prefer rhodamine because its orange-red fluorescence is not easily confused with autofluorescence. To make a stock solution of the fluorescently labelled lectin, add 0.5–1.0ml of the following solution to the vial.

0.05M TRIS buffer, pH7.2:	99ml
Trace metal solution:	1ml
Sodium azide (NaN_3):	65mg

(The "trace metal solution" contains calcium, magnesium and manganese chlorides, each at 0.01M concentration)

The Ca^{2+}, Mg^{2+} and Mn^{2+} satisfy the requirements of lectins for these metals. The NaN_3 is to inhibit growth of microorganisms in the solution. The type of buffer is not critical, and many people use a phosphate-buffered saline rather than TRIS. The same solution is used for further dilutions of the lectin, and for washing sections before and after staining.

The stock solution of the fluorescently labelled lectin should contain 1.0mg of protein per ml. A tiny drop of this should be tested by staining a section (see below). Probably the fluorescence will be too intense, with non-specific background staining. Other sections should then be stained with further dilutions until the observed fluorescence is confined to structures that are known or expected to contain stainable material. Such structures commonly include collagen fibres, basement membranes and mucus-secreting cells. When the optimum concentration has been found, an appropriate volume of the buffer is added to the initially dissolved lectin in its container, to give a working solution. This solution should be kept at 4°C, **not frozen**. It is stable for several weeks.

Staining procedure

Fixation is not critical, but ethanol and mercuric chloride were found by Allison (1987) to be somewhat superior to Carnoy, Bouin, neutral buffered formalin-saline and calcium acetate-paraformaldehyde. Glutaraldehyde should be avoided because it causes non-specific binding of lectins to tissues (Rittman & Mackenzie, 1983). Paraffin or cryostat sections may be used.

1. Bring sections to water, rinse in buffer containing trace metals (see above) and blot dry.
2. Place the slides, sections up, on a piece of wet filter paper on a flat bench top.
3. Carefully apply the smallest possible drop of fluorescent lectin solution to cover the sections to be stained.
4. Cover the slides with a suitable airtight lid, such as a petri dish. The air inside must be saturated with water vapour, to prevent drying of the sections.
5. Wait for 30min. (Sometimes 10min will be enough; rarely 2h may be needed.)
6. Rinse the slide in 3 changes of buffer containing trace metals.
7. Dehydrate in 95% and 2 changes of absolute alcohol, clear in xylene and cover, using a non-fluorescent resinous mounting medium. *Alternatively*, mount from water into buffered glycerol (see Chapter 4, p.45). The latter method is useful if the same section is to be stained again by some other technique.

Result
Lectin binding sites are fluorescent.

Notes
1. Lectins can also be bought labelled with horseradish peroxidase. When these are used, the working solution can be more dilute (typically 0.01–0.05mg/ml) than that of a fluorescent lectin. After Stage 5 above, the sections are subjected to a histochemical method for peroxidase, usually with DAB as the chromogen (see Chapter 16).
2. Controls are important in lectin histochemistry. They are discussed in the next section of this chapter.

11.5.3. Interpretation of results

Two considerations are important in assessing the significance of observed binding of a lectin to a component of a tissue.

1. The specificity of the binding must be established by means of suitable controls. **Inhibition** with appropriate sugars or glycosides is the most important control procedure. This is done by staining control slides in the presence of high (0.2–1.0M) concentrations of sugars or glycosides that will occupy binding sites on the lectin

molecules before the latter can attach to the tissue. For example, staining with concanavalin A should not occur from a solution containing α-methyl-D-glucoside (but should occur normally in the presence of β-methyl-D-glucoside). **Chemical blocking** methods destroy or modify the glycosyl residues to which the lectin is expected to bind. Acetylation, for example, prevents the binding of concanavalin A, but not of aprotinin, and methylation prevents staining by fluorescently labelled aprotinin or limulin but not by concanavalin A and most other lectins. The reasons for these effects can be determined by consulting Table 11.1. Oxidation by periodic acid prevents the binding of concanavalin A at most sites in tissues, but some structures paradoxically acquire the ability to bind this lectin, even though all glucosyl and mannosyl residues would be expected to be destroyed. The reasons for this effect are not yet understood, but it has been suggested that certain mannosyl units of glycoproteins, previously inaccessible to the large molecules of concanavalin A, are "unmasked" but not oxidized as a result of the treatment with periodic acid (Katsuyama & Spicer, 1978).

2. It may not be assumed that the amount of lectin bound at any site (as judged by intensity of staining) is proportional to the local concentration of the appropriate monosaccharide in the tissue. The binding of any lectin is profoundly affected by neighbouring sugar residues in the oligosaccharide chains of glycoproteins, even though these may not be ones which attach themselves directly to the lectin. The histochemical recognition of individual monosaccharide units forms only a small part of the study of the interactions between lectins and glycoproteins (see Nicolson, 1974).

11.6. Enzymes as reagents in carbohydrate histochemistry

Three enzymes are commonly used in the histochemical study of mucosubstances. These are amylase, hyaluronidase, and neuraminidase. The precautions discussed in Chapter 9 (p. 139) apply equally to the use of these enzymes.

11.6.1. Amylase

(A collective name for α-1,4-glucan hydrolases, E.C. 3.2.1.1, 3.2.1.2 and 3.2.1.3) Amylase catalyses the hydrolysis of the glucosidic linkages of starch and glycogen. The enzyme usually used in histochemistry is β-amylase (E.C. 3.2.1.2), the action of which yields the soluble disaccharide maltose. Glycogen (or starch) can therefore be identified as a substance whose stainability by the PAS method is prevented by digestion of the sections with amylase.

Procedure

Use a solution containing 1.0mg of enzyme per ml of water. Ideally a pure preparation of β-amylase should be used. An effective alternative is human saliva, though this also contains various proteolytic enzymes and a ribonuclease. An adequate supply of saliva may be obtained by thinking of lemons and drooling into a small beaker. Bubbles should be removed by stroking the surface of the collected liquid with filter paper before applying it to the sections.

Incubate for 30min at 37°C. The treatment is somewhat destructive to the sections, which may need to be covered with a film of celloidin after applying the enzyme but before staining. A control section should be incubated with water.

11.6.2. Hyaluronidase

(Hyaluronate lyase, E.C. 4.2.99.1.) Hyaluronidase attacks the glycosidic linkages of some proteoglycans, including hyaluronic acid, causing depolymerization of these substances. Hyaluronidases of different types are available, the one most often used by histochemists being that extracted from ovine testes. The specificities of the enzymes have been reviewed by Pearse (1985) who considers that the testicular enzyme can remove hyaluronic acid, chondroitin-4-sulphate, and chondroitin-6-sulphate from fixed tissues. Streptococcal hyaluronidase is specific for hyaluronic acid. Chondroitinase AC and chondroitinase ABC are enzymes of bacterial origin that degrade chondroitin sulphates*. They are occasionally used histochemically, in the same way as the hyaluronidases.

Most samples of hyaluronidases and of chondroitinases are contaminated with proteolytic enzymes, but these probably do not affect the specific degradation of glycosaminoglycans (Harrison, Van Hoof & Vanroelen, 1986).

Procedure

Both hyaluronidases are used as solutions containing 1.0mg of enzyme per ml of 0.9% aqueous NaCl.

Incubate overnight at 37°C.

11.6.3. *Neuraminidase*

(*N*-acetylneuraminate glycohydrolase, E.C. 3.2.1.18; the enzyme from *Vibrio cholerae* is usually employed.) Neuraminidase removes the terminal NANA residues of glycoproteins. It is therefore possible to determine whether staining is due to the presence of these sugars. However, not all sialic acid groups are removed by the enzyme, so no significance can be attached to a failure to prevent staining by pre-treating the sections with the enzyme. Prior saponification makes some previously neuraminidase resistant sialic acid residues susceptible to the action of the enzyme, probably by converting O-acetylated sialic acids to NANA. The type of fixative used does not usually affect digestibility by neuraminidase, but cyanuric chloride, which cross-links hydroxyl and amino groups, is exceptional in that it prevents the subsequent action of the enzyme (Sorvari & Laurén, 1973). The identity of the sugar to which NANA is glycosidically bound can affect the susceptibility of the linkage to attack by neuraminidases from different sources. The *Vibrio* enzyme, however, catalysed the hydrolysis of all the types of linkage studied by Drzeniek (1973). Acid hydrolysis (see p.195) will detach all sialic acid residues from glycoproteins.

---

* The formal names of these enzymes are chondroitin AC lyase (E.C. 4.2.2.5) and chondroitin ABC lyase (E.C. 4.2.2.4).

Procedure

0.1M acetate buffer, pH5.5:	1.0ml
Calcium chloride (CaCl$_2$):	1.0mg
	(or 0.1ml of a 1% aqueous solution)
Neuraminidase:	0.025 international units

The sections may be saponified (see below) before exposure to neuraminidase. This will make some otherwise resistant sialic acid residues susceptible to removal by treatment with the enzyme. Incubate overnight at 37°C.

11.6.4. *Other glycosidases*

There are many enzymes that split glycosidic linkages, in addition to those mentioned in the preceding paragraphs (see Hughes, 1976). Some of these are histochemically useful, especially in conjunction with lectins (Whyte *et al.*, 1978). Enzymatic removal of the terminal residue of an oligosaccharide makes the next sugar in the chain accessible to lectin molecules. For example, removal of terminal fucose by an α-L-fucosidase will abolish the affinity of the tissue for UEA$_1$ (see Table 11.1), and may unmask binding sites for other lectins. Sequential degradation with glycosidases is extensively used in the biochemical study of glycoproteins. Caution in the interpretation of results is necessary, however, because linkages other than those to the terminal sugar unit may be attacked, and also because commercially available preparations usually contain more than one enzyme (Stoddart, 1984).

11.7. Chemical blocking procedures

The reactive groups of carbohydrates are hydroxyls, carboxyls, and sulphate esters. These can be chemically "blocked" to prevent their subsequent histochemical demonstration. The significance of a negative result produced in this way is dependent upon the selectivity of the blocking procedure. The blocking reactions are discussed more fully in Chapter 10 and the following notes apply only to their applications in the histochemical study of carbohydrates.

11.7.1. Acetylation of hydroxyl groups

Sections are treated with acetic anhydride in dry pyridine. Hydroxyl groups are converted to acetyl esters:

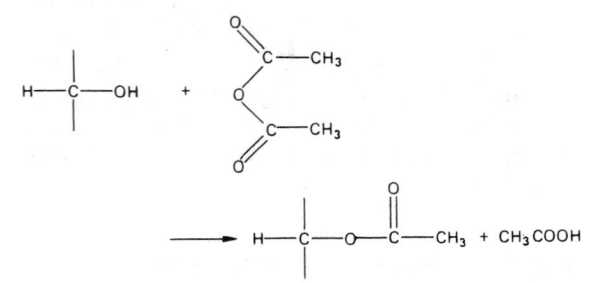

Histochemical reactions due to the hydroxyl groups of carbohydrates are prevented. Acetic anhydride also causes N-acetylation of amines and can react with carboxyl groups (of proteins, but probably not of carbohydrates) to form carbonyl compounds, though the optimum conditions for these reactions are different from those used for acetylating sugars. The acetyl esters of carbohydrates are hydrolysed by treatment with alkali (saponification, p. 166), which restores the reactivity of the hydroxyl groups.

Procedure
1. Take sections to absolute alcohol.
2. Immerse overnight at room temperature in a mixture of acetic anhydride (2 volumes) and pyridine (3 volumes) in a tightly closed container. Exercise care in handling acetic anhydride, which reacts with water to form acetic acid.
3. Wash in two changes of absolute alcohol and then pass through 70% alcohol to water.

It is often said that the pyridine must be anhydrous, but ordinary reagent-grade pyridine is perfectly satisfactory. This is not surprising

because the traces of water in it will react rapidly with acetic anhydride to form acetic acid, though not in quantities sufficient to interfere with the function of the pyridine as a basic catalyst.

11.7.2. Methylation and desulphation

When sections are treated with a solution of either HCl or thionyl chloride in methanol (or by any of several other procedures), methyl esters are formed with carboxyl groups:

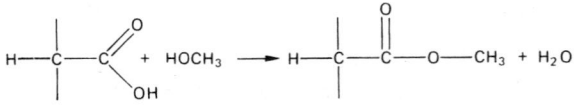

(In the case of the uronic acids it is likely that other reactions also occur—see Sorvari & Stoward, 1970.)

The same reagent, if applied for sufficient time, also **removes** O-sulphate and N-sulphate groups and methylates amino groups. Methylation prevents the staining of nucleic acids with cationic dyes, but this effect is not due to esterification of their phosphate anions. Prento (1980) attributes the abolition of basophilia to (a) depolymerization and extraction of RNA, and (b) the induction of a high concentration of positively charged groups in the chromatin in which the DNA is located. This positive charge is due to the excess of amino groups following methylation of the carboxyl groups of the nucleoproteins. Another effect of the methylating agent is that it causes acid hydrolysis of most of the glycosidic linkages of sialic acid residues. The loss of basophilia of sialic acid-containing glycoproteins therefore cannot, for the most part, be restored by saponification. This last effect is rather unpredictable, so methylation is more useful for the study of uronic acids of proteoglycans than for the investigation of glycoproteins.

As with acetylation, the esters can be saponified with restoration of carboxyl groups, but this treatment does not, of course, restore any tinctorial properties which were due to the presence of sulphate esters. Hydroxyl groups are not blocked by methanolic HCl under the conditions in which the reagent is normally used. Free hydroxyls at position C1 of a monosaccharide unit would be converted to methyl glycosides, but in mucosubstances almost all the sugars are already linked at this position. The formation of methyl ethers with the alcoholic hydroxyl groups of carbohydrates requires methyl iodide or dimethyl sulphate with a basic rather than an acid catalyst and does not occur in histochemical methylations.

Procedures

Methylation

1. Take sections to absolute methanol.
2. Immerse in methanol containing 1% by volume of concentrated hydrochloric acid in a tightly closed container at 58–60°C for 48h.
3. Rinse in absolute methanol and pass through 70% alcohol to water.

Methylation for only 4 h is usually adequate for blocking carboxyl groups and removing sulphate-ester groups. The completeness of the desired consequences of methylation is easily determined by staining with alcian blue at pH1.0 and 2.5.

Saponification

This unblocking procedure can be used after acetylation or methylation.
1. Take sections to absolute ethanol.
2. Cover with a film of celloidin (optional), drain, dry, and harden the film in 70% ethanol.
3. Immerse for 20min in a freshly prepared 0.5% solution of potassium hydroxide (KOH) in 70% ethanol.

11.7.3. Mild acid hydrolysis

The α-2-glycosidic linkage by which sialic acids are attached to the subjacent sugars of the oligosaccharide chains of glycoproteins is easily broken by acid-catalysed hydrolysis. Sections are treated with an aqueous acid (pH2.5 or lower) for 1 or 2h at 70–80°C. All sialic acid residues are removed, including those that resist digestion by

neuraminidase, but most other glycosidic linkages remain intact.

Procedure
1. Take sections to water.
2. Heat some 0.05M sulphuric acid (add 2.8ml of concentrated sulphuric acid, S.G. 1.84, 96% H_2SO_4, to about 800ml of water. Mix thoroughly and add water to obtain 1000ml) to 80°C. Place the slides in the hot, dilute acid and maintain at 80°C for 1h.
3. Wash in running tap water for 2min, rinse in distilled or de-ionized water, and proceed with the histochemical testing. (Sections often come off their slides and are lost in this procedure.)

11.8. Combinations of techniques

It is often informative and convenient to apply two or more different methods for carbohydrates to the same section. A counterstain for nuclei is often added, to provide general morphological information. Notes on two such procedures follow. Many other combinations are possible.

11.8.1. Alcian blue and PAS

Procedure
1. Take sections to water.
2. Stain with alcian blue, either at pH1.0 or at pH2.5).
3. Wash in running tap water, then rinse in distilled water.
4. Stain by the PAS method. The periodate oxidation may be the standard one (p.186) or the mild oxidation that is selective for sialic acids (p.187).
5. Wash thoroughly in rapidly running tap water.
6. Counterstain nuclei with an alum-haematein (pp.96–97) or with an iron-chromoxane cyanine R (p.99) mixture.
7. Wash in water, dehydrate, clear, and mount in a resinous medium.

Result

Alcian blue-positive materials turquoise-blue. (With the dye at pH1.0, only sulphated glycoproteins and proteoglycans are stained.) PAS-positive material magenta. (With the mild oxidation, sialic acids are the only PAS-positive substances.) A purple colour is seen in material stained by both alcian blue and PAS. Nuclei are dark blue.

Note

Usually the alcian blue is used before the PAS staining. If the sequence is reversed, staining by alcian blue is stronger, especially in mucus that is also PAS-positive. The additional blue staining has been attributed by Johannes & Klessen (1984) to the formation of aldehyde bisulphite addition compounds, by reaction of the periodate-engendered aldehyde groups with the bisulphite in Schiff's reagent. Aldehyde bisulphite compounds are sulphonic acids, and can therefore bind alcian blue even at pH1.0 (see Chapter 10, p. 161). The purple and blue colours produced in sialoglycoproteins by the PAS-alcian blue sequence are not uniformly distributed. Yamabayashi (1987) has suggested that the variations may be due to differences in the positions of the sialic acid groups. Steric hindrance by parts of other macromolecules might keep the alcian blue cations away from some anionic sites. Alcian blue also binds to the coloured tissue-Schiff compound (Reid & Owen, 1988), which contains a sulphonic acid group (see p. 160). For microspectrophotometric studies, it is necessary to apply alcian blue and PAS procedures to separate sections (Roe et al., 1989).

11.8.2. A two-colour periodate method for sialoglycoproteins

This method (Reid et al., 1984) is based on principles explained in Section 11.4.1 of this chapter. Aldehydes formed from neutral hexoses are irreversibly blocked by phenylhydrazine, and those derived from oxidizable sialic acids are stained with a blue Schiff-like reagent (thionine-Schiff). Saponification then releases more hydroxyl groups by removing O-acyl groups from esterified sialic acid residues. The newly formed glycols are then demonstrated by the ordinary PAS reaction. The borohydride reduction (Step 11) blocks any aldehyde groups that failed to combine with the thionine-Schiff reagent, and also changes the colour imparted by the thionine-Schiff from purple to blue.

Solutions required

Thionine-Schiff reagent (van Duijn, 1956)

Add 0.5g thionine (C.I. 52000) to 250ml of water. Boil for 5min. Leave to cool, then add water to restore the volume to 250ml. Add 250ml t-butanol (freezes at 26°C; may need to be warmed). Do not filter at this stage, even though some of the dye is undissolved. Add 37.5ml 1.0M hydrochloric acid, followed by 5.0g of sodium metabisulphite ($Na_2S_2O_5$), shake, and leave in a stoppered bottle for 24h at room temperature, then transfer to a refrigerator for storage. May be used after 48h at 4°C. Filter what is needed into a coplin jar, and use only once.

Phenylhydrazine

Phenylhydrazine hydrochloride: 0.25g
Water: 50ml
    Prepare as needed, and use only once.
    Caution: Phenylhydrazine is poisonous.

Other solutions needed are described with the other components of the method (see page references below)

Procedure

1. Take sections to water.
2. Oxidize with 1% periodic acid (p. 186) for 2h, room temperature.
3. Wash in running tap water, 10min.
4. Transfer to 0.5% phenylhydrazine hydrochloride, 2h.
5. Wash in running tap water, 10min, then leave overnight in distilled water.
6. Immerse in thionine-Schiff for 4h.

7. Wash in running tap water, 10m.
8. Transfer to 70% alcohol.
9. Saponify in alcoholic KOH (see p. 167).
10. Wash in running tap water, 10min.
11. Immerse in sodium borohydride solution (see p. 47) for 20min.
12. Wash in running tap water, 10min.
13. Oxidize with 1% periodic acid (p. 167) for 30–60min, room temperature.
14. Wash in running tap water, 10min, then rinse in distilled water.
15. Immerse in ordinary Schiff's reagent for 1h.
16. Wash in running tap water, 10min.
17. Dehydrate, clear and cover, using a resinous mounting medium.

Results

Sialic acids with free hydroxyls at C7, C8 and C9: blue.

Sialic acids acylated at C7 are stained blue.

Sialic acids acylated at C8 *or* C9 are stained magenta.

Sialic acids acylated at two or three of positions C7, C8, C9 are stained magenta.

Mixtures are stained purple.

Mucosubstances that do not contain sialic acids are unstained.

11.9. Exercises

Theoretical

1. Which mucosubstances are stained by (a) the periodic acid-Schiff reaction, (b) fluorescent-labelled concanavalin A?

2. Hydroxyl groups of sugars can be **sulphated** (i.e. converted to sulphate esters) by treating sections with sulphuric or chlorosulphonic acid. What effect would you expect this procedure to have on the pattern of staining by thionine or toluidine blue?

3. What substances (if any) would you expect to be PAS-positive in a section which had been (a) acetylated, (b) acetylated and then saponified, (c) methylated, (d) methylated and then saponified?

4. What substances (if any) would you expect to stain with alcian blue (pH1.0 or 2.5) after (a) acetylation, (b) acetylation followed by saponification, (c) methylation, (d) methylation followed by saponification, (e) mild acid hydrolysis?

5. Which proteoglycans and glycoproteins can bind fluorescent-labelled aprotinin? How would the pattern of staining be affected by prior treatment of the sections with (a) amylase, (b) hyaluronidase, (c) neuraminidase?

6. Which of the histochemical techniques discussed here would be suitable for the demonstration of glycogen? Describe exactly how you would proceed to demonstrate glycogen in the cells of a piece of liver freshly removed from an animal.

7. To what extent would chemical blocking procedures be useful in confirming the specificity of binding of a fluorescently labelled lectin believed to be specific for (a) the α-L-fucosyl configuration, (b) N-acetylgalactosamine? How else one can ascertain the specificity of a labelled lectin used as a histochemical reagent?

8. With which (if any) of the methods described in this chapter would you expect to obtain a positive reaction with chitin? How do you account for the fact that the exoskeletal tissues of invertebrates are commonly PAS-positive?

9. A glycoprotein secreted by mucous cells of mammalian submandibular glands bears oligosaccharide chains with the structure

$$NANA(\alpha\text{-}2\rightarrow6)\text{D-Gal}(\beta\text{-}1\rightarrow)PROTEIN$$

How much of the information contained in this formula could be obtained by histochemical analysis?

10. Reid *et al.* (1988) have developed the following sequence of reactions for the histochemical demonstration of glycoproteins with *O*-acyl substitution at C2 or C3 of hexose sugars:

Periodate oxidation—Borohydride—Saponification—Mild periodate oxidation—Borohydride—Periodate oxidation—Schiff's reagent.

If the section initially contains glycoproteins with PAS-positive neutral sugars and variously substituted sialic acids, what is oxidized by each of the three treatments with periodate?

11. Fluorescent "Neoglycoproteins" can be made by covalently conjugating sugar residues and fluorochrome molecules to a protein. How could such a reagent be used to determine the distribution of a lectin in the tissues of a plant? (see Latge *et al.*, 1988).

Practical

Stain paraffin sections of liver for glycogen and establish the specificity of the result by the judicious use of amylase.

Take some paraffin sections of skin, tongue, salivary gland, and intestine. Treat some of the sections by the procedures described for methylation and acetylation. Stain methylated, acetylated, and control sections with (a) PAS, (b) alcian blue, pH2.5, (c) alcian blue, pH1.0. What mucosubstances can be identified by these procedures and in what locations?

Stain a section of the external ear to demonstrate metachromasia. Which structures are metachromatically stained and why?

Using sections of skin or tongue, demonstrate that the cytoplasmic granules of mast cells contain a heavily sulphated proteoglycan with some uronic acid residues, but that sialic acids are absent.

12

Lipids

The term "lipid" is applied to a chemically heterogeneous group of substances that are extracted from tissues by non-polar organic solvents such as chloroform and ether. These substances vary greatly in structural complexity but are built from a limited number of simpler molecules joined together in different ways. These component substances do not occur in large amounts in the free state in living tissue but are present as metabolic precursors of the lipids themselves.

12.1. Components of lipids

In the following brief account the symbols R, R' and R" indicate alkyl radicals, mostly of 16–20 carbon atoms. Simplified structural formulae are used for whole lipids: most carbon atoms are represented only as junctions of bonds, and hydrogen atoms attached to carbon are omitted.

(a) Aliphatic alcohols

ROH. Present in waxes as their esters. An example is cetyl alcohol, $CH_3(CH_2)_{14}CH_2OH$. Long-chain alkyl groups are also present as glyceryl ethers in the ether phosphatides.

(b) Fatty acids

RCOOH. Present as acyl groups in all those lipids that are esters or amides. Most of them have unbranched chains of even numbers of carbon atoms. Olefinic (unsaturated; CH=CH) linkages may or may not be present. In the following list of the common fatty acids of animals, the length of chain is shown as C16, C18, etc., and the number of double bonds is indicated by the symbol $\triangle$.

Saturated:	Myristic acid	C14
	Palmitic acid	C16
	Stearic acid	C18
	Lignoceric acid	C24
Unsaturated:	Palmitoleic acid	C16, $\triangle1$
	Oleic acid	C18, $\triangle1$
	Linoleic acid	C18, $\triangle2$
	Linolenic acid	C18, $\triangle3$
	Arachidonic acid	C20, $\triangle4$
	Clupanodonic acid	C22, $\triangle5$

In normal animals, all lipids (with the exception of some cholesterol esters and sulphatides) contain at least one unsaturated acyl group per molecule.

(c) Glycerol

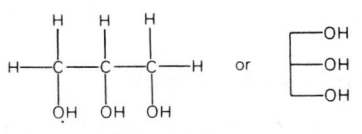

A trihydric alcohol, present as its esters in most lipids.

(d) Phosphoric acid

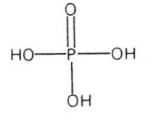

This formulation, though not strictly in accord with modern chemical notation, is the one usually used by biochemists. In phospholipids the H_3PO_4 is esterified through one or two of its hydroxyl groups. The unesterified hydroxyls ionize as acids.

(e) Choline, ethanolamine and serine

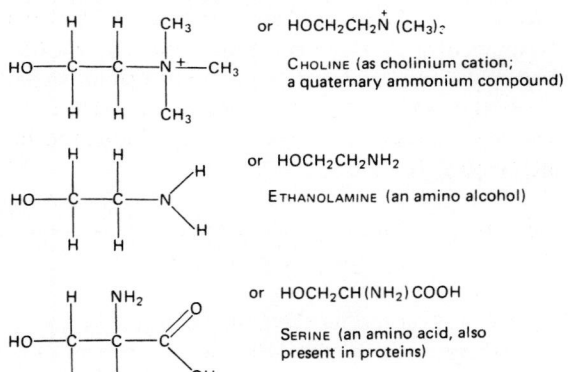

These three bases are esterified through their hydroxyl groups with phosphoric acid in phosphoglycerides and sphingomyelins.

(f) Myo-Inositol

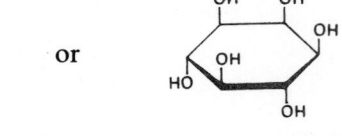

A cyclic alcohol, esterified through phosphoric acid to glycerol in the phosphoinositides, in which some other hydroxyl groups of the inositol are also phosphorylated.

(g) Sugars

Galactose, *N*-acetylgalactosamine, *N*-acetylglucosamine, glucose, *N*-acetylneuraminic acid, and other monosaccharides, sometimes carrying sulphate-ester groups, are present in the glycosphingolipids. See Chapter 11 for structures and abbreviated formulae.

(h) Sphingosine

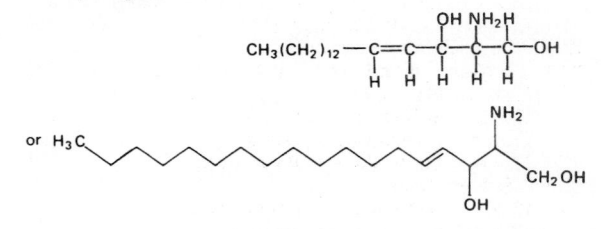

This long-chain (C18) unsaturated amino alcohol is joined in amide linkage to fatty acids and is esterified with phosphoric acid through the hydroxyl group shown at the right-hand side of the formula. In the related compound dihydrosphingosine the bond between C4 and C5 is saturated. In the dehydrosphingosines, more than one unsaturated linkage is present in the hydrocarbon chain.

(i) Cholesterol

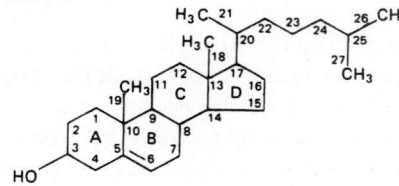

This is a **sterol** (a steroid alcohol). It is insoluble in water or cold ethanol, but freely soluble in ether or acetone, so it is classified as a hydrophobic lipid (see Table 12.1).

The fused ring system (cyclopentanoperhydrophenanthrene), with a wide variety of substituents, is characteristic of all **steroids**. The numbering system and the letters identifying the rings are applicable to all steroids.

(j) Isoprene

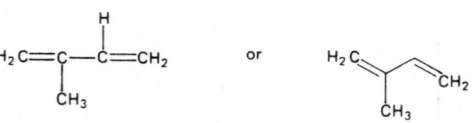

This unsaturated hydrocarbon is the monomer from which the carbon skeletons of most of the terpenes are constructed.

(k) Proteins

These are frequently conjugated with phospholipids in proteolipids and lipoproteins.

12.2. Classification of lipids

The following scheme, which includes only the lipids of vertebrate animals, differs slightly in its arrangement from the classifications used by biochemists. Substances with similar histochemical properties are, as far as possible, grouped together.

The classification is summarized in Table 12.1. Brief descriptions of the main groups of lipids follow.

12.2.1. Free fatty acids

These occur only in traces in normal tissues, but the amounts are increased in some pathological states. Crystal-like deposits of free fatty acids may also form in lipid-rich tissues that have stood for long periods of time in formaldehyde solutions. The fatty acids form water-soluble salts with sodium and potassium and insoluble salts, known as soaps, with calcium and many other metal cations.

The prostaglandins are polyunsaturated fatty acids (C20, $\triangle$ 2–3) produced by all animal cells, but in minute quantities, which are unlikely to be histochemically detectable.

12.2.2. Terpenes

Numerous terpenes occur in plants. The only important one in higher animals is squalene:

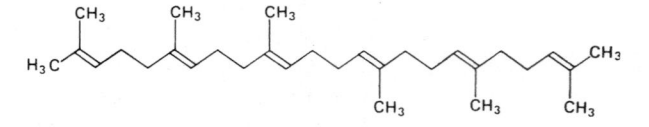

This hydrocarbon occurs in sebum and is also a metabolic precursor of cholesterol.

12.2.3. Steroids

Cholesterol (see p. 199) is a waxy solid with a much higher melting point (150°C) than most other lipids. Inspection of its structural formula shows that it is an unsaturated alcohol. It dissolves easily in most organic solvents but is insoluble in water.

The steroid hormones are numerous but it is doubtful whether they can be identified histochemically. They are stored only in minute quantities in the glands that secrete them, though metabolic precursors of similar chemical structure exist in detectable concentrations in steroid-secreting endocrine cells. Many of the steroid hormones are ketones.

The corticosteroids have two ketone groups per molecule, as well as three alcoholic hydroxyl groups. An example is cortisol (hydrocortisone):

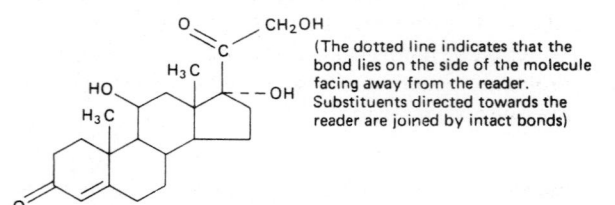

(The dotted line indicates that the bond lies on the side of the molecule facing away from the reader. Substituents directed towards the reader are joined by intact bonds)

Androgens are also ketones but lack the side-chain at position 17. Examples are:

TABLE 12.1. *A classification of the major groups of lipids*

A. LIPIDS WHICH ARE NOT ESTERS OR AMIDES
      1. Free fatty acids
      2. Terpenes
      3. Steroids
         (a) Cholesterol
         (b) Steroid hormones and related compounds   **Hydrophobic lipids**

B. LIPIDS WHICH ARE ESTERS OR AMIDES

Esters
      1. Neutral fats
      2. Waxes
      3. Cholesterol esters
      4. Phosphoglycerides
         (a) Phosphatidyl cholines
         (b) Phosphatidyl ethanolamines
         (c) Phosphatidyl serines
         (d) Ether phosphatides   **Phospholipids**
         (e) Phosphatidylinositols
         (f) Diphosphatidyl glycerols   **Hydrophilic lipids**

Amides (sphingo-lipids)
      5. Sphingomyelins
      6. Ceramides
      7. Glycosphingolipids
         (a) Cerebrosides
         (b) Gangliosides   **Glycolipids**
         (c) Sulphatides

Note. Phospholipids and sphingolipids are often covalently bound to proteins in proteolipids and lipoproteins.

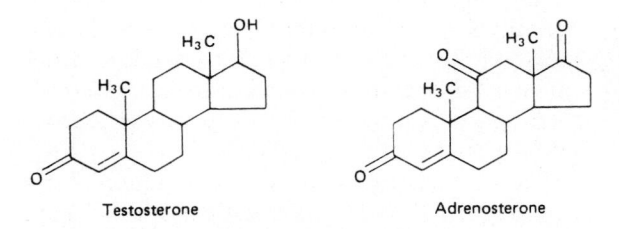

Testosterone   Adrenosterone

Progesterone, the principal progestogen secreted by the corpus luteum, is also a metabolic precursor of other steroid hormones. It is chemically quite similar to the corticosteroids:

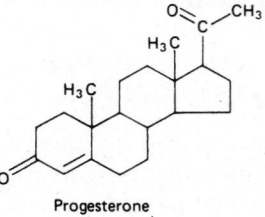

Progesterone

Oestrogens differ from other steroids in that ring A is aromatic, so that the hydroxyl group attached to it confers a phenolic character on the molecule. Ketone and secondary alcohol groups may also be present.

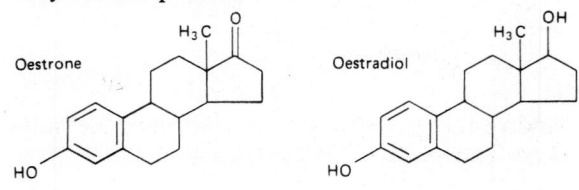

Oestrone   Oestradiol

12.2.4. Neutral fats

These may be mono-, di-, or triglycerides:

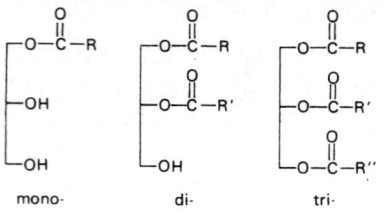

mono·   di·   tri·

The last-named type is the most abundant. Neutral fats occur principally in adipose connective tissue.

12.2.5. Waxes

Natural waxes are esters of long-chain aliphatic alcohols with fatty acids:

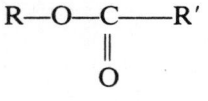

Waxes occur in a wide variety of organisms (spermaceti and beeswax are well-known examples) but are unlikely to be encountered in the tissues of laboratory animals or of man.

12.2.6. Cholesterol esters

These compounds resemble waxes, but the alcohol half of the ester is derived from cholesterol. The acyl groups are unusual in that they are predominantly those of saturated fatty acids.

12.2.7. Phosphoglycerides

These lipids are hydrophilic on account of their content of polar groups such as phosphate ester, hydroxyl, and primary or quaternary amino. They fall into six groups:

(a) Phosphatidylcholines

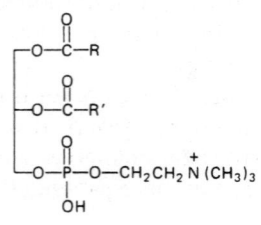

These phospholipids are commonly known as "lecithins". They are soluble in all lipid solvents, including ethanol, with the notable exception of acetone.

(b) Phosphatidyl ethanolamines

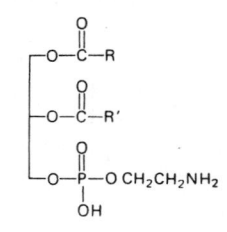

An old synonym is "cephalins". The ethanolamine has sometimes been called "cholamine" or "colamine". In the pure state, phosphatidyl ethanolamines are, like the lecithins, soluble in ethanol but insoluble in acetone. However, the crude mixture of cephalins isolated from tissues, which is a mixture of phosphatidyl ethanolamines, serines, and inositols, is insoluble in ethanol. Extractions with solvents are of little value when attempting to distinguish histochemically between different phosphoglycerides in sections of tissues.

(c) Phosphatidyl serines

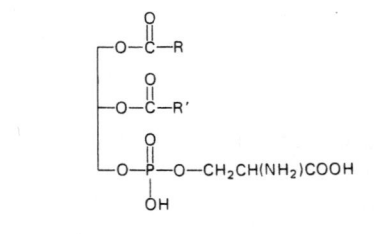

(d) Ether phosphatides

In these phospholipids one of the oxygen atoms of glycerol is joined to a long-chain alkyl group to give an ether. The most important ether phosphatides to the histochemist are those in which there is an unsaturated linkage adjacent to the ether oxygen. These are the **plasmalogens:**

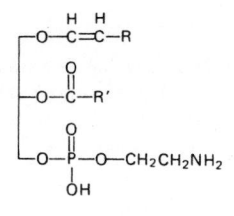

It was once thought that plasmalogens were acetals rather than ethers, hence the old name "acetal lipids". The basic component is most commonly ethanolamine but sometimes choline. Plasmalogens are extracted rapidly from sections of tissue by 80% ethanol but only slowly by 60% ethanol.

(e) Phosphatidyl inositols

The simplest lipids of this type, which are also known as "inositol cephalins", have the structure

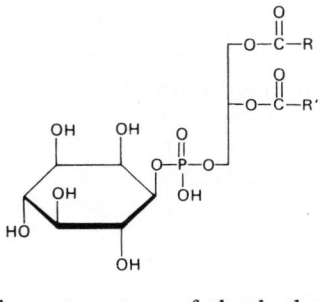

Commonly one or two of the hydroxyl groups shown in the *myo*-inositol ring are esterified by phosphoric acid.

(f) Diphosphatidyl glycerols

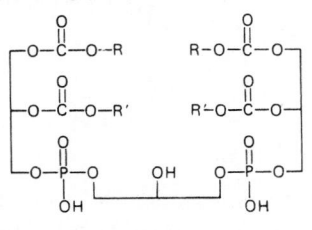

The most important lipid of this type is cardiolipin, a component of the inner mitochondrial membrane. Cardiolipin is unusual among lipids in that it can serve as an antigen.

12.2.8. Sphingomyelins

These choline-containing phospholids are amides derived from sphingosine, choline and fatty acids, the last named being connected by an amide linkage:

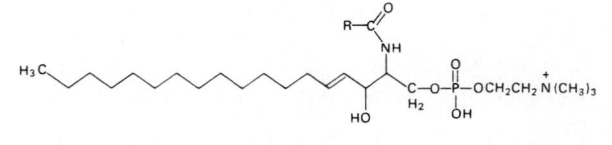

The sphingosine may be replaced by dihydrosphingosine or dehydrosphingosine. The amide linkage of sphingomyelins is much more resistant to alkali-catalysed hydrolysis than are the ester linkages of other phospholipids.

Sphingomyelins, which dissolve in hot ethanol but not in ether or acetone, have more hydrophilic character than do the phosphoglycerides, which are all soluble in ether.

12.2.9. Ceramides

These lipids, which are simple amides of sphingosine containing no phosphorus, are widely distributed but are never present in high concentration in tissues.

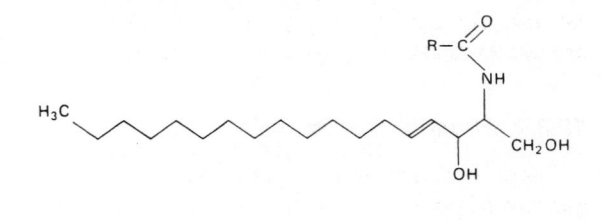

12.2.10. Glycosphingosides (glycolipids)

Because of their carbohydrate content, these lipids are strongly hydrophilic and are easily lost from tissues owing to their solubility in water. There are three types:

(a) Cerebrosides. These are ceramides in which the terminal hydroxyl group of sphingosine is joined by a glycosidic linkage to a hexose sugar, which is most commonly a β-D-galactosyl residue:

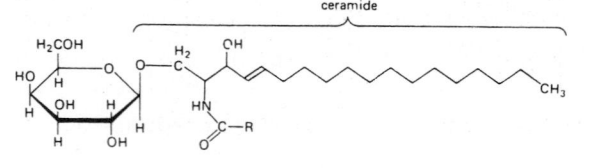

(b) Gangliosides resemble the cerebrosides, but have an oligosaccharide chain in place of the single hexose residue. Sialic acids and *N*-acetyl hexoses are always present. They have structures such as: NANA(α-2$\rightarrow$4)DGalNAc(β-1$\rightarrow$4)D-Gal(β-1$\rightarrow$Ceramide).

(c) Sulphatides are cerebrosides in which one of the alcoholic hydroxyl groups of the hexose is esterified by sulphuric acid. In the commonest sulphatides of mammalian nervous tissue a β-galactosyl residue is sulphated at position C3.

12.2.11. Lipids conjugated to protein

Proteolipids are compounds in which each protein molecule is combined with several of lipids, so that the complete molecule is soluble in non-polar solvents.

Lipoproteins are large protein molecules with some bound lipid. They are insoluble in non-polar solvents and are also insoluble in most polar solvents, so they are not extracted from tissues by the process of embedding in paraffin wax. The lipid moieties of lipoproteins, which are phospholipids, can be dissolved out of sections only if the bonds to protein are broken. The cleavage can be brought about by adding a strong acid to a suitable solvent. The chemical nature of the bonds between lipid and protein is imperfectly understood.

12.3. Histochemical methodology

The many techniques of lipid histochemistry have been thoroughly reviewed by Adams (1965) and Bayliss High (1984). Some older methods, valuable especially for revealing structural features of invertebrate tissues, are reviewed by Wigglesworth (1988). Very few of the available procedures selectively demonstrate any of the classes of lipid described above. It is possible,

however, to obtain information about several physical properties and chemical constituents of these substances. Methods are available that will detect with reasonable certainty the presence of:

1. Any lipids.
2. The hydrophobic or hydrophilic nature of a lipid.
3. Unsaturation.
4. Carbohydrates.
5. Free fatty acids.
6. The 1,2-unsaturated ether groups of plasmalogens.
7. Cholesterol and its esters.
8. Choline in phospholipids.
9. Amide rather than ester linkages.

Both the physical and chemical properties have to be taken into consideration when attempting to identify and localize lipids in tissues. The specificities of techniques for the demonstration of lipids have been determined mainly by experiments using pure lipids incorporated into thin paper, which is then treated as if it were a section being stained. Baker (1946) and Adams (1965) have carried out thorough studies of this kind.

Lipids are mostly unaffected by chemical fixation of tissue (Heslinga & Deierkauf, 1961; Deierkauf & Heslinga, 1962), but the presence of calcium ions in the fixative (as in Baker's formal–calcium) increases the preservation of the hydrophilic phospholipids, possibly by forming insoluble calcium–phosphate–lipid complexes.

Bayliss High (1977) has shown that fixation in formal–calcium should not exceed 2–3 days' duration for blocks of tissue or 1 h for cryostat sections of fresh tissue. Longer fixation impairs the staining of phosphoglycerides. Calcium ions derived from the fixative form salts (calcium soaps), insoluble in non-polar solvents, with free fatty acids. Treatment with a strong acid (e.g., 1.0 M HCl for 60 min) will regenerate the fatty acids.

Frozen sections are used for the histochemical examination of lipids, but appreciable quantities of phospholipids persist in sections of paraffin-embedded tissue. They occur mainly in myelin sheaths of nerve fibres and in erythrocytes. These lipids are stained in some of the histological techniques for the demonstration of myelin (see Chapter 18). They can be extracted from paraffin sections by suitable mixtures of solvents.

Phospholipids are rendered completely insoluble in organic solvents by prolonged treatment with potassium dichromate. This technique, known as **"chromation"**, has been known for over a century (Elftman, 1954) but the chemistry of the process is not yet understood. Lillie (1969) has shown that the chromating reagent reacts with double bonds in lipids, though other functional groups such as hydroxyls may also be involved in the binding of chromium, especially if chromation is carried out at 60°C rather than at 3° or 24°C. He has suggested that a cyclic ester is produced:

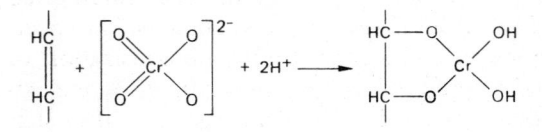

The oxidation number of the chromium in the proposed product is +4. Compounds of Cr(IV) are rare, though some stable organometallic complexes are known. The free hydroxyl groups of such a complex would be expected to participate in the formation of dye–metal complexes (e.g. with haematein) if the coordination number of the chromium atom were 6. Coordination numbers 4 and 6 exist in known Cr(IV) complexes. Lillie's speculations do not take into account the fact that the chromating reagent, between pH 2 and pH 6, exists as an equilibrium mixture of hydrogen chromate and dichromate ions:

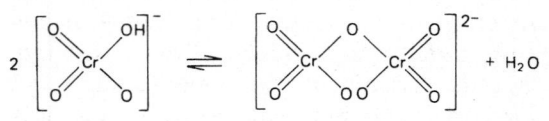

The chromate ion $[CrO_4]^{2-}$ required by Lillie's equation is formed when the pH of the solution is above 6 (see Cainelli & Cardillo, 1984 for chemistry of chromium). A dichromate ion might be able to combine with the two unsaturated sites to give a Cr(VI) *bis*-ester. Such cross-linking could account for the insolubilization of chromated lipids.

12.4. Extraction and chemical degradation

12.4.1. Rationale

The hydrophobic lipids, of which the neutral fats of adipose tissue are the most abundant, are extracted by **cold acetone**. It is important that the acetone be anhydrous. If traces of water are present, partial extraction of hydrophilic lipids occurs (Elleder & Lojda, 1971).

Either **pyridine** or a mixture of **chloroform and methanol** will extract all lipids except those that are firmly bound to protein. Proteolipids are soluble in methanol-chloroform but lipoproteins are not. In order to extract the firmly bound lipids of lipoproteins it is necessary to use an **acidified solvent**, which will hydrolyse the protein–lipid linkages and then dissolve the lipids (Adams & Bayliss, 1962).

Extractive methods may be used in conjunction with any staining method in order to confirm the specificity for the demonstration of lipids.

With some lipoproteins it is necessary to employ acid hydrolysis in an aqueous medium to release the lipid components for staining with solvent dyes (see below). Without such treatment, the hydrophilic protein prevents access of hydrophobic dye molecules to the lipids.

The ester linkages between fatty acids and glycerol in fats and phosphoglycerides can be broken by **alkaline hydrolysis (saponification)**. The fatty acids are liberated as their soluble sodium soaps. Amide linkages (with sphingosine) are not hydrolysed under the same conditions of time, temperature, and concentration of alkali. Consequently the fatty acid moieties of ceramides and sphingomyelins remain insoluble in water after saponification of the other lipids in the tissue. Glycosphingolipids may also be presumed to resist saponification, though they are largely extracted from tissues by aqueous reagents.

12.4.2. Solvent extraction procedures

Times given are for frozen sections. Use solvents in tightly closed containers. The sections are mounted onto slides and allowed to dry before extraction.

Cold acetone. Acetone (4°C, 1 h, but sometimes may need to be left overnight) extracts hydrophobic lipids only. It is important that no water be present in the acetone. Ordinary acetone is dehydrated by adding anhydrous calcium chloride, one-fifth of the total volume of $CaCl_2$ + acetone, and allowing to stand for 2 days. Use a bottle of at least 500 ml capacity, since much acetone is lost by absorption into the desiccant. Glassware must be dry and the sections must be allowed to dry in air before and after immersion of the slides in the anhydrous acetone (Elleder & Lojda, 1971).

Hot methanol-chloroform. A mixture of methanol (2 volumes) and chloroform (1 volume), at 58–60°C for 18 h (2 h is sometimes sufficient) removes all lipids except those firmly bound to protein.

Methanol-chloroform-HCl. Methanol, 66 ml; chloroform, 33 ml; concentrated hydrochloric acid, 1.0 ml; 18 h at room temperature (25°C) or 2 h at 58°C. Removes all lipids, including the phospholipid moieties of lipoproteins (Adams & Bayliss, 1962).

12.4.3. Hydrolysis of esters

Esters (but not amides) are hydrolysed by treating free-floating sections for 1 h at 37°C with 2 M (8%) sodium hydroxide (NaOH). The sodium soaps of the liberated fatty acids are dissolved out. After hydrolysis the sections are fragile and should be gently washed in water, followed by 1% aqueous acetic acid, and then returned to water.

12.4.4. Unmasking masked lipids

Lipids firmly bound to protein are released by acid hydrolysis prior to staining with a solvent dye. Sections are treated with 25% aqueous acetic acid for 2 min and are then washed in five changes of water. The released lipids are then amenable to extraction by solvents and to staining with lysochromes.

12.5. Solvent dyes

12.5.1. Solvent dyes (lysochromes)

Coloured, non-polar substances dissolve in lipids and render them visible under the microscope. Because coloured substances are not necessarily dyes, Baker (1958) prefers to call them **lysochromes**. These coloured substances stain fat because they are more soluble in it than in the solvents from which they are applied. The first solvent dyes to be used for staining fat were Sudan III and Sudan IV. Later, oil red O, which is more hydrophobic, was introduced. (See Chapter 5, pp. 66–67 for chemistry of these and other solvent dyes.) These dyes dissolve in hydrophobic lipids that are liquid at the temperature used for staining: neutral fats and those esters of cholesterol whose acyl groups are unsaturated. They give only weak colour to the hydrophilic phospholipids.

A more useful reagent for colouring all types of lipid is Sudan black B. This is a mixture of two main components and 8–11 coloured contaminants (Marshall, 1977). Several other unwanted dyes are formed in solutions that are more than one month old. Some of these products of deterioration stain proteins and nucleic acids (Frederiks, 1977). Even in fresh solutions, contaminating anionic dyes can give rise to false-positive identification of lipids (Malinin, 1977).

Lysochromes can dissolve in lipids only at temperatures above the melting points of the latter. This is a point of some importance, since lipids in which all the fatty acid chains are saturated melt above 60°C. Unsaturated lipids are mostly liquid at room temperature for a reason given in Section 12.6. Cholesterol melts at 150°C and its saturated esters also have high melting points. However, it is possible to stain these lipids with Sudan black B if the sections have first been treated with bromine water. This reagent reacts with unsaturated linkages in fatty acids (see Section 12.6.3) and Bayliss and Adams (1972) have shown that it also reacts with cholesterol to form an oily derivative which is liquid at room temperature. This derivative is not 5,6-dibromo-cholesterol (which has a higher melting point) and its chemical nature is unknown. The bromine-Sudan black method stains virtually all lipids. The only ones not stained would be fully saturated triglycerides or saturated free fatty acids.

Adams (1965) points out that the staining of lipids is affected by the solvent used for the lysochrome. Thus phospholipids are more likely to be stained from a solution in 70% than from one in absolute alcohol. Phosphoglycerides and free fatty acids are generally supposed to be more prone to extraction by the solvent than are lipids of other types. Some lipoproteins cannot be stained by solvent dyes without prior acid hydrolysis (see p. 205). In such substances the lipid is said to be "masked" by the protein.

12.5.2. Sudan IV method

Preparation of stain

Prepare a saturated solution of Sudan IV in 70% ethanol in a tightly stoppered bottle. Allow to stand for 2 or 3 days before using the supernatant solution. Keeps indefinitely.

Procedure

1. Cut frozen sections and rinse them in 70% ethanol.
2. Stain in Sudan IV solution for 1 min.
3. Transfer to 50% ethanol for a few seconds, until no more clouds of dye leave the section.
4. Wash in two changes of water.
5. (Optional.) Counterstain nuclei progressively with alum-haematoxylin.
6. Wash in water.
7. Mount in an aqueous medium.

Result

Lipids (especially neutral fats)—orange to red. Small intracellular droplets cannot be resolved and hydrophilic lipids are only weakly stained, but the method is quite adequate for adipose tissue and for the detection of isolated fat cells.

12.5.3. Sudan black B methods

This is the bromine-Sudan black B method of Bayliss & Adams (1972) for the demonstration of all types of lipid. If the bromination is omitted, free cholesterol is not stained and phospholipids are less strongly coloured. (See *Note 1* below.)

Solutions required

A. Bromine water

Bromine:	5.0 ml
Water:	200 ml

Bromine is caustic, and its brown, pungent vapour is injurious to the respiratory system. It should be kept in a fume cupboard and must not be pipetted by mouth. The aqueous solution is no more hazardous than most other laboratory chemicals. Bromine water is stable for about 3 months at room temperature in a glass-stoppered bottle and may be used repeatedly. It should be replaced when its colour becomes noticeably weaker.

B. Sodium metabisulphite

0.5% aqueous $Na_2S_2O_5$. Dissolve on the day it is to be used.

C. Sudan black B

Add 600 mg of Sudan black B (C.I. 26150) to 200 ml of 70% ethanol. Place on a magnetic stirrer for 2 h, then pour into a screw-capped bottle. Leave to stand overnight. To use the solution, filter it into a coplin jar and try not to disturb the sediment of undissolved dye in the bottom of the bottle.

The solution of Sudan black B may be kept (and used repeatedly) for 4 weeks.

D. A counterstain

Alum-brazilin and Mayer's carmalum (see Chapter 6) are suitable nuclear counterstains because they act progressively and are stable in aqueous media.

Procedure

1. Cut frozen sections, mount them onto slides, and allow them to dry.
2. Immerse slides in bromine water (solution A) for 1 h.
3. Rinse in water.
4. Immerse in sodium metabisulphite (solution B) for about 1 min until the yellow colour of bromine has been removed from the sections. (See *Note 2* below.)
5. Wash in four changes of water.
6. Rinse in 70% ethanol.
7. Stain in Sudan black B (solution C) for 10 min with occasional agitation of the slides.
8. Rinse in 70% ethanol, 5–10s (see *Note 3* below), with agitation, then wash in water.
9. Apply counterstain if desired.
10. Wash in water and mount in an aqueous medium.

Result

Lipids appear in shades of deep grey, very dark blue, and black.

Notes

1. Stages 2–5 may be omitted if it does not matter that cholesterol will be unstained.
2. In the original account of this method, "sodium bisulphate" was specified as the reagent for decolorizing the brominated sections: obviously an error. Dilute aqueous solutions of sodium or potassium sulphite, bisulphite, or thiosulphate may be used instead of sodium metabisulphite.
3. This differentiation is the only critical step in the method. Ideally a lipid-extracted control section should be included with those being stained. When the lipid-free section is completely decolorized the end-point of the differentiation has been reached.

12.6. Tests for unsaturation

Olefinic linkages occur in isoprene, cholesterol, and sphingosine, and in the very widely distributed unsaturated fatty acids. Consequently, the histochemical demonstration of the carbon-carbon double bond is tantamount to the staining of all lipids. Unsaturated linkages in fatty acids can be of the *cis* or *trans* type. The *cis* configuration is the more abundant in fatty acids of animal tissues.

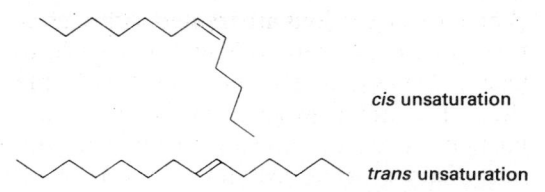

cis unsaturation

trans unsaturation

It can be seen that a *cis* double bond produces a bend in the chain of carbon atoms. This impairs the packing of the molecules in the solid state and in consequence *cis*-unsaturated lipids have lower melting points than do saturated or *trans*-unsaturated ones. Thus glyceryl tripalmitate melts at 65.1°C while glyceryl trioleate melts at –4°C. Since lysochromes dissolve only in lipids that are in the liquid phase, these dyes impart their colours only to structures containing unsaturated lipids. The ordinary lysochromes, such as Sudan III, Sudan IV, and oil red O, stain only those unsaturated lipids that are hydrophobic. Sudan black B (see p.67) is different and can also enter hydrophilic domains. The methods based on the use of oil-soluble dyes demonstrate lipids by virtue of their physical properties and may reasonably be called "histophysical" techniques (Adams, 1965; Bayliss High, 1977). There are, however, several genuinely histochemical reactions for the demonstration of unsaturation. Some of these will now be discussed.

12.6.1. Osmium tetroxide

The reactions of osmium tetroxide with various substances present in tissues were discussed in some detail in Chapter 2. There it was shown that this compound oxidizes the —CH=CH— bond and is reduced to a black substance, probably osmium dioxide. Since osmium tetroxide is soluble in both polar and non-polar solvents, it serves as a stain for both hydrophilic and hydrophobic lipids. It is possible, however, to distinguish between the two types by mixing the osmium tetroxide with an oxidizing agent that dissolves only in polar liquids. Potassium chlorate is a suitable oxidizing agent for this purpose. When frozen sections are treated with such a mixture, the osmium tetroxide will form cyclic esters (see p. 17) at both hydrophilic and hydrophobic sites. The other product of the reaction, the unstable osmium trioxide, will then disproportionate to give osmium tetroxide (soluble) and dioxide (insoluble and black). However, at hydrophilic sites in the tissue, the newly formed osmium dioxide will immediately be oxidized by chlorate ions:

$$3OsO_2 + 2ClO_3^- \rightarrow 3OsO_4 + 2Cl^-$$

Hydrophilic unsaturated lipids will therefore remain converted to cyclic osmium esters, which may be colourless or brown and are barely visible in the sections. At hydrophobic sites, the precipitation of osmium dioxide will be unimpeded, so that intense black staining will be observed. The cyclic esters of the unsaturated hydrophilic lipids can be stained subsequently by treating the sections with α-naphthylamine, which forms a red or orange complex with the bound osmium. These reactions form the basis of the OTAN (osmium tetroxide-α-naphthylamine) method for differential staining of hydrophobic and hydrophilic lipids. The Marchi method for degenerating myelin (see Chapter 18) is similar. The chemical reactions described above follow Adams *et al.* (1967) and Adams & Bayliss (1968), but have been slightly modified from these accounts in view of the work of Korn (1967).

In practice, osmium tetroxide seems to be a highly sensitive and specific reagent for the demonstration of unsaturation, though the OTAN reaction may sometimes fail to distinguish correctly between hydrophilic and hydrophobic lipids (see Bayliss High, 1977). The reactions of osmium tetroxide with phenolic compounds and with proteins (Nielson & Griffith, 1978, 1979; see also Chapter 2) should not be ignored, however. The specificity can be checked by staining control sections in which unsaturated linkages have been blocked by bromination (see p. 211) and others from which the lipids have been extracted by solvents.

Solution required

Osmium tetroxide solution

Water:	100 ml
Osmium tetroxide:	1.0 g

To prepare this solution, carefully clean the outside of the sealed glass ampoule in which the OsO_4 is supplied, removing all traces of the label and any gum with which it was attached. Score the glass with a diamond scriber, clean off sebum from your skin with acetone, and allow the ampoule to dry. Drop the scored, cleaned ampoule into a very clean bottle containing 100 ml of the purest available water. Insert the glass or rubber stopper (no grease may be used) and shake the bottle until the ampoule breaks. If necessary, the ampoule may be broken by striking it with a clean, degreased glass rod. The OsO_4 often takes several hours to dissolve completely.

In a clean, tightly stoppered bottle, this solution keeps for a few months at 4°C. The solution may be used several times provided that all glassware is very clean. Debris derived from sections may be removed by filtration (in fume cupboard) when the solution is poured back into its bottle. The used filter paper should be soaked overnight in 10% alcohol to reduce OsO_4 to $OsO_2.2H_2O$, before throwing it away.

Osmium tetroxide should always be used in a fume cupboard.

Deterioration is indicated by the presence of a blue-grey colour in the solution, due to colloidal osmium dioxide. Old solutions should be pooled in a screw-capped container and kept for recycling (see Kiernan, 1978).

Procedure

1. Wash frozen sections in water. They may be mounted onto slides, preferably without any adhesive.
2. Transfer to the OsO_4 solution and leave in a tightly closed container for 1 h.
3. Wash free-floating sections in five changes of water (at least 10 ml per section), 2 min in each change. Mounted sections may be washed in running tap water if they are still firmly adherent to their slides.
4. Dry free-floating sections onto slides or blot mounted sections with filter paper.
5. Mount in an aqueous medium. Alternatively, dehydrate in dioxane (two changes, each 4 min with occasional agitation), clear in carbon tetrachloride (two changes, each 1 min), and mount in a resinous medium.

Result

Unsaturated lipids—black.

Notes

1. The hydrophobic lipids can be extracted with acetone (see p. 206). This may reveal hydrophilic lipids in the same sites.
2. Thorough washing is necessary to remove excess OsO_4.
3. Alcohols are avoided for dehydration because they would reduce any OsO_4 not removed by aqueous washing. The clearing agent recommended is one in which OsO_4 is extremely soluble. **Caution:** toxic vapours from dioxane and carbon tetrachloride.

12.6.2. Palladium chloride

Unsaturated hydrophilic lipids can be demonstrated by virtue of their ability to reduce the chloropalladite ion $[PdCl_4]^{2-}$ to metallic palladium. The chloropalladite ion is formed when palladious chloride is dissolved in aqueous hydrochloric acid:

$$PdCl_2 + 2H^+ + 2Cl^- \rightarrow 2H^+ + [PdCl_4]^{2-}$$
$$(H_2PdCl_4 = \text{chloropalladious acid})$$

An unstable complex is formed with olefinic compounds and the complex is reduced to the metal, which is visible as a black deposit. Since the reagent is used in aqueous solution, it is reduced only by hydrophilic unsaturated lipids. The chloropalladite ion also behaves as an anionic dye, imparting a yellow background colour to non-lipid substances. This non-specific staining can be largely removed by treating the sections with pyridine, with which are formed complexes of the type $[Pd(pyr)_2Cl_2]$. The chemistry of the technique is discussed in greater detail by Kiernan (1977a).

Palladium chloride has been used in empirical staining procedures for the nervous system (Paladino, 1890) and in electron microscopy to impart electron density to elastin (Morris *et al.*, 1978). The reaction with elastin has not been studied chemically, but it does not result in the formation of black products, so there is no possibility of confusion with hydrophilic lipids in light microscopy.

In addition to its use as a histochemical test, this method (Kiernan, 1977a) is suitable for the histological demonstration of myelinated nerve fibres in the peripheral nervous system. In the central nervous system, myelin is also coloured , but stained lipids in the grey matter reduce the contrast.

Solutions required

A. Chloropalladious acid solution

Stock solution

Palladium chloride ($PdCl_2$):	1.0 g
Concentrated hydrochloric acid:	1.0 ml
Water:	5.0 ml

Mix thoroughly to disperse the $PdCl_2$, then add:

Water:	to 50 ml

Keeps for several months

Working solution

Stock solution:	1.0 ml
Water:	9.0 ml

This diluted solution can be used repeatedly until it becomes cloudy.

B. 20% aqueous pyridine

Pyridine:	20 ml ⎱ Keeps
Water:	80 ml ⎰ indefinitely

Procedure

1. Wash frozen sections in water. (See *Note 1*)
2. Transfer to working solution of chloropalladious acid for 2 h at 37°C. (See *Note 2* below.)
3. Rinse in two changes of water, 1 min in each.
4. Immerse sections in 20% aqueous pyridine for 1 min.
5. Rinse in water.
6. Dehydrate through graded alcohols, clear in xylene, and mount in a resinous medium.

Result

Unsaturated hydrophilic lipids—dark brown or black. Background pale yellow. (See *Note 3* below.)

Notes

1. Either free-floating sections or sections dried onto slides may be used.
2. Alternatively, leave the sections in chloropalladious acid overnight at room temperature, or for 30 min at 58°C.
3. Bromination or extraction of hydrophilic lipids prevents the reaction but the yellow background coloration is unaffected.

12.6.3. Bromination

Unsaturated linkages are **blocked** by bromination, either by exposure to bromine vapour

or by treatment with bromine water or an aqueous solution of bromine in potassium bromide $(KBr + Br_2 = KBr_3)$

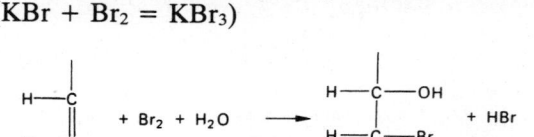

This test, though not entirely specific, may be used to confirm the unsaturated nature of substances stained by the methods discussed above.

The bromo derivatives of unsaturated lipids are decomposed with liberation of bromide ions when treated with a dilute mineral acid. The bromide ions can be precipitated as silver bromide, which can then be reduced to black metallic silver. These reactions constitute the bromine-silver method for the histochemical detection of unsaturation (Norton *et al.*, 1962):

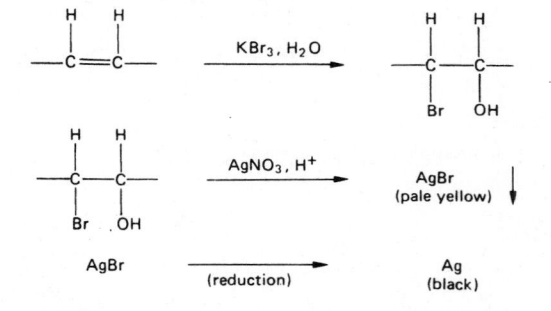

The specificity of this technique is marred, however, by the occasional non-specific deposition of silver at sites in the tissue that do not contain lipids.

Procedure for bromination

Potassium bromide, 6 g; water, 300 ml; bromine, 1.0 ml (**caution**: use fume cupboard; see p. 207). Keeps for a few months. Replace when the colour has faded.

Treat sections with this solution for 5 min at room temperature. Rinse in a bisulphite or thiosulphate solution to remove yellow stain of bromine and wash thoroughly with water.

This treatment prevents reactions due to unsaturated (—CH=CH—) linkages.

12.7. Glycolipids

Carbohydrates in glycosidic combination with lipids are demonstrated by the techniques of carbohydrate histochemistry. It should be remembered that inositol, though not a sugar, has an arrangement of hydroxyl groups similar to that found in some hexoses. Phosphatidyl inositols will contain the glycol configuration unless three or more of their hydroxyl groups are phosphorylated. Distinction between glycolipids and mucosubstances is made by using extractive procedures for the lipids. It is not possible to draw satisfactory conclusions when the two types of substance are present in the same place.

When the PAS reaction is used with frozen sections, allowance must be made for the possible presence of aldehydes generated by atmospheric oxidation of unsaturated fatty acids. Direct positive staining with Schiff's reagent from this cause is known as the **pseudoplasmal** reaction. The periodate ion also oxidizes a proportion of the unsaturated linkages to pairs of aldehyde groups, resulting in a truly positive PAS reaction that is not due to carbohydrate.

A modified PAS method, devised to circumvent these reactions of olefinic bonds, is described by Bayliss High (1977, 1984). Primary amine groups are first oxidatively deaminated (producing aldehydes), unsaturated linkages are next oxidized to aldehydes, with performic acid. All the free aldehyde groups are then blocked with 2,4-dinitrophenylhydrazine. A conventional PAS procedure is then carried out. Since all the aqueous reagents used extract gangliosides, the only lipids stained by this method are cerebrosides (and possibly phosphatidyl inositols).

Lipids containing acid sugars (i.e. gangliosides and sulphatides) are conspicuous only in the tissues of patients with certain lipid-storage diseases. Special methods, not used in ordinary carbohydrate histochemistry, are usually employed for the histopathological diagnosis of these conditions.

12.8. Free fatty acids

The soaps formed from fatty acids and heavy metals are insoluble in water. The cationic component of such a soap can be demonstrated by any suitable chromogenic reaction.

In **Holczinger's method**, sections are treated with a dilute solution of cupric acetate. Loosely bound copper is then removed by a brief treatment with a chelating agent, EDTA. The residual metal, which has been shown by Adams (1965) to be associated only with fatty acids, is made visible by forming an insoluble dark-green compound with dithiooxamide (see Chapter 13).

Elleder & Lojda (1972) were less impressed with the specificity of this technique than was Adams (1965). They found that free fatty acids were sometimes weakly stained when known to be present in considerable quantities in some tissues and that there was false-positive coloration of some phospholipids and of calcified material as well as background staining of proteinaceous and carbohydrate-containing material. They showed that pre-treatment with 1.0M hydrochloric acid enhanced the staining of free fatty acids, probably by causing hydrolysis of the calcium soaps formed during fixation in formal-calcium. This pre-treatment also dissolved calcium phosphates and carbonate. The differentiation in EDTA could usefully be extended beyond the time suggested in the original method, to minimize the staining of copper bound to "background" structures, but extraction of control sections with cold acetone was necessary in order to ensure that positive results were due to free fatty acids and not to phospholipids. The modifications recommended by Elleder & Lojda (1972) are incorporated into the practical instructions for Holczinger's method given here.

Use unfixed cryostat sections or frozen sections of tissue fixed in formal-calcium. Acetone-extracted control sections must also be examined.

Solutions required

A. 1.0M hydrochloric acid

Concentrated hydrochloric acid (S.G. 1.19):	40 ml	Keeps indefinitely
Water:	to 500 ml	

B. Copper acetate

Cupric acetate $(CH_3COO)_2Cu.H_2O$:	5.0 mg	Prepare before using
Water:	100 ml	

C. 0.1% EDTA

Disodium ethylenediamine tetraacetate $(Na_2(EDTA).2H_2O)$:	100 mg	Prepare before using
Water:	80 ml	
Adjust to pH 7.1 with drops of 1.0 M (=4%) NaOH, then add:		
Water:	to 100 ml	

D. Dithiooxamide solution

Dithiooxamide:	100 mg	Keeps for several months
Ethanol:	70 ml	
Dissolve and then add:		
Water:	30 ml	

Procedure

1. Affix frozen sections to slides.
2. Immerse in 1.0M HCl (solution A) for 1 h.
3. Rinse in three changes of water. Drain and allow sections to dry. Carry out acetone-extraction of control sections (see p. 206).
4. Immerse sections in copper acetate (solution B) for 3–4 h.
5. Transfer to two changes of 0.1% EDTA (solution C), 30 s in the first, 60 s in the second.
6. Wash in two changes of water.
7. Immerse in dithiooxamide solution (D) for 30 min. A shorter time (e.g. 10 min) will usually suffice. This stage need not be extended after the sections have stopped darkening.

8. Rinse in 70% ethanol, two changes, 1 min in each.
9. Wash in water and mount in aqueous medium.

Result

Free fatty acids—dark green to black. Any staining seen in acetone-extracted control sections is **not** due to free fatty acids.

12.9. The plasmal reaction

In frozen sections, plasmalogens yield aldehydes following a brief treatment with a 1% aqueous solution of mercuric chloride. The aldehydes are then demonstrated by means of Schiff's reagent (see Chapter 10, p. 160). The chemistry of the plasmal reaction has been worked out by Terner & Hayes (1961), whose paper should be consulted for the experimental evidence upon which the following account is based.†

Mercuric chloride adds to the double bond of the vinyl (1,2-unsaturated) ether linkage of plasmalogens:

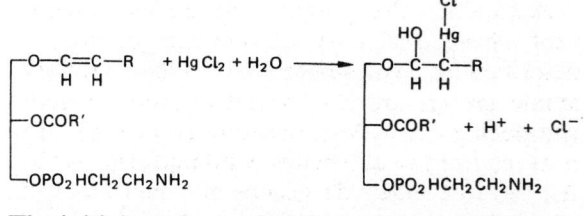

The initial product, a hemiacetal, is unstable and immediately dissociates into an alcohol and an aldehyde:

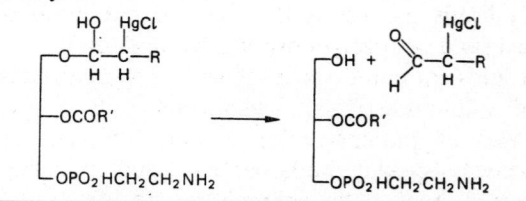

---

†In most publications dealing with plasmalogens, the double bond of the vinyl ether is said to be between the α- and β-carbon atoms. This is an incorrect usage of the Greek letters (see "Conventions and Abbreviations", page x of this book). These two carbons should be designated as 1 and 2, as in the present account. The letters α and β, used correctly, would refer to carbon atoms 2 and 3.

The mercury atom remains attached to carbon 2 of the aldehyde; its presence there can be demonstrated histochemically.

Vinyl ethers are much more easily hydrolysed in the presence of acids than are ordinary ethers, but the pH of the mercuric chloride solution used in the plasmal reaction (3.5) is not low enough to catalyse the hydrolysis. However, 6 M hydrochloric acid is just as effective as 1% mercuric chloride in generating aldehydes from plasmalogens:

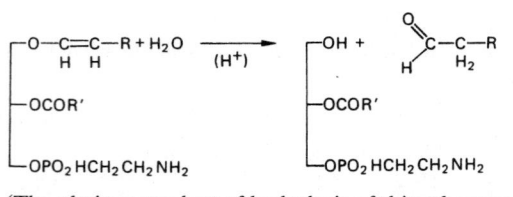

(The obvious product of hydrolysis of this ether would be an enolic "vinyl alcohol", RCH=CHOH. These compounds do not exist as such but as the tautomeric aldehydes.)

In practice, 6 M HCl is not used because it is more injurious to the sections than aqueous mercuric chloride. However, acid-catalysed hydrolysis of plasmalogens can occur in Schiff's reagent (pH 2.5), especially if the sections are immersed in it for more than 20 min (Elleder & Lojda, 1970). Consequently, omission of the treatment with mercuric chloride is not always an adequate control procedure for the plasmal reaction. Staining with Schiff's reagent alone could be due to plasmalogens as well as to a "pseudoplasmal" reaction (see p. 212). However, if there is no staining with Schiff's reagent alone, a positive reaction after treatment with mercuric chloride certainly indicates the presence of plasmalogens.

It is important to control for aldehydes already present in the tissue. If found these must be chemically blocked with sodium borohydride (see p. 164). If any staining occurs when there are no aldehyde groups in the tissue, the Schiff's reagent is not working properly and must be replaced. Use either cryostat sections of unfixed tissue or frozen sections of small specimens fixed in formaldehyde for no more than 6 h. The sections should be used within an hour of being cut.

Solutions required

A. 1% mercuric chloride

Mercuric chloride ($HgCl_2$)   1.0 g   } Keeps
Water: 100 ml   } indefinitely

B. Schiff's reagent

See under the Feulgen reaction (p. 13). Allow the Schiff's reagent to warm to room temperature before use.

C. Bisulphite water

Potassium metabisulphite
($K_2S_2O_5$):   5.0 g   } Prepare
Water: 1000 ml   } before
Concentrated hydrochloric   } using
acid (S.G. 1.19):   5.0 ml

Procedure

1. Wash sections (mounted on slides or coverslips) in three changes of water. This step is omitted for sections of unfixed tissue.
2. Immerse in 1% $HgCl_2$ (solution A), 1 min.
3. Transfer slides directly to Schiff's reagent for 5 min.
4. Transfer to bisulphite water (solution C), three changes, 2 min in each.
5. Wash in three changes of water.
6. A progressive aluminium-haematein counterstain (e.g., Baker's haematal-16; see p. 97) may be applied at this stage, if desired. Wash in running tap water after counterstaining.
7. Mount in an aqueous medium.

Result

Plasmalogens pink to purple, provided that a positive reaction in the absence of treatment with $HgCl_2$ is not obtained. Nuclei blue if counterstained as suggested.

12.10. Cholesterol and its esters

Of the various histochemical tests for steroids which have been devised, the one with the most

certain specificity is the perchloric acid-α-naph-
thoquinone reaction. Adams (1965) has shown
that this method gives a positive result with no
lipids other than cholesterol, its esters, and a few
other closely related sterols. The sections are
heated in a solution containing perchloric acid,
1,2-naphthoquinone-4-sulphonic acid, formaldehyde, and ethanol.

Perchloric acid is thought to convert cholesterol to a conjugated diene:

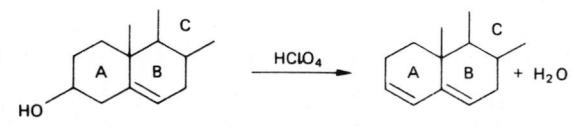

(Only the A and B rings of the cholesterol molecule are
shown. The remainder of the structure does not take part
in the reaction.)

This reaction is analogous to the well-known
preparative technique whereby ethylene is
formed by elimination of water from ethanol in
the presence of concentrated sulphuric acid.
Esters of cholesterol may be hydrolysed in the
strongly acid reagent and the resultant cholesterol then converted to the diene. In the next
phase of the method, the diene reacts with 1,2-
naphthoquinone-4-sulphonic acid (see p. 149) to
form a blue compound. The chemistry of this latter reaction and the nature of the end-product
are not understood. Neither are the roles of the
ethanol and the formaldehyde contained in the
reagent. The method does not work with pure
cholesterol. Prior oxidation either by air or by
ferric chloride is necessary (Bayliss-High, 1984).

Free cholesterol may be distinguished from its
esters by treating sections with a solution of digitonin before staining. Digitonin is a glycoside of
a plant sterol and it forms with cholesterol an
adduct which is insoluble in cold acetone. Esters
of cholesterol remain soluble in cold acetone,
along with other hydrophobic lipids. Digitonin
has also been incorporated into fixatives for the
purpose of making cholesterol insoluble and also
osmiophilic for histochemical studies with the
electron microscope. However, Vermeer *et al.*
(1978) have found that digitonin added to aqueous fixatives fails to immobilize cholesterol on
filter paper and that it accelerates the diffusion
of cholesterol esters. The specificity of histochemical tests involving the use of digitonin to
insolubilize cholesterol is therefore uncertain.

Fixation in formal-calcium is recommended for
this method. Nicholson & Monkhouse (1985)
have shown that much cholesterol is retained in
sections of tissue that have been fixed in glutaraldehyde and embedded (without alcoholic dehydration) in glycol methacrylate.

Preparation of reagents

*A. Perchloric acid-naphthoquinone (PAN) sol-
ution*

Ethanol:	20 ml
60% aqueous perchloric acid (S.G. 1.54):	10 ml
(Handle with care)	
Formalin (40% HCHO):	1.0 ml
Water:	9.0 ml
1,2-naphthoquinone-4-sulphonic acid:	40 mg

B. 60% aqueous perchloric acid (S.G. 1.54)

Procedure

1. Cut frozen sections and leave in 4% aqueous formaldehyde (from formalin) for 1
   week. Alternatively, leave the sections in a
   freshly prepared 1% (w/v) aqueous solution
   of ferric chloride ($FeCl_3.6H_2O$) for 4 h.
   Before starting the staining procedure, pre-
   heat a hotplate to approximately 65°C.
2. Wash sections in water and dry them onto
   slides.
3. Paint the sections with a thin layer of the
   PAN (solution A) and place on the hotplate
   for 10 min. Use the brush to apply more
   PAN solution at intervals to prevent drying.
   Colour changes from red to blue.
4. Place a drop of 60% $HClO_4$ (reagent B) on
   the section and apply a coverslip. Carefully
   remove excess perchloric acid from the
   edges of the coverslip, with filter paper.

Result

Cholesterol, its esters and a few closely related steroids are stained dark blue. (See *Note* below.) Pink background colours are not due to lipids. The colour is not stable in water or in ordinary mounting media.

Note

To demonstrate free cholesterol alone, proceed as follows:

(a) Carry out stages 1 and 2 of the above method.

(b) Place slides in a 0.5% solution of digitonin in 40% ethanol for 3 h.

(c) Immerse slides in acetone for 1 h at room temperature.

(d) Rinse in water.

(e) Proceed with stages 3 and 4 of the above method. Esters of cholesterol are extracted by the acetone, but free cholesterol is rendered insoluble by combination with digitonin.

The perchloric acid-naphthoquinone method is damaging to sections, so the stained preparations show poor structural preservation. There are other histochemical tests for steroids, but most are just as destructive and have less chemical specificity. Cholesterol can be stained by the bromine-Sudan black method (p. 207) and by the fluorescent Nile red method described in Section 12.11.2 of this chapter. These techniques are not injurious to the tissue, but they have little specificity for cholesterol, even when used in conjunction with solvent extractions and digitonin.

12.11. Miscellaneous techniques

Two useful methods for lipids remain to be considered. Baker's acid-haematein test is a valuable staining method for mitochondria, myelin and other structures rich in phospholipids. Nile blue and Nile red have been in occasional use for many years, but have recently enjoyed a resurgence of popularity with the rediscovery and further investigation of the fluorescence of Nile red.

12.11.1. Acid-haematein for choline-containing lipids

Various techniques are available for the selective demonstration of phosphatidyl cholines and sphingomyelins. Some of them (e.g. Bottcher & Boelsma-van Houte, 1964; Hadler & Silveira, 1978) are derived from analytical chemical methods for choline and its derivatives. They will not be discussed here because they are not very frequently used. A more popular technique is Baker's (1946) acid-haematein test. Although the chemistry of this technique is poorly understood, model experiments with lipids and other substances on paper have revealed that it is specific for choline-containing phospholipids (Adams, 1965) if the instructions are followed faithfully. Baker's (1946) investigation indicated a lower degree of specificity, but it is likely that his specimens of phosphatidyl ethanolamines ("cephalin") and cerebrosides ("brain galactolipine") were less pure than those available to Adams (1965).

In the first stage of the acid-haematein procedure, sections fixed in either formal-calcium or a mixture similar to Bouin's fluid are exposed to a solution containing potassium dichromate. As explained earlier (p. 205), this reagent probably combines covalently with unsaturated linkages and also with nearby hydroxyl groups. The chromated sections are then stained with an acidic solution of haematein (see Chapter 5), which forms a dark blue dye-metal complex with the bound chromium. Differentiation in an alkaline solution of potassium ferricyanide leaves the strong colour only in structures such as mitochondria and myelin sheaths, which contain phosphatidyl cholines and sphingomyelin. This coloured product is insoluble in alcohols and xylene, so the preparations can be mounted permanently in resinous media. Aside from its histochemical value, the acid-haematein method is one of the best techniques for staining mitochondria. Other methods for the demonstration of these organelles by light microscopy of fixed tissues (see Gabe, 1976, for detailed descriptions) are purely empirical. A yellow background coloration is probably due to unmordanted haematein acting as a simple anionic dye.

The mode of action of the differentiator is unknown. Baker (1958) suggested that it might oxidize haematein to more faintly coloured compounds (see Chapter 5, p. 73).

Baker (1946) prescribed an extraction with pyridine as a control procedure to confirm that stained structures were lipids. Nowadays other solvents are preferred and it is also possible to employ a saponification procedure (see p. 206) so that the method becomes specific for sphingomyelins.

In the original method of Baker (1946) the chromation was carried out before sectioning, and the procedure took 4 days to complete. The current practice is to treat sections with dichromate, and to use much shorter times for staining and differentiation.

Solutions required

A. Dichromate-calcium

Potassium dichromate
$(K_2Cr_2O_7)$:   15 g ⎫ Keeps
Calcium chloride $(CaCl_2)$: 3.0 g ⎬ indefinitely
Water: 300 ml ⎭

B. Acid-haematein

Haematoxylin (C.I. 75290):   50 mg
Water: 48 ml
1% aqueous sodium iodate $(NaIO_3)$: 1.0 ml

Heat until it just boils, then allow to cool to room temperature and add:
Glacial acetic acid: 1.0 ml
This reagent is used on the day it is prepared.

C. Borax-ferricyanide differentiator

Potassium ferricyanide, $K_3Fe(CN)_6$: 0.75 g
Borax (sodium tetraborate,
$Na_2B_4O_7.IOH_2O$):   0.75 g
Water: 300 ml

Stable indefinitely at 4°C. Use only once.

Procedure

1. Cut frozen (or cryostat) sections of tissue that has been fixed in formal-calcium, and mount on slides.
2. Transfer the slides to the dichromate-calcium solution (A), and leave overnight at about 60°C. (4 h is usually enough.)
3. Wash in 5 changes of water, 1 min in each.
4. Stain for 2 h in acid-haematein (Solution B), at about 37°C.
5. Wash in water until excess dye is removed.
6. Differentiate in Solution C for 2 h at 37°C.
7. Wash in water (3 or 4 changes, taking care to avoid detachment of the sections after treatment with the alkaline differentiator).
8. **Either** mount in an aqueous medium **or** dehydrate through graded alcohols, clear in xylene, and mount in a resinous medium.

Result

Certain phospholipids (phosphatidyl cholines and sphingomyelins) are coloured blue, blue-black, or grey. Background yellow. Blue-stained objects may only be assumed to contain phospholipid if they are unstained after appropriate solvent extraction. (See *Note 1* below.) Mitochondria are shown well by the acid-haematein method.

Notes

1. The extraction test used by Baker (1946) involved fixing a second piece of the same tissue in a "weak Bouin" mixture, extracting it with hot pyridine, rehydrating, and then proceeding with the acid-haematein method. Current practice favours extraction of frozen sections with methanol-chloroform or acidified methanol-chloroform (see Section 12.4.2) prior to chromation.
2. Sections may be subjected to alkaline hydrolysis (see Section 12.4.3) before chromation. Sphingomyelins will then be the only lipids stained by acid-haematein.

12.11.2. *Nile blue and Nile red*

The chemistry of Nile blue and its oxidation product Nile red has been reviewed in Chapter 5 (Section 5.9.14). The staining solution used in lipid histochemistry is a mixture of the two dyes, made in the laboratory by boiling an acidified aqueous solution of Nile blue. Although Nile red is insoluble in pure water, it dissolves easily in mixtures of alcohol or acetone with water, or in an aqueous solution of Nile blue. Presumably the hydrophobic molecules of the red dye are held by van der Waals forces to the similarly shaped Nile blue cations. Nile red can be extracted from the mixture of dyes by shaking with a non-polar solvent. The purified dye is used alone as a fluorochrome.

The fluorescence of Nile red is quenched in the presence of water, so the dye is a probe of hydrophobic environments. The wavelengths of maximum excitation and emission are affected by the polarity of the solvent. Thus, in a strongly hydrophobic medium Nile red is excited by blue (450–500 nm) light and emits yellow-orange (maximum at 528 nm). In an organic solvent miscible with water, the excitation maximum is around 550 nm (green), with an emission maximum at 628 nm (Greenspan & Fowler, 1985). Equivalent differences in fluorescence are seen in hydrophobic and hydrophilic lipids stained with Nile red.

When a frozen section of a tissue is treated with a solution of Nile red, this dye behaves like any other lysochrome and dissolves in hydrophobic lipids: principally the neutral fats. These are coloured red. If a very dilute solution of Nile red is used (Greenspan *et al.*, 1985), there is fluorescence at the sites in which the dye is present. Unlike the red colour seen by ordinary microscopy, this fluorescence is visible even in inclusions composed of cholesterol, which is solid at the temperature of staining. The visibility of fluorescence in cholesterol is due to the fact that fluorescence microscopy is more sensitive than bright-field microscopy. Even minute amounts of a fluorochrome stand out conspicuously against a dark background.

In frozen sections stained with an aqueous solution of Nile blue and Nile red, the cations of the former dye stain some of the hydrophilic lipids. These are the ones that are acidic due to the presence of ionized phosphate (or sulphate) ester groups. Free fatty acids, if present, are stained blue or purple, because they are somewhat hydrophilic on account of their carboxyl groups.

The staining of phospholipids and fatty acids by Nile blue cannot be explained simply by attraction of oppositely charged ions. The pH of the reagent is approximately 2, so most of the non-lipid macromolecular cations, such as nucleic acids, acquire little of the blue colour (see Chapters 6 and 9 for reasons). Sulphated carbohydrates, which can bind cationic dyes even at pH 1 (see Chapter 11), are stained by Nile blue, and the resulting colour is metachromatic (i.e. red), so it must not be confused with that of Nile red. When using Nile blue and Nile red, it is necessary to stain suitable control sections from which lipids have been extracted.

Nile blue and red staining solution

Nile blue sulphate (C.I. 51180):	1.0 g
Water:	200 ml
Sulphuric acid (10% v/v aqueous):	10 ml

Boil for 4 h (use a reflux condenser to prevent loss of water, or make up the volume manually from time to time). Cool and filter. Keeps for several months.

Nile red stock solution

Shake 100 ml of the Nile blue and red solution (above) in a separating funnel with three changes of 100 ml of xylene. Collect the xylene (with extracted Nile red) and evaporate to dryness in a rotary evaporator, to obtain about 0.4 g of solid dye. The powder is probably stable indefinitely. A stock solution is made by dissolving it in acetone to a concentration of 0.5 mg/ml. The solution in acetone can be kept for several months in a tightly capped container in darkness (e.g. in a refrigerator).

Nile red working solution (for fluorescence)

Nile red stock solution (above):   0.05 ml
Glycerol-water (75:25 by volume) mixture: 50 ml

Stir vigorously. Removal of bubbles is facilitated by exposing the mixture to reduced atmospheric pressure for a few minutes. This is the highest concentration of Nile red likely to be needed for fluorescence microscopy. Dilution (5 times) is necessary if over-staining occurs.

(Greenspan & Fowler, 1985, started with Nile blue chloride, rather than the sulphate. The chloride is less soluble in water, and is not used in staining methods for ordinary light microscopy.)

Nile blue sulphate method *(Cain, 1947)*

1. Cut frozen sections and attach to slides.
2. Stain in the Nile blue and red solution at 37°C, for 30 min.
3. Differentiate in 1% acetic acid, 2 min.
4. Wash in water, and mount in an aqueous medium.

Result

Unsaturated hydrophobic lipids pink to red; free fatty acids and phospholipids blue to purple. There is sometimes some blue staining of nuclei. Mast cell granules and other glycosaminoglycan-containing structures are red to purple (metachromasia). Free fatty acids can be extracted (cold acetone), leaving phospholipids as the only stainable substances; they then stain blue (see Bayliss-High, 1984).

Nile red fluorescence method

This procedure (Fowler & Greenspan, 1985) is applicable to frozen or cryostat sections or to monolayers or suspensions of cultured cells. They may be unfixed or fixed in a mixture suitable for lipid histochemistry.

1. Place a drop of the Nile red working solution (see above) on the preparation, apply a coverslip, and wait for 5 min.
2. Examine in a fluorescence microscope. Two filter sets are needed: one for blue excitation, as used for fluorescein, and one for green excitation (as used for rhodamine derivatives).

Result

With blue exciting light, hydrophobic lipids (neutral fat or cholesterol) show orange-yellow fluorescence. This can be prevented by prior extraction with acetone or alcohol. With green excitation, phospholipids show red fluorescence.

Notes

1. Nile blue sulphate (presumably containing Nile red) has been used as a fluorochrome for the examination of muscle biopsies (Bonilla & Prelle, 1987).
2. Anilinonaphthalene sulphonic acid (ANSA) was used by Cowden & Curtis (1974) as a supravital fluorochrome for hydrophobic cytoplasmic inclusions in the cells of invertebrates. ANSA can also impart fluorescence to hydrophobic proteins such as elastin (Vidal, 1978).

12.12. Exercises

Theoretical

1. How would you determine whether a structure stained by Sudan black B in 70% ethanol was composed of triglycerides or of phosphoglycerides?

2. **Tristearin** is a neutral fat in which glycerol is esterified by three molecules of stearic acid. Which of the following methods would and which would not give a positive reaction with tristearin? Give reasons for your answers.
  (a) Sudan IV at room temperature.
  (b) Sudan IV at 70°C.
  (c) Osmium tetroxide.
  (d) The plasmal reaction.

3. Answer question 2 for **triolein**, in which glycerol is esterified by three molecules of oleic acid.

4. Myelin, in frozen sections, commonly gives a direct-positive reaction with Schiff's reagent. Why? Describe exactly how you would proceed to demonstrate the presence in myelin of (a) plasmalogens, (b) fatty acid amides (ceramides and sphingomyelins).

5. It has been shown by Malinin (1980) that different types of intracellular lipid inclusions may be distinguished from one another by red staining with Nile blue at various temperatures from 19 to 65°C. What does this tell you about the mechanism of coloration?

6. What histochemical evidence could be obtained to

support the contention that lipid droplets in the adrenal cortex contain cholesterol and steroid hormones rather than neutral fats or phospholipids?

7. In metachromatic leucodystrophy there are accumulations in the brain of **sulphatide** in which the fatty-acid residues are largely saturated. Using histochemical methods, how would you distinguish these pathological deposits from (a) proteoglycans, (b) glycoproteins, (c) cerebroside, (d) ganglioside, (e) neutral fats, (f) any unsaturated lipids?

8. If wax is not adequately removed from paraffin sections, birefringent crystal-like bodies are seen in the tissue, within nuclei. Using histochemical methods, how would you show that these objects are composed of a hydrocarbon rather than of a naturally occurring lipid, lipoprotein, or mineral material? See Nedzel (1951) for further information concerning this interesting artifact.

13

Methods for
Inorganic Ions

13.1. General considerations

Inorganic ions are present in all parts of all tissues, but owing to their solubility they are not generally amenable to detection by histochemical methods. It has been possible to demonstrate some soluble ions, such as potassium and chloride, by making use of freeze-dried material or by including in the fixative a reagent which forms an insoluble salt with the ion concerned. It is, however, difficult to determine the extent to which a soluble ion has diffused during the course of the procedure from its location *in vivo* to the site at which the end-product of the histochemical reaction is seen.

When inorganic substances are present in tissues as insoluble compounds, be these salts such as calcium carbonate or complexes with protein, such as haemosiderin, they remain *in situ* while the specimens are being processed and can therefore be localized accurately by histochemical methods. Some metals, such as calcium, iron, and zinc, are present in animal tissues in sufficient quantity to permit their demonstration in normal material. Others are normally present but can only be detected histochemically when large amounts accumulate as the result of disease or experimental manipulation. Yet other metals are not involved in normal mammalian metabolism but may be found in the tissues following intoxication or excessive environmental exposure. Of the inorganic anions, phosphate and carbonate of mineralized tissue are the only ones normally present in an insoluble form.

Histochemical techniques for inorganic ions fall into two categories: those in which the visible products are black or coloured inorganic substances, and those in which coloured compounds are formed by chelation of metals by organic ligands. The chemistry of chelation has already been introduced in connection with mordant dyeing (Chapter 5). It is important to realize that the methods described in this chapter will only detect metals present as ions in the tissue. Metals already bound by strong coordinate bonds to organic ligands cannot react with the reagents used to demonstrate their ions. For example, the iron in erythrocytes is so firmly combined with the haem group of haemoglobin that it will not give a positive reaction in histochemical tests for iron.

It should be remembered that histochemistry is only one approach to the examination of inorganic elements in tissues. There are valuable physical methods too, including electron probe microanalysis, cathodoluminescence, mass spectrometry and laser microprobe emission spectroscopy. (For reviews, see Moreton, 1981; Schmidt & Moenke-Blankenburg, 1986.)

As examples of histochemical techniques for inorganic substances, we will consider methods for calcium, phosphate (and carbonate), iron, zinc, copper, and lead.

13.2. Calcium

This metal is demonstrated by virtue of its ability to form coloured chelates with dyes and other organic compounds. Dyes are chosen that do not form complexes of similar colour with other naturally occurring metals such as magnesium and iron.

13.2.1. Alizarin red S method

Suitable reagents for calcium are **alizarin red S**, a synthetic anthraquinone dye, and **morin**, a coloured (yellow) polyphenolic ketone obtained from certain tropical plants. Morin is used as a fluorescent stain for calcium and it also forms fluorescent complexes with many other metals not normally present in animal tissue.

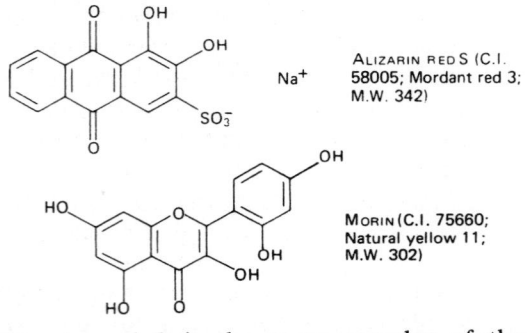

ALIZARIN RED S (C.I. 58005; Mordant red 3; M.W. 342)

MORIN (C.I. 75660; Natural yellow 11; M.W. 302)

Alizarin red S is the more popular of these reagents. Like other hydroxyanthraquinone dyes, alizarin red S forms chelates in which a quinone oxygen and a phenolic oxygen serve as electron donors to the metal atom, so that a stable six-membered ring is formed. Calcium, with coordination number 4, combines with two molecules of the dye:

The structure shown is widely accepted but has been challenged by Lievremont *et al.* (1982), who found that when the complex was precipitated *in vitro*, the ratio of metal:dye was close to 1.0. They suggested chelation by the ionized phenol and sulphonic acid groups.

Alizarin red S is also an anionic dye in its own right and, as such, imparts non-specific pink "background" coloration to the tissue.

The calcium ions must be released from the insoluble deposits in which they occur before they can combine with the dye to form an insoluble chelate. Consequently, some diffusion from the original site of deposition of calcium is inevitable. The histological picture is greatly influenced by the pH of the staining solution. An acid solution will extract much calcium and produce a conspicuous coloured deposit, but this will be diffusely localized. An alkaline solution will liberate fewer calcium ions, so that although the accuracy of localization will be greater the sensitivity of the method will be less. Different authors have recommended pH values ranging from 4 to 9 for solutions of alizarin red S.

An alcoholic or a neutral aqueous fixative should be used. Maximum preservation of calcium is probably achieved by fixation in 80% ethanol, though this is a rather poor fixative for histological purposes.

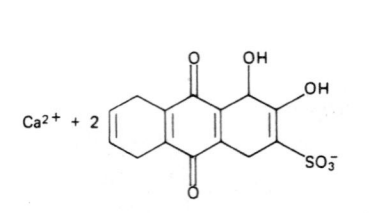 Ca^{2+} + 2 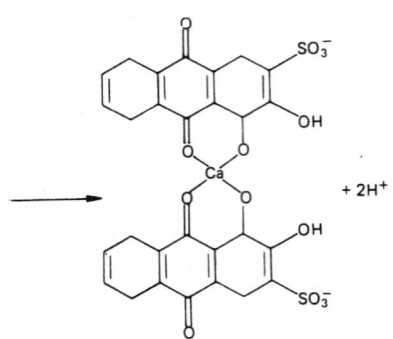 + $2H^+$

Solutions required

A. *Alizarin red S*

> Alizarin red S (C.I. 58005): 1.0 g
> Water: 90 ml
> Dilute NH₄OH (28% ammonia
>   diluted 100 times with water)
> Add in small aliquots until the
>   pH is 6.4—approximately
>   10 ml
>
> } Keeps for 4 weeks

B. *Differentiating fluid*

> Ethanol (95%):   500 ml
> Concentrated hydro-
>   chloric acid: 0.05 ml
>
> } Mix before using

Procedure

1. De-wax and hydrate paraffin sections. Stain in solution A for 2 min.
3. Wash in water for 5–10 s.
4. Differentiate in solution B for 15 s.
5. Complete dehydration in two changes of absolute ethanol, clear in xylene, and mount in a resinous medium.

Result

Calcium—orange to red. Background—dull pink.

13.2.2. GBHA method

A more sensitive histochemical method for calcium involves a chelating agent that is not a dye. This is glyoxal-*bis*-(2-hydroxyanil), also known as GBHA:

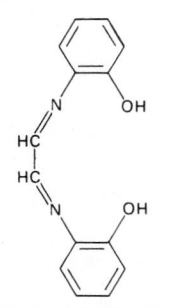

In the presence of a strong base such as sodium hydroxide, the phenolic hydroxyl groups ionize.

The anion of GBHA forms insoluble chelates with several metals, including calcium:

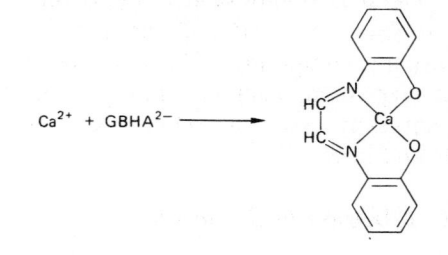

$$Ca^{2+} + GBHA^{2-} \longrightarrow$$

Metal chelates are also formed with strontium, barium, cadmium, cobalt, and nickel. If any of these metals are likely to be present in the tissue their chelates can be decomposed by treating the stained sections with an alkaline solution containing cyanide ions.

The GBHA method is very sensitive and can be used to detect calcium ions other than those of insoluble salts of the metal, provided that the tissue has been freeze-dried or freeze-substituted. After fixation in an aqueous reagent, the soluble calcium diffuses and some of it can be detected in nuclei of cells by the GBHA method. This artifact is presumably due to electrostatic attraction of calcium ions by the phosphate groups of nucleic acids. As with alizarin red S, the solution of GBHA used for staining deposits of insoluble calcium salts must contain some water and must not be too strongly alkaline. "Soluble" calcium salts, however, are best stained with a solution of GBHA in absolute ethanol containing some sodium hydroxide.

For this technique (Kashiwa & Atkinson, 1963; Kashiwa & House, 1964) it is essential to prepare the tissues as described and to avoid the use of any reagents (including water) that might contain traces of calcium.

Solutions required

A. *GBHA stock solution*

> Glyoxal *bis*-(2-hydroxy-
>   anil):   200 mg
> Absolute ethanol: 50 ml
>
> } Keeps for several weeks at 4°C

B. 10% NaOH solution

Sodium hydroxide (NaOH):	10 g
Water:	to 100 ml

Keeps for several weeks

C. GBHA working solution I

GBHA stock solution (A):	2.0 ml
10% NaOH (solution B):	0.15 ml
Water:	0.15 ml

Mix just before using

D. GBHA working solution II

GBHA stock solution (A):	2.0 ml
10% NaOH (solution B):	0.6 ml

Mix just before using

E. Alkaline cyanide solution

Absolute ethanol:	45 ml
Water:	5.0 ml
Sodium carbonate (Na_2CO_3): to saturation (approx. 1.0 g)	
Potassium cyanide (KCN): to saturation (approx. 1.0 g)	

Stable for a few weeks at room temperature

(**Caution:** potassium cyanide is poisonous. Do not allow it to come into contact with acids. Before throwing away add an excess of sodium hypochlorite solution and wait for 10 min. Then flush down sink with plenty of water.)

F. Fast green FCF counterstain

Fast green FCF (C.I. 42053):	0.24 g
95% ethanol:	300 ml

Keeps indefinitely and may be used repeatedly

Preparation of specimens

Freeze small blocks, no more than 2.0 mm³, in isopentane cooled in liquid nitrogen. **Either** freeze dry and embed in paraffin wax **or** freeze substitute in acetone at −80°C, clear in xylene, and embed in paraffin wax. Wolters *et al.* (1979) state that superior results are obtained by freeze substitution for 10 days at −80°C in acetone containing 1% oxalic acid (presumably as $H_2C_2O_4.2H_2O$). Cut sections at 7 µm and mount them onto slides without using water or any adhesive. The sections are not de-waxed before staining. (See also *Note 2* below.)

Staining procedure

1. Place the slides bearing the sections on a horizontal staining rack. **Do not remove wax.**
2. **Either:** flood with GBHA working solution I (C) for 3 min, **or** apply one or two drops of GBHA working solution II(D) per section and allow to evaporate to dryness.
3. Rinse in 70% ethanol, then in 95% ethanol.
4. (Optional, see *Note 1* below.) Immerse for 15 min in alkaline cyanide solution (E).
5. Rinse in three changes of 95% ethanol.
6. Apply a counterstain, if desired: rinse in absolute ethanol; de-wax in xylene; rinse in absolute, followed by 95% ethanol. Stain in alcoholic fast green FCF (solution F) for 3 min. Rinse in three changes of 95% ethanol, then proceed to stage 7 below.
7. Dehydrate in four changes of absolute ethanol, clear in xylene, and mount in a resinous medium.

Results

When GBHA working solution I is used, a red colour is seen at sites of soluble calcium salts, but insoluble calcified material is largely unstained. With GBHA working solution II, the red colour is seen predominantly at sites of insoluble calcium salts. If the fast green FCF counterstain is used, the background is bluish green. The red colour fades after 2 or 3 days, so the sections should be photographed to obtain a permanent record.

Notes

1. The alkaline cyanide reagent decolorizes the chelates of metals other than calcium (Sr^{2+}, Ba^{2+}, Cd^{2+}, Co^{2+}, Ni^{2+}). Stage 4 may be omitted if the presence of other metals is unlikely.

2. It is necessary to use freeze-dried or freeze-substituted tissue in order to minimize the diffusion of the calcium ions dissolved in the cytoplasm and extracellular fluid. If the method is applied to sections of conventionally fixed tissue, calcified material is stained but there is also red staining of nuclei, cytoplasmic RNA, and sites of proteoglycans (see Chapter 11). These artifacts are due to electrostatic attraction of diffused calcium ions to macromolecular anions.

3. A green filter in the illuminating system of the microscope enhances the contrast of the red-stained calcium deposits and is recommended for photomicrography.

13.3. Phosphate and carbonate

The **von Kossa technique** is often designated as a histochemical method for calcium, but it is really a method for phosphate and carbonate, the anions with which the metal is associated in normal and pathological calcified tissues.

The sections are treated with silver nitrate. The calcium cations are replaced by silver:

$$CaCO_3 + 2Ag^+ \rightarrow Ag_2CO_3 + Ca^{2+}$$

$$Ca_3(PO_4)_2 + 6Ag^+ \rightarrow 2Ag_3PO_4 + 3Ca^{2+}$$

(The phosphates are really more complicated compounds than those indicated in this simple equation.)

The reaction is carried out in bright light, which promotes the reduction to metal of silver ions in the crystals of the insoluble silver phosphate and carbonate. Alternatively, the silver nitrate may be applied in darkness or subdued light and the reduction accomplished chemically with a solution of hydroquinone, metol or another photographic developing agent. Finely divided metallic silver is black. Any unreduced ionized silver is finally removed by treatment with sodium thiosulphate, which dissolves otherwise insoluble salts by forming complex anions such as $[Ag(S_2O_3)_3]^{5-}$.

The von Kossa procedure provides sharp and accurate localizations of calcified material in tissues, but it is less sensitive than the alizarin red S or the GBHA methods for calcium.

Solutions required

A. *1% aqueous silver nitrate (AgNO₃)*

B. *5% aqueous sodium thiosulphate (NA₂S₂O₃.5H₂O)*

These can be used repeatedly until precipitates form in them.

C. *A counterstain (0.5% aqueous safranine or neutral red)*

Procedure

1. De-wax and hydrate paraffin sections.
2. Immerse in silver nitrate (solution A) in bright sunlight or directly underneath a 100 W electric light bulb for 15 min.
3. Rinse in two changes of water.
4. Immerse in sodium thiosulphate (solution B) for 2 min.
5. Wash in three changes of water.
6. Counterstain nuclei (solution C), 1 min.
7. Rinse briefly in water.
8. Dehydrate (and differentiate counterstain) in 95% and two changes of absolute alcohol.
9. Clear in xylene and mount in a resinous medium.

Result

Sites of insoluble phosphates and carbonates—black. Nuclei—pink or red.

13.4. Iron

The major deposits of iron in mammals are in the red blood cells as haemoglobin, and in the phagocytes of the reticulo-endothelial system as ferritin and haemosiderin. In these compounds, the metal is present in the ferric (oxidation state +3) form, but, being complexed to protein, it is not available as ions to participate in simple chemical reactions.

The iron of ferritin and haemosiderin is readily released as Fe^{3+} ions by treatment with a dilute mineral acid. In the presence of ferrocyanide ions, Prussian blue is immediately precipitated:

$$4Fe^{3+} + 3[Fe(CN)_6]^{4-} \rightarrow Fe_4[Fe(CN)_6]_3 \downarrow$$

Prussian blue is a more complicated compound than the simple ferric ferrocyanide implied above (see Chapter 10, p. 153). This is the simplest histochemical test for iron; it is known as the Prussian blue reaction of Perls.

An alternative strategy is to reduce all the ferric ions in the section to ferrous by treatment with a dilute aqueous solution of ammonium sulphide:

$$2Fe^{3+} + 3S^{2-} \rightarrow 2FeS + S$$

The sulphur dissolves in the excess of aqueous ammonium sulphide. Ferrous sulphide is soluble in dilute mineral acids, so when the sections are treated with acidified potassium ferricyanide solution:

$$FeS + 2H^+ \rightarrow Fe^{2+} + H_2S$$

$$3Fe^{2+} + 2[Fe(CN)_6]^{3-} \rightarrow Fe_3[Fe(CN)_6]_2 \downarrow$$

As explained in Chapter 10, p. 153, the precipitate of "Turnbull's blue" is not really ferrous ferricyanide, but is identical to Prussian blue. This method would, of course, detect any ferrous ions that were present in a tissue.

The simple Perls method is adequate for the detection of iron in higher vertebrates, but according to Gabe (1976) the Turnbull blue technique is more sensitive and is preferred for the demonstration of iron in the tissues of cold-blooded vertebrates and invertebrates. The Perls reaction is also suitable for the identification of exogenous ferritin introduced into animals as an experimental tracer protein (Parmley *et al.*, 1978).

It should be noted that the iron of haemoglobin cannot be released as ions by any method that does not totally destroy the section, so it cannot be demonstrated histochemically.

The stainable iron can be **removed** from the sections by treatment with acid in the absence of a precipitant anion. Aqueous solutions of oxalic acid or sodium dithionite are also suitable for this purpose (Morton, 1978) and act more rapidly than dilute mineral acids.

13.4.1. *Perls' Prussian blue method*

The fixative must not be acidic and must not contain chromium. Hadler *et al.* (1969) have compared several fixatives and found that the largest amounts of histochemically detectable iron were preserved by immersion of tissues for 24 h in 6% aqueous formaldehyde containing 0.27 M calcium chloride, with the pH adjusted to 4.0 by addition of 0.1 M NaOH or 0.1 MHCl. Alcoholic fixatives or buffered aqueous formaldehyde solutions were somewhat less satisfactory.

Solutions required

A. *Acid ferrocyanide reagent*

Potassium ferrocyanide:	2.0 g	Prepare just before using
Water:	100 ml	
Dissolve and add:		
Concentrated hydrochloric acid:	2.0 ml	

B. *Counterstain for nuclei*

0.5% aqueous safranine or neutral red.

Procedure

1. De-wax and hydrate paraffin sections.
2. Immerse in acid ferrocyanide reagent (solution A) for 30 min.
3. Wash in four changes of water.
4. Counterstain nuclei (solution B), 1 min.
5. Rinse briefly in water.
6. Dehydrate (and differentiate counterstain) in 95% and two changes of absolute alcohol.
7. Clear in xylene and mount in a resinous medium.

Result

Blue precipitate with Fe^{3+} liberated from ferritin and haemosiderin. Nuclei pink or red. Haemoglobin is not stained. See *Note 1* below for a control procedure.

Notes

1. Iron can be removed, before staining, by treatment of the hydrated sections with 5% aqueous oxalic acid ($H_2C_2O_4.2H_2O$) for 6 h or with freshly prepared 1% sodium dithionite (=sodium hydrosulphite, $Na_2S_2O_4$) in acetate buffer, pH 4.5, for 5 min. Perls' method is also applicable to sections of plastic-embedded tissue, but the times required for extraction of iron are longer: 12 h for oxalic acid or 15 min for sodium dithionite (Morton, 1978). A negative result after this extraction shows that the sections were not coloured artifactually as a consequence of the presence of unwanted traces of iron in the acid ferrocyanide reagent.

2. The stained slides fade after a year or two of storage. One way to avoid fading is based on the ability of Prussian blue to catalyse the oxidation of 3,3'-diaminobenzidine by hydrogen peroxide (Nguyen-Legros *et al.*, 1980). The product of the catalysed reaction is an insoluble brown polymer that does not fade. To make preparations permanent in this way, carry out the DAB reaction for peroxidase (Chapter 16, p. 285) between Steps 3 and 5 of Perls' method.

13.5. Zinc

Zinc is associated with insulin in the β-cells of the pancreatic islets and with various enzymes (notably carbonic anhydrase in exocrine glands) and also occurs in the cytoplasmic granules of leukocytes and of the Paneth cells of the small intestine. The proteins that bind zinc are easily denatured, with loss of the metal ions, so for histochemical demonstration it is desirable to use freeze-dried material, cryostat sections of unfixed tissue or air-dried smears. If a liquid fixative must be used, cold absolute ethanol is probably the least objectionable one. Histochemical methods certainly do not detect all the zinc that can be shown by chemical analysis to be present in animal tissues (Elmes & Jones, 1981).

A useful reagent for the detection of zinc is **dithizone** (diphenylthiocarbazone; 3-mercapto-1,5-diphenylformazan):

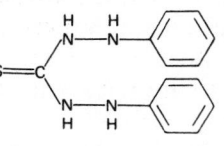

This compound forms a red chelate with zinc. Many other metal ions form differently coloured chelates with dithizone. The reagent has also been used for the histochemical detection of mercury, which forms a reddish-purple complex with the structure (See Harris & Livingstone, 1964):

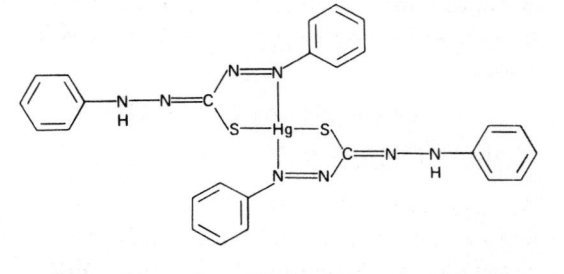

The zinc chelate probably has a similar structure (De *et al.*, 1970). Dithizone and metal-dithizone complexes are much more soluble in non-polar solvents than in water, so it is necessary to use an aqueous mounting medium for the stained preparations. However, it is also necessary to use a non-polar solvent to wash off the excess of dithizone (which is itself coloured). The unusual procedure in the last three steps of the method described below is an attempt to reconcile the removal of the reagent and the desired product.

13.5.1. Dithizone method

This method should be used with paraffin sections of freeze-dried tissue or with cryostat sections of unfixed tissue. Cryostat sections may be fixed for 10 min in absolute ethanol at 4°C for improvement of morphological preservation. If a liquid fixative is unavoidable, fix small pieces for 1 h at 4°C in absolute methanol or ethanol, clear in benzene or xylene, and prepare paraffin sections. Extraction of zinc may be reduced if sections are not flattened on water when mounting onto slides. Dinsdale (1984) fixed in absolute ethanol, cleared in propylene oxide, embedded in an epoxy resin, and used dithizone successfully on 1 μm sections that were collected directly into the staining solution.

Solutions required

A. Dithizone stock solution

Dithizone:	100 mg
Absolute acetone (see Chapter 12, p. 198):	100 ml

Keeps for 1 or 2 months in darkness at 4°C

B. Complexing solution

Sodium thiosulphate (Na$_2$S$_2$O$_3$.5H$_2$O):	55 g
Sodium acetate (anhydrous):	5.4 g
Potassium cyanide (KCN):	1.0 g
(**Caution**: poisonous. See p. 224)	
Water:	100 ml

Dissolve the salts in the water. Dissolve a little dithizone in 200 ml of carbon tetrachloride. Shake the aqueous solution in a separatory funnel with successive 50 ml aliquots of the solution of dithizone in CCl$_4$ until the CCl$_4$ layer is a clear green colour. This manipulation extracts traces of zinc from the reagents. The carbon tetrachloride fractions are discarded. The vapour of CCl$_4$ is toxic. Ideally, this solvent should be collected into a metal drum or canister designated for the disposal of "chlorinated solvents", not a drum for ordinary "waste solvents" such as alcohol and acetone. The purified aqueous complexing solution is stable for a few months.

C. 1.0 M acetic acid

Glacial acetic acid:	60 ml
Water:	to 1000 ml

Keeps indefinitely

D. Sodium potassium tartrate solution

NaKC$_4$H$_4$O$_6$.4H$_2$O:	20 g
Water:	to 100 ml

Keeps indefinitely

E. Working dithizone solution (see also Note 1 below)

This is mixed when needed and used immediately.

Solution A:	24 ml
Water:	18 ml
1.0 M acetic acid (solution C): add in 0.1 ml aliquots until the pH is 3.7 (about 2 ml required)	
Solution B:	5.8 ml
Solution D:	0.2 ml

F. Chloroform

Required for rinsing.

Procedure

1. Cryostat sections are allowed to dry on coverslips or slides. Paraffin sections are dewaxed in xylene (three changes) and allowed to dry by evaporation.
2. Immerse the slides or coverslips bearing the sections in freshly mixed working dithizone solution (E) for 10 min.
3. Rinse the slides or coverslips in two changes of chloroform (F), with agitation, for 30s.
4. Pour off the chloroform and allow the slides or coverslips to drain but do not let the solvent evaporate completely (to avoid cracking of the sections).
5. Rinse in water, allow most of the water to drain off onto filter paper, and put a drop of an aqueous mounting medium onto each section. Apply slides or coverslips according to the type of preparation.

Result

A red to purple colour is formed where zinc is present in the tissue. See *Notes 1* and *2* below.

Notes

1. The complexing solution is mixed with the dithizone to prevent the formation of coloured chelates with metals other than zinc. 24 ml of solution A may be diluted with

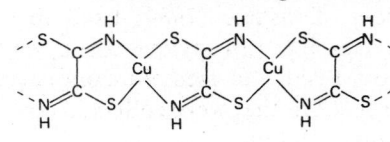

16 ml of water instead of with solutions B, C, and D as described above. The simple dithizone solution will form coloured complexes with Ag, Au, Bi, Cd, Co, Cu, Fe, Hg, In, Mn, Ni, Pb, Pd, Pt, Sn, and Tl, as well as with Zn.

2. Zinc is removed by treating the sections for 5 min with 1% aqueous acetic acid before staining.

3. The sulphide-silver method (see end of this chapter) will detect concentrations of zinc that are too low to give a visible colour with dithizone.

13.6. Copper

Normal mammalian tissues do not contain enough copper to allow histochemical detection of the metal, but in Wilson's disease (hepato-lenticular degeneration) large quantities of the metal accumulate in phagocytic cells in the brain, liver, and cornea. It is also possible, of course, to demonstrate copper which has been artificially introduced into an animal. In some arthropods and molluscs the haemocyanins, respiratory pigments equivalent to the haemoglobin of vertebrate animals, are copper-containing proteins dissolved in the plasma. Copper ions can be released from combination with protein by exposing sections to the fumes of concentrated hydrochloric acid.

The most satisfactory histochemical method for the detection of copper (as the cupric ion, oxidation state +2) makes use of a chelating agent, dithiooxamide (also called rubeanic acid):

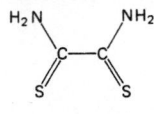

This reagent forms an insoluble dark-green chelate with copper in the presence of ethanol and acetate in alkaline solution. Other metals, such as nickel, cobalt, and silver, also form coloured complexes with dithiooxamide, but under these conditions they are unreactive or give soluble complexes. The copper chelate is believed to be a polymer with the structure

(Harris & Livingstone, 1964).

Dithiooxamide combines only with cupric ions. A reagent that forms a coloured complex with cuprous ions is p-dimethylaminobenzylidenerhodanine:

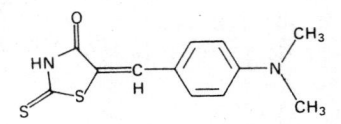

Copper probably replaces the hydrogen attached to the nitrogen atom in the five-membered ring to form a red product. Brown, red, and purple compounds are formed with several other metals. Irons et al. (1977) have shown that while dithiooxamide and p-dimethylaminobenzylidenerhodanine are equally sensitive reagents for the histochemical detection of copper, only the latter gives an intensity of staining proportional to the concentration of the metal in the tissue.

Little attention has been paid by histochemists to the oxidation states of copper in tissues. The invertebrate haemocyanins contain Cu(I), while coeruloplasmin, the copper-binding globulin of mammalian blood plasma, contains both Cu(I) and Cu(II). The copper-containing deposits in the tissues of patients with Wilson's disease are stainable with dithiooxamide, so they must consist partly or entirely of a compound capable of releasing cupric ions. Further study of the interconversion of Cu(I) and Cu(II) in tissues, which is readily brought about by treatment with oxidizing and reducing agents, might lead to broadening of the applicability of histochemical techniques for the demonstration of copper.

13.6.1. Dithiooxamide method

This technique is applicable to frozen or paraffin sections of specimens fixed in phosphate-buffered formaldehyde, to cryostat sections of unfixed tissue, and to de-waxed paraffin sections of freeze-dried material. Fixatives other than formaldehyde are probably acceptable but have not been investigated. But it is probably desirable for the fixative to contain anions such as phosphate, hydroxide, or carbonate, which form insoluble cupric salts.

Solution required

Dithiooxamide (rubeanic
acid):   0.2 g
70% aqueous ethanol: 200 ml
Sodium acetate
(CH_3COONa):   0.4 g

} Keeps for several months

Procedure

1. Take frozen or paraffin sections to water. (See *Note 1*.)
2. Immerse in the dithiooxamide reagent for 30 min.
3. Rinse in two changes of 70% ethanol. (See *Note 2*.)
4. Dehydrate, clear, and mount in a resinous medium.

Result

Sites of copper deposits—dark green to black.

Notes

1. According to Pearse (1985), protein-bound copper can be released by placing the slides (after de-waxing) face downwards over a beaker of concentrated hydrochloric acid for 15 min followed by washing for 15 min in absolute ethanol.
2. A counterstain may be applied after stage 3.
3. In another histochemical method, bound copper is released by treatment of cryostat sections with trichloroacetic acid, and then precipitated with the magnesium salt of dithizone. The deposit is intensified by a treatment with silver nitrate. It is claimed that this method demonstrates the copper normally present in several tissues of the rat (Szerdahelyi & Kása, 1986a,b).

13.7. Lead

Lead has no known physiological function, but can be demonstrated histochemically in the tissues of animals poisoned by its salts. The element is accumulated in intranuclear inclusion bodies, especially in the kidney (Goyer & Cherian, 1977).

The simplest histochemical method for lead is based on the formation of the insoluble chromate, $PbCrO_4$, which is yellow. In order to ensure precipitation of lead ions, it is probably prudent to include sulphate ions in the fixative. Lead sulphate is insoluble in water (solubility product = 1.6×10^{-8}), but the chromate is even less soluble (solubility product = 2.8×10^{-13}), so the reaction

$$PbSO_4 + CrO_4^{2-} \rightarrow PbCrO_4 + SO_4^{2-}$$

can proceed, though slowly, in the direction indicated.

Though this is a fairly specific test for lead (Ba, Sm, Sr, Tl, and Zn also have insoluble yellow chromates but are unlikely to cause confusion), the product is not strongly coloured. A bright coloration is obtained by using a chelating agent, sodium, or potassium rhodizonate:

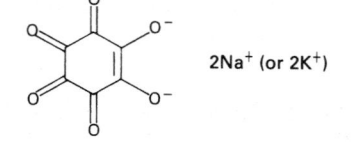

Under mildly acid conditions this forms a pink to red chelate with lead. Neutral solutions of the reagent give a brown product (Molnar, 1952). Barium, strontium, and mercury can also form red compounds with rhodizonate, and a blue-black product is formed with iron (Pearse, 1985).

13.7.1. Rhodizonate method

This method is applicable to frozen or paraffin sections of formaldehyde-fixed specimens. An anion such as sulphate or phosphate should be present in the fixative to ensure precipitation of salts of lead. If the technique is to be applied to bone, sodium sulphate ($Na_2SO_4.10H_2O$; 5 to 10% w/v) should be added to the decalcifying fluid.

Solution required

Sodium (or potassium) rhodizonate:	0.2 g	Prepare
Water:	99 ml	just before
Glacial acetic acid:	1.0 ml	using

Procedure

1. Take sections to water.
2. Immerse in rhodizonate solution for 30 min.
3. Wash in water.
4. Mount in a water-soluble medium.

Result

Sites of salts of lead—pink to red, or brown.

Note

Several other metals form coloured rhodizonate complexes, notably Ag, Ba, Bi, Ca, Cd, Hg_2^{2+}, Sn, Sr, Tl and UO_2^{2+}. Confusion with Pb is unlikely in tissues from experimental animals. See Feigl & Anger (1972) and Lillie & Fullmer (1976) for suggested methods for increasing the specificity of the method.

13.8. Methods of high sensitivity but low specificity

Some histochemical techniques for metal ions have greater sensitivity than those described in the preceding sections of this chapter. Such methods, however, do not identify the detected element with any certainty. The specificity can sometimes be improved by the use of control procedures, or the investigator may already know what the metal is, and be interested only in its localization in the tissue. Two of these methods will now be discussed.

13.8.1. Detection of metals with haematoxylin

The chemistry of haematoxylin has been reviewed in Chapter 5. Its ability to form darkly coloured complexes with many metals makes the dye useful in the histochemistry of inorganic ions. For greatest sensitivity, the solution should be freshly made: it is haematoxylin itself, not haematein, that is the reagent in this histochemical test.

Procedure

1. Take the sections to water (which must not be contaminated with traces of metal salts).
2. Dissolve 50 mg haematoxylin (C.I. 75290) in 1 ml of absolute ethanol. Add the solution to 99 ml of the purest available water. The solution is almost colourless at first and becomes pink on standing.
3. Stain the sections in the freshly prepared haematoxylin solution. Examine after 2 h. If nothing is seen, leave overnight.
4. Do not wash in water. Transfer from the staining solution to 95%, then 2 changes of 100% ethanol.
5. Clear in xylene and mount in a synthetic resin.

Results

Colours are obtained with the following metals (cited from Lillie & Fullmer, 1976):

Al	Blue-black	Ni	Blue
Be	Blue	Os	Blue or Greenish-brown
Bi	Purple		
Cr	Blue-black	Pb	Blue
Cu	Greenish-blue	Pt	Blue
Dy	Blue	Rh	Blue
Fe	Blue-black	Sn	Purplish-red
Ga	Blue-black	Ta	Brownish-red
Hf	Blue-black	Tb	Blue
Ho	Blue	Ti	Brown
In	Blue-black	Tl	Purplish-red
Ir	Blue	U	Blue
Mn	Blue	Yb	Blue
Mo	Blue	Zn	Blue
Nb	Brown	Zr	Blue-black
Nd	Blue		

Notes

1. All the above metals react with unbuffered haematoxylin (pH 5 to 6). Optimum pH levels for staining a few metals are given by Lillie & Fullmer (1976), who also recommend a mounting medium with a low refractive index, so that the architecture of the unstained parts of the section can be seen.
2. The sensitivity of this method can be increased by prolonging the staining time, which may be as long as 48 h.
3. Another chromogenic chelating agent is bromopyridylazo-diethylaminophenol (bromo-PADAP):

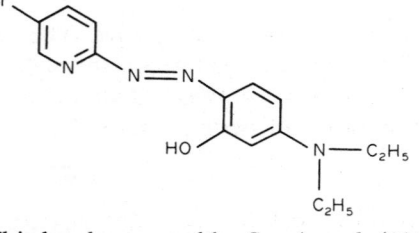

This has been used by Sumi *et al.* (1985) as an extremely sensitive histochemical reagent for several metals. By treating the sections with suitable mixtures of colourless complexing agents, it is possible to "mask" the colour reactions of some metals with bromo-PADAP while preserving that of others. Masking is derived from a comparable practice widely used in analytical chemistry, but chelating agents for histochemical use cannot be selected only on the basis of their chemical properties. The reactivities of chelating agents with metal salts in tissues have been shown to differ significantly from the reactions *in vitro* (Sumi *et al.*, 1983).

13.8.2. *Timm's sulphide-silver method*

In this technique, the fresh specimen is fixed in a liquid containing sulphide ions, either as hydrogen sulphide (bubbled through the fluid before use) or as sodium sulphide (more convenient and less smelly, but usable only with aqueous fixatives). Metals are precipitated:

$$M^{2+} + S^{2-} \rightarrow MS \downarrow$$

In the above equation, M^{2+} is a cation of any metal that has an insoluble sulphide. The method has been used histochemically to detect Ag, Au, Ca, Co, Cu, Fe, Hg, Ni, Pb, Pt and Zn. Of these, copper, lead and zinc have been most often studied.

The next stage of Timm's method is the preparation of sections of the fixed and sulphide-treated specimen. The sections are treated with **physical developer**, which is a solution containing silver ions, a reducing ("developing") agent, and a stabilizing agent that retards the reduction of silver ions to metal. Deposition of silver occurs first at sites where there are deposits of metal sulphide, which have a catalytic action. Deposited silver also acts as a catalyst, so the black deposits of the metal continue to accumulate, thus increasing the sizes and optical densities of the original sulphide precipitates. The slides must be removed from the physical developer and washed in water before there is appreciable reduction of Ag^+ to Ag at non-catalytic sites. For more information about physical developers, see Chapter 18.

The method now described is based on the modification of Timm's original method by Danscher & Zimmer (1978).

Solutions required

A. *Aqueous fixative (for perfusion)*

Sodium sulphide (Na_2S):	11.7 g
Sodium phosphate, monobasic ($NaH_2PO_4.H_2O$):	11.9 g
Water:	1000 ml

Mix in fume cupboard, immediately before use. Sodium sulphide is deliquescent, and cannot be weighed accurately unless it is new and dry.

About one litre of a phosphate-buffered 3% glutaraldehyde solution (see Chapter 2) will also be needed.

B. *Alcoholic fixative (for immersion)*

Bubble hydrogen sulphide (from a Kipp's apparatus or similar generator) through 200 ml of *either* 70% ethanol *or* Carnoy's fluid (see Chapter 2), for 5 min in a fume cupboard.

C. Physical developer

It is important to add the last ingredient ($AgNO_3$) immediately before using.

50% (w/v) aqueous gum arabic (see *Note 1* below):	60 ml
Citrate buffer (see *Note 2*):	10 ml
Hydroquinone (5.67% w/v, aqueous):	30 ml
Silver nitrate (17% w/v, aqueous):	0.5 ml

Procedure

1. Arrange to fix an anaesthetized rat by vascular perfusion (see Chapter 2). Inject 5 ml of the sodium sulphide solution (A) into the tubing, then perfuse the buffered glutaraldehyde for 3 min. The sodium sulphide (Solution A) is then perfused, for about 10 min. This uses up some 250 ml of the solution. Remove the organs of interest, and continue their fixation by immersion of small pieces in buffered glutaraldehyde for 1 h.

    Alternatively, fix pieces of tissue by immersion in an alcoholic fixative that is saturated with hydrogen sulphide (Solution B), overnight at room temperature.
2. Dehydrate in graded alcohols.
3. Either embed in plastic (after the aqueous fixative) or clear (48 h in toluene recommended) and embed in paraffin wax (after either fixative). Cut sections and mount them on slides.
4. Bring the sections to water. (Wax must be removed, but plastic need not be removed.)
5. Immerse the slides in the freshly prepared physical developer (Solution C) for 30–90 min. *Remove a slide from time to time, rinse in water, and check with a microscope.* If they are left too long in the developer, there will be non-catalytic reduction of silver, with generalized non-specific brown staining of the sections.
6. Rinse in water, then wash for 5 min in running tap water.
7. Counterstain, if desired. Any method that does not give dark colours is suitable.
8. Wash, dehydrate, clear, and cover, using a resinous mounting medium.

Result

Sites of heavy metals black. Background according to counterstain. If no counterstain is used, a faint yellow-brown background colour may be discernible.

Notes

1. Gum arabic is slow to dissolve. The viscous solution should be filtered through cloth, and stored frozen in 60 ml aliquots.
2. Preparation of stock citrate buffer:

Citric acid ($H_3C_6H_5O_7.H_2O$):	22.5 g	The pH is 3.9.
Sodium citrate ($Na_3C_6H_5O_7.2H_2O$):	23.5 g	Keeps for several
Dissolve in water and make up to 100 ml		months

3. Timm's sulphide-silver method can also be used for electron microscopy. Practical instructions are given by Danscher & Zimmer (1978) and Danscher & Norgaard (1985).

13.9. Exercises

Theoretical

1. No satisfactory method has ever been devised for the histochemical localization of sodium ions. Why should this be?

2. Devise a simple technique for the histochemical demonstration of chloride in animal tissues, based on the use of the reactions:

(a) $Ag^+ + Cl^- \rightarrow AgCl \downarrow$
(b) $2AgCl \xrightarrow{\text{(light)}} 2Ag + Cl_2$

What factors would limit the accuracy of localization of chloride ions by this method?

3. How would you proceed to show that a brown pigment found in sections of a tissue contains iron?

4. Classify the types of technique for histochemical demonstration of metallic ions according to the types of reagents used. What factors limit the chemical specificity and the accuracy of localization of metals?

5. Examine the structural formula of bromo-PADAP (p. 232). Why is this compound coloured? Suggest a structure for its metal complexes. (Information that will help the reader to answer these questions can be found in Chapter 5).

Practical

6. Mix 100 ml of 0.5 M NaH_2PO_4 with 400 ml of 0.5 M Na_2HPO_4 (to give pH 7.4). Inject a rat intraperitoneally, once daily, with 1.0 ml per 100 g body weight of this solution for 8 days. This procedure induces calcification of renal basement membranes (Haase, 1975).

Twelve hours after the last injection, kill the rat with an intraperitoneal injection of pentobarbitone. Remove the kidneys and fix slices about 2 mm thick in 80% ethanol overnight. Trim the specimens and prepare paraffin sections 7 μm thick.

Stain some sections with alizarin red S and some by the von Kossa technique. Counterstain some of the von Kossa sections with the PAS method and alum-haematoxylin. Observe the distribution of the calcified deposits. Stain some more of the sections with GBHA and compare the result with those obtained by the other two methods. Account for the differences.

7. Using the histochemical methods described in this chapter, demonstrate the normal presence of (a) iron in the liver; (b) zinc in the pancreatic islets.

8. Stain a paraffin section of a decalcified bone for calcium and for phosphate/carbonate, and check the completeness of decalcification.

9. Incubate paraffin sections of formalin-fixed skin in 10^{-5} M aqueous solutions of salts of iron(III), copper(II), chromium(III) and zinc, for 1–2 h at 37°C. Wash in several changes of pure water, then demonstrate the metals histochemically. The haematoxylin method (p. 231) is suitable. (The keratohyaline granules of cells in the epidermis are demonstrated by this procedure.)

14

Enzyme Histochemistry: General Considerations

This chapter introduces the large subject of enzyme histochemistry. A short account of the general properties of enzymes is followed by a discussion of the physical and chemical principles underlying the methods used for their localization in tissues. These techniques are exemplified by more detailed treatments of the histochemistry of some hydrolytic enzymes (Chapter 15) and some oxidative enzymes (Chapter 16).

14.1. Some properties of enzymes

An enzyme is a protein which functions as a catalyst. The catalysis is brought about as a result of combination of a molecule of the enzyme with a molecule of one of the reactants, known as the **substrate**. The substrate, which is nearly always an organic compound or ion, is thereby made more chemically active than it would otherwise be towards another reactant, which may be an inorganic ion or molecule or another organic compound. The products of the reaction are released from the enzyme molecule, which will then be free to bind another molecule of the substrate. It is an important feature of enzymatic catalysis that reactions are enabled to occur at ambient or body temperature, usually in a medium whose pH is near to neutrality. In most instances the same chemical reactions could, in the absence of enzymes, occur only at unphysiologically high temperatures, in strongly acid or alkaline solutions, or in non-aqueous solvents.

Individual enzymes are highly specific for their substrates and for the types of reaction they catalyse. However, the substrate specificity is not always absolute and this is fortunate for the biochemist or histochemist. By providing an artificial substrate, similar but not identical to the natural one, it is often possible to study the enzymatic catalysis of a reaction that yields products more easily detectable than those generated from the physiological substrate.

For more information about enzymes and their physiological functions the reader should consult a textbook of biochemistry or enzymology. A comprehensive work of reference is that of Dixon & Webb (1979). The histochemistry of enzymes is treated in great detail by Burstone (1962), Pearse (1968, 1977) and Lojda *et al.* (1979). For plant enzymes, see Vaughn (1987), and for enzyme histochemistry with the electron microscope the multi-author work edited by Hayat (1973–77).

14.2. Names of enzymes

Knowledge of enzymes and their properties has, like all other scientific information, been acquired gradually. The recognition of more and more enzymes has resulted in a bewildering profusion of names for these substances. The International Union of Biochemistry has introduced a scheme of nomenclature in which each enzyme is named according to the reaction it catalyses. The system also embraces a classification of enzymes and allows the use of approved trivial names when the full systematic names are cumbersome. Each enzyme is given a number (the Enzyme Commission, E.C. number), which establishes its place in the classification. For example, E.C.3.1.1.3 is one of the major group (Group 3) of hydrolases. It acts on ester bonds (3.1) of carboxylic acids (3.1.1) and is the third member (3.1.1.3) of the list of carboxylic ester hydrolases. Its systematic name is triacylglycerol ester hydrolase, and its recommended trivial name is triacylglycerol lipase. It catalyses the reaction:

$$A \text{ triglyceride} + H_2O \rightarrow a \text{ diglyceride} + a \text{ fatty acid}$$

In this and the following two chapters, the approved trivial names of enzymes will be used. The E.C. numbers and systematic names will be given at the first mention in the text. Some non-approved names will also have to be used and discussed. This is an unfortunate consequence of the fact that the histochemical study of enzymes is still a much less precise science than is bio-chemical enzymology.

14.3. Scope and limitations of enzyme histochemistry

The accurate identification of the cellular and subcellular locations of many enzymes has been one of the most conspicuous achievements of modern histochemistry. When using a method for the detection of a particular enzyme, a histo-chemist seeks answers to two questions.

1. *Is the activity detected in the tissue the same as the activity of the purified enzyme identified by biochemical methods?*

This is a re-statement of the requirement for chemical specificity associated with every histo-chemical technique. It is important to remember that the biochemist purifies enzymes from homo-genized tissue and usually studies their properties in solution. In a section of tissue, many enzymes are present, and all of them may be available to

act upon the histochemical reagents chosen for the purpose of demonstrating just one of their number.

The sequestration of an enzyme in an organelle can prevent or suppress the activity of that enzyme in a section. When this is the case, the failure of a histochemical reaction may be a more informative indication of enzymatic activity *in vivo* than the biochemical detection of plentiful activity in an extract of the same tissue, in which all the cells and their organelles have been thoroughly destroyed by homogenization. In practice, however, histochemical techniques are usually devised in such a way as to detect as much enzymatic action as possible.

2. *Where in the tissue is the enzyme located* in vivo?

This is usually the more difficult question to answer. The problems associated with accuracy of localization are different for the three principal types of method used for the histochemical dem-onstration of enzymes. These will now be dis-cussed.

14.4. Methods not based on enzymatic activity

Enzymes are proteins and are antigenic. It is therefore possible to prepare antibodies to pur-ified enzymes and to use these antibodies in immunofluorescent and other **immunohisto-chemical techniques**. For accurate localization it is necessary that the antigenic (though not necessarily the enzymatic) properties of the enzyme be unaffected by fixation or other pre-treatment of the tissue and that the enzyme does not diffuse away from its normal position before forming an insoluble complex with its antibody. Increasing numbers of enzymes are being loca-lized in this way, but the method is immunolog-ical rather than strictly histochemical, so it will not be considered here. Immunohistochemical techniques are discussed in Chapter 19.

Another technique, which may be used more extensively in the future, is that of **affinity label-ling with specific inhibitors** of enzymes. The first method of this type to be described involved the binding of radioactively labelled DFP (see p.245) to sections. The bound inhibitor was then detected by autoradiography. Unfortunately this inhibitor attaches to and inhibits many enzymes that have serine residues at or near their active sites, so the specificity of the technique is low. Sodium, potassium-adenosine triphosphatase (**Na, K-ATPase**; a variant of E.C. 3.6.1.3, ATP phosphohydrolase) has been detected by virtue of its binding of ouabain, a highly specific inhibitor of the enzyme. The ouabain was covalently coupled to a peptide with peroxidase activity. The peptide could then be identified by a histochemical method for peroxidase (see Chapter 16). This gave an osmiophilic product, visible with the electron microscope (Mazurkiewicz *et al.*, 1978).

Carbonic anhydrase (E.C. 4.2.2.1) has been demonstrated by means of a fluorescent inhibitor, dimethylaminonaphthalene-5-sulphonamide, by Pochhammer *et al.* (1979). This inhibitor was given orally to animals, which were killed 7–12 h later. Cryostat sections of unfixed tissue were examined and it was possible, with appropriate filters, to distinguish the fluorescence of the enzyme-bound from that of the unbound inhibitor. Dermietzel *et al.* (1985) applied the same fluorescent inhibitor directly to unfixed cryostat sections. Controls for specificity included mixing the fluorescent compound with competitive non-fluorescent inhibitors.

Some irreversible enzyme inhibitors can be covalently tagged with a fluorochrome or other label, without impairment of the specificity. Thus, α-difluoromethylornithine is firmly and specifically bound by **ornithine decarboxylase** (L-ornithine carboxylyase; E.C. 4.1.1.17) after labelling with rhodamine or with biotin (Gilad & Gilad, 1981). Rhodamine is fluorescent. The significance of labelling with biotin is discussed in Chapter 19.

14.5. Methods based on enzymatic activity

14.5.1. Substrate-film methods

In this type of technique, the enzyme acts upon a substrate carried in a film of some suitable material which is closely applied to the section of tissue. After incubation for an appropriate time, the change in the substrate is detected by any expedient means. The sites of change in the film can then be seen to correspond to sites in the underlying (or overlying) section.

While substrate-film techniques are potentially very versatile, they have two important limitations. Firstly, **the enzyme must diffuse** from its original locus to the film. Since diffusion takes place in all directions, not just straight up or down, the change in the film will occur at some distance from the place in which the enzyme was located in the section. The inaccuracy from this cause will be minimized if the contact between section and film is uniformly very close, and if the film is as thin as is compatible with seeing the changes in it. Secondly, **the change produced in the substrate must be immediate** and irreversible, otherwise lateral diffusion of products of enzymatic action within the film would render the method useless. Probably the highest resolution that can be expected with a substrate-film technique is the identification of the cells in which an enzyme is contained.

A method related to the substrate-film procedure is the **enzyme transfer technique**, developed by Schulz-Harder & Graf von Keyserlingk (1988) for the localization of ribonuclease activity. The enzyme is transferred electrophoretically from a cryostat section to a polyacrylamide gel containing RNA. The gel is incubated for a suitable time, and the remaining RNA is then fluorescently stained with ethidium bromide (see Chapter 9). The original section is stained, and its microscopic appearance is compared with that of the gel. This approach may be suitable for other enzymes too.

14.5.2. *Dissolved substrate methods*

In these methods all reagents are used in solution and the end-products of enzymatic action are deposited within the sectioned tissue. The great majority of histochemical methods for enzymes fall into this category. In all such techniques, the enzyme in the section acts upon a substrate (often not the natural substrate of the enzyme), which is provided in an **incubation medium**. Depending upon the enzyme, the substrate may be hydrolysed, or it may be oxidized or reduced by another substance provided in the incubation medium or already present in the section. Other chemical reactions are also possible. The enzymatic reaction results in the formation of **products** (of hydrolysis, oxidation, etc.), one of which must be immobilized at its site of production and eventually made visible under the microscope. This immobilization of the product is accomplished by means of a **trapping agent**, usually a compound that reacts rapidly with the product to form an insoluble, coloured **final deposit**. Sometimes the precipitate produced by the trapping agent is colourless and must be made visible by a third chemical reaction. Every histochemical method in which a dissolved substrate is used entails, therefore, at least two chemical processes: the enzyme-catalysed reaction and the trapping of a product. Techniques of this type are subject to five possible sources of artifact.

(i) Diffusion of the enzyme may occur, either during preparation of the tissue or during incubation with the substrate and trapping agent. The diffusion of many soluble enzymes can be greatly reduced by incorporating an inert synthetic polymer, such as polyvinyl alcohol, in the incubation medium. Fixation of the tissue will also immobilize enzymes, but most fixatives also cause sufficient denaturation to destroy the catalytic properties. If it is possible the tissue should always be fixed: many histochemical methods will work even when only a small fraction of the original enzymatic activity survives.

(ii) The product of the enzymatic reaction may diffuse away from its site of production before it is precipitated by the trapping agent. The diffused product may then attach itself to nearby structures that are different from the sites in which the enzyme is located. For example, if the product is a positively charged ion, it may be bound to anionic sites in the nucleus of the cell in whose cytoplasm it was formed. In order to minimize diffusion of the primary product of the enzyme-catalysed reaction, it is important that: (a) the trapping reaction occurs very rapidly, (b) the final deposit has a very low solubility in the incubation medium. The latter condition depends, in some cases, on the solubility of the final deposit. For example, if phosphate ions, released by a phosphate ester hydrolase, are to be trapped by precipitation of calcium phosphate, the incubation must contain as high a concentration of calcium ions as possible. In this way, the product $[Ca^{2+}]^3[PO_4^{3-}]^2$ is likely to exceed the very low solubility product of $Ca_3(PO_4)_2$ in the presence of only minute traces of phosphate. (See also Exercise 4 at the end of this chapter.)

(iii) The final deposit is necessarily precipitated somewhere in the section. If the deposit is insoluble in water but somewhat soluble in lipids, there may be erroneous localization of the enzyme in the latter. Protein is ubiquitous in cells, so it is preferable for the final deposit to be bound to protein once it has been precipitated. The term "substantivity" is used to denote the affinities of different types of final deposit for lipid or protein. High substantivity for protein is clearly a desirable property.

(iv) If a third reaction is necessary to make the trapped product visible, diffusion can occur at this stage. This is rarely a cause of artifact in light microscopy, but may assume greater significance in enzyme histochemistry at the higher levels of resolution of the electron microscope.

(v) At any stage of the procedure, coloured deposits may be formed that do not result from activity of the enzyme being studied. False-positive reactions of this type may be due to (a) other enzymes acting upon constituents of the incubation medium, (b) spontaneous occurrence of the reaction catalysed by the enzyme, or (c) unwanted non-enzymatic reactions of other kinds. As **controls** against this type of artifact it is necessary to try out the method (a) in the absence of the substrate, and (b) on sections in which specific enzymatic activity is prevented (e.g. by heating or, preferably, by a specific inhibitor of the enzyme). Large numbers of compounds that inhibit enzymes are classified, reviewed and indexed by Jain (1982).

14.6. Types of enzymes

Biochemists classify enzymes according to the types of chemical reaction they catalyse. The systematic classification of enzymes is, however, too thorough for the histochemist, whose repertoire of techniques is much more limited than that of the enzymologist. Histochemically demonstrable enzymes fall into the following rather broad categories:

(a) Hydrolytic enzymes

These catalyse the reactions of their substrates with water. The substrate molecule is split into two parts, one of which can be trapped.

1. Phosphatases. Catalyse the hydrolysis of esters and amides of phosphoric acid.
2. Sulphatases. Catalyse the hydrolysis of sulphate esters.
3. Carboxylic esterases. Catalyse the hydrolysis of ester linkages between carboxylic acids and any of a variety of alcohols or phenols.
4. Glycosidases, which catalyse the hydrolysis of glycosidic linkages.
5. Proteolytic enzymes, which are of several types and catalyse the splitting of peptide and amide bonds.

(b) Oxidoreductases

These enzymes, which catalyse oxidation-reduction reactions, are discussed in Chapter 16.

(c) Transferases

A transferase is an enzyme which removes part of a molecule to form a new compound. A number of chemical reactions (degradation, synthesis, changes in coenzymes) are associated with the activities of transferases, and there is a varied assortment of histochemical methods for enzymes of this type.

(d) Lyases

This group includes enzymes that catalyse the decomposition of larger into smaller molecules. Commonly one of the products is a small molecule such as water or ammonia.

14.7. Technical considerations

14.7.1. Preparation of tissue

The preservation of an enzyme in a tissue may be achieved in one of three ways.

(a) The tissue may be rapidly frozen and sectioned with a cryostat. The unfixed sections can then be incubated to detect enzymatic activity. This procedure is strongly recommended by Chayen *et al.* (1973) for all histochemical methods for enzymes. However, most histochemists prefer to fix the tissue if possible (see Pearse, 1968, 1972).

(b) A chemical fixative that does not inhibit or denature the enzyme may be used. This procedure is convenient for those enzymes that resist fixation. For example, most carboxylic ester hydrolases can be demonstrated in frozen sections of formaldehyde-fixed tissue, and some phosphatases survive fixation in acetone followed by paraffin embedding. Cryostat sections of unfixed tissue may be fixed for a few minutes (usually in cold formaldehyde or acetone) before incubation, and several enzymes will survive this treatment. Fixation limits diffusion of enzymes.

(c) The tissue may be freeze-dried and then embedded in glycol methacrylate. The resin must be kept cold while it polymerizes. Many enzymes are demonstrable in sections of the embedded tissue. For details, see Murray, Burke & Ewen (1989).

14.7.2. *Conditions of reaction*

Many incubations are carried out at 37°C, though room temperature or even 4°C is preferred in some methods. The enzymes of homoiothermic animals generally function optimally at 37°C; those of poikilotherms and of plants at 25°C. However, there is less diffusion of the enzymes and of the products of reaction at lower temperatures. Sections mounted on slides or coverslips are most conveniently incubated in coplin jars. When only a small volume of medium is available (e.g., if an expensive ingredient is used), drops may be placed over individual sections. The slides are then enclosed in a humid container, such as a petri dish containing wet gauze, in order to prevent evaporation. Free-floating frozen sections are incubated in small petri dishes, glass cavity-blocks, or haemagglutination trays.

Clean glassware is essential—even more so when dealing with enzymes than in other branches of histochemistry, because contaminating substances may inhibit the enzymes or react with constituents of the incubation media.

Some enzyme inhibitors (e.g., potassium cyanide, diethyl-*p*-nitrophenyl phosphate and many others) are toxic and must be handled with care. It should also be noted that some reagents (e.g., benzidine and related compounds) are carcinogenic and that the possible hazards associated with the use of many substances are unknown.

14.8. Exercises

Theoretical

1. What factors limit the accuracy of localization of an enzyme by a technique that is known to be chemically perfectly specific?

2. The final coloured deposit formed in a histochemical method for an enzyme is sometimes seen to be in the form of well-defined intracytoplasmic granules. Explain why this result, though aesthetically satisfying, does not necessarily prove that the enzyme of the living cell is identically distributed.

3. In histochemical methods for dehydrogenases, hydrogen and electrons are transferred from the physiological substrate to a tetrazolium salt, which is thereby reduced to an insoluble pigment. Identify, in general terms, the substrate, the products of enzymatic action, the trapping agent, and the final product.

4. A section is incubated in 0.1 M calcium chloride together with a substrate, which releases phosphate ions under the influence of an enzyme. How many phosphate ions would have to be released in order to precipitate calcium phosphate, $Ca_3(PO_4)_2$: (a) in an organelle with a volume of 1.0 μm^3; (b) in a spherical cell 8 μm in diameter. The solubility product $[Ca^{2+}]^3[PO_4^{3-}]^2$ is 2.0×10^{-29}. Avogadro's number, $N_A = 6.022 \times 10^{23}$ molecules per mole. Assume that the enzymatically liberated phosphate ions do not diffuse outside the organelles or cells in which they are formed and ignore any effects of pH and of formation of other phosphates of calcium.

15

Hydrolytic Enzymes

Methods for three groups of enzymes will be discussed in this chapter. The techniques have been selected with a view to illustrating a wide variety of histochemical principles. It should be remembered that many other methods are available for these and related hydrolytic enzymes.

15.1. Phosphatases
(Phosphoric monoester hydrolases; E.C.3.1.3.)

These enzymes catalyse the hydrolysis of esters of phosphoric acid. Acid and alkaline phosphatases are distinguished by their widely separated pH optima. The former are typical lysosomal constituents, and the latter group includes enzymes that hydrolyse specific substrates (e.g. adenosine triphosphate, thiamine pyrophosphate) as well as ones that can attack a variety of substrates.

15.1.1. Acid phosphatase
(Orthophosphoric monoester phosphohydrolase-acid pH optimum; E.C.3.1.3.2.)

This name is applied to a family of lysosomal enzymes that catalyse the hydrolysis of phosphate esters optimally at around pH 5.0.

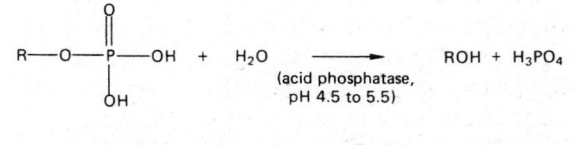

R may be an alkyl or aryl radical. Only one hydroxyl group of phosphoric acid may be esterified in a substrate for this enzyme.

In the lysosomes acid phosphatase is associated with several other enzymes, all with acid pH optima, which catalyse the hydrolysis of a wide variety of esters, amides, and proteins. These enzymes occur in high concentrations in cells of the renal tubules and the prostate gland and in phagocytes and cells undergoing degeneration. Smaller numbers of lysosomes are found in most other types of cells.

In Gomori's lead phosphate method, which has been improved upon by many later investigators, the phosphate ions released by hydrolysis of the substrate (sodium β-glycerophosphate) are trapped by lead ions, with which they combine to form insoluble lead phosphate. The latter is white, but the colour can be changed to black by treatment with hydrogen sulphide or a solution of sodium or ammonium sulphide.

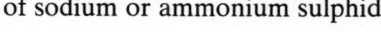

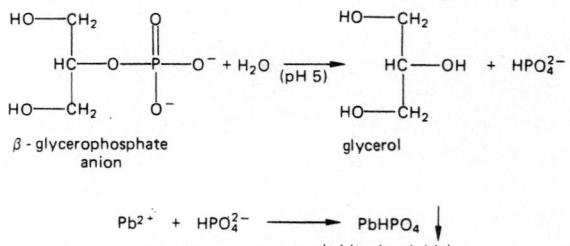

Although lysosomal enzymes are enclosed in membranous organelles, the substrate can penetrate the membranes at the optimum pH of acid phosphatase.

The concentration of Pb^{2+} ions is critical and the incubation medium has to be mixed well in advance in order to allow any spontaneous precipitation of insoluble lead salts to occur before the sections are introduced. The effects of variations in concentration of Pb^{2+} and substrate, the pH, and the overall ionic strength of the incubation medium have all been examined (see Pearse, 1968; Chayen et al., 1973, for references to original literature). The mixture used is formulated so as to minimize the binding of lead ions to nuclei and other structures, including the lysosomes that contain the enzyme.

Lead phosphate precipitates are electron-dense, but are coarsely crystalline and adhere to membranes. A fine precipitate is obtained, within lysosomes, if aluminium ions are substituted for lead in the incubation medium (Berry et al., 1982). The electron-opacity is surprising in view of the low atomic number of aluminium (see Lewis, 1987). Possibly there is deposition of osmium, from the post-fixation solution, upon the precipitated aluminium phosphate. Lanthanide metals, notably cerium, have also been used to precipitate phosphate ions in ultrastructural histochemical methods for phosphatases (Halbhuber et al., 1988).

Acid phosphatase in most sites is inhibited by fluoride ions (e.g. 10^{-3}–10^{-2} M NaF in incubation medium). Certain neurons in sensory ganglia, however, contain a fluoride-resistant acid phosphatase. These neurons also contain a peptide, substance P, and are believed to be involved in the perception of pain (see Jancsó et al., 1985). Other inhibitors include cupric ions (5×10^{-4} M) and tartrate ions (10^{-2} M) (Lojda et al., 1979). Phosphate ions are inhibitory, but cannot, of course, be added to the lead-containing incubation medium. It is desirable to **avoid phosphate buffers in fixatives or washing solutions** for specimens in which phosphatases are to be histochemically demonstrated.

Fixation and processing

Acid phosphatase survives brief fixation in cold (4°C) acetone or aqueous formaldehyde. Decalcification in formic acid inhibits the enzyme in osteoclasts, but this can be reactivated by incubating the sections, before staining, for 1 h at 37°C in the buffer that will be used for the histochemical incubation (Liu et al., 1987). Frozen or cryostat sections retain the greatest amounts of enzymatic activity, but the histochemical method will work with glycol methacrylate-embedded material, and even on paraffin sections if a wax with a low melting point has been used.

Solutions required

A. *Incubation medium (Waters & Butcher, 1980)*

Dissolve 132 mg of lead nitrate, $Pb(NO_3)_2$, in 25 ml of 0.2M acetate-acetic acid buffer, pH 4.7.

Dissolve 315 mg sodium β-glycerophosphate in 75 ml water.

Combine the two solutions and warm to 37°C before use.

This medium is more stable than earlier formulations and it does not form a precipitate on standing. It may be used immediately or stored at room temperature for several days.

B. *Sulphide solution*

Add about 0.5 ml of yellow ammonium sulphide to about 100 ml of water immediately before use. Do not allow fumes from the ammonium sulphide bottle to come into contact with the incubation medium (solution A). Ammonium sulphide is malodorous and toxic and should be used in a fume cupboard. Discard the used solution by washing it down the sink with plenty of running tap water.

Procedures

1. Incubate sections in solution A for 30 min at 37°C.
2. Wash in four changes of water, 1 min in each.

3. Immerse in dilute ammonium sulphide (solution B) for about 30 s.
4. Wash in three changes of water.
5. Mount in a water-miscible medium. (See *Note 1*.)

Result

Black deposits of lead sulphide indicate sites of acid phosphatase activity. (See also *Note 2* below.)

Notes

1. An aqueous mountant is preferred because small amounts of PbS may be dissolved during dehydration and clearing. Saturation of the alcohols and xylene with PbS is sometimes recommended when it is desired to use a resinous mounting medium.
2. Control sections should be incubated (a) without substrate, (b) in the full incubation medium containing, in addition, 0.01 M sodium fluoride, which inhibits the enzyme in most sites. For other inhibitors, see above.

15.1.2. *Alkaline phosphatase*
   (Orthophosphoric monoester phosphohydrolase-alkaline pH optimum; E.C.3.1.3.1.)

The reaction catalysed by this group of enzymes (which excludes, in the present context, those phosphomonoesterases that have specific substrates) is the same as for acid phosphatase except that the pH optimum lies around 8.0. Alkaline phosphatase occurs in many types of cell, especially in regions specialized for endocytosis and pinocytosis.

Although it is possible to demonstrate alkaline phosphatases by a method similar to the one described above, which produces an eventual precipitate of a metal sulphide, we shall consider instead another technique in which the organic product of hydrolysis is trapped.

The substrate is the monobasic sodium salt of α-naphthyl phosphate, the monoester of α-naph-

thol, and phosphoric acid. The α-naphthol freed by hydrolysis is a phenolic compound and can therefore couple with a diazonium salt, which is included in the incubation medium. Coupling occurs rapidly at an alkaline pH and an insoluble, coloured azoic dye is produced.

It will be noticed that the azo compound is also phenolic and is therefore an acid dye, albeit one which is insoluble in water and not appreciably ionized at pH 8.

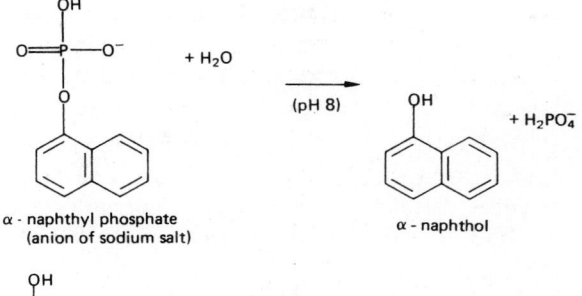

α - naphthyl phosphate
(anion of sodium salt)

α - naphthol

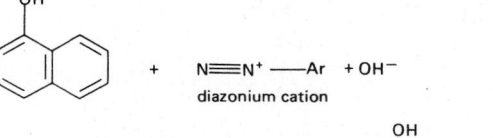

diazonium cation

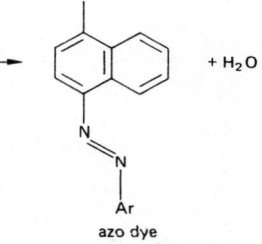

azo dye

This property may confer some substantivity for protein. However, the azo dyes formed from naphthols are usually soluble in non-polar organic liquids. A phosphate ester of naphthol-AS (see p.70) can also be used as a substrate. Naphthol-AS forms more intensely coloured azoic dyes than does α-naphthol, but its rate of coupling with diazonium salts is slower (Burstone, 1962). The diazonium salt used as a trapping agent should be as stable as possible and must not inhibit the enzyme. The dye formed as the end product of the histochemical reaction should precipitate as exceedingly small particles, with substantivity for protein. The perfect diazonium salt has not yet been found, but fast blue RR salt (see p. 69) is one of the best available. The relative merits and faults of many stabilized diazonium salts used in enzyme histochemistry are discussed by Pearse (1968) and Lillie (1977).

There are no inhibitors of high specificity for alkaline phosphatase, but enzymatic activity is prevented by prior treatment of the sections with a solution of iodine in potassium iodide. Some alkaline phosphatases are inhibited when cysteine is added to the incubation medium. Mammalian alkaline phosphatases other than that of the intestinal epithelium are inhibited by including 10^{-3} M levamisole or tetramisole* in the incubation medium. The latter two compounds are useful for inhibiting endogenous alkaline phosphatase when antisera labelled with the intestinal enzyme are used as reagents in immunohistochemical techniques (Ponder & Wilkinson, 1981; Appenteng et al., 1986; see also Chapter 19).

Fixation and processing

Like acid phosphatase, alkaline phosphatase can be detected after brief fixation in cold acetone or cold aqueous formalin followed by careful embedding in paraffin (see Bancroft & Cook, 1984) or nitrocellulose. The latter is valuable for demonstrating the enzyme in capillary blood vessels in thick sections of large organs (see Bell & Scarrow, 1984 for details of technique). Formaldehyde-fixed blocks may be stored for a few weeks at 4°C in gum-sucrose. Frozen or cryostat sections are preferred for most purposes. The latter are fixed on their slides or coverslips by immersion for 2 min in cold (4°C) acetone.

The following instructions apply to free-floating frozen sections, which must be handled carefully to minimize physical damage in the alkaline incubation medium.

---

*Levamisole is L[–]-2,3,5,6–Tetrahydro-6-phenyl-imidazo[2,1-b]thiazole. Tetramisole is the slightly cheaper racemic ([±]) form of the same compound.

Solutions required

A. Incubation medium

0.05M TRIS buffer, pH 10.0 (see Chapter 20):	10 ml
α-naphthyl acid phosphate (sodium salt):	10 mg
Magnesium chloride ($MgCl_2.6H_2O$):	10 mg
Fast blue RR salt (C.I. 37155):	10 mg

Mix in the order stated. Filter and use immediately.

B. 1% aqueous acetic acid

Procedure

1. Carry sections through two changes of water to remove fixative, then incubate them in solution A for 20 min at room temperature.
2. Transfer to water (in a large dish) for 1 min.
3. Transfer to 1% acetic acid (solution B) for 1 min.
4. Rinse in water.
5. Mount sections onto slides and cover, using a water-miscible mounting medium.

Result

Sites of alkaline phosphatase activity purple to black.

Notes

1. Incubate sections in solution A without the substrate as a control for non-specific staining by the diazonium salt.
2. To control for spontaneous hydrolysis of the substrate, treat some sections with an iodine solution for 3 min followed by a rinse in $Na_2S_2O_3$ (as for removal of mercurial deposits, see p. 46) and four rinses in water, before incubating in the substrate-containing solution A. The treatment with iodine inhibits the enzyme.
3. The magnesium salt may enhance enzymatic activity, but it is often omitted. The rinse in dilute acetic acid neutralizes the alkaline buffer and renders the sections less fragile.

4. Many methods have been developed for ultrastructural visualization of alkaline phosphatase activity. Those based on precipitation of cerium phosphate are probably the most satisfactory (Hulstaert *et al.*, 1983; Halbhuber *et al.*, 1988).

15.2. Carboxylic esterases
(Carboxylic ester hydrolases; E.C.3.1.1.)

The carboxylic esterases catalyse the general reaction

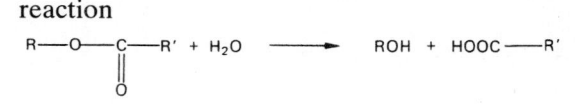

The alkyl or aryl radical R may be derived from one of many possible alcohols (including glycerol), or phenols, or from a hydroxylated base such as choline. The acyl group

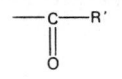

may be derived from a simple carboxylic acid such as acetic or from one of the long-chain fatty acids.

15.2.1. Classification

Many names have been used for the histochemically demonstrable enzymes in this group. The synonymy is explained in Table 15.1. The substrates of the different enzymes and the products of their hydrolysis are listed in Table 15.2, on p. 247.

15.2.2. Substrates and inhibitors

In this text only the first five enzymes in Table 15.2 will be considered. These enzymes are detected histochemically by providing them with synthetic substrates and trapping the products of hydrolysis. Unfortunately, no artificial substrate is available that is acted upon exclusively by any one of the individual enzymes. All five catalyse the hydrolysis of the substrates for carboxylesterase, arylesterase, and acetylesterase. The hydrolysis of some other substrates, which are not attacked by these three enzymes, is catalysed by both acetylcholinesterase and cholinesterase. Unlike the two phosphatases discussed earlier in this chapter, the carboxylic esterases cannot be distinguished from one another by taking advantage of their pH optima: they all function efficiently over the range of pH 5.0 to 8.0. It is therefore necessary to make use of toxic substances that specifically inhibit some of the enzymes.

The two choline esterases are inhibited by **eserine** (also known as physostigmine). This alkaloid competes with esters of choline for substrate-binding sites on the enzyme molecules. Because it is a competitive inhibitor, eserine must be included in the incubation medium together with the substrate. In the presence of eserine, a substrate for all five histochemically detectable carboxylic esterases will be hydrolysed only as a result of the activities of carboxylesterase, arylesterase, and acetylesterase.

The first of these three is inhibited by low concentrations of **organophosphorus compounds**. These potent inhibitors, which also inactivate both the choline esterases and some proteinases, act by blocking serine residues at or near to the active sites on the enzyme molecules, thus preventing access of the substrate. Typical organophosphorus inhibitors are diisopropylfluorophosphate (DFP) and diethyl-*p*-nitrophenyl phosphate (E600).

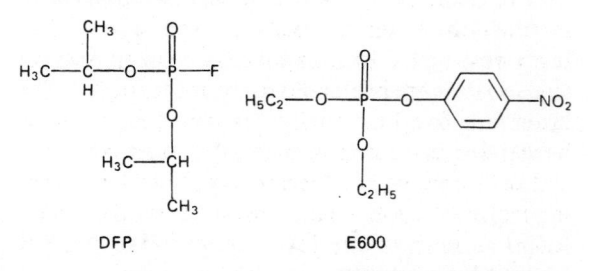

The hydroxyl group of serine displaces the substituent shown on the right-hand side of the phosphorus atom in each of the above formulae. Organophosphorus compounds are dangerously toxic, mainly because they inhibit acetylcholinesterase. Inhibition is irreversible, so when sections have been pre-incubated in DFP or E600 it

TABLE 15.1. *Nomenclature of some carboxylic esterases*

Recommended trivial name	E.C. number and systematic name	Synonyms which should no longer be used	
Carboxylesterase	3.1.1.1 Carboxylic ester hydrolase	Ali-esterase B-esterase Organophosphate-sensitive esterase	⎫
Arylesterase	3.1.1.2 Aryl ester hydrolase	Arom-esterase A-esterase Organophosphate-resistant esterase E600-esterase DFP-ase	⎬ Non-specific esterases
Acetylesterase	3.1.1.6 Acetic ester acetyl-hydrolase	C-esterase Organophosphate-resistant, sulphydryl inhibitor-resistant esterase	⎭
Acetylcholinesterase (AChE)	3.1.1.7 Acetylcholine acetyl-hydrolase	Specific cholinesterase True cholinesterase Cholinesterase (term still used by physiologists and pharmacologists)	⎫
Cholinesterase (ChE)	3.1.1.8 Acylcholine acyl-hydrolase	Pseudocholinesterase Butyrylcholinesterase Non-specific cholinesterase	⎬ Choline esterases ⎭
Lipase	3.1.1.3 Glycerol ester hydrolase		
Phospholipase B	3.1.1.5 Lysolecithin acyl-hydrolase		

is not necessary to add these inhibitors to the incubation medium.

Only arylesterase and acetylesterase are still active after exposure to an organophosphorus compound. The activity of arylesterase is dependent upon the integrity of a free sulphydryl group at the substrate-binding site of the enzyme molecule. It can therefore be inhibited by blocking this sulphydryl group with an organic mercurial compound. The inhibitor usually chosen is the *p*-chloromercuribenzoate ion (PCMB).

If the sections are first incubated in a dilute solution of E600, then in a solution of PCMB, and then in a substrate-containing medium with added PCMB, acetylesterase will be the only enzyme to give a positive histochemical reaction.

Selective inhibitors of acetylcholinesterase and cholinesterase will be mentioned later in connection with the histochemical methods for these enzymes. For reviews of the carboxylic esterases and their inhibitors, the reader is referred to Pearse (1972) and Luppa & Andrä (1983).

Instructions for making solutions of esterase inhibitors are given at the end of this chapter.

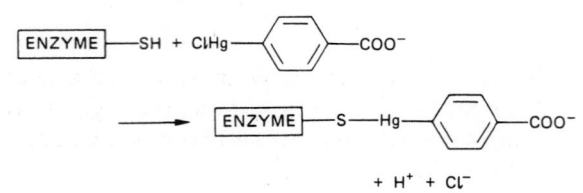

TABLE 15.2. *Substrates and actions of some carboxylic esterases*

Enzyme	Substrate whose hydrolysis is catalysed[a]	Products of reaction
Carboxylesterase	A carboxylic ester (may be of an aliphatic or aromatic alcohol or a phenol)	An alcohol or a phenol and a carboxylic acid
Arylesterase	An ester formed from acetic acid and a phenol	A phenol and acetic acid
Acetylesterase	An ester of acetic acid	An alcohol or a phenol and acetic acid
Acetylcholinesterase	Acetylcholine[b]	Choline and acetic acid
Cholinesterase	An acylcholine[b]	Choline and an acid
Lipase	A triglyceride[c]	A diglyceride and a fatty acid
Phospholipase B	A lysolecithin	Glycerolphosphocholine and a fatty acid

[a]These are the substrates which serve to define the biochemical specificities of the enzymes in conjunction with effects of specific inhibitors. The enzymes also act upon other substrates, including those used in histochemistry.
[b]Histochemical substrates include phenolic esters and thioesters of choline.
[c]Histochemical substrates include esters of polyhydric alcohols other than glycerol and phenolic esters of higher fatty acids.

15.2.3. Indigogenic method for carboxylic esterases

Histochemical methods for these enzymes are available in which acetyl esters of naphthols and of indoxyls are used as substrates. The former methods, in which the substrates are naphthyl esters, are similar in principle to the one for alkaline phosphatase discussed earlier, so only the latter type of technique will be described.

The substrate may be one of a selection of halogenated derivatives of indoxyl acetate. 5-bromo-indoxyl acetate is a typical example. The unsubstituted indoxyl ester is not used because the end product of its hydrolysis forms unduly large crystals.

Hydrolysis yields the halogenated indoxyl:

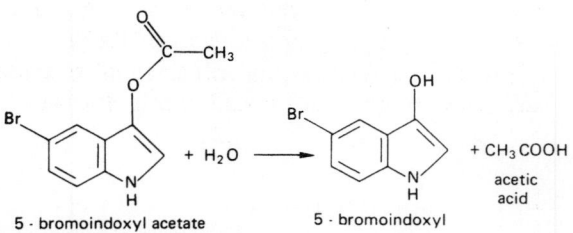

5 - bromoindoxyl acetate
(5 - bromo - *O* - acetyl indoxyl)

5 - bromoindoxyl

The indoxyl exhibits keto-enol tautomerism:

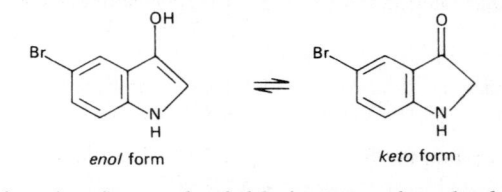

enol form   *keto* form

It is colourless and soluble in water, but the *keto* form is rapidly oxidized by a balanced mixture of ferrocyanide and ferricyanide ions present in the incubation medium. The continuous removal of the *keto* tautomer causes the reversible reaction shown above to be driven from left to right, virtually to completion. The product of oxidation of two molecules of the halogenated indoxyl is 5,5′-dibromoindigo:

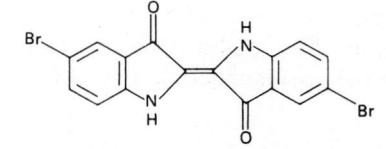

This compound is insoluble in water and also in histological dehydrating and clearing agents. It contains the indigoid chromophore (see Chapter 5) and is blue.

All five of the carboxylic esterases under consideration are able to catalyse the hydrolysis of halogenated indoxyl acetates, so the individual enzymes must be identified by the judicious use of inhibitors, as outlined in the preceding section (15.2.2) of this chapter. No inhibitors are available that spare only the choline esterases. Consequently, the indigogenic method is used mainly for carboxylesterase, arylesterase, and acetylesterase. It is also useful, however, for the demonstration of acetylcholinesterase at motor endplates in skeletal striated muscle (see Chapter 18), where other carboxylic esterases are absent.

Indigogenic methods for esterases have been in use for many years, and the original technique (Holt & Withers, 1952) is very reliable. It works well with frozen sections of tissues fixed in neutral, buffered formaldehyde for 12–24 h at 4°C. The fixed blocks may be stored for several months at 4°C if they are transferred directly from the fixative to **gum-sucrose** (gum acacia (arabic), 2 g; sucrose, 60 g; water, 200 ml). Several hydrolytic enzymes can be immobilized by soaking tissues in gum-sucrose. The large molecules of the gum, which is a polysaccharide, may hinder the diffusion of the enzyme molecules.

Stock solutions

A. TRIS-HCl buffer, pH 7.2 (see Chapter 20)

B. 0.1 M calcium chloride ($CaCl_2$; 1.11%)

} Keeps for several months at 4°C, unless infected

C. 0.05 M potassium ferricyanide ($K_3Fe(CN)_6$; 1.65%)

D. 0.05 M potassium ferrocyanide ($K_4Fe(CN)_6.3H_2O$; 2.11%) } Keeps for about 4 weeks at 4°C

It is slowly oxidized by air to the ferricyanide, with consequent change of colour from almost colourless to yellow.

Incubation medium

Dissolve 5 mg of O-acetyl-5-bromoindoxyl (= 5-bromoindoxyl acetate) in about 0.2 ml of ethanol in a small beaker, then add:

Solution A (buffer):	8.0 ml	Mix just before using
Solution B ($CaCl_2$):	4.0 ml	
Solution C ($K_3Fe(CN)_6$):	4.0 ml	
Solution D ($K_4Fe(CN)_6$):	4.0 ml	

(See also *Note 1*.)

Procedure

1. Cut frozen sections and transfer them without washing to the incubation medium. Leave at room temperature for about 30 min. The time is not critical; some tissues will show an adequate reaction in 5 min while others may require 2 h.
2. Rinse section in two changes of water.
3. (Optional.) Counterstain nuclei with neutral red or safranine. van Gieson's stain (p. 118) is also suitable.
4. Rinse in water, mount onto slides, allow to dry, dehydrate, clear and mount in a resinous medium.

Result

Sites of carboxylic esterase activity—blue. The deposit is finely granular. See *Note 2* for comments on specificity and controls.

Notes

1. Indoxyl acetate is a poor substrate, giving a conspicuously crystalline, diffused end product, but it has been used. O-acetyl-4-chloro-5-bromoindoxyl is said to be superior to the substrate suggested here, in that an even more finely granular blue precipitate is formed.
2. This method detects all carboxylic esterases, but AChE and ChE are inhibited by eserine. E600 and PCMB should be used to identify the other enzymes, but in critical histochemical studies it is necessary to determine the effects of other inhibitors as

well, especially with species other than the rat.

Instructions for making solutions of esterase inhibitors are given at the end of this chapter.

3. Bromoindoxyl phosphate and various bromoindoxyl glycosides are available as substrates for phosphatases and glycosidases. In some of the methods, a diazonium salt is included in the incubation medium. This couples with the liberated indoxyl, and the final reaction product is therefore an azo dye, not a derivative of indigo. See Lojda *et al*. (1979) for more information.

15.2.4. Choline esterases

Mammalian tissues contain two enzymes that catalyse the hydrolysis of esters of choline. Both are inhibited by the alkaloid eserine.

Acetylcholinesterase (AChE; see Table 15.1 for synonyms) is present in erythrocytes and in some neurons. This is the enzyme that terminates the action of acetylcholine at cholinergic synapses and neuromuscular junctions, though it must have other functions as well.

Cholinesterase (ChE; also frequently called "pseudocholinesterase", see Table 15.1) can also catalyse the hydrolysis of acetylcholine, though more slowly than AChE. The preferred substrates are esters of choline with acyl groups containing more carbon atoms than acetyl. ChE occurs in serum, in neuroglia, and in some neurons. It is also present in the endothelial cells of cerebral capillaries in some species, notably the rat.

Histochemical methods for AChE and ChE are valuable in the histological study of the nervous system, because they selectively demonstrate certain groups of neuronal somata and axons. The uses of the methods in neuroanatomy are reviewed by Kiernan and Berry (1975) and Butcher (1983).

The substrate in the most widely employed technique (introduced by Koelle & Friedenwald, 1949) is **acetylthiocholine** (AThCh), the thio-ester analogous to acetylcholine.

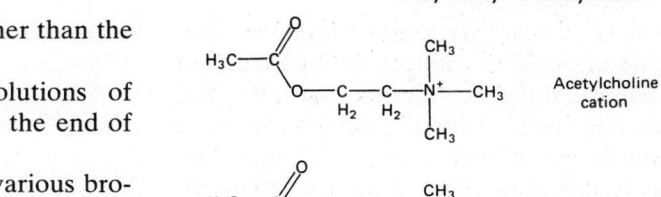

Enzymatic cleavage of AThCh yields acetic acid and the thiocholine cation:

$$CH_3COS(CH_2)_2\overset{+}{N}(CH_3)_3 \; + \; H_2O$$

AThCh cation

$$\longrightarrow CH_3COOH \; + \; HS\!-\!\!-\!(CH_2)_2\!-\!\!-\!\overset{+}{N}(CH_3)_3$$

Thiocholine cation

Thiocholine has a free sulphydryl group. The incubation medium also contains copper (complexed with glycine, so that the AChE is not inhibited by a high concentration of Cu^{2+}) and sulphate ions. Thiocholine combines with copper and iodide ions to form an insoluble crystalline product, copper-thiocholine iodide, which is probably

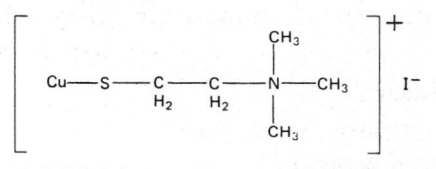

(Tsuji, 1974). This product is usually made visible in the microscope by treatment with a soluble sulphide (H_2S, Na_2S or $(NH_4)_2S$), which results in the formation of brown cuprous sulphide. The crystals of copper-thiocholine sulphate are visible in light microscopy, but the deposits of cuprous sulphide are amorphous (Malmgren & Sylven, 1955) even under the electron microscope. This redistribution of the final product of the reaction must be remembered when the supposed localizations of AChE and ChE are studied at the ultrastructural level.

In another widely used technique (Karnovsky & Roots, 1964) the incubation medium contains a thiocholine ester, copper (as a citrate complex), and ferricyanide ions. The latter are reduced to ferrocyanide by the sulphydryl group of the thiocholine released by enzymatic hydrolysis of the substrate. Brown copper ferrocyanide (Hatchett's brown) is formed and is immediately precipitated at the site of the enzymatic activity. This technique is known as a "direct-colouring" method to distinguish it from the procedures in which the product of the reaction has to be converted to cupric sulphide.

The efficiency of the direct-colouring method is reduced by simultaneous precipitation of copperthiocholine iodide, which does not contribute colour to the final product (Tewari *et al.*, 1982). In one modification of the technique (Tsuji & Larabi, 1983), the acetylthiocholine substrate is used as its chloride rather than iodide, so that Hatchett's brown is the only precipitated product of the reaction.

The deposits can be darkened by treatment with diaminobenzidine (DAB) and hydrogen peroxide. $Cu_2Fe(CN)_6$ resembles the enzyme peroxidase (see Chapter 16) in that it catalyses the oxidation of aromatic amines by H_2O_2. The dark colour of oxidized DAB is made even darker if the reaction of formation occurs in the presence of nickel or cobalt ions.

Cholinesterase substrates and inhibitors

AThCh is hydrolysed, though slowly, under the influence of catalysis by ChE, so it is necessary to include a selective inhibitor of this enzyme in the incubation medium if AChE is to be demonstrated in isolation. A more suitable substrate for the deliberate demonstration of ChE is **butyryl-thiocholine** (BuThCh). This is, however, slowly attacked by AChE, so it is then necessary to inhibit the latter enzyme. The most generally useful selective inhibitor of AChE is a quaternary ammonium compound, **B.W. 284C51** (1,5-*bis*-(4-allyldimethylammoniumphenyl)pentan-3-one di-bromide). **Ethopropazine** hydrochloride is a convenient inhibitor of ChE, but this enzyme is also inhibited by concentrations of **DFP**, which are too low to inhibit both the choline esterases. Another organophosphorus compound useful for inhibiting ChE is tetraisopropylpyrophosphoramide (**isoOMPA**). It should be remembered that there is considerable variation among different species of animals in the susceptibilities of choline esterases and other carboxylic esterases to different inhibitors. Except in well understood species (such as the rat), it is necessary to investigate the effects of several inhibitors before reaching a decision as to the identity of an enzyme. Interspecific variation is discussed by Kiernan (1964) and Pearse (1972). These references and a paper by Pepler and Pearse (1957) may be consulted for lists of substances that inhibit the different carboxylic esterases.

Instructions for making solutions of esterase inhibitors are given at the end of this chapter (Section 15.4).

15.2.5. *A method for acetylcholinesterase and cholinesterase*

This method is that of Karnovsky & Roots (1964), with modifications based on the work of Hanker *et al.* (1973) and Tago *et al.* (1986).

Frozen sections are cut from specimens fixed at 4°C for 12–24 h in neutral, buffered formaldehyde. Glutaraldehyde (1.0–3.0%, pH 7.2–7.6) is also a suitable fixative and formal–sucrose–ammonia (formalin (40% HCHO), 100 ml; sucrose, 150 g; strong ammonia (NH_4OH, S.G. 0.880), 10 ml; water to 100 ml) is a mixture devised specially for the preservation of these enzymes (Pearson, 1963).

Two incubation media, C.(1) and C.(2), are prescribed, each for a different purpose.

Reagents required

A. (i) Acetylthiocholine iodide (substrate for AChE)
   (ii) Butyrylthiocholine iodide (substrate for ChE)
   (iii) Inhibitors, according to the requirements of the investigation.

B. *Stock solution for incubating medium*

   0.1 M acetate buffer, pH 6.0 (see
   Chapter 20):   65 ml

Sodium citrate (trisodium salt;
dihydrate): 147 mg
Cupric sulphate (anhydrous): 48 mg
Water: to 100 ml
Then add:
Potassium ferricyanide ($K_3Fe(CN)_6$): 17 mg

This pale green solution keeps for about 1 week at room temperature. It must not be used if it contains a brown precipitate.

C.(1) Incubation medium (working solution, full strength)

Dissolve 5.0 mg (approximately) of **either** acetylthiocholine iodide **or** butyrylthiocholine iodide in a drop of water and add 10 ml of solution B. This medium is stable for several hours and may be used repeatedly if not cloudy, but should be discarded at the end of the day.

This medium is used for tissues containing high enzymatic activity, including skeletal muscle (motor end-plates).

C.(2) Incubation medium (working solution, 108 × dilution)

Mix 10 ml of Solution C. (1) above with 90 ml of 0.1 M acetate buffer, pH 6.0. *This medium is used for tissues containing fine AChE-positive nerve fibres, or cell-bodies with low enzymatic activity.*

Inhibitors should be added to the medium C.(1) or C.(2) as needed. See Section 15.4 for instructions on use of inhibitors.

D. TRIS buffer, pH 7.2 (see Chapter 20)

E. DAB solution

3,3′-diaminobenzidine
tetrahydrochloride: 10 ml ⎫ Use within
TRIS buffer (pH 7.6): 20 ml ⎬ one hour
Nickel ammonium sulphate ⎭ of making
($Ni(NH_4)_2(SO_4)_2.6H_2O$): 150 mg

F. Hydrogen peroxide

A stock solution of 30% H_2O_2 (= "100 volumes available oxygen"; handle carefully) is diluted to 0.03% with water, less than 15 minutes before it is needed (see *Procedure* below). 2 ml of the diluted solution will be needed.

Procedure

1. Collect frozen sections into water in which they may remain for up to 1 h. Pre-incubate any control sections in appropriate inhibitors. Wash irreversibly inhibited sections in four changes of water. Do not wash after pre-incubation with competitive inhibitors.
2. Transfer sections to the incubation medium, C.(1) *or* C.(2), for 10–30 min at room temperature. When regions of enzymatic activity go reddish brown, incubation is adequate. Check under a microscope for isolated sites of activity such as motor end-plates.
3. Wash in 3 changes of TRIS buffer, each about 2 min. The sections may be left in the last change, but total time in TRIS buffer should not exceed 30 min.
4. Transfer sections to the 20 ml of DAB solution (E), and wait for 5 min.
5. Add 2 ml freshly diluted hydrogen peroxide (Solution F) to the sections and DAB solution. Mix well by stirring with a glass rod. Make sure the sections are not collapsed into little balls or knots.
6. Wait for 10 min, then wash in 3 changes of water, each at least 1 min.
7. Mount sections onto slides and allow to dry.
8. *Either* mount in an aqueous medium, *or* dehydrate through graded alcohols, clear in xylene, and mount in a resinous medium.

Result

Sites of enzymatic activity black. Brown if steps 4 and 5 are omitted.

Notes

1. Stages 4 and 5 may be omitted when only sites of strong activity, such as AChE at motor end plates or ChE in cerebral capillaries and some central neurons, are of

interest. Stages 4 and 5 are necesary when medium C.(2) is used.

2. Although the substrates (especially butyryl-thiocholine) show partial specificity, the use of inhibitors is imperative for certain identification of the two enzymes. A control with eserine should also be provided, though non-enzymatic precipitation of the end-product is unusual with this technique. If the DAB-nickel-H_2O_2 intensification is carried out, there may be a non-specific grey background, and sites of endogenous peroxidase activity (erythrocytes, granular leukocytes, occasional groups of neurons in brain) will be stained.

3. Endogenous peroxidase activity can be blocked by treating the sections with 0.1% hydrogen peroxide for 30 min, before Step 1 of the method.

15.3. Peptidases and proteinases

The proteolytic enzymes catalyse the hydrolysis of peptide bonds between amino acids in polypeptides and proteins:

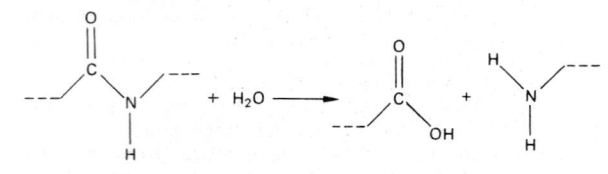

The different enzymes attack peptide linkages adjacent to particular amino acids or at one or other end (the C- or N-terminal amino acid) of a polypeptide chain.

15.3.1. Classification of proteolytic enzymes

The scheme in Table 15.3 is abstracted from Dixon & Webb (1979) and slightly modified to accommodate histochemically demonstrable enzymatic activities.

Some of the peptidases catalyse the hydrolysis of synthetic substrates containing the peptide configuration, such as naphthyl amides. Histo-chemical methods are available in which the naphthyl amines liberated from substrates of this type are trapped by diazonium salts in a similar way to that in which α-naphthol is trapped in the method for alkaline phosphatase described earlier in this chapter. Some proteinases are able to act upon esters and this property has also been utilized histochemically. Unfortunately, considerable uncertainty exists as to the correspondence between the histochemically demonstrated enzymes and those identified by biochemists.

15.3.2. Method for enteropeptidase

Enteropeptidase (E.C. 3.4.21.9; also called enterokinase) is secreted in the duodenum. It catalyses the hydrolysis of a peptide bond between lysine and leucine in pancreatic trypsinogen, thus forming a hexapeptide and trypsin:

$$\text{TRYPSINOGEN} + H_2O \rightarrow \text{TRYPSIN} + \text{PEPTIDE}$$

The incubation medium for the histochemical detection of enteropeptidase (Lojda & Malis, 1972) contains trypsinogen, an artificial chromogenic substrate for trypsin, and a diazonium salt. Three reactions occur:

1. Enteropeptidase catalyses the hydrolysis of its natural substrate, trypsinogen, releasing trypsin. At this stage, the newly formed trypsin is in the medium within and near to the cells that contain enteropeptidase.

2. Trypsin catalyses the hydrolysis of N-benzoylarginine-2-naphthylamide (BANA):

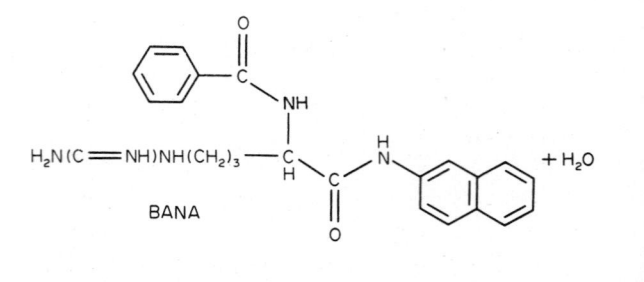

TABLE 15.3. *Peptidases and proteinases*

Peptidases (= exopeptidases) E.C. 3.4.11–3.4.17
(Release single amino acids from peptide chain)

Aminopeptidases (= α-aminoacylpeptide hydrolases) E.C. 3.4.11
(Release amino acids from N-terminus)
Many mammalian and bacterial enzymes. Some have little specificity; others are specific for the N-terminal amino acid. Most require metal ions (Zn^{2+}, Mg^{2+}, Mn^{2+}) for activity.

Dipeptidases E.C. 3.4.13
(Act on dipeptide substrates)
Most of these are specific for one of the amino acids of the dipeptide. Related enzymes act at the penultimate peptide linkage, and liberate dipeptides from larger polypeptides.

Carboxypeptidases E.C. 3.4.16–3.4.17
(Release amino acids from C-terminus)

Serine carboxypeptidases E.C. 3.4.16
(Acid pH optimum; inhibited by organophosphorus compounds that bind to serine residues)
Includes a group of enzymes of broad specificity, and enzymes that release C-terminal proline and tyrosine.

Metallocarboxypeptidases E.C. 3.4.17
(Need a metal: Zn^{2+}, Co^{2+}, for activity)
Includes the pancreatic enzymes Carboxypeptidase A (E.C. 3.4.17.1.; releases C-terminal amino acids other than arginine, lysine and proline), and Carboxypeptidase B (E.C. 3.4.17.2.; releases lysine or arginine), and some enzymes that remove individual specific C-terminal amino acids.

Proteinases (= endopeptidases) E.C. 3.4.21–3.4.24
(Attack non-terminal peptide linkages)

Serine proteinases E.C. 3.4.21
(Have histidine and serine at active site; inhibited by organophosphorus compounds that bind to serine residues)
Includes Chymotrypsin (E.C. 3.4.21.1), which cleaves peptide linkages at the carbonyl end of phenylalanine, tyrosine, tryptophan or leucine; Trypsin (E.C. 3.4.21.4), which cleaves peptide linkages at the carbonyl end of lysine or arginine, and many enzymes with specialized metabolic functions.

Thiol proteinases E.C. 3.4.22
(Have cysteine at active site; inhibited by compounds that react with –SH groups)
Include the plant proteinase Papain (E.C. 3.4.22.2), which cleaves peptide bonds at the carbonyl end of lysine or arginine or next-but-one to the carboxyl group of phenylalanine. Cathepsin B (E.C. 3.4.22.1) is an enzyme with similar specificity that occurs in many tissues of vertebrate animals.

Carboxyl (acid) proteinases E.C. 3.4.23
(Acid pH optimum; unionized carboxyl group at active site)
Include Pepsins A, B & C (E.C. 3.4.23.1–3), which attack peptide linkages on the carbonyl side of leucine and phenylalanine, and Cathepsin D (E.C. 3.4.23.5), and intracellular proteinase with similar specificity.

Metalloproteinases E.C. 3.4.24
(Need a metal: Zn^{2+}, Ca^{2+}, Mg^{2+}, Fe^{2+}, for activity; inhibited by chelating agents)
Include various Collagenases (from bacteria and from vertebrate animals), which attack linkages joining glycine to proline, and numerous enzymes of invertebrates and microorganisms.

N – benzoylarginine   2 – naphthylamine

3. One of the products of the reaction is 2-naphthylamine (= β-naphthylamine). This couples with the diazonium salt. Fast blue B salt has two $-N_2^+$ groups, so it can combine with two molecules of the aromatic amine, to form a dis-azo dye:

0.05 M. A 0.1 M buffer is prepared by making up the final volume to only half that stated in the table. Adjustment of the pH may be necessary.

Agar solution

 Dissolve 0.5 g of agar in 50 ml of water, by

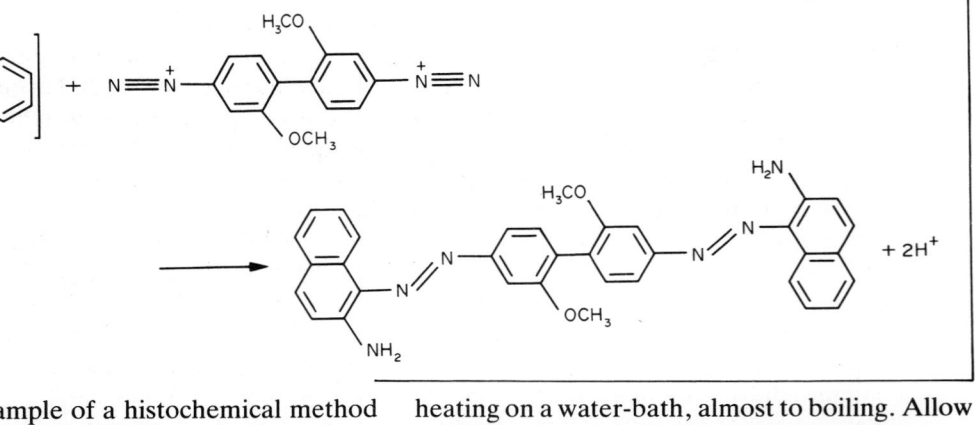

This is an example of a histochemical method in which two enzymatically catalysed reactions occur sequentially. The trypsin is formed in solution, and can be expected to diffuse away from the sites in which it is formed by the enteropeptidase-catalysed hydrolysis of trypsinogen. The accuracy of localization depends on rapid hydrolysis of BANA by the released trypsin and immediate production of the azo dye. The ingredients of the incubation medium are dissolved in a gel made from agar, and this is allowed to set in contact with the sections. The high viscosity of the gelled agar retards diffusion of trypsin and of 2-naphthylamine, thereby improving the accuracy of localization of enteropeptidase.

 Enteropeptidase is present at the luminal ends of the absorptive epithelial cells of the duodenum. Control specimens can be taken from the distal ileum. The tissue may be fixed for a few hours in neutral buffered formaldehyde at 4°C, but Lojda *et al.* (1979) preferred to use unfixed cryostat sections.

Solutions required

0.1 M TRIS buffer, pH 6.5, with 0.01 M Ca²⁺ (0.1% CaCl₂)

 The TRIS buffer in Chapter 20, p. 358 is

heating on a water-bath, almost to boiling. Allow to cool to about 60°C, and maintain at this temperature until ready to use. Make on the day it is to be used.

Copper sulphate solution

 2%(w/v) aqueous $CuSO_4.5H_2O$. Keeps indefinitely.

Incubation medium

 This medium is made up immediately before it is to be used.

  N-benzoyl-L-arginyl-2-naphthylamide: 8 mg
   (If the pure L-form of BANA is not
   available, use the DL- form)
  N,N-dimethyl formamide (DMF): 0.5 ml
   Dissolve the BANA in the DMF, then
   add to 5 ml of the TRIS buffer with Ca²⁺.
   Mix well, then add:
  Fast blue B salt (C.I. 37235): 5–10 mg

Filter the solution if it is not transparent. Add 2 mg of trypsinogen, swirl or stir to dissolve, then add 2 ml of the warm agar solution. Quickly check the pH, and adjust to 6.5 if necessary. Proceed immediately to stage 2 of the procedure described below.

 For a **control medium**, omit the trypsinogen.

Procedure
1. Mount the sections on slides and allow to dry. Place on a horizontal staining rack.
2. Pour the warm incubation medium onto the slides. It should spread to form a thin layer covering the sections, and set within about 1 min.
3. Transfer the slides to an incubator with a moist atmosphere at 37°C, and inspect at 30 min intervals. Optimum staining is usually seen between 1 and 2 h.
4. Immerse the slides in 2% copper sulphate for 5 min at room temperature. Handle very carefully to avoid damaging the agar layer and the sections.
5. Transfer the slides to water, and leave for 5 min to remove copper sulphate and any other residual reagents.
6. Examine and photograph, then allow the slides to dry before storing. If they are to be examined later, a coverslip may be applied, using an aqueous mounting medium or a drop of immersion oil.

Result

The product of the reaction is purple; it is seen in the section, and in the overlying film of agar. There is also a brownish-yellow background, due to azo-coupling of the fast blue B salt with aromatic amino acid side-chains of proteins (see Chapter 10).

Enteropeptidase occurs at the luminal brush-borders of the epithelial cells of the duodenum, and is absent from the distal ileum. Purple colour produced in the control medium (without trypsinogen) may be due to trypsin-like lysosomal enzymes, or to a peptidase present in mast cells.

15.3.3. Substrate film methods

We now consider a group of techniques that have low specificity but are based on principles quite different from those in which soluble substrates and trapping agents are used. In the first method of this kind to be described (Adams & Tuqan, 1961), the sections are placed on the emulsion of an over-exposed, developed, fixed and washed photographic plate. The section and the film are both slightly moist (but not wet) and are equilibrated with a suitable buffer. This preparation is incubated for about an hour in a humid atmosphere and then dried and mounted for examination.

The emulsion on the film contains tiny particles of metallic silver suspended in gelatin (which is a protein of high molecular weight made by boiling collagen in water). Proteolytic activity derived from the section changes the subjacent gelatin into soluble peptides of lower molecular weight which diffuse into the surrounding regions of the emulsion, carrying the suspended silver particles with them. After a suitable time of incubation, the blackened photographic emulsion will have holes in it which correspond to the sites of proteinases in the tissues. The section itself is obscured by the black emulsion, so the positions of the digested holes have to be identified by comparison of the substrate-film preparation with an adjacent, stained section of the same specimen. Colour films have also been used as substrates, but most are unsuitable because their emulsions are made of gelatin that has been excessively cross-linked in the manufacturing process (Hasegawa & Hasegawa, 1977).

A proteinaceous substrate of more certain composition than a photographic emulsion may be prepared by covalently linking a dye to a layer of gelatin on a microscope slide (Cunningham, 1967; Kiernan, 1981). A dyed gelatin film is sufficiently transparent to allow examination of the section by phase-contrast microscopy, thereby permitting more accurate localization of the holes produced by the action of proteolytic enzymes. An alternative approach is to use an undyed gelatin film and, after incubation, to stain both the film and the section by a general method for protein (Fried et al., 1976). The colour is then strongest where both section and film are present, so that digested areas are easily seen only at the edges of the tissue.

Substrate-film methods are not applicable only to proteolytic enzymes. It is possible to make gelatinous films containing other macromolecular substrates, such as nucleic acids or glycosamino-glycans, and to demonstrate local areas of degradation by appropriate staining. In practice, none of these methods are much used, because of the poor resolution of the localization of enzyme activity.

15.4. Use of inhibitors

The carboxylic esterases are defined partly on the basis of their susceptibilities to different inhibitors. The following instructions relate to the practical uses of such compounds. *Remember: Most substances that block the active sites of enzymes are poisonous, and the pure substances and stock solutions must be handled with great care.* Incubation media contain greatly diluted inhibitors, and do not pose any danger.

15.4.1. Preparation of inhibitor solutions

A. **Eserine.** Used as eserine (= physostigmine) sulphate, M.W. 649, usually as a 10^{-5} M solution. Dissolve 6.5 mg of eserine sulphate in 10 ml of water. Add 0.1 ml of this to 9.9 ml of buffer (for pre-incubation) or incubation medium (for simultaneous incubation and inhibition). 10^{-5} M eserine inhibits AChE and ChE.

B. **E600.** Diethyl-*p*-nitrophenyl phosphate is sold as an oily liquid in an ampoule. It is poisonous and care must be taken to avoid ingestion, contact with the skin, or inhalation of vapour. Ampoules contain 1.0 ml (M.W. 275; S.G. 1.27). Drop an ampoule into 462 ml of propylene glycol in a stoppered reagent bottle under a fume hood. Break the ampoule with a glass rod and insert the stopper. This gives 10^{-2} M stock solution of E600, which keeps indefinitely at 4°C. The bottle should be stood in a container of absorbent material such as kieselgühr or vermiculite.

For use, dilute with buffer to obtain the desired concentration (usually 10^{-7} to 10^{-3} M). Dispose of old solution as described for DFP.

10^{-5} M E600 inhibits AChE, ChE, and carboxylesterase. Arylesterase and acetylesterase are inhibited only by very high concentrations (e.g. 10^{-2} M).

C. **DFP**. Diisopropylfluorophosphate is a liquid, supplied in a glass ampoule. On account of its volatility, it is more dangerous than E600. Ampoules contain 1.0 g of DFP (M.W. 208). In a fume hood, drop an ampoule into 480 ml of propylene glycol in a glass-stoppered reagent bottle. Break the ampoule with a glass rod and insert the stopper. This stock solution is 10^{-2} M DFP. Store at 4°C with the bottle standing in a container of absorbent material. Keeps indefinitely. Do not pipette by mouth. For use, dilute with buffer to obtain the desired concentration (usually 10^{-7} to 10^{-5} M). Solutions containing 10^{-2} M DFP should be discarded by pouring into an excess of 4% aqueous NaOH and leaving in the fume cupboard for 4 days before washing down the sink with plenty of water. More dilute solutions, in quantities less than 100 ml, may be discarded without special precautions.

AChE and ChE are inhibited by 10^{-5} to 10^{-4} M DFP. ChE is inhibited by 10^{-7} to 10^{-6} M DFP.

D. **Ethopropazine.** Available as its hydrochloride (M.W. 349), or as its methosulphate (M.W. 359), which is used clinically as a drug for relieving symptoms of Parkinson's disease. Ethopropazine is usually used as a 10^{-4} M solution. Dissolve 34.9 mg (hydrochloride) or 35.9 mg (methosulphate) in 10 ml of water. Add 0.1 ml of this to 9.9 ml of buffer (for pre-incubation) and to 9.9 ml of incubation medium (for simultaneous incubation and inhibition). ChE is inhibited selectively.

E. **B.W. 284C51** (1,5-*bis*-(4-allyldimethyl-ammoniumphenyl)pentan-3-one dibromide) (M.W. 560). Dissolve 140 mg in 25 ml of water to give a stock solution which is 10^{-2} M. Keeps for a few months at 4°C. Dilute with buffer and with incubation medium to the desired concentration (10^{-5} to 10^{-4} M). B.W. 284C51 inhibits AChE but not ChE.

F. **_p_-Chloromercuribenzoate** (PCMB). Available as *p*-chloromercuribenzoic acid (M.W. 357) or its sodium salt (M.W. 379). Dissolve 36 mg of the acid in the smallest possible volume (1.0–2.0 ml) of 2 N (8%) NaOH. Add 5 ml of buffer, then adjust to the desired pH by careful addition of 1.0 M HCl. Make up to 10 ml with buffer to obtain a 10^{-2} M solution of the inhibitor.

The sodium salt (38 mg) can be dissolved directly in 10 ml of buffer. Dilute with buffer and with incubation medium to 10^{-4} to 10^{-6} M PCMB. Acetylesterase is not inhibited.

G. *iso*-OMPA. Tetraisopropylpyrophosphoramide (M.W. 342.4) is a somewhat sticky solid. Store in a metal can in a freezer. For use, weigh out approximately 10 mg, dissolve this in 5 ml of alcohol, and make up to about 1000 ml with water. The final volume should be 1 ml for every 1.0 mg of *iso*-OMPA, giving a solution that is close to 3×10^{-5} M.

For use, dilute with buffer to obtain the desired concentration. The principal use of *iso*-OMPA is in the histochemistry of choline esterases. A $10^{-7} – 10^{-6}$ M solution selectively inhibits cholinesterase but spares acetylcholinesterase. Higher concentrations (e.g. 10^{-4} M) inhibit AChE and carboxylesterase, in much the same way as E600 and DFP. Disposal of old solutions: as for DFP.

15.4.2. Methods of application

The **irreversible inhibitors** (i.e. the organophosphorus compounds E600, DFP and *iso*-OMPA) are used by **pre-incubation** in a solution with the same pH as the substrate-containing mixture. The sections are placed in a buffered solution containing the desired concentration of the inhibitor for 30 min at 37°C. The sections are rinsed in four changes of buffer before being transferred to the incubation medium. The rinsing is a particularly important part of the procedure when inhibited and uninhibited sections are to be incubated side by side in the same batch of medium.

Other inhibitors work by **competition** with the substrate for the active sites of the enzymes and must be **incorporated in the incubation medium**. Usually it is desirable also to pre-incubate the sections with a solution of the inhibitor, so that there will be no chance of enzymatic hydrolysis of the histochemical substrate during the earliest stages of incubation. Pre-incubation in a competitive inhibitor should **not** be followed by washing.

In any attempt to identify carboxylic esterases it is necessary to use several inhibitors over a range of concentrations extending at least two orders of magnitude either side of that generally thought to be optimal for the inhibition of any particular enzyme. Inhibitors may not be necessary when a histochemical procedure is used only as a staining method for some structure which happens to contain one of the enzymes. Inhibitors other than the ones described here are listed by Pearse (1972, pp. 796 and 798).

15.5. Exercises

1. Explain in general terms the methods by which the cellular localizations of hydrolytic enzymes are determined. What factors limit (a) the chemical, and (b) the topographical accuracy of such techniques?

2. What are the differences between reversible and irreversible inhibitors of carboxylic esterases? How are the differences reflected in the histochemical utilization of inhibitors?

3. In sections of nervous tissue it was found that the cells of a certain nucleus had the following histochemical properties:
   (a) Strong reaction with 5-bromoindoxyl acetate as substrate in the presence of 10^{-5} M diethyl-*p*-nitrophenyl phosphate (E600).
   (b) Strong reaction with butyrylthiocholine as substrate; inhibited by 10^{-5} M eserine, 10^{-5} M E600, 10^{-5} M ethopropazine, or 10^{-7} M diisopropylfluorophosphate (DFP).
   (c) Weak reaction with acetylthiocholine as substrate, inhibited by the same agents which inhibited the hydrolysis of butyrylthiocholine.

   Which enzyme (or enzymes) is (are) likely to be responsible for these histochemical findings? What further investigations are indicated?

4. **Thiamine pyrophosphatase** (TPPase), an enzyme found in the Golgi apparatus, releases phosphate ions from the substrate, thiamine pyrophosphate (TPP). The phosphate can be trapped by calcium ions. TPP is also hydrolysed to some extent by alkaline phosphatases, but the latter are inhibited by cysteine. TPPase can function over a wide pH range. Calcium salts are colourless, but the Ca^{2+} ion can be replaced by cobalt, which has an almost colourless phosphate but a black sulphide. Co^{2+} cannot, however, be included in the substrate mixture, since it would inhibit the enzyme.

   On the basis of the above facts, devise an incubation medium for TPPase and explain how the sites of enzymatic activity could be made visible under the microscope.

5. It is possible to determine the histological localization of the enzyme **deoxyribonuclease** by placing cryostat sections on a film made from a solution of DNA in aqueous gelatin and incubating for 1 h. Because of the ways in which the section and film are mounted, it is then possible to separate them and obtain two slides, one bearing the section and the other bearing the substrate-film.

   How could the sites of DNase activity be demonstrated? What control procedures would be necessary in order to establish the specificity of the method?

16

Oxidoreductases

The following account is a simplified one, and several controversial issues are not taken into consideration. Pearse (1972) and Wohlrab, Seidler & Kunze (1979) discuss the subject in great detail, and shorter treatments are given by Chayen *et al.* (1973) and Lojda *et al.* (1979). The metabolic functions of the oxidoreductases are described in textbooks of biochemistry.

16.1. Oxidation and reduction

An atom or molecule is said to be oxidized when it loses one or more electrons and to be

reduced when it gains one or more electrons. A simple example of a reaction of this type is a change in the oxidation state of a metal ion:

$$Fe^{2+} \xrightleftharpoons[\text{reduction}]{\text{oxidation}} Fe^{3+} + \epsilon^-$$

The position of the equilibrium is determined by the presence of other substances which can accept or donate electrons (**oxidizing** and **reducing agents** respectively). For example, the ferric ion is reduced by hydroquinone in acid conditions:

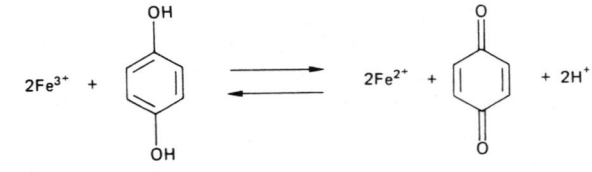

Here, the ferric ion is the oxidizing agent and is itself reduced to the ferrous state. It is equally valid to consider that hydroquinone is a reducing agent which acts upon the ferric ions and is itself oxidized to *p*-quinone.

The reaction between ferric ions and hydroquinone is the algebraic sum of two half-reactions:

$$Fe^{3+} + \epsilon^- \rightleftharpoons Fe^{2+} \tag{16.1}$$

$$H_2Q \rightleftharpoons Q + 2H^+ + 2\epsilon^- \tag{16.2}$$

(Q = *p*-quinone; H_2Q = hydroquinone)

Both sides of equation (16.1) are multiplied by 2 before adding, in order that the number of electrons participating in each half-reaction will be the same. Thus:

$$\begin{array}{c} 2Fe^{3+} + 2\epsilon^- \rightleftharpoons 2Fe^{2+} \\ H_2Q \rightleftharpoons Q + 2H^+ + 2\epsilon^- \\ \hline 2Fe^{3+} + H_2Q \rightleftharpoons 2Fe^{2+} + Q + 2H^+ \end{array}$$

It will be noticed that the electrons have been cancelled out and do not appear in the equation for the complete reaction. It will also be noticed that the reaction is reversible. The oxidation of hydroquinone in neutral and alkaline solutions is more complicated and leads to the formation of polymeric products.

The net effect of the oxidation of hydroquinone has been the loss of two atoms of hydrogen. The gain or loss of hydrogen is a feature of most oxidation-reduction reactions of biological importance. According to an earlier definition, a substance was said to be oxidized when it gained oxygen, lost hydrogen, or increased its positive charge and to be reduced when oxygen was lost, hydrogen gained, or negative charge increased. The modern definition, in terms of the transfer of electrons, simplifies the multiple requirements of the older one.

The number of electrons gained or lost by an organic molecule is not always as obvious as it is in the case of an inorganic ion. It may be determined (see Hendrickson et al., 1970, for more information) from the change in the oxidation number of the carbon atom at which oxidation or reduction takes place. The oxidation number of a carbon atom is found by adding the following values for each of its four bonds:

-1 for each —H
 0 for each —C
+1 for each bond to an element other than H or C

Thus, for the carbon atom at the top of the p-quinone molecule there are two bonds to other carbon atoms $(2 \times 0 = 0)$ and two to an oxygen atom $(2 \times +1 = +2)$, so the oxidation number is $+2$. At the equivalent position in hydroquinone there are three bonds to carbon $(3 \times 0 = 0)$ and one to oxygen $(1 \times +1 = +1)$, giving the oxidation number $+1$. This carbon atom therefore gains one electron in the process of being reduced. The carbon on the opposite side of the ring behaves identically, so the reduction of the whole molecule of p-quinone involves the acquisition of two electrons. Conversely, the oxidation of hydroquinone is accomplished by its losing two electrons. Application of the rule given above will show that each of the other four carbon atoms has oxidation number -1 in both p-quinone and hydroquinone.

16.2. Oxidation-reduction potentials

It is possible to make an electrical cell in which one electrode is gaseous hydrogen and the other is an inert metal such as platinum, immersed in an electrolyte which is a solution of a substance capable of being oxidized or reduced. Completion of the external circuit by a wire will result in the passage of electrons from one electrode to the other as a consequence of the gain or loss of electrons by the electrolyte at the inert electrode. The potential difference between the electrodes can be measured, in volts, and its magnitude, under standardized conditions of temperature and concentration, is the **oxidation-reduction potential** (E_0) of the electrolyte. It is a measure of the ease with which the electrolyte is oxidized or reduced or, conversely, of the strength of the electrolyte as an oxidizing or reducing agent.

Hydrogen is arbitrarily assigned $E_0 = 0$. The oxidation-reduction potentials of other substances are, by the most widely used convention, applied to half-reactions in which electrons are gained. Thus the half-reaction in which a strong oxidizing agent (which takes up electrons avidly) is reduced will have a high positive value of E_0. The half-reaction in which a weaker oxidizing agent than the hydrogen ion is reduced will have a negative E_0. These half reactions are reversible. If they are written the other way round, the negative or positive sign of E_0 is changed.

Consider, for example, the following half-reactions, arranged in descending order of their oxidation-reduction potentials:

$$MnO_4^- + 8H^+ + 5\varepsilon^- \rightleftharpoons Mn^{2+} + 4H_2O \quad (E_0 = +1.51\,V)$$
$$Fe^{3+} + \varepsilon^- \rightleftharpoons Fe^{2+} \qquad (E_0 = +0.77\,V)$$

$$2CO_2 + 2H^+ + 2\varepsilon^- \rightleftharpoons (COOH)_2 \quad (E_0 = -0.49\,V)$$
$$Na^+ + \varepsilon^- \rightleftharpoons Na \qquad (E_0 = -2.71\,V)$$

Of the substances shown, permanganate ion is the strongest oxidizing agent, and sodium metal is the strongest reducing agent. A substance on the left-hand side of one of the above half-reactions can be expected to oxidize a substance on the right-hand side only if the latter has a more negative E_0 than the former. Permanganate ions in acid solution therefore react with oxalic acid to give manganous ions and carbon dioxide, but sodium ions will not react with ferrous ions. The equation for the overall reaction is obtained by reversing the half-reaction that contains the reducing agent and adding it to the half-reaction containing the oxidizing agent. It may be necessary to multiply both sides of one or both of the equations by appropriate integers in order to obtain equal numbers of electrons on the two sides of the final equation. For the reaction between permanganate ion and oxalic acid, one reaction must be multiplied by 2 and the other by 5:

$$2MnO_4^- + 16H^+ + 10\varepsilon^- \rightleftharpoons 2Mn^{2+} + 8H_2O$$
$$5(COOH)_2 \rightleftharpoons 10CO_2 + 10H^+ + 10\varepsilon^-$$

$$2MnO_4^- + 6H^+ + 5(COOH)_2 \rightleftharpoons 2Mn^{2+} + 10CO_2 + 8H_2O$$

Although this reaction is theoretically reversible, it proceeds from left to right, virtually to completion, because there is a large difference between the oxidation-reduction potentials of the two component half-reactions. (The continuous removal of carbon dioxide from the system, as gas or by combination with water to form carbonic acid, also helps to drive the reaction from left to right, in accordance with Le Chatelier's principle and the law of mass action.)

Tables of oxidation-reduction potentials are valuable for showing which oxidations are likely to occur and which are not, but they must be used with caution (see Latimer, 1952 and textbooks of inorganic and general chemistry). Other chemical properties of the reactants are not taken into account and may complicate the overall reaction. The mixing of ferric ions with oxalic acid, for example, will result in the formation not only of ferrous oxalate (a sparingly soluble salt) but also of soluble complexes in which one, two, or three oxalate ions are coordinately bound to iron ions in both oxidation states +2 and +3. These complications could only be predicted by taking into account the chemistry of complex formation as well as the oxidation-reduction potentials.

Of greater importance in biochemical oxidation-reduction reactions is the fact that the value of the potential for any system varies with temperature, with the pH of the medium in which the reactants are dissolved, and with the proportions of the oxidizing and reducing agents present. A constant more useful than E_0 is E'_0, the oxidation-reduction potential of the half-reduced system at specified temperature and pH. Table 16.1 (p. 261) gives values of E'_0 for some biochemically and histochemically important half-reactions at pH7 and 25°C.

16.3. Biological oxidations

The life of every cell depends upon the coordinated oxidation and reduction of many organic compounds. These chemical reactions, collectively known as **cellular respiration**, involve a great number of substrates and enzymes and a much smaller number of **electron carriers**. One electron carrier can function as coenzyme or prosthetic group to many different substrate-specific respiratory enzymes. A **coenzyme** is a soluble substance (itself enzymatically inert and not a protein) which can diffuse in the cytoplasm and attach itself, reversibly, to the proteinaceous **apoenzyme**. A **prosthetic group** is a non-protein organic compound which is covalently bound to the protein molecule constituting the enzyme.

When attached to the apoenzyme, the coenzyme accepts electrons from the substrate and protons from either the substrate or the surrounding medium. Thus the coenzyme is reduced and the substrate is oxidized. Effectively the coenzyme removes one or two atoms of hydrogen from the substrate, so it is equally valid to call the electron carrier a **hydrogen acceptor**.

The reduced form of the coenzyme can act as an electron donor for the enzymatic reduction of some other substrate. The original coenzyme is thus regenerated. Enzymes that catalyse the re-oxidation of reduced coenzymes are often called **diaphorases**. They are more correctly named as the enzymes that catalyse the reduction of their specific substrates. The diaphorase activity (catalysis of the oxidation of reduced coenzyme) is incidental to the main function of such an enzyme. The histochemist, however, is often primarily interested in localizing the sites of re-oxidation of reduced coenzymes. Since more than one enzyme may catalyse this reaction for a single reduced coenzyme, the conveniently vague term "diaphorase" will continue to be used when the enzymes concerned cannot be accurately identified and named.

Before discussing some of the enzymes that catalyse oxidation and reduction and the histochemical methods for their identification, it is necessary to review the system that transfers electrons and protons from oxidized metabolites to the ultimate oxidizing agent, which is atmospheric oxygen. The transport takes place in several stages and involves the repetitive oxidation and reduction of various prosthetic groups, coenzymes, and cytochromes. The last-named substances are proteins with iron-containing haem groups. The oxidation number of the iron atom is $+3$ or $+2$, when the molecule is in the oxidized or reduced state, respectively. Cytochromes are not, strictly speaking, enzymes, but one of them, cytochrome aa_3, is commonly known as **cytochrome oxidase**. It transfers electrons and protons to molecular oxygen. It is not itself consumed in this reaction. Thus, cytochrome aa_3 catalyses the reduction of molecular oxygen to water. The reactions of the electron-transport chain occur in an order dictated largely by the oxidation-reduction potentials of the various half-reactions that make up the system. These, together with the names of the more important coenzymes, prosthetic groups and cytochromes, are set out in Table 16.1. The data in the table are taken mainly from Loach (1976).

Not all the substances shown in Table 16.1 are involved in the oxidation of all metabolites. Electrons most often pass from NADH or NADPH to one of the flavoproteins and thence via ubiquinone and the cytochromes to molecular oxygen.

All the components of the electron-transport system are present in the mitochondria of eukaryotic cells and some occur also in the general cytoplasmic matrix, associated with soluble enzymes. The oxidation of any metabolite involves the enzymatically catalysed transfer of protons and electrons from **substrate** to an **acceptor**:

TABLE 16.1. *Oxidation-reduction potentials in the electron-transport system (abbreviations explained at foot of table)*

Half-reaction			E_0' (V) (at pH7.0; 25°C)
$H_2O_2 + 2H^+ + 2\varepsilon^-$	=	$2H_2O$	$+1.35$
$O_2 + 4H^+ + 4\varepsilon^-$	=	$2H_2O$	$+0.82$
cyt. $aa_3^{3+} + \varepsilon^-$	=	cyt. aa_3^{2+}	$+0.29$
cyt. $a^{3+} + \varepsilon^-$	=	cyt. a^{2+}	$+0.29$
cyt. $c^{3+} + \varepsilon^-$	=	cyt. c^{2+}	$+0.25$
cyt. $c_1^{3+} + \varepsilon^-$	=	cyt. c_1^{2+}	$+0.22$
$UQ + 2H^+ + 2\varepsilon^-$	=	UQH_2	$+0.10$
cyt. $b^{3+} + \varepsilon^-$	=	cyt. b^{2+}	$+0.08$
$FMN + 2H^+ + 2\varepsilon^-$	=	$FMNH_2$	-0.21
$FAD + 2H^+ + 2\varepsilon^-$	=	$FADH_2$	-0.22
$NAD^+ + 2H^+ + 2\varepsilon^-$	=	$NADH + H^+$	-0.32
$NADP^+ + 2H^+ + 2\varepsilon^-$	=	$NADPH + H^+$	-0.32

Coenzymes: UQ = ubiquinone (coenzyme Q); UQH_2 = reduced form. NAD^+ = nicotinamide adenine dinucleotide (formerly known as coenzyme I or DPN); NADH = reduced form. $NADP^+$ = nicotinamide adenine dinucleotide phosphate (formerly known as coenzyme II or TPN); NADPH = reduced form.

Prosthetic groups: FMN = flavin mononucleotide; $FMNH_2$ = reduced form. FAD = flavin adenine dinucleotide; $FADH_2$ = reduced form. The value of E_0' for a prosthetic group may differ from that given above for different apoenzymes. For example, the FAD of succinic dehydrogenase has $E_0' = -0.03$.

Cytochromes: Designated "cyt." followed by a letter (a, aa_3, b, c, c_1) with superscript indicating oxidation state of the iron atom. Thus cyt. c^{3+} and cyt. c^{2+} are the oxidized reduced forms of cytochrome c.

$$\text{substrate} + \text{acceptor} \underset{(\text{enzyme})}{\rightleftharpoons}$$

$$\text{oxidized substrate} + \text{reduced acceptor}$$

Commonly the acceptor is a coenzyme such as NAD^+ or $NADP^+$ or a prosthetic group such as FMN or FAD. The enzyme catalysing the reaction is then known as a **dehydrogenase**. Such an enzyme (consisting of apoenzyme + oxidized form of the coenzyme) combines specifically with its substrate and renders it highly reactive towards the coenzyme. When the reaction has taken place, the oxidized substrate and the reduced coenzyme part company with the apoenzyme and are then free to enter into other chemical reactions. Specificity for the substrate in the apoenzyme, though the latter cannot bind to the substrate unless it has first combined with the coenzyme.

A simple example of a reaction catalysed by a dehydrogenase is the oxidation of the lactate ion. The enzyme concerned is known as lactate dehydrogenase.

The lactate ion first combines with an enzyme molecule:

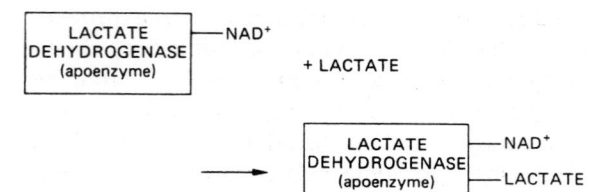

Then, on the surface of the enzyme molecule:

$$LACTATE + NAD^+ \rightarrow PYRUVATE + NADH + H^+$$

and finally,

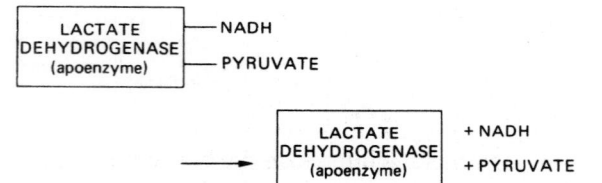

The NADH will subsequently be re-oxidized to NAD^+, when it serves as coenzyme in the enzymatically catalysed reduction of some other substrate. It will be noticed that the acceptor in the oxidation of lactate was NAD^+. The systematic name for lactate dehydrogenase, which identifies both the substrate and the acceptor, is L-lactate: NAD oxidoreductase (E.C. 1.1.1.27). The same enzyme will also catalyse the reduction of pyruvate ions. The direction in which the reversible reaction proceeds is determined by the relative concentrations of the reactants: lactate, pyruvate, NAD^+, NADH, and H^+. Oxidation of lactate occurs when this ion is present in excess and when NADH and H^+ are continuously removed, either by other metabolic activities or by deliberate manipulation of the conditions of the reaction.

The electrons removed from a substance such as lactate, when it has been oxidized, are incorporated into a reduced coenzyme. When this is re-oxidized, the electrons will be transferred to another acceptor, which has a higher oxidation—reduction potential than the coenzyme. Usually the flow of electrons passes via flavoprotein enzymes, ubiquinone, and the cytochrome system to molecular oxygen. For the histochemist, the importance of all this lies in the fact that the flow of electrons may be interrupted by the introduction of an artificial electron-acceptor with an oxidation-reduction potential intermediate between those of any two of the members of the electron transport chain. The tetrazolium salts, to be discussed below, are suitable for this purpose since on reduction they are converted to insoluble pigments. Thus, whenever a substrate is oxidized in the presence of a tetrazolium salt, the released electrons will not be transported through the usual sequence of cytochromes, etc., but will be trapped in the formation of a stable, coloured substance.

Not all oxidoreductases make use of the coenzymes NAD^+ and $NADP^+$; some dehydrogenases are flavoproteins. The most familiar of these is succinate dehydrogenase, whose prosthetic group is FAD. The physiological acceptors associated with the flavoprotein dehydrogenases are not certainly known, though ubiquinone is a likely candidate. Other oxidoreductases use molecular oxygen as an acceptor, thus bypassing all the intermediate components of the electron-transport system. These enzymes are known as **oxidases**. Tetrazolium salts cannot be used to detect the activity of oxidases unless they can be made to act as substitutes for oxygen. Other methods are therefore usually needed for these enzymes. The only other oxidoreductases considered in this chapter are the **peroxidases**. These catalyse the oxidation of many substances by hydrogen peroxide and are discussed in Section 16.6.

16.4. Histochemistry of dehydrogenases

The dehydrogenases catalyse the general reaction

$$\text{(S)}H_2 + \text{(A)} \rightleftharpoons \text{(S)} + \text{(A)}H_2$$

(S) and (S)H_2 represent oxidized and reduced forms of the substrate. (A) and (A)H_2 represent oxidized and reduced forms of the acceptor, which is a substance other than O_2 or H_2O_2.)

When the reaction proceeds from left to right, the net effect is the removal of hydrogen (usually two atoms of it) from the substrate. The acceptor is the coenzyme (NAD^+ or $NADP^+$) in the case of coenzyme-linked dehydrogenases. These enzymes are detected histochemically by substituting an artificial electron acceptor for the naturally occurring substances constituting the electron transport chain. The artificial substance chosen is one which becomes insoluble and coloured in its reduced state. Consequently, a visible precipitate forms at sites where hydrogen is given up by an electron carrier whose oxidation-reduction potential is negative with respect to that of the artificial hydrogen acceptor.

In the histochemical methods for dehydrogenases, the substrates are the physiological ones, and no attempts are made to trap the products of their oxidation. Instead, a special kind of indicator (the artificial hydrogen acceptor or electron carrier) is used to detect the place in which a biological oxidation is taking place. The substrate is provided in large amounts in the incubation medium. An adequate quantity of the acceptor must also be present, either as coenzyme added to the medium or as intermediate electron carriers already present in the tissue.

16.4.1. Tetrazolium salts

The artificial hydrogen acceptors of greatest value to the histochemist are the tetrazolium salts. These are heterocyclic compounds (derivatives of tetrazole, CH_2N_4) which are changed by reduction into insoluble, coloured **formazans**.

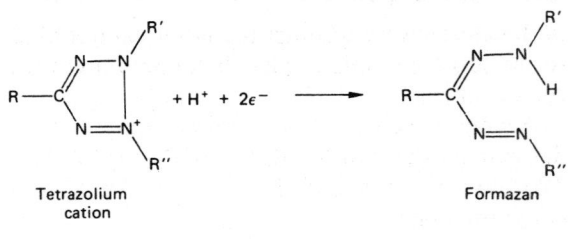

Tetrazolium
cation

Formazan

This reaction is irreversible because the formazan is insoluble. Several tetrazolium salts have been used in histochemical methods for dehydrogenases. The ideal one would be stable, not chemically altered by exposure to light, and would be reduced very rapidly to yield a formazan with exceedingly small crystals that were insoluble in lipids and had some substantivity for protein. These properties are most closely approached by some of the ditetrazolium salts, which have the general structure

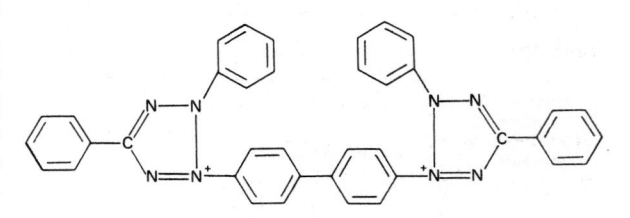

with various substituents on the benzenoid rings. Monotetrazolium salts, which have only one tetrazole ring in the molecule, are generally less suitable, though some have been used as histochemical reagents. When a ditetrazolium salt is reduced, the product may be either a monoformazan in which only one of the tetrazole rings has been opened or a diformazan in which both tetrazole rings have been opened. Monoformazans are usually red, and diformazans are blue, purple, or black. Both the coloured products may be formed in histochemical reactions, though the diformazan is the one desired. Other colours (usually reds) may also result from the presence of monotetrazolium salts as contaminants in samples of ditetrazolium salts.

A list of tetrazolium salts, with some of their properties, is given in Table 16.2.

The most generally useful tetrazolium salt for use in light microscopy is **nitro blue tetrazolium**

TABLE 16.2. *Properties of some tetrazolium salts and their formazans**

Trivial name, abbreviation, and M.W.	E'_0(V) (pH 7.2, 22°C)	Properties of formazan	Carriers from which electrons are accepted in histochemical usage
MONOTETRAZOLIUM SALTS			
Triphenyltetrazolium (chloride). TTC (M.W. 335)	−0.49	Large red crystals. High lipid solubility. Reduction is slow	cyt. a; cyt. a_3
2,3-*p*-dinitrotriphenyltetrazolium (chloride). 2,3-*p*-DNTTC (M.W. 425)	??−0.05	Small blue crystals. High lipid solubility, but also binds to protein. Reduction is rapid.	Flavoproteins; UQ; cyt.b
m-3-trinitrophenyltetrazolium (chloride). *m*-3-TNTTC (M.W. 470)	??−0.05	Small red crystals. Lipid-soluble, but also binds to protein. Reduction is very rapid.†	Flavoproteins; UQ; cyt.b
Tetrazolium violet (chloride). TV (M.W.384)	?−0.20	Large dark blue crystals (pink contaminant). High lipid solubility. Reduction is slow.	
Methylthiazolyldiphenyltetrazolium (bromide). MTT (M.W. 414)	−0.11	Co^{2+}-chelate has small, black crystals. Lipid-soluble but also binds to protein. Reduction is rapid.	UQ; cyt. b; cyt.c_1
Iodonitrotetrazolium (chloride). INT (M.W. 505)	−0.09	Large dark red crystals (orange contaminant). High lipid solubility. Reduction is slow.	UQ: cyt. b; cyt.c_1
2-(2-benzthiazolyl-5-styryl-2-(4-phthalhydrazidyl)tetrazolium (chloride). BSPT (M.W. 502)	?−0.10	Purple, amorphous. Osmiophilic. Used in EM histochemistry. Reduction is rapid.	
DITETRAZOLIUM SALTS			
Neotetrazolium (dichloride). NT (M.W. 668)	−0.17	Dark purple, small crystals, lipid-soluble (red monoformazan or contaminant). Reduction is rapid.	UQ; cyt. b; cyt. c
Blue tetrazolium (dichloride). BT (M.W. 728)	−0.16	Small deep blue crystals, lipid-soluble (red monoformazan). Reduction is slow.	
Nitro blue tetrazolium (dichloride). Nitro-BT (M.W. 818)	−0.05	Dark blue, amorphous, slight lipid-solubility. Binds to protein. Resists organic solvents (red monoformazan and contaminant: lipid- and alcohol-soluble). Reduction is rapid.	Flavoproteins; UQ; cyt. b
Tetranitro blue tetrazolium (dichloride). TNBT (M.W. 908)	?−0.05	Brown, amorphous, insoluble in lipids and organic solvents. Binds to proteins. (Pink monoformazan or contaminant). Reduction is rapid.	(Probably closely similar to nitro-BT)
Distyryl nitro blue tetrazolium (dichloride). DS-NBT (M.W. 870)	?−0.10	Amorphous, osmiophilic. Used in EM histochemistry. Reduction is rapid.	

*For more information, see Burstone (1962), Lillie & Fullmer (1976), Lillie (1977) and Seidler (1980). The oxidation-reduction potentials E'_0 are taken from Pearse (1972). These values of E'_0 may not be accurate, and cannot be compared in a meaningful way with the potentials of systems in which the oxidizing and reducing agents are soluble, for reasons given by Jámbor (1954) and Clark (1972). They are useful, however, for comparing one tetrazolium salt with another. (Values of E'_0 marked ? are guessed, on the basis of comparison of histochemical properties with those of tetrazolium salts with known oxidation-reduction potentials.)

†A blue-black formazan deposit is formed in injured and diseased cells containing enzymes that bring about reduction of *m*-3-TNTTC and some other monotetrazolium salts. Seidler (1980) has suggested that the blue colour is due to alignment of formazan molecules on abnormal proteins. This explanation is similar to that offered for metachromasia of basic dyes (see Chapter 11).

(nitro-BT), which has the advantage of forming a formazan which is not visibly crystalline, is insoluble in lipids, and is substantive for protein.

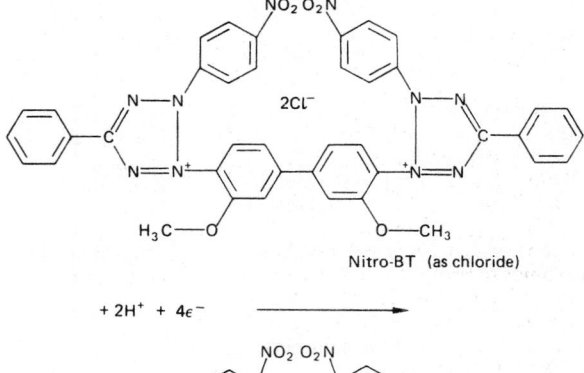

Nitro-BT (as chloride)

+ 2H⁺ + 4ε⁻ ⟶

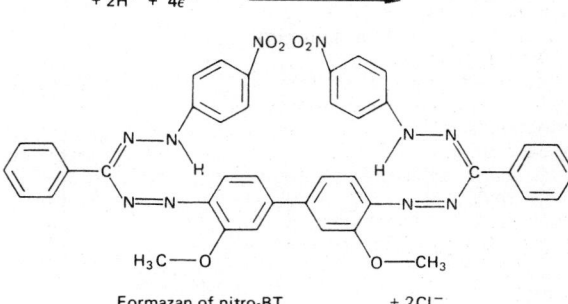

Formazan of nitro-BT   + 2Cl⁻

It will be seen from the above formula that nitro-BT is a ditetrazolium salt in which the radical R″ of the general formula is joined to two substituted tetrazole rings. The oxidation-reduction potential of nitro-BT is −0.05 V, which lies between that of FAD and that of ubiquinone. This tetrazolium salt can therefore be expected to accept electrons from NADH, NADPH, or $FADH_2$, but not from dihydroubiquinone or from any of the cytochromes. However, studies in which metabolic inhibitors of various components of the electron-transport chain have been used indicate that within mitochondria tetrazolium salts collect electrons from UQH_2 and even from reduced cytochromes (see Burstone, 1962; Seidler, 1979).

Reduction of tetrazolium salts by systems with higher oxidation-reduction potentials would occur if there were large differences between the concentrations of the products and of the reactants. The values of E'_0 in Tables 16.1 and 16.2 pertain when [products] = [reactants]. The

actual potential for the reduction, at a given temperature and pH, is

$$E = E'_0 + \frac{RT}{nF} \log_e \frac{[\text{oxidizing agent}]}{[\text{reducing agent}]}$$

where R is the gas constant, T is the absolute temperature, n is the number of electrons gained by the reduced molecule or ion (usually two), and F is the faraday. $RT/F = 0.026$ at 25°C.

Thus for a tetrazolium salt E will be higher than E'_0 when the concentration of this reagent in the medium exceeds that of the formazan. The concentration of formazan is, of course, always very low on account of its very low solubility. It has been pointed out by Clark (1972) that artificial electron acceptors used in biochemical studies of dehydrogenases commonly react at rates completely out of line with their oxidation-reduction potentials. The values of E'_0 apply to systems in true thermodynamic equilibrium, which is not likely to be the state of an experimental system, with an excess of oxidizing agent present, or indeed that of the substances in a living cell.

The tetrazolium salt in a histochemical incubation medium is in competition with the naturally occurring electron carriers of the cell. In order to divert electrons from the oxidized substrate to the tetrazolium salt, it is sometimes necessary to inhibit the flow of electrons to oxygen. This may be accomplished either by incubating under strictly anaerobic conditions or, more easily, by adding cyanide ions to the medium. The cyanide inhibits cytochrome oxidase (cyt. a_3). Azide ions act similarly. Cyanide ions can also combine with aldehydes or ketones that are formed in some dehydrogenations, thus enhancing the enzyme-catalysed reaction by removing one of its products.

16.4.2. Diaphorases

When a tetrazolium salt is reduced by NADH or NADPH, the reaction is catalysed by an enzyme, either NADH-diaphorase or NADPH-diaphorase.

These enzymes catalyse the reaction

Reduced coenzyme	tetrazolium salt	Oxidized coenzyme
(NADH + or NADPH)	salt	$\rightarrow$ (NAD$^+$ + formazan or NADP$^+$)

in which the reduced coenzyme is the substrate and the tetrazolium salt is the acceptor. When a tetrazolium salt is mixed with NADH or NADPH in the absence of a diaphorase apoenzyme, the reaction is very slow. Thus, the coloured product of a histochemical method for a coenzyme-linked dehydrogenase is formed by the catalytic action of another enzyme, the diaphorase. **Consequently, a coenzyme-linked dehydrogenase will be accurately localized only if it occurs in the same place as the diaphorase.** Fortunately, the diaphorases are present in all cells, in mitochondria, and sometimes also in the cytoplasmic matrix. They are rather "tough" enzymes, unlikely to be inhibited by short fixation in formaldehyde or by other preparative manipulations.

The natural acceptors associated with the two histochemically recognized diaphorases are not known with certainty, and the activities may be shared by several enzymes. The enzymes are believed to be flavoproteins. NADH-diaphorase may well be lipoamide dehydrogenase (NADH: lipoamide oxidoreductase; E.C. 1.6.4.3), which contains FAD as its prosthetic group and catalyses the reaction

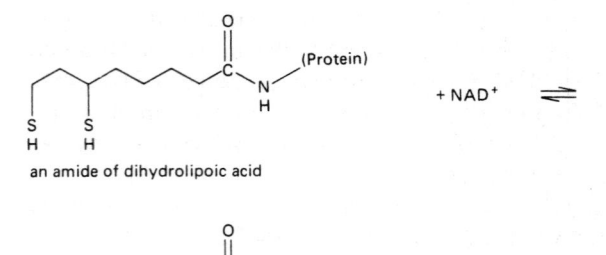

an amide of dihydrolipoic acid

an amide of lipoic acid

Tetrazolium salts are able to serve as acceptors in the place of lipoic acid when the reaction pro-

ceeds from right to left. The properties of NADP-diaphorase are shared by "Warburg's old yellow enzyme" (NADPH: (acceptor) oxidoreductase: E.C. 1.6.99.1), which contains FMN. The physiological acceptor is unknown. Until more is known of the identities of the enzymes that catalyse the oxidation of reduced coenzymes by tetrazolium salts, it is probably best to continue to call them NADH- and NADPH-diaphorases. Names such as "NADH-tetrazolium reductase" are also used and are acceptable.

The deliberate histochemical localization of the diaphorases is a very simple matter. Sections are incubated in a suitably buffered medium containing a tetrazolium salt and the appropriate **reduced form** of the coenzyme. The general methodological principles applicable to dehydrogenase histochemistry (see below) should also be observed.

In the case of the flavoprotein dehydrogenases, which have prosthetic groups rather than coenzymes, diaphorases cannot be responsible for the production of the formazan deposits. The mechanisms of electron-transfer from enzyme-bound FADH$_2$ or FMNH$_2$ to the tetrazolium salt are poorly understood, but may well involve ubiquinone and cytochromes, as discussed earlier. Some tetrazolium salts are able to accept electrons directly from reduced flavin nucleotides.

16.4.3. Technical considerations

16.4.3.1. Tissue preparation

Oxidoreductases are generally much more easily inactivated than the hydrolytic enzymes discussed in Chapter 15 (Chalmers & Edgerton, 1989). It is not possible to fix the tissues thoroughly enough to allow the cutting of sections on an ordinary freezing microtome and embedding in wax is out of the question. Small blocks of tissue may be fixed for 5–10min in neutral, buffered formaldehyde at 4°C and then sectioned in a cryostat, or fresh frozen sections from the cryostat may be similarly fixed. Unfixed cryostat sections are often used, though meticulous attention to technique (see Chayen et al., 1973) is necessary if the cells and their mitochondria are to remain recognizable after incubation. The two diaphorases and lactate dehydrogenase are notable in that they will survive fixation for several hours in neutral formaldehyde solutions. For the reasons given in Chapter 14, minimal fixation should be employed if possible. Fixation of the sections is more easily controlled than that of blocks. An alternative to formaldehyde is acetone (5–10min at 4°C), which also extracts some of the cytoplasmic lipids, to which certain of the formazans may be artifactually bound. Acetone also extracts ubiquinone, which may be a necessary intermediate for the reduction of tetrazolium salts in methods for the flavoprotein enzymes (see above). It has been shown in the case of succinate dehydrogenase that it is necessary to apply UQ to sections that have been treated with acetone in order to be able to detect the enzyme at all sites of activity (Contestabile & Andersen, 1978).

16.4.3.2. Composition of incubation medium

The incubation medium for histochemical localization of a dehydrogenase includes the following:

(i) **Buffer.** The pH of the medium should be 7.0–7.2, even if this is not optimum for the enzyme. At pH values more alkaline than this, non-enzymatic reduction of NAD^+ or $NADP^+$ occurs. The NADH or NADPH so produced serves as substrate for its appropriate diaphorase, with consequent meaningless deposition of formazan within the section. This artifact, known as "nothing dehydrogenase", is probably due to reduction of the coenzyme by sulphydryl groups of proteins containing cysteine. Nothing dehydrogenase activity is maximal at pH9.

Since incubation media for dehydrogenases often contain cations of divalent metals, TRIS buffer is usually used. Phosphate buffer is suitable when no metal ions that form insoluble phosphates are present.

(ii) **Substrate.** Commonly the substrate is an organic anion and is used as its sodium salt, at a concentration of 0.1M. Addition of the substrate usually changes the pH of the buffer, which must therefore be adjusted to the correct value by adding a few drops of 1.0M NaOH or 1.0M HCl.

(iii) **Coenzymes.** The amount of coenzyme contained in a section is usually very small, so for coenzyme-linked dehydrogenases it is necessary to provide an excess of NAD^+ or $NADP^+$ in the incubation medium at a concentration of approximately 0.003M. For demonstration of diaphorases, the reduced forms of the coenzymes (NADH, NADPH) are used.

(iv) **Cofactors.** Many dehydrogenases have requirements for traces of divalent metal cations. It is usual to include magnesium chloride (0.005M) in the medium. This does no harm, but is not necessary for all the enzymes. Magnesium ions probably also help to prevent rupture of mitochondria during incubation (see also (vii) below).

(v) **Tetrazolium salt.** The concentration of the tetrazolium salt is not very critical and may range from 10^{-4} to 10^{-3}M. Some samples of nitro-BT and TNBT are difficult to dissolve in water, so they are dissolved in a small volume of an organic solvent before being added to the aqueous medium. The solvent must be one that is not a substrate for dehydrogenases. Ethanol would not be suitable; acetone or N,N,-dimethylformamide is satisfactory.

(vi) **Electron transport inhibitors.** To suppress aerobic cellular respiration, sodium or potassium cyanide or sodium azide (0.005–0.01M) is incorporated in the incubation medium. An alternative, but inconvenient technique is to incubate in the complete absence of oxygen. For many enzymes, these precautions are unnecessary when a rapidly reducible tetrazolium salt such as nitro-BT is used.

(vii) **Protective agents.** The inclusion of a chemically unreactive synthetic polymer in the medium prevents osmotic damage to mitochondria during incubation and, by increasing the viscosity of the medium, limits the diffusion of soluble enzymes. A protective agent is not always needed if the tissue has been partially fixed, but is desirable when sections of unfixed tissues are used. The polymers employed for this purpose are polyvinylpyrrolidone (PVP) (7.5% w/v) and polyvinyl alcohol (PVA) (20% w/v). The molecular weight of PVP used for this purpose is not critical. PVA should have a molecular weight of 30,000. Addition of these polymers often acidifies the medium and the pH must be adjusted accordingly.

(viii) **Intermediate electron acceptors**. It is a common practice to add phenazine methosulphate (PMS) (10^{-5}–10^{-3}M) to incubation media for dehydrogenases. This easily reduced substance transfers electrons directly from reduced coenzymes or other acceptors to tetrazolium salts. The addition of PMS accelerates the reaction and gives more intense staining, but sometimes also causes non-specific deposition of formazan in the sections. This is due to spontaneous, non-enzymatic reduction of the tetrazolium salt by the reduced form of PMS. Menadione has also been used for the same purpose, though less often. An

oxazine dye, Meldola's blue (C.I. 51175; Basic blue 6) has also been proposed as an intermediate electron acceptor in dehydrogenase histochemistry (Kugler & Wrobel, 1978). It is used at a concentration of 10^{-4}M. The effects of Meldola's blue are the same as those of PMS, but the dye, unlike PMS, is not rapidly decomposed by light and causes only slight spontaneous reduction of tetrazolium salts. In a comparison of intermediate electron carriers, van Noorden & Tas (1982) found that menadione was ineffective, and that Meldola's blue imparted some of its own colour to the cells. They preferred PMS and a related compound, 1-methoxyphenazine methosulphate.

Intermediate electron acceptors cannot completely replace the naturally occurring diaphorases, and they may work mainly by enhancing the activity of these enzymes (Raap *et al.*, 1983a,b). Inclusion of PMS in the incubation medium can also lead to non-enzymatic formation of form-

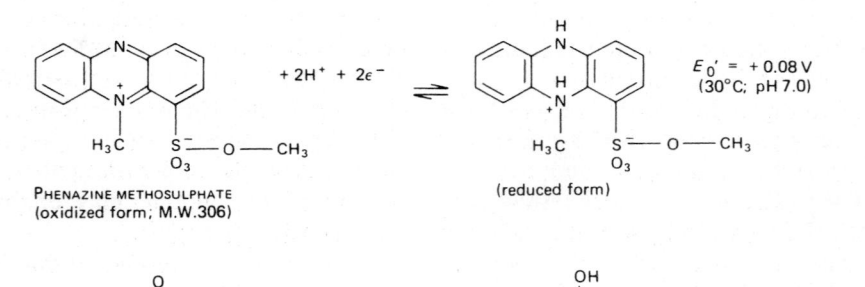

PHENAZINE METHOSULPHATE
(oxidized form; M.W.306)

(reduced form)

$E_0' = +0.08$ V
(30°C; pH 7.0)

MENADIONE M.W.172
(2-methyl-1,4-naphthoquinone)

(2-methyl-α-naphthohydroquinone)

$E_0' = +0.42$ V
(25°C; pH 7.0)

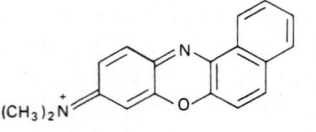

MELDOLA'S BLUE (M.W.311)

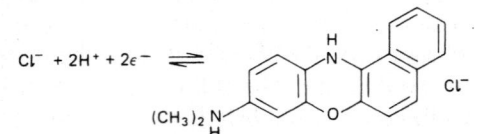

(a reduction product of Meldola's blue)

E_0' at 30°C, pH 7.0 is − 0.12 V for the closely related dye Nile blue. Other oxazines have E_0' close to zero (Clark, 1972)

FIG. 16.1. Intermediate electron acceptors.

azan. The reduced form of PMS can reduce oxygen (from the atmosphere) to the superoxide ion, O_2^-, which is able to reduce tetrazolium salts (Raap, 1983). False localization from this cause can be prevented by excluding oxygen from the incubation vessel when PMS is used.

16.4.3.3. Conditions of reaction

Sections, carried on slides or coverslips, are incubated at room temperature or 37°C for 10–20 min. Chieco *et al.* (1984) recommend treating the sections with a solution of the tetrazolium salt in acetone for 1–2 min before incubating in the complete medium. This manoeuvre gives the least diffusible reactant a chance to permeate into all parts of the section, but is only permissible for enzymes that resist brief fixation in acetone. The histochemical reaction is terminated by transferring the sections to neutral buffered formalin, which also completes the fixation and stabilizes the tissue for any further manipulations. Counterstains may be applied if desired in colours that contrast with that of the formazan.

When the tetrazolium salt is nitro-BT or TNBT, the preparations may, with advantage, be dehydrated, cleared, and mounted in a resinous medium. The formazans from other tetrazolium salts are extracted by alcohol, so water-miscible mounting media are necessary. The type of mounting medium will, of course, influence the choice of a counterstain.

16.4.3.4. Controls

When a histochemical method for a dehydrogenase is performed, it is necessary to show that the production of the coloured end-product is brought about as a result of enzymatic oxidation of the substrate and that the product is present in the same place as the enzyme. The following control procedures will help to establish the biochemical specificity and the accuracy of localization:

(i) Omit the substrate from the incubation medium. No formazan should be produced.

(ii) Inhibitors are available for some dehydro-genases. They are usually competitive and are used by short pre-incubation followed by addition of the inhibitor to the substrate-containing incubation medium. Fixation of the section in formaldehyde for a few hours inactivates most dehydrogenases but usually spares the diaphorases, but this is a test of low specificity.

(iii) In the case of a coenzyme-linked dehydrogenase, carry out the technique for the appropriate diaphorase. This enzyme should be present at the same sites as the dehydrogenase, and will usually be seen in other places too. If there is deposition of formazan from the dehydrogenase medium at sites where there is no diaphorase, the formazan must have diffused away from its place of production. Lipid-soluble formazans are often falsely localized in cytoplasmic lipid droplets. For this reason, the tetrazolium salts whose formazans are substantive for protein are preferred.

(iv) Before accepting a negative result, try the method with PVA or PVP added to the medium, with cyanide or azide added (if not done the first time) and with PMS or Meldola's blue added. Try also with unfixed as well as with briefly fixed material. In the case of flavoprotein dehydrogenases (which have prosthetic groups and do not use coenzymes), replenishment of the section's content of ubiquinone may enable a positive reaction to be obtained.

(v) It must be remembered that in histochemical demonstrations of dehydrogenases, the production of the final product is a consequence of at least three different chemical reactions. It is optimistically assumed that the intermediate reactants, especially the reduced forms of coenzymes, do not diffuse appreciably during the progress of the incubation. This assumption appears to be justified in the case of some mitochondrial enzymes at the level of resolution of the light microscope. It is not justifiable to draw conclusions concerning the fine structural localization of soluble enzymes other than perhaps to identify the cells in which they occur.

16.4.4. *Method for succinate dehydrogenase*

(Succinate: (acceptor) oxidoreductase; E.C. 1.3.99.1.)

Fresh, unfixed tissue should be rapidly frozen and then sectioned at 4–10µm in a cryostat. The sections, carried on slides or coverslips, may be fixed for 10min at 0–4°C in neutral, buffered formaldehyde, or in acetone. The formaldehyde fixative is washed off by rinsing in three changes of 0.06M phosphate buffer, pH7.0. Acetone is allowed to evaporate. See also *Note 1* below.

Succinate dehydrogenase is a flavoprotein whose electron-carrying prosthetic group, FAD, is part of the enzyme molecule, so an exogenous coenzyme does not have to be added to the substrate mixture.

Solutions required

A. Incubation medium

0.06M phosphate buffer, pH7.0:	50ml
Nitro blue tetrazolium (nitro-BT):	20mg
Disodium succinate (hexahydrate):	0.68g

Prepare just before using. Warming and stirring are sometimes needed in order to dissolve the nitro-BT. The final solution should be filtered if it contains any undissolved material.

B. Neutral, buffered 4% formaldehyde

Procedure

1. Incubate sections, prepared as described above, in the medium (solution A) for 10-30min at room temperature or at 37°C. Cellular regions of the sections should become blue or purple to the unaided eye. Check under a microscope for intracellular deposition of formazan.
2. Transfer to fixative (solution B) for 10min. This will stop the reaction and provide morphological fixation of the tissue.
3. Wash in water, apply a counterstain if desired (e.g. a pink nuclear stain), dehydrate through graded alcohols, clear in xylene, and cover, using a resinous mounting medium.

Result

Sites of enzymatic activity (mitochondria) dark blue to purple. Any red formazans (see p. 263) that form are extracted during dehydration.

Notes

1. Fixation in acetone extracts ubiquinone from the sections. This electron-acceptor may be restored by depositing a thin layer of a 0.1% solution of ubiquinone$_{10}$ (coenzyme Q_{10}) in a mixture of equal volumes of ether and acetone on the coverslip or slide (Wattenberg & Leong, 1960) or onto the fixed section (Contestabile & Andersen, 1978) and allowing the solvent to evaporate away. Alternatively, an intermediate electron carrier may be added to the incubation medium (see *Note 3*).
2. *Controls.* (a) Omit the substrate. A positive reaction in the absence of succinate ions can not be due to succinate dehydrogenase. (b) Pre-incubate the sections for 5min in 0.05M sodium malonate (0.37g of the anhydrous disodium salt in 50ml of buffer) and add sodium malonate at the same concentration to the complete incubation medium. Malonate is a competitive inhibitor of succinate dehydrogenase.
3. The incubation time can be shortened by adding 2.0mg of phenazine methosulphate (PMS) to 50ml of the incubation medium. The incubation must be carried out in darkness if PMS is used. Meldola's blue (see p. 268), 1.5mg per 50ml of medium, may be preferable to PMS on account of its greater stability. Do not incubate for more than 10min if any intermediate electron-acceptor is used, or there may be non-specific deposition of formazan. The addition of electron-transport inhibitors (CN^- or N_3^-) to media for succinate dehydrogenase is not necessary when a rapidly reducible tetrazolium salt such as nitro-BT is used.

16.4.5. General method for coenzyme-linked dehydrogenases

The following procedure, which is suitable for the demonstration of several dehydrogenases, is based on the techniques described by Pearse (1972), Lojda *et al.* (1979) and van Noorden & Butcher (1984). For individual enzymes the incubation media are made up by adding the tetrazolium salt, the substrate, and the appropriate coenzyme to a previously prepared stock solution containing the stable ingredients. The substrates and coenzymes required by various enzymes are set out in Table 16.3.

Stock solution

0.2M TRIS-HCl buffer, pH7.2:	65 ml
Magnesium chloride (MgCl$_2$.6H$_2$O):	200 mg
Sodium azide (NaN$_3$):	15 mg
Water:	85 ml
Add **either:** Polyvinyl alcohol (M.W. 30,000):	40 g
or: Polyvinylpyrollidone (M.W. about 20,000, but not critical):	15 g

(Let the PVA or PVP float on the surface with a magnetic stirrer bar revolving slowly in the bottom of the beaker or flask in which the solution is being prepared. If the powder sinks it will form lumps, which take longer to dissolve.)

Adjust to pH7.0–7.2 if necessary by adding 1.0N sodium hydroxide (4% NaOH).

Add water to bring the volume up to 200ml.

This solution is stable for several weeks at 4°C. The sodium azide, included as an electron-transport inhibitor, also serves to check bacterial and fungal growth. Potassium or sodium cyanide (26 or 20mg) may be substituted for sodium azide, but the cyanides are less stable, and solutions containing them should be used on the day they are made. All three of these substances are poisonous and must be handled carefully, but the small amounts contained in 200ml of this solution can safely be discarded by flushing down the sink with plenty of water.

Incubation media

Some of the ingredients, especially the coenzymes, are expensive, so it is the usual practice to prepare only small amounts of incubation media. For each section of tissue, 0.1–0.2ml of medium will be needed. The following instructions are for the preparation of 5ml volumes of the media.

Since the numbers of atoms of the cation and of molecules of water of crystallization may vary with some of the substrates, always check the M.W. shown by the supplier and ensure that the correct number of moles of substrate is taken.

TABLE 16.3. *Some coenzyme-linked dehydrogenases*

Trivial name	E.C. number	Systematic name (indicating substrate and coenzyme)	Product(s) of oxidation of substrate
Alcohol dehydrogenase	1.1.1.1	Alcohol: NAD oxidoreductase	An aldehyde or ketone
Glycerolphosphate dehydrogenase	1.1.1.8	L-glycerol-3-phosphate: NAD oxidoreductase	Dihydroxyacetone phosphate
UDPG dehydrogenase	1.1.1.22	UDP-glucose: NAD oxidoreductase	UDP-glucuronate
Lactate dehydrogenase	1.1.1.27	L-lactate: NAD oxidoreductase	Pyruvate
Glucose-6-phosphate dehydrogenase	1.1.1.49	D-glucose-6-phosphate: NADP oxidoreductase	D-glucono-8-lactone 6-phosphate
Glutamate dehydrogenase	1.4.1.2	L-glutamate: NAD oxidoreductase (deaminating)	2-oxoglutarate + NH$_3$
Glutamate dehydrogenase	1.4.1.3	L-glutamate: NADP oxidoreductase (deaminating)	2-oxoglutarate + NH$_3$

Alcohol dehydrogenase

Stock solution:	5.0ml
Absolute ethanol:	0.03ml
$(5 \times 10^{-4}$ mole)	
Nitro-blue tetrazolium:	1.0mg
Nicotinamide adenine	
dinucleotide:	1.0mg

(The small volume of ethanol is more easily measured out by diluting 10ml of ethanol to 33ml with water and adding 0.1ml of the diluted alcohol to the stock solution.)

Glycerolphosphate dehydrogenase

Stock solution:	5.0ml
Glycerol-3-phosphate,	
disodium salt:	158mg
$(5 \times 10^{-4}$ mole)	
Nitro-blue tetrazolium:	1.0mg
Nicotinamide adenine	
dinucleotide:	1.0mg

Check that the pH is 7.0–7.2.
Adjust with drops of 1.0M HCl if necessary.

UDPG dehydrogenase

Stock solution:	5.0ml
Uridine-5-diphosphate	
glucose trisodium salt:	1.0mg
(approx. 1.5×10^{-6} mole)	
Nitro-blue tetrazolium:	1.0mg
Nicotinamide adenine	
dinucleotide:	1.0mg

Lactate dehydrogenase

Stock solution:	5.0ml
Sodium DL-lactate	
$(NaC_3H_5O_3)$:	56mg
$(5 \times 10^{-4}$ mole)	
Nitro-blue tetrazolium:	1.0mg
Nicotinamide adenine	
dinucleotide:	1.0mg

Glucose-6-phosphate dehydrogenase

Stock solution:	5.0ml
Glucose-6-phosphate,	
disodium salt. $3H_2O$:	179mg
$(5 \times 10^{-4}$ mole)	
Nitro-blue tetrazolium:	1.0mg
Nicotinamide adenine	
dinucleotide phos-	
phate, sodium salt:	1.0mg

Check that the pH is 7.0–7.2.
Adjust with drops of 1.0M HCl if necessary.

Glutamate dehydrogenases

Stock solution:	5.0ml
Sodium-L-glutamate:	85mg
$(5 \times 10^{-4}$ mole)	
Nitro-blue tetrazolium:	7.0mg
Either nicotinamide	
adenine dinucleotide **or**	
nicotinamide adenine	
dinucleotide phosphate,	
sodium salt:	1.0mg

Check that the pH is 7.0–7.2.
Adjust with drops of 1.0M HCl if necessary.

(Note that a higher than usual concentration of the tetrazolium salt is needed for the demonstration of these enzymes.)

Procedure

Fresh tissue is rapidly frozen and sectioned on a cryostat (4–10μm), the sections being collected onto coverslips or slides. The sections may be fixed for 5–10min in pre-chilled (4°C) acetone or neutral, phosphate-buffered formaldehyde. Unfixed sections may also be used: the reactions will occur more rapidly but the integrity of the tissue will suffer.

Rinses are carried out in small coplin jars or beakers. The incubation takes place in a closed petri dish with a piece of moist filter paper in the bottom to ensure a humid atmosphere and prevent evaporation of the medium. Read the *Notes* below before carrying out this method.

1. (Fixed sections.) Allow acetone to evapor-

ate or rinse the formaldehyde-fixed section in buffer (10–15s, with agitation). Drain.

2. Place slides or coverslips, section uppermost, on the damp filter paper in the bottom of the petri dish. Cover each section with a generous drop of freshly prepared incubation medium. Put the lid on the dish and carefully place it in an oven at 37°C. (Often room temperature is satisfactory.)

3. Inspect the sections at 10-min intervals for the formation of blue, intracellular deposits. The time of incubation should not exceed 1h.

4. When staining is judged to be optimum, *Either* Rinse in buffer or saline, blot, and allow to dry, *or* place in neutral, buffered formaldehyde (see Chapter 2). This arrests the histochemical reaction and provides further morphological fixation. The time in formaldehyde is not critical: minimum 10min; can be left overnight.

5. Air-dried sections are mounted in an aqueous medium just before being examined and photographed. Fixed sections are rinsed in water and may then be counterstained if desired. A pink nuclear stain is suitable (see Chapter 6). They may then be dehydrated, cleared and mounted in a resinous medium.

Result

Sites of enzymatic activity—purple to dark blue. With an aqueous mounting medium, the alcohol-soluble red monoformazan of nitro-BT is not extracted, and it contributes to the observed colour.

Notes

1. Lactate dehydrogenase is noteworthy in that it is more resistant to fixation in formaldehyde than the other enzymes. Small blocks may be fixed at 4°C in 2.5–4.0% neutral, buffered formaldehyde and the histochemical method performed on ordinary frozen sections.

2. Always remember that the formazan deposit is formed as a result of the activity of a diaphorase, not of the dehydrogenase whose substrate was included in the incubation medium. See also *Note 3*.

3. In order to by-pass the diaphorase, an intermediate electron-acceptor may be added to the incubation medium (see p. 268). Immediately before applying the medium to the sections, dissolve in it 0.15mg of phenazine methosulphate (PMS) per 5.0ml and incubate in darkness. The time of incubation should not exceed 10min if PMS is used, or spontaneous reduction of the tetrazolium salt in solution may cause non-specific precipitation of formazan on the sections. Meldola's blue (0.15mg per 5.0ml of medium) may be preferable to PMS, for reasons given on p. 270.

4. *Controls.* It is necessary to control for non-enzymatic deposition of formazan and for production of formazan as a result of the activity of enzymes other than the dehydrogenase in which one is interested. The following control procedures are recommended:

(a) Incubate in a solution containing all the ingredients of the incubation medium except the substrate. No staining should occur. If colour does develop, it may be due to "nothing dehydrogenase" (see p. 267 and check that the medium is not too alkaline).

(b) Omit the coenzyme from the incubation medium. If staining is seen in the absence of the coenzyme, the oxidation of the substrate is being catalysed by a dehydrogenase with a prosthetic group. For example, in addition to the glycerolphosphate dehydrogenase shown in Table 16.3 there is also a mitochondrial flavoprotein (E.C. 1.1.2.1) which catalyses the reaction:

L-glycerol-3-phosphate + (acceptor) ⇌ dihydroxyacetone phosphate + (reduced acceptor)

No coenzyme is involved, but the

reduction of the acceptor (once believed to be cytochrome c, but now thought to be ubiquinone) will trigger the transport of electrons to other intermediates of the respiratory chain and to a tetrazolium salt. The mitochondrial flavoprotein glycerolphosphate dehydrogenase can be demonstrated histochemically by using a medium without a coenzyme, though the inclusion of an intermediate electron-acceptor is desirable.

(c) Inhibitors of high specificity are not available for most dehydrogenases. Some of the enzymes (e.g. alcohol, glycerolphosphate, and UDPG dehydrogenases) have sulphydryl groups at their active sites and are inhibited by SH-blocking agents such as PCMB and N-ethylmaleimide (10^{-4} to 10^{-3}M). Metal ions (Mg^{2+}, Mn^{2+}, Zn^{2+}) are cofactors for many dehydrogenases, so that chelating agents such as EDTA and 8-hydroxyquinoline (10^{-3} to 10^{-2}M) are inhibitory. A few of the enzymes (e.g. soluble glycerolphosphate dehydrogenase) display increased activity in the presence of chelators.

(d) If an unexpected negative result is obtained, try again with up to ten times the concentration of the coenzyme. Enzymes that catalyse the hydrolysis of NAD^+ and $NADP^+$ are present in some tissues. These coenzymes can also deteriorate on storage in the laboratory.

5. The histochemical detection of an enzyme requires the penetration of cellular and mitochondrial membranes by all the reagents. Usually freezing and thawing will damage the membranes sufficiently to make them permeable. The glutamate dehydrogenases show increased activity in mitochondria that have been traumatized by rough handling of the tissue (Chayen *et al.*, 1973).

6. Although the intensity of the colour of the final reaction product provides an approximate indication of the activity of the enzyme, the concentration of the latter does not vary in direct proportion with the amount of formazan deposited.

16.4.6. Method for diaphorases (tetrazolium reductases)

The presence of these systems of enzymes in tissue is essential for the production of coloured end-products in methods for coenzyme-linked dehydrogenases, except when intermediate electron-acceptors such as PMS are used. In any examination of dehydrogenases whose activities involve NAD^+ or $NADP^+$, the distribution of the appropriate diaphorases should also be ascertained. NADH-diaphorase is located predominantly in mitochondria, whereas NADPH-diaphorase is mainly found elsewhere in the cytoplasm.

Incubating medium

Stock solution (see p. 271):	2ml
Nitro-blue tetrazolium:	0.5mg
Either nicotinamide adenine dinucleotide, reduced form (disodium salt):	4 mg
Or nicotinamide adenine dinucleotide phosphate, reduced form (tetrasodium salt):	4 mg

For NADH-diaphorase, the "stock solution" may be replaced by TRIS buffer, pH 7.0–7.2. The sodium azide included in the stock solution is also unnecessary, though it does no harm. The incubation medium should be made up immediately before using.

Procedure

1. Cryostat sections are prepared as for histochemical methods for dehydrogenases.

2. Incubate in the above medium for 10–30 min at room temperature or at 37°, as described in stages 2 and 3 of the general method for coenzyme-linked dehydrogenases (p. 273).

3. Drain off incubating medium and transfer

the slides or coverslips bearing the sections to neutral buffered formaldehyde for 10–15 min.

4. Rinse in water, dehydrate through graded alcohols, clear in xylene, and mount in a resinous medium.

Result

Purple to blue–black deposits indicate sites of formazan deposition due to diaphorase activity.

16.5. Histochemistry of oxidases

Oxidases catalyse the general reaction

$$2 \begin{pmatrix} \text{reduced} \\ \text{substrate} \end{pmatrix} + O_2 \underset{\text{(oxidase)}}{\rightleftharpoons}$$

$$2H_2O + 2 \text{ (oxidized substrate)}$$

(It is assumed for convenience that the oxidation of each molecule of this generalized substrate entails the abstraction from it of two hydrogen atoms.)

The equation can be derived from two half-reactions (see Section 16.1):

$$O_2 + 4H^+ + 4\varepsilon^- \rightleftharpoons 2H_2O \qquad (E'_0 = +0.82)$$
$$\begin{pmatrix} \text{oxidized} \\ \text{substrate} \end{pmatrix} + 2\varepsilon^- \rightleftharpoons \begin{pmatrix} \text{reduced} \\ \text{substrate} \end{pmatrix} (E'_0 < +0.82)$$

If the value of E'_0 for the second half-reaction is lower (or only slightly higher) than that for the reduction of a tetrazolium salt, it is possible to use a histochemical method similar to those used for the detection of dehydrogenases. The tetrazolium salt will act as a substitute for oxygen and will be reduced to its formazan. However, the substrates for many oxidases have oxidation–reduction potentials appreciably higher than those of the tetrazolium salts, so it is necessary to use different histochemical techniques. Some oxidases catalyse the reaction:

$$\frac{\text{Reduced}}{\text{substrate}} + H_2O + O_2 \longrightarrow \frac{\text{Oxidized}}{\text{substrate}} + H_2O_2$$

The "oxidized substrate" often consists of two molecules. Histochemical methods for such enzymes are based on detection of the hydrogen peroxide that is produced.

Methods for the localization of three oxidases are discussed in this chapter.

16.5.1. Cytochrome oxidase

(Cytochrome c:O_2 oxidoreductase; E.C.1.9.3.1)

The terminal members of the electron-transport chain are cytochromes a and aa₃, from which electrons are transferred to oxygen. The electrons are derived from cyt. c^{2+}, the reduced form of cytochrome c. Cytochrome oxidase catalyses the reaction

$$4 \text{ cyt. } c^{2+} + O_2 + 4H^+ \rightarrow 4 \text{ cyt. } c^{3+} + 2H_2O$$

Cytochrome oxidase, which is probably identical to cytochrome a or aa₃, contains iron atoms, tightly bound in haem-like prosthetic groups, and copper atoms. The enzyme is inhibited by cyanide and azide ions and by several other toxic substances, including hydrogen sulphide and carbon monoxide. Cytochrome oxidase occurs in all cells of aerobic organisms and is present in mitochondria. The histochemical demonstration of cytochrome oxidase is useful in several fields of research, as a way to demonstrate populations of functionally active cells.

The earliest histochemical reaction for cytochrome oxidase was the "NADI" (naphtholdiamine) technique, in which the formation of an indoaniline ("indophenol") dye from α-naphthol and N-dimethyl-p-phenylenediamine is catalysed in the presence of atmospheric oxygen and cytochrome c. The last-named substance is naturally present in the tissue. Two oxidation–reduction reactions are involved:

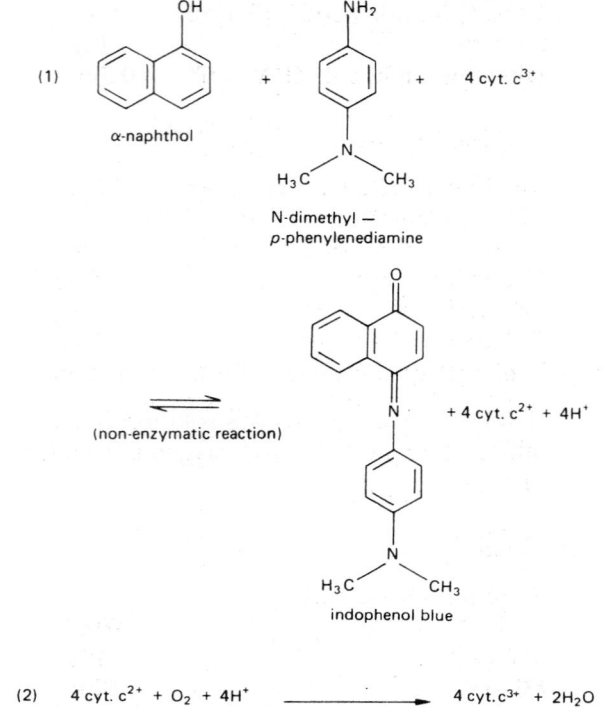

(1)

α-naphthol

N-dimethyl — p-phenylenediamine

+ 4 cyt. c^{3+}

(non-enzymatic reaction)

+ 4 cyt. c^{2+} + 4H⁺

indophenol blue

(2) 4 cyt. c^{2+} + O_2 + 4H⁺ $\longrightarrow$ 4 cyt. c^{3+} + 2H_2O

(cytochrome oxidase)

Reaction (2) serves to remove cyt. c^{2+} and H⁺ from the products of reaction (1), thereby promoting formation of the indophenol blue and assuring a continued supply of oxidized cytochrome c. It is the diamine, not the naphthol, that is oxidized by cytochrome c. The unstable product of oxidation of the diamine oxidizes and couples with the naphthol to form the dye. The discovery of cytochrome oxidase was intimately linked with the elucidation of the mechanism of the NADI reaction (Keilin & Hartree, 1938). Inhibition of the NADI reaction by inhibitors of cytochrome oxidase confirms its specificity. The oxidized form of cytochrome c (cyt. c^{3+}) does not, by itself, cause the oxidation and coupling of α-naphthol and N-dimethyl-p-phenylenediamine to occur as rapidly as is observed in the histochemical NADI reaction. It is probable, therefore, that the substrate, cyt. c^{3+}, is made more active (by being bound to cytochrome oxidase) in the tissue than it would be in solution.

The NADI method in its original form is un-satisfactory as a histochemical technique for various reasons. A positive reaction is seen in some sites, especially leukocyte granules, when cytochrome oxidase is not active, as in fixed tissues. The product of the reaction, indophenol blue, fades quite rapidly on exposure to light. It is highly soluble in lipids and has no substantivity for protein. The production of indophenol blue in myeloid leukocytes (known as the M-NADI reaction, in contrast to the cytochrome oxidase-dependent or G-NADI reaction) is now known to be catalysed by a peroxidase. Improved methods for cytochrome oxidase have been developed from the original NADI technique (see Burstone, 1962, for a detailed account). They are based on the production of indoaniline, indamine, and indophenol dyes, of uncertain composition, from a variety of naphthols, amines, quinones, and quinolines. The most satisfactory amine for NADI-type methods is N-phenyl-p-phenylenediamine (=p-aminodiphenylamine), and the most convenient coupler is 1-hydroxy-2-naphthoic acid (Burstone, 1959). These give a brownish-blue product. Cytochrome c may be added to the incubation medium, or there may be enough of this substance in the section. Catalase may also be added. This enzyme accelerates the decomposition of any H_2O_2 that might be formed by metabolic processes. In the absence of H_2O_2 there can be no peroxidase-catalysed oxidation of the substrate. Organelles that contain peroxidase (notably leukocyte granules) would otherwise give false-positive staining for cytochrome oxidase.

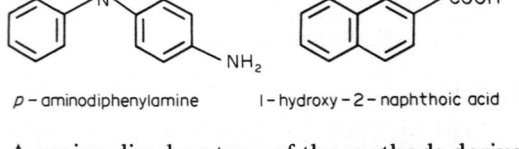

p – aminodiphenylamine   l – hydroxy – 2 – naphthoic acid

A major disadvantage of the methods derived from the NADI procedure is their sensitivity, which is too low to permit even minimal fixation of the tissue (see Pearse, 1972; Lojda *et al.*, 1979).

The modern methods for cytochrome oxidase

make use of diaminobenzidine (DAB), a reagent that is valuable in many histochemical methods. The oxidation of DAB by cytochrome c is catalysed by cytochrome oxidase, and the product is an insoluble brown polymer, whose colour can be darkened by incorporation of transition metal ions in the incubation medium. The oxidation of DAB is discussed in more detail later in this chapter, in connection with methods for peroxidases. In methods for cytochrome oxidase, exogenous cytochrome c is needed, and the medium must also contain catalase, for the reasons given in the previous paragraph. The following method is based on that of Silverman & Tootell (1987), which was devised for application to the central nervous system.

Preparation of tissue

An anaesthetized rat is perfused (see Chapter 2) for about 15s with about 50ml of 0.1M phosphate buffer, pH 7.4, containing 10% sucrose and 1% sodium nitrite ($NaNO_2$, as a vasodilator), to displace the blood. The fixative, which is 3.5% formaldehyde in phosphate-buffer, is then run through for 15s. The phosphate buffer with sucrose is then perfused for about 15s, to displace the fixative. The brain is removed, and pieces are frozen and cut at 20–40μm with a cryostat. The sections are collected on slides or coverslips that have been subbed with chrome-gelatin (see Chapter 4), and quickly dried by placing on a hotplate (50°C) for 10–15s. The dry sections should be refrigerated if they are not to be stained immediately.

Solutions required

A. *Cold (10°C) acetone*

B. *Rinsing solution*

   0.1M phosphate buffer, pH 7.4, containing 10% (w/v) sucrose.

C. *Pre-incubation solution*

   0.05M TRIS buffer, pH 7.6
   (see Chapter 20):   99.5ml

Cobalt chloride ($CoCl_2.6H_2O$):	28mg
Sucrose:	10g
Dimethylsulphoxide (DMSO):	0.5ml

D. *Reaction medium*

0.1M phosphate buffer, pH 7.6:	100ml
Diaminobenzidine tetrahydrochloride:	50mg
Cytochrome c:	7.5mg
Sucrose:	5g
Catalase:	2mg
Dimethylsulphoxide (DMSO):	0.25ml

Before and during use, this solution is maintained at about 40°C, and oxygen is bubbled through it.

E. *Final fixative*

0.1M phosphate buffer, pH 7.2–7.6:	90ml
Sucrose:	10g
Formalin:	10ml

Procedure

1. Immerse the slides (or coverslips) in cold acetone for 5 min. (This is said to improve adhesion; it would also cause further fixation, and extract some lipids.)
2. Rinse in 3 changes of the rinsing solution (B).
3. Immerse for 10 min in the pre-incubation solution (C).
4. Rinse briefly in Solution B.
5. Place in oxygenated reaction medium (Solution D) at 40°C, for 30 min to 6h, examining from time to time.
6. When staining is adequate, transfer to final fixative (Solution E) for 30 min, then dehydrate, clear, and mount in a resinous medium.

Result

Sites of cytochrome oxidase activity greyish-brown to blue-black.

Note

To inhibit cytochrome oxidase, include 10^{-3}M potassium or sodium cyanide or azide in the incubation medium (Solution D).

16.5.2. Monophenol monooxygenase

(Monophenol, DOPA: oxygen oxidoreductase; E.C. 1.14.18.1. Almost indistinguishable from **catechol oxidase**, which is *o*-diphenol: O_2-oxidoreductase; E.C. 1.10.3.1.)

Several names have been applied to the copper-containing enzymes that catalyse the oxidation of *o*-diphenols by oxygen to yield *o*-quinones. These include monophenol monooxygenase, catechol oxidase, tyrosinase, phenol oxidase, polyphenol oxidase, and DOPA-oxidase. All occur in the histochemical literature. Related enzymes catalyse the oxidation of *p*-diphenols and aromatic amines. The importance of catechol oxidase in histochemistry derives from the function of the enzyme in the synthesis of melanin. This pigment is produced in melanocytes and some other types of cells by a series of reactions, some enzymatically catalysed and others spontaneous. The initial metabolite in the sequence is the amino acid tyrosine. This is first slowly oxidized to dihydroxyphenylalanine (DOPA):

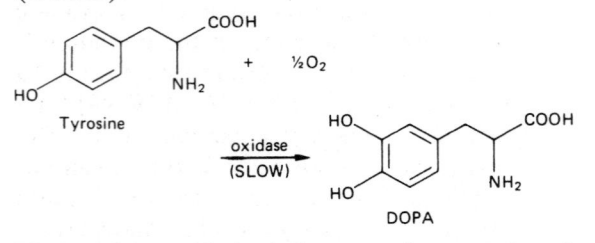

Under the catalytic influence of monophenol monooxygenase the DOPA is now rapidly oxidized to DOPA quinone:

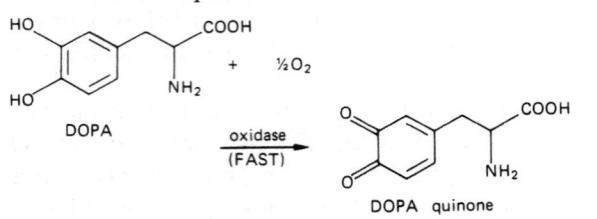

The remaining reactions are believed to occur spontaneously:

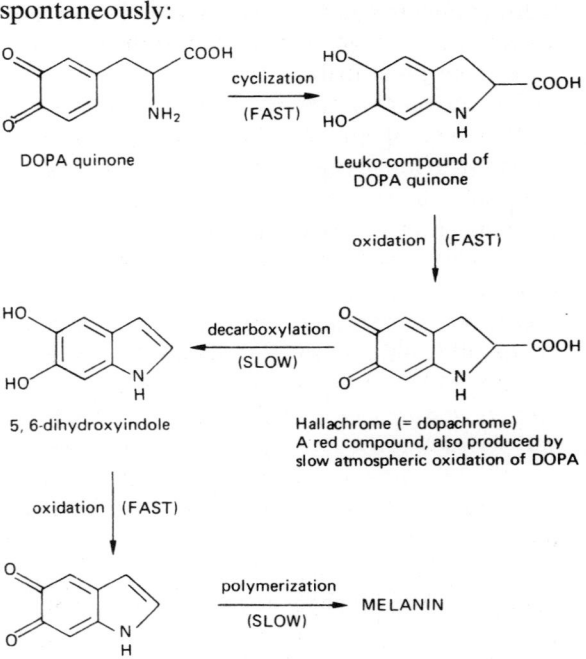

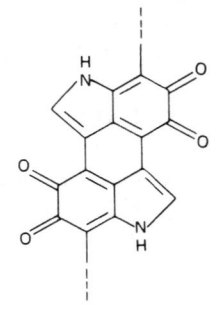

The composition of melanin is not known with certainty, but it is probably a polymer of the form derived from indole-5,6-quinone and bound to the proteinaceous matrix of the granules in which it occurs. It is a stable, black, insoluble substance.

Monophenol oxygenase is demonstrated histochemically by virtue of its catalysis of the rapid oxidation of DOPA by oxygen, with the ultimate formation of melanin. Several potentially diffusible intermediates are produced in this series of reactions, so it is possible that the final deposits of melanin are not formed in exactly the same sites as those at which the substrate was oxidized.

However, the formation of a finely granular pigment in melanocytes, where melanin is normally synthesized, suggests that diffusion does not occur over great distances.

The chemical specificity of the reaction is certainly not complete. Positive staining of erythrocytes and of leukocyte granules is probably due to peroxidase activity, with tissue-derived hydrogen peroxide as the substrate and DOPA as the electron-donor (see Section 16.6). Inhibitors are of little value since both monophenol monooxygenase and the peroxidases are inhibited by cyanide, azide, and sulphide ions, though lower concentrations (10^{-4}–10^{-3}M) are effective with the former enzyme. These inhibitors also block the activity of cytochrome oxidase, but this enzyme is unlikely to be involved in the histochemical oxidation of DOPA when formaldehyde-fixed tissue is used. Lillie & Fullmer (1976) state that sodium dithionite ($Na_2S_2O_4$, 5×10^{-3}M) and cysteine (10^{-3}M) enhance the reaction. They might do this by reducing hallachrome to 5,6-dihydroxyindole, thus speeding up the slowest non-enzymatic reaction in the series leading from DOPA to melanin.

The technique for the histochemical demonstration of monophenol monooxygenase is unusual in that the substrate and the visible product are those naturally used and formed by the enzyme. The method works with cryostat sections of unfixed tissue, with frozen sections of tissue fixed for 6–24h at 4°C in neutral, buffered 4% formaldehyde and with paraffin sections of freeze-dried material. Small specimens can be stained whole by a slight modification of the technique (see Pearse, 1972, and other texts) and then fixed and embedded in wax.

Incubation medium

0.06M phosphate buffer, pH 7.4:   100ml
DL-β-dihydroxyphenylalanine
   (DOPA):   100mg

Pre-warm the buffer to 37°C. Add the DOPA and place on a magnetic stirrer or shake vigorously for 10–15 min. Any of the solute that has not dissolved after this time should be removed by filtration.

Procedure

1. Incubate sections in the medium, in darkness, for 60 min at 37°C. After the first 45 min, prepare a fresh batch of incubation medium.
2. Replace the incubation medium with the new batch and incubate for another 60 min at 37°C.
3. Wash in three changes of water.
4. Apply a counterstain (e.g. neutral red or safranine) if desired, and rinse in water.
5. Dehydrate, clear, and mount in a resinous medium.

Result

A dark brown deposit of melanin forms at sites of enzymatic activity (see *Notes* below).

Notes

1. It is important to incubate control sections in buffer without added DOPA in order to detect pigments already present in the tissue. See also Exercise 13 at the end of this chapter.
2. Catechol oxidase is inhibited by cyanide ions, though specificity of the inhibition is low. Sections may be pre-incubated in 10^{-3}M KCN (M.W. 65) or NaCN (M.W. 49) for 5 min. The same concentration should also be included in the incubating medium. (**Caution.** Sodium and potassium cyanides are poisonous. They must be handled carefully and not allowed to come into contact with acids. Their solutions must not be pipetted by mouth. Quantities smaller than 100mg may safely be discarded by washing down the sink with copious running tap water.)
3. Incubation may be prolonged to 6h if necessary, changing the medium every 45–60 min. White *et al.* (1983) recommended a pH of 6.8 for detecting the enzyme in cell cul-

tures. They used L-DOPA as the substrate, and D-DOPA for the controls.

16.5.3. Amine oxidase

(Amine:O_2 oxidoreductase (deaminating) (flavin-containing); E.C. 1.4.3.1)

This flavoprotein enzyme is commonly known as **monoamine oxidase** or **MAO**, because it is involved in the metabolism of such compounds as dopamine, noradrenaline, serotonin, tryptamine and tyramine, which contain one amino group. The overall reaction catalysed by MAO is

$$RCH_2NH_2 + H_2O + O_2 = RCHO + NH_3 + H_2O_2$$

The oxidation is due to removal of two electrons from the carbon atom of —CH_2NH_2. (Application of the rules summarized at the beginning of this chapter shows a change in oxidation number from -1 to $+1$.) The electron acceptor is FAD, the prosthetic group of the enzyme. From the reduced FAD, electrons are transferred to the cytochromes and to molecular oxygen.

In early histochemical methods for MAO, the aldehyde product was trapped by a naphthoic hydrazide (see Chapter 10), and the resulting hydrazone converted to a dye by coupling with a diazonium salt. Such methods were unsatisfactory because of diffusion of intermediate products of the reactions and inhibition of the enzyme by some of the reagents.

More satisfactory histochemical localization of MAO is obtained by providing a tetrazolium salt to accept electrons from the reduced FAD, thus substituting for the cytochromes and for oxygen. The formazans formed from reduced tetrazolium salts are coloured and insoluble (see Section 16.4.1 of this chapter). Technical instructions for methods of this type are given by Pearse (1972), Lojda et al. (1979) and Shannon (1981). It is necessary to use cryostat sections of unfixed tissue for these methods.

More sensitive methods for detecting MAO (and other enzymes catalysing reactions that produce H_2O_2) are based on detection of the released hydrogen peroxide. The sensitivity of these methods is high enough to permit the detection of MAO even when much of the enzymatic activity has been inhibited by brief fixation. The accuracy of localization permits their application to slices of tissue cut with a vibrating microtome and subsequently processed for electron microscopy. The simplest way to detect released H_2O_2 is to have cerium(III) ions in the incubation medium. Cerium(IV) perhydroxide, $Ce(OH)_3(OOH)$, is precipitated (Sneed & Brasted, 1955; Briggs et al., 1975). This is electron dense and can be detected within mitochondria and other organelles (Christie & Stoward, 1982; Angermuller & Fahimi, 1987). The product is not visible in light microscopy, however. The other strategy for detection of newly produced H_2O_2 is to let it be the substrate of another enzyme, peroxidase. The final reaction products in histochemical methods for peroxidase are visible and electron-dense (see Section 16.6).

The following method (from Maeda et al., 1987) is one of several "coupled peroxidatic oxidation" techniques for MAO and for other oxidases that form hydrogen peroxide. As described, the procedure requires free-floating sections cut on a vibrating microtome. However, it should also work with cryostat sections mounted on slides or coverslips.

Solutions required

Phosphate-buffered saline (PBS)
       (See Chapter 20)
       100 ml at room temperature, and
       100 ml at 0°C.

Fixative

2–4% formaldehyde with 2–3% glutaraldehyde, in 0.1M phosphate buffer, pH 7.4 (see Chapter 2).

Incubation medium

0.05M TRIS buffer, pH 7.6, at 4°C
       (see Chapter 20):   10 ml

Diaminobenzidine
   tetrahydrochloride: 0.5 mg
Horseradish peroxidase: 10 mg
Tyramine hydrochloride: 7.5 mg
Nickel ammonium sulphate: 60 mg
Sodium azide: 6.5 mg

Make up less than 1 h before it is needed. The medium is used at 4°C.

Procedure

1. Perfuse an anaesthetized rat with 100 ml of PBS (room temperature), followed by 250 ml of the fixative (at 0–5°C). About 250 ml of fixative should be perfused, over the course of 6–7 min.
2. Remove the brain or other organs to be studied, and cut sections 30 μm thick on a vibrating microtome.
3. Collect the sections into ice-cold PBS.
4. Transfer the sections to the cold incubation medium. Agitate occasionally. Examine at hourly intervals for the first 4 h. Longer times (up to 48 h) are sometimes needed.
5. When staining is satisfactory, rinse the sections in 3 changes of ice-cold PBS.
6. Mount sections onto slides; leave to dry.
7. Dehydrate, clear, and mount in a resinous medium.

Result

Sites of MAO activity blue-black. Controls for specificity (see *Note 2* below) are needed.

Notes

1. Tyramine is the substrate. The sodium azide is necessary to inhibit other enzymes (notably cytochrome oxidase) that would cause oxidation of DAB. The nickel salt must be $Ni(NH_4)_2(SO_4)_2.6H_2O$. It serves to darken the product of the oxidation of DAB (which would otherwise be brown) and to shorten the time required for the development of adequate colour in the sections.
2. **Controls.** (a) Omit the substrate. (b) Inject a rat with **pargyline hydrochloride**, or **nial-amide**, 50 mg/kg body weight intraperitoneally, 2 h before killing it. These drugs inhibit MAO. They may be added to the incubation medium ($10^{-2}M$); if this is done, the pH will require adjustment. (c) To inhibit *Type A* MAO, incubate the sections, for 15 min between stages 3 and 4 of the procedure, in 10^{-7}–$10^{-5}M$ **deprenyl** in PBS. To inhibit *Type B* MAO, pre-incubate in 10^{-6}–$10^{-4}M$ **clorgyline**. The *Type A enzyme* occurs in neuroglia (mainly astrocytes). *Type B* is found within the cell-bodies, axons and dendrites of neurons that use amines as synaptic transmitters.
3. This method can also be used in conjunction with electron microscopy. The final reaction product is seen to be in the cytoplasmic matrix and also in mitochondria. It cannot be stated with any confidence that these are the true subcellular sites of monoamine oxidase activity however, because biochemical studies indicate that this is a mitochondrial enzyme. Some diffusion of hydrogen peroxide would be expected to occur prior to the occurrence of the histochemical reaction catalysed by the added peroxidase.

16.6. Histochemistry of peroxidases

The peroxidases catalyse the oxidations of various substances, including reduced co-enzymes, fatty acids, amino acids, reduced cytochromes, and many other substances by hydrogen peroxide.

16.6.1. Actions and occurrence

The name "peroxidase" (in the singular; donor: H_2O_2 oxidoreductase; E.C. 1.11.1.6) embraces several enzymes of plant and animal origin. They are all iron-containing haemoproteins and they catalyse the reaction

$$donor + H_2O_2 \rightarrow oxidized + 2H_2O$$
$$donor$$

in which the net effect is the removal of two atoms of hydrogen from each molecule of the donor.

Many organic compounds, including amines, phenols, and the leuco-compounds of dyes, can serve as donors. The substrate is hydrogen peroxide which, when it is bound to the enzyme, is made to oxidize other substances much more rapidly than it would if it were acting alone.

In mammals, peroxidase activity is present in the granules of myeloid leukocytes and in some neurons and secretory cells. A positive histochemical reaction is also given by the haemoglobin of erythrocytes, though this is not considered to be due to truly enzymatic catalysis. The animal enzyme is inhibited by cyanide or azide ions at 10^{-2}M, a concentration higher than that which will inhibit cytochrome oxidase. It is also inhibited by treatment of unfixed cryostat sections with methanol (Streefkerk & van der Ploeg, 1974) but not by fixation of tissues in 70% ethanol or 4% formaldehyde. Hydrogen peroxide irreversibly inhibits peroxidase, if applied to sections at a concentration of 0.3% (approximately 0.1M). Complete inhibition of the peroxidase activity of leukocytes and erythrocytes can be achieved by treating formaldehyde- or acetone-fixed sections or smears with *either* 0.024M HCl in ethanol (Weir *et al.*, 1974) *or* a solution containing 0.3% H_2O_2 and 0.1% NaN_3 in phosphate-buffered saline, for 10 min (Li *et al.*, 1986). The peroxidase activity of leukocytes survives paraffin embedding, but only if the formaldehyde fixative has been completely washed out of the specimen prior to dehydration.

The importance of histochemical methods for peroxidase has increased in recent years because the enzyme extracted from the root of the horseradish (*Armoracia rusticana*) is extensively used as an intravital tracer protein in studies of vascular permeability, and in neuroanatomy for both light and electron microscopy. Horseradish peroxidase (HRP) is also used as a reagent in immunohistochemistry (Chapter 19). Inhibition of the endogenous peroxidase of animal tissues is often necessary before applying methods in which HRP is used as a reagent. The blocking procedures mentioned above cannot, however, be used on tissue containing exogenous HRP, because this enzyme is also inhibited by them.

16.6.2. *Histochemical localization*

Histochemical methods for peroxidase are based on the catalysed reactions of hydrogen peroxide with substances that yield insoluble coloured products upon oxidation. In one of the oldest techniques, the donor is benzidine, which is oxidized to a blue substance. The incubation medium contains benzidine and hydrogen peroxide. In this and other methods, the concentration of hydrogen peroxide, the substrate, should not exceed 0.03M, because higher concentrations inhibit the enzyme.

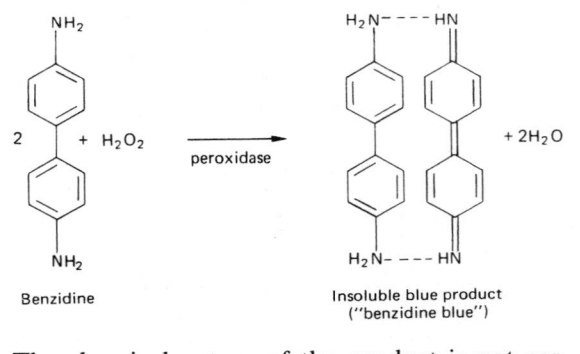

Benzidine   Insoluble blue product ("benzidine blue")

The chemical nature of the product is not certainly known, but it is generally believed to have the quinhydrone-like structure shown above. The simple benzidine technique is not often used because the blue product often forms as unduly large crystals and its colour soon fades to a less conspicuous brown. Under some conditions of reaction (pH>7, temperature >4°C) a brown product is formed in the first instance. It is probably a polymer derived from condensation of benzidine with its unstable quinone–imine. Various methods are available for the stabilization of benzidine blue, the best-known being treatment of the stained preparations with a concentrated aqueous solution of sodium nitroprusside (sodium nitroferricyanide, $Na_2Fe(CN)_5 NO.2H_2O$) Straus, 1964). However, donors which give more stable products are preferred to benzidine.

The donor most widely applicable to the histochemical localization of peroxidases is 3,3'-diaminobenzidine tetrahydrochloride (DAB), introduced by Graham & Karnovsky (1966):

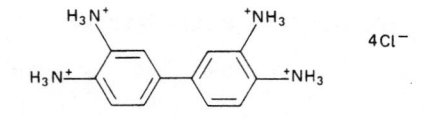

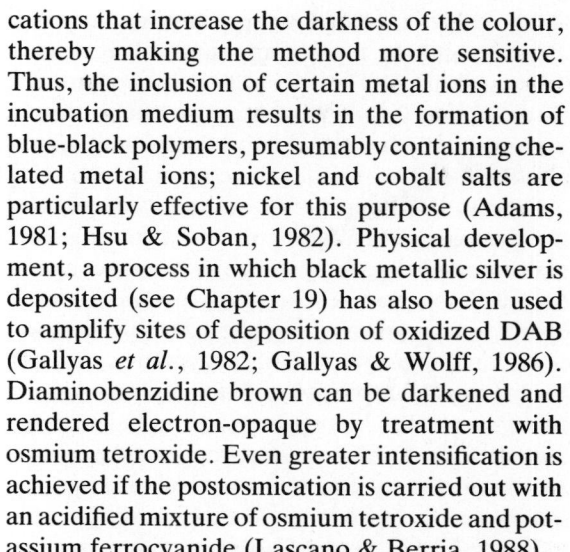

The spontaneous oxidation of this amine by hydrogen peroxide is quite slow, but in the presence of peroxidase an insoluble, amorphous, brown substance is rapidly precipitated. The initial products of oxidation are presumed to be quinone–imines:

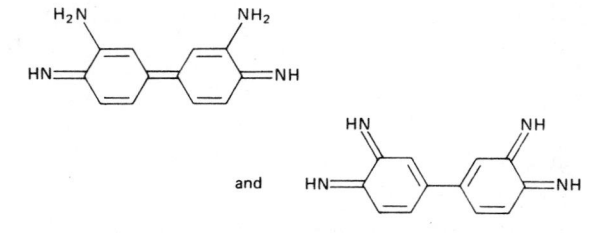

These unstable compounds immediately react with DAB to give polymers, which contain the quinonoid and indamine chromophores. The polymerization, which involves the elimination of hydrogen atoms attached to aromatic rings, is also an oxidation reaction and may also be brought about by hydrogen peroxide and catalysed by peroxidase. The polymers are thought to contain such structural formations as

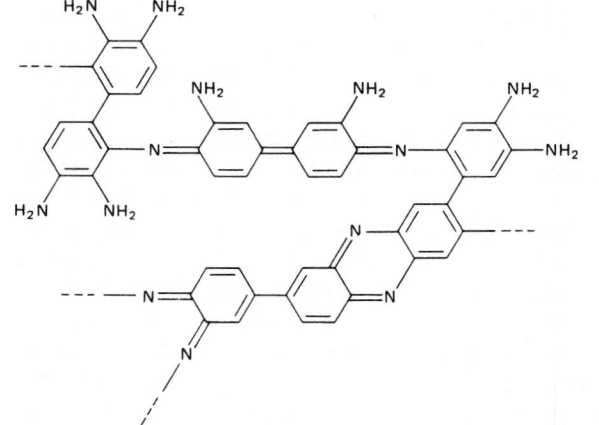

The visibility of this brown, amorphous product can be greatly enhanced by including appropriate blue filters in the light path of the microscope (Gordon, 1988), and there are chemical modifi-

cations that increase the darkness of the colour, thereby making the method more sensitive. Thus, the inclusion of certain metal ions in the incubation medium results in the formation of blue-black polymers, presumably containing chelated metal ions; nickel and cobalt salts are particularly effective for this purpose (Adams, 1981; Hsu & Soban, 1982). Physical development, a process in which black metallic silver is deposited (see Chapter 19) has also been used to amplify sites of deposition of oxidized DAB (Gallyas et al., 1982; Gallyas & Wolff, 1986). Diaminobenzidine brown can be darkened and rendered electron-opaque by treatment with osmium tetroxide. Even greater intensification is achieved if the postosmication is carried out with an acidified mixture of osmium tetroxide and potassium ferrocyanide (Lascano & Berria, 1988).

Other methods for the localization of peroxidase are based on the oxidation and coupling of amines with phenols, quinones, and other substances. The methods for cytochrome oxidase (Section 16.5.1) will demonstrate peroxidase if hydrogen peroxide is added to the incubation medium (Burstone, 1962; Lojda et al., 1979). Related methods use combinations of p-phenylenediamine with catechol (Hanker et al., 1977), o-dianisidine or o-tolidine with catechol (Segade, 1987), and DAB with p-cresol (Streit & Reubi, 1977). The peroxidase-catalysed oxidation of α-naphthol by H_2O_2 yields an anionic polymer that gives metachromatic staining (see Chapter 11) with cationic dyes (Mauro et al., 1985). The insoluble products of these methods are probably similar to the "oxidation colours" used in the dyeing of fur and hair (see Chapter 5). Though not the most sensitive method, this approach gives the histochemist a choice of colours for the final reaction product.

The use of HRP as an intravital tracer in neuroanatomical studies has resulted in the development of several techniques for the demonstration of this plant enzyme in animal tissues. The activity of HRP is optimally preserved by vascular perfusion of a cold glutaraldehyde–formaldehyde mixture followed by immersion of the excised tissues in a buffered solution of sucrose (Rosene & Mesulam, 1978). The method using DAB has been widely employed, but much greater sensitivity is attained if tetramethylbenzidine (TMB) is the donor. A blue product, stabilizable by sodium nitroprusside, is formed (Mesulam, 1978). The incubation medium for the TMB method has to be quite strongly acidic (pH 3.3). It disrupts antigen–antibody complexes and cannot therefore be used in immunohistochemical techniques (see Chapter 19). The TMB technique is the most sensitive one yet developed for the demonstration of exogenous HRP in fixed tissues (Mesulam & Rosene, 1979). Preparations made by this method are, however, sometimes marred by the deposition of large blue crystals, both at the sites of enzymatic activity and elsewhere (Reiner & Gamlin, 1980). The product of oxidation of TMB can be stabilized by treatment with ammonium molybdate, which is used instead of sodium nitroprusside (Jhaveri *et al.*, 1988), or by treatment after staining with a solution containing DAB, Co^{2+} ions and H_2O_2 (Rye *et al.*, 1984). The latter technique changes the blue product to a black substance that resists counterstaining and treatment with solvents.

16.6.3. Specificity and accuracy of localization

The activities of various enzymes may be expected to cause false-positive reactions in sections stained by histochemical methods for peroxidase. Those most likely to cause confusion are cytochrome oxidase and catalase.

Donors such as DAB can be oxidized by cytochrome c in the presence of cytochrome oxidase and oxygen. However, cytochrome oxidase is inactivated by fixatives such as formaldehyde, which are usually employed in the preparation of tissues for the demonstration of peroxidase. With unfixed tissue intended for the localization of cytochrome oxidase, false-positive results can be due to peroxidase, as in the M-NADI reaction. Enough hydrogen peroxide is generated within the tissue to act as substrate for the enzyme. The endogenous substrate can be destroyed, thus

eliminating artifacts of the M-NADI type, by adding purified catalase to the histochemical incubation medium.

Catalase (H_2O_2: H_2O_2 oxidoreductase; E.C. 1.11.1.6) catalyses the reaction in which hydrogen peroxide functions as both an oxidizing and a reducing agent:

$$2H_2O_2 \rightarrow 2H_2O + O_2$$

The enzyme occurs in nearly all cells, in organelles known as "microbodies" or "peroxisomes". Its physiological function is probably to prevent the accumulation of hydrogen peroxide, a potentially toxic metabolite. The decomposition of hydrogen peroxide occurs exceedingly rapidly in the presence of catalase. This enzyme can, however, also function as a peroxidase and catalyse the oxidation of chromogenic donors by hydrogen peroxide. Catalase is specifically inhibited by 3-amino-1,2,4-triazole, but attempts to distinguish between this enzyme and peroxidase are rarely made.

Peroxidase and catalase may also be distinguished by varying the concentration of their substrate. Silveira & Hadler (1978) have shown that catalase cannot be detected (by benzidine) when the concentration of hydrogen peroxide in the incubation medium is less than about 3×10^{-3} M, but remains active when [H_2O_2] is as high as 4.0 M. Peroxidase, on the other hand, is fully active in the presence of 1.5×10^{-3} M H_2O_2, but is inhibited by concentrations greater than about 0.05 M. Silveira & Hadler found, however, that 4.0 M H_2O_2 was necessary for complete inhibition of all peroxidases. The peroxidase-like activity of haemoglobin resembles catalase in that it is not detectable when [H_2O_2] is very low. Cyclopropanone hydrate is a specific inhibitor of the endogenous peroxidases of animal tissues and does not damage sections of tissue (Schmid *et al.*, 1989), but this reagent is not commercially available.

16.6.4. DAB methods for peroxidase

The following procedure is the original method of Graham & Karnovsky (1966), with some optional modifications (Adams, 1981) to increase the sensitivity. It is suitable for the demonstration of endogenous peroxidases of animal tissues and for horseradish peroxidase used as a tracer or as a reagent in immunohistochemical and other techniques. More sensitive methods using TMB (see previous paragraph) are preferred when HRP is used as a tracer in neuroanatomical studies.

The enzyme survives brief fixation in formaldehyde or glutaraldehyde. For optimum preservation of exogenous HRP, the animal should be perfused for 30 min with 0.1M phosphate buffer, pH 7.4, containing 1% paraformaldehyde and 1.25% glutaraldehyde. Glutaraldehyde alone (1–2% in the same buffer) is also suitable. Excess fixative is then washed out by perfusing cold (4°C) 10% sucrose in 0.1M phosphate buffer, pH 7.4, for a further 30 min. The specimens may be stored in phosphate-buffered sucrose solution for up to 7 days at 4°C. Frozen or cryostat sections should be used, either free-floating or mounted on slides or coverslips.

Incubation medium

This solution is prepared just before use. Dissolve 25mg of 3,3′-diaminobenzidine tetrahydrochloride (DAB) in 50ml of either 0.1M phosphate buffer or 0.05M TRIS buffer, pH 7.3. See *Note 1* below.

Optionally, add 0.5ml of the following solution:

Cobalt chloride ($CoCl_2.6H_2O$):	2.5g
Nickel ammonium sulphate ($Ni(NH_4)_2(SO_4)_2.6H_2O$):	2.0g
Water:	to make 100ml

Keeps indefinitely

(The effect of either of the above salts used alone at 2% is almost identical to that of the mixture.)

Dilute a stock solution of hydrogen peroxide (e.g. "100 volumes" = 30% w/w) to give a 1% w/v solution of H_2O_2 in water.

(**Caution:** Avoid contact of the strong H_2O_2 with skin or clothing. H_2O_2 is unstable, so do not use old stock. Decomposition is accelerated by chemical contamination, especially by contact with metals.)

Immediately before use (see step 2 of the procedure, and *Note 2*), add 0.5ml of the 1% hydrogen peroxide to the 50ml of DAB solution.

Procedure

1. Incubate the sections in the buffered DAB solution (with or without Co^{2+} and Ni^{2+}), without hydrogen peroxide, for 15 min at room temperature.
2. Add the hydrogen peroxide to the medium. Mix well and wait for a further 5–15 min.
3. Wash in 3 changes of water, each 1 min.
4. (Optional) Apply a counterstain if desired.
5. Dehydrate, clear, and cover, using a resinous mounting medium.

Result

Sites of peroxidase activity are brown if the metal salts are not included in the medium; blue-black if Co^{2+} and/or Ni^{2+} is included.

Notes

1. The choice of phosphate or TRIS buffer is not critical. Samples of DAB vary in quality. The compound should be very pale, almost white. Darkly coloured material gives nonspecific background staining and weaker specific staining. Brown DAB solutions can be cleaned and made usable by shaking with activated charcoal, 1 mg per ml for 1 min, and filtering (Ros Barcelo *et al.*, 1989). The most reliable DAB is supplied in rubber-capped vials, each containing a pre-weighed amount. DAB is handled carefully, though it is probably not carcino-

genic (see Burns, 1982). Residual DAB can be destroyed (oxidized) by adding a few ml of 5% sodium hypochlorite (household bleach) to the used solution and waiting for an hour before discarding.

2. If the sections are thin, they may be put directly into the complete incubation medium. For thicker ($>40\mu m$) sections or whole mounts, the pre-incubation is necessary to allow penetration of the DAB, whose larger molecules diffuse more slowly than those of hydrogen peroxide.

3. Control sections should be incubated with DAB in the absence of H_2O_2. In unfixed sections a positive reaction in the absence of H_2O_2 can be due to cytochrome oxidase.

4. Catalase may be inhibited by adding 3-amino-1,2,4-triazole (10^{-2}M) to the incubation medium. Halving the recommended concentration of hydrogen peroxide in the medium should also prevent the formation of coloured products due to catalase.

5. Some pre-treatments to inhibit endogenous peroxidases of animal tissues are listed in Section 16.6.1, p. 282).

16.7. Exercises

Theoretical

1. In the following chemical reactions, what is oxidized and what is reduced?
(a) The reaction at the cathode in the electrolysis of molten sodium hydroxide:

$$Na^+ + \varepsilon^- \rightarrow Na$$

(b) The copper-plating of an iron nail immersed in a solution of cupric sulphate:

$$Cu^{2+} + Fe \rightarrow Fe^{2+} + Cu$$

(c) The conversion of an unsaturated lipid to a saturated one:

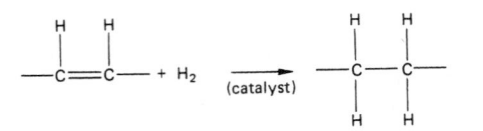

2. Examine the formula for the formazan of nitro blue tetrazolium. Why is this compound coloured? Is it a dye?

3. The physiological action of **catalase** is to accelerate the reaction

$$2H_2O_2 \xrightarrow{\text{(catalase)}} 2H_2O + O_2$$

What are the oxidizing and reducing agents in this reaction?

4. If, in the histochemical method for succinate dehydrogenase, one used a tetrazolium salt with an oxidation–reduction potential of + 1.0V, the specificity and accuracy of localization might be lower than when nitro blue tetrazolium was used. Why?

5. Compare the rationales of histochemical methods for (a) NAD^+-linked dehydrogenases; (b) flavoprotein dehydrogenases. Justify the inclusion of all constituents of the incubation media and indicate how the specificity for the substrate can be determined in each case.

6. In a histochemical test for enzymatic oxidation of a substrate XH_2 to X by an enzyme requiring NAD^+, the following results were obtained in a tissue containing two cell types, A and B.

(a) With substrate, NAD^+ and nitro-BT: reaction in mitochondria of all cells of both types.
(b) With NADH and nitro-BT: strong reaction in mitochondria of both cell types.
(c) With substrate and nitro-BT but no coenzyme: moderate reaction in mitochondria of type A cells; nothing in type B.
(d) With nitro-BT but no substrate or coenzyme: no reaction in mitochondria of type A or B cells.

What are the probable localizations of (i) XH_2: NAD^+ oxidoreductase, (ii) NADH-diaphorase? What other enzyme has been detected, and in which cell type?

7. In ultrastructural cytochemical methods for dehydrogenases, the ferricyanide anion is sometimes preferred to a tetrazolium cation as an artificial electron acceptor (see Benkoel *et al.*, 1976). The phosphate-buffered incubation medium contains, in addition to the substrate and coenzyme, potassium ferricyanide, sodium citrate and cupric sulphate. What are the functions of these ingredients? (*Hint*: See Chapter 15, Section 15.2.4.) What is the electron-dense product of the histochemical reaction?

Practical

8. Dissolve 10mg of *N*-dimethyl-*p*-phenylenediamine (=*p*-amino-*N*-dimethylaniline) or its hydrochloride, and 10mg of α-naphthol in a few drops of alcohol, and add to 10ml of phosphate buffer (pH between 7 and 8). Pour 5ml of the solution into a petri dish or watchglass, and put some small pieces of freshly removed tissue into it. Do a control experiment with about 1mg of sodium azide or potassium cyanide added to the other 5ml of incubation medium. After a few minutes, the specimens in the medium without the N_3^- or CN^- go blue. Explain the observations.

9. Carry out the method for monophenol oxidase on sections of skin, with appropriate controls. The skin must not be from an albino animal. Account for the different distributions of the enzyme and of the melanin already present in the skin.

17

Methods for Soluble Substances of Low Molecular Weight

There is a shortage of histochemical techniques for the demonstration of soluble organic compounds of low (<1000) molecular weight. This is due partly to the fact that such substances diffuse rapidly and partly to a lack of suitable chemical reactions. For one group of small molecules, however, there are several satisfactory methods; this is the series of substances known as biogenic amines.

Most of the present chapter is concerned with the histochemical study of amines, though several other substances are discussed.

17.1. Nature and occurrence of biogenic amines

Of the amines occurring in mammalian tissues, it is possible to localize dopamine, noradrenaline, adrenaline, serotonin, and histamine in sections of tissue. The structures of these substances are shown below.

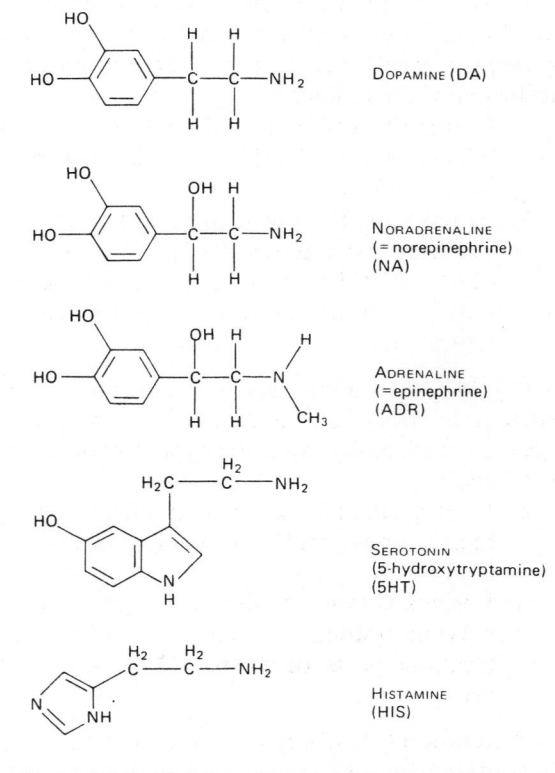

Dopamine (DA) and noradrenaline (NA) are primary monoamines derived from phenylethylamine ($C_6H_5CH_2CH_2NH_2$), whereas adrenaline (ADR) is a secondary amine formed by N-methylation of NA. Serotonin (5HT) is another primary monoamine, being an indolylethylamine derivative. Histamine (HIS), an imidazolylethylamine, is usually classified as a diamine on account of the basicity of the imidazole ring. All these compounds have fully aromatic rings attached to the ethylamine moieties, and all except HIS are phenols as well as amines.

All the above-mentioned amines are readily soluble in water and would therefore be expected to diffuse rapidly from their cellular sites of storage under the ordinary physical conditions of a histochemical technique. Diffusion can be minimized by using freeze-dried material or by the application of a chemical fixative which reacts rapidly with the amine to produce an insoluble compound. Both these approaches are used. The amines are found in two types of site. These are:

(a) In large quantities, in secretory granules of endocrine (or similar) cells. Here, binding to a protein or carbohydrate matrix greatly reduces diffusion of the amines.

1. Chromaffin cells of the adrenal medulla and related paraganglia (NA and ADR in separate cell-types).
2. Argentaffin (=enterochromaffin) cells of the intestinal mucosa (5HT).
3. Mast cells (HIS in all mammalian species; also 5HT in rodents and DA in the lungs of ruminants).

(b) In small quantities as transmitter substances in aminergic neurons. The amines in these sites are easily extracted by water and other solvents.

1. Most postganglionic sympathetic neurons, their axons and terminal aborizations (NA).
2. Various systems of neurons in the central nervous system, most conspicuously in the terminal parts of axons (DA, NA, and 5HT).

In situation (a), where the diffusion of amines is hindered by the associated structural macromolecules, the sensitivity of a histochemical method does not need to be as high as for the demonstration of amines in situation (b). Furthermore, the amines of the endocrine cells are fairly easily preserved by chemical fixation, but those of neurons are much more labile. This difference is probably also due to stronger binding of the amines to the proteinaceous matrices of the granules of secretory cells than to the synaptic vesicles of neurons.

The histochemical techniques for the biogenic amines (for review, see Hahn von Dorsche *et al.*, 1975; Björklund, 1983) depend on the detection either of the phenolic groups or of ethylamine with an aromatic substituent on the β-carbon atom.

17.2. Histochemical methods for amines in secretory granules of argentaffin, chromaffin, and mast cells

17.2.1. Serotonin: azo coupling methods

The serotonin-containing cells of the alimentary tract are named "argentaffin" on account of their ability to reduce ammoniacal or similarly complexed solutions of silver nitrate to the metal. Other presumably endocrine cells found in the gastrointestinal epithelium reduce complex silver ions more weakly and can only be seen if treatment with the silver reagent is followed by the application of a developer, which increases the sizes and densities of the initially deposited metal particles. The latter types of cell are termed "argyrophil" and the methods for demonstrating them are reminiscent of the methods used for staining axons in the nervous system (see Chapter 18). Unfortunately, the argentaffin and argyrophil reactions, though useful to histopathologists, have very low histochemical specificities.

The most convenient histochemical techniques for argentaffin cells are based on the detection of the phenolic function of serotonin. Phenols couple with diazonium salts in alkaline solution to form coloured azo compounds. Azo-coupling with phenols occurs preferentially *para* to the hydroxyl groups. Since this position is already occupied in 5HT, one of the *ortho* sites will be used:

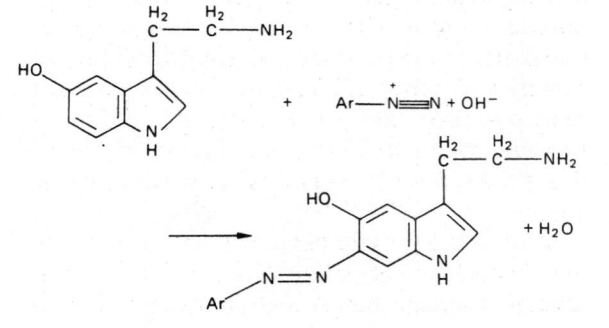

Similar azo-coupling reactions are also to be expected with the aromatic amino acids of all proteins and with any other phenolic compounds present in the tissues. Consequently, generalized "background" staining occurs if the time of exposure to the diazonium salt is unduly prolonged. The secretory granules of argentaffin cells (and of rodent mast cells) are recognized by virtue of their more intense and more rapidly developing coloration.

The argentaffin cells also give a positive chromaffin reaction, though this is not generally used for their identification.

Fixation and preparation

Use paraffin sections of formaldehyde-fixed material.

Solutions required

A. *Diazonium salt solution*

Fast red B salt (stabilized diazonium
salt derived from 5-nitroanisidine)
(see Chapter 5):   100mg
TRIS buffer, pH 9.0:   100ml
    Prepare just before using.

B. *A counterstain*

(An alum-haematoxylin, used progressively, is suitable.)

Procedure

1. De-wax and hydrate sections (see *Note* below).
2. Immerse in the diazonium salt solution for 30s.
3. Carefully rinse in five changes of water, each 20–30s, without agitation.
4. Counterstain nuclei. If alum-haematoxylin is used for this purpose, do not wash in running water but "blue" the sections in water containing a trace of $Ca(OH)_2$.
5. Dehydrate, clear, and cover, using a resinous mounting medium.

Result

Argentaffin cell granules—orange–red; background (protein)—yellow; nuclei (if counterstained with alum–haematoxylin)—blue.

Note

It may be necessary to cover the sections with a film of celloidin in order to prevent their detachment from the slides after treatment with the alkaline solution A. The film should be removed before clearing. The instructions for stages 3 and 4 allow for more gentle handling of the preparations than is usually necessary.

17.2.2. *The chromaffin reaction*

Cytoplasmic granules in the cells of the adrenal medulla assume brown colours after fixation in a solution containing potassium dichromate. Although such cells are said to be "chromaffin", the coloured product is derived not from the fixative but from the adrenaline and noradrenaline present in the cells. The amines are oxidized by dichromate to coloured quinones. The reaction involves oxidation of the catechol (*o*-diphenol) moiety to a quinone and oxidative coupling of the amino group to one of the carbon atoms of the ring:

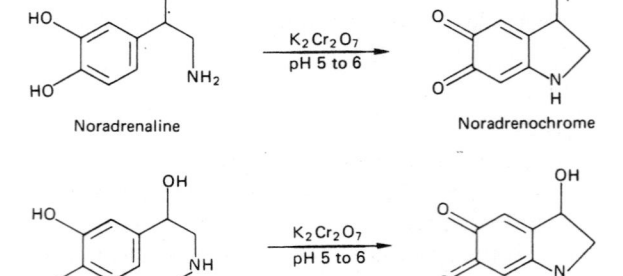

A similar reaction occurs with DA. It is likely that further oxidations and polymerization take place in the formation of the final brown products, which are insoluble in water and in organic solvents. The chromium atoms in the reagent are reduced from oxidation state +6 to +3. The resulting Cr^{3+} ions are bound at the sites of the amine (Lever *et al.*, 1977), and cause increased electron density of the chromaffin granules.

This technique is applied to small pieces or thin slices of fresh tissue (e.g. a rat's adrenal gland cut in half).

Solution required

Potassium dichromate ($K_2Cr_2O_7$):	5.0 g
Potassium chromate (K_2CrO_4):	0.45 g
Water:	to 100 ml

  Keeps indefinitely

Procedure

1. Fix small pieces of tissue in the above solution (see *Note 1* below) for 24h.
2. Wash in running tap water overnight. (See also *Note 2* below.)
3. Dehydrate, clear, embed in paraffin wax, and cut sections at 4 or 7μm.
4. De-wax and clear (three changes of xylene) and mount in a resinous medium.

Results

The cytoplasm of cells containing adrenaline (ADR) and noradrenaline (NA) is coloured brown. The product formed from ADR is darker than that from NA.

Notes

1. A chromate-dichromate mixture is used in order to obtain a solution of pH 5.5–5.7. Alternatively, one may fix the tissue in a fixative mixture based on potassium dichromate provided that it does not contain acids or other heavy metal ions. However, these mixtures are more acid than the optimum pH for the chromaffin reaction.
2. Stage 2 may be followed by a secondary fixation for 12–24h in 4% formaldehyde if desired. This is necessary if frozen rather than paraffin sections are to be cut.

3. A counterstain may be applied to the sections if desired. Alum-haematoxylin is suitable for nuclei. Many cationic dyes stain the chromaffin cells, thereby increasing the intensity of their colour in addition to demonstrating nuclei.
4. The **iodate ion** can oxidize catecholamines in the same way as dichromate though it acts much more rapidly upon NA than upon adrenaline. Consequently it is possible, using sodium iodate, to stain noradrenaline-containing cells selectively. Thin slices of tissue are immersed for 24h in saturated (10%) aqueous sodium iodate prior to fixation in formaldehyde and sectioning on a freezing microtome.

17.2.3. Catecholamines in glutaraldehyde-fixed tissue

Glutaraldehyde reacts with amino groups, probably to form imines:

$$OHC(CH_2)_3CHO + H_2NR \longrightarrow OHC(CH_2)_3 \overset{\overset{H}{|}}{C}\!=\!\!=\!N\!-\!\!-R + H_2O$$

Ordinarily an imine such as that shown above will be unstable when, as in the biogenic amines, R is not an aromatic ring.[†] However, in a tissue the other end of the glutaraldehyde molecule is likely to combine similarly with a protein-bound amino group. Extensive cross-linking of the proteinaceous matrix of the tissue probably increases the resistance of the linkages to hydrolysis. Primary catecholamines (DA and NA) are bound by their amino groups and are therefore immobilized, but their catechol groups remain free to react with other reagents. Adrenaline, which is a secondary amine, reacts with glutaraldehyde more slowly than NA or DA and is not usually retained in the tissue.

The catechols are strong reducing agents and may be detected by several simple histochemical methods. A simple procedure is to post-fix the

---

[†]In aromatic amines, NH_2 is attached **directly** to an aromatic ring, not to an aliphatic side-chain.

glutaraldehyde-fixed specimens in osmium tetroxide (Coupland *et al.*, 1964), which is reduced to a black material (see also Chapter 2, p. 17). Alternatively, the catechol-containing site may be made visible by virtue of its ability to reduce silver diammine ions to the metal (Tramezzani *et al.*, 1964; see also Chapter 10, p. 162). Methods of this type impart optical blackness and electron density to granules containing NA or DA. If adrenaline is to be demonstrated as well, it is necessary to fix the specimens in a solution containing both glutaraldehyde and potassium dichromate at empirically determined optimum concentrations and pH (Coupland *et al.*, 1976; Tranzer & Richards, 1976).

Solutions required

A. *Fixative*

25% aqueous glutaraldehyde:	20 ml
0.1 M phosphate buffer, pH 7.3:	80 ml
Prepare on day of use	

B. *Buffered osmium tetroxide*

2% aqueous OsO$_4$ (stock solution):	5.0 ml
0.1 M phosphate buffer, pH 7.3:	5.0 ml

Stable for about 6 h after addition of the phosphate buffer. A simple 1% aqueous solution of osmium tetroxide, which is stable for several weeks, may be substituted.

Procedure

1. Fix freshly removed pieces of tissue no more than 3 mm thick (e.g. halved adrenal gland of a rat) in solution A for 4 h.
2. Rinse specimen in water. Cut frozen sections 10–40 μm thick and collect them into water. Mount onto slides, blot, and allow to dry for 5–10 min. Position the sections near the ends of the slides so that only a small volume of solution B will be needed.
3. Immerse the sections in solution B in a closed coplin jar for 30 min.
4. Wash slides in running tap water for 30 min.
5. Dehydrate, clear, and mount in a resinous medium.

Result

Noradrenaline-containing cells—black to grey. Osmiophilic lipids (see Chapter 12) are also blackened.

Control

Substitute neutral, buffered formaldehyde for solution A in stage 1 of the method. The osmium tetroxide will now give black products only with lipids.

17.2.4. *Histamine*

Various histochemical methods for histamine have been described, but none are very specific. The least unsatisfactory is based on the formation of a fluorescent adduct with *o*-phthaldialdehyde (OPT). The method is usually applied to paraffin sections of freeze-dried tissue. Unfortunately, the reagent also forms fluorescent products with many amino acids, and with some proteins (Hakanson *et al.*, 1972) and peptides (Murray *et al.*, 1986). The OPT method has revealed the presence of histamine in cells already suspected of containing this amine, as in the gastric glands (Hakanson *et al.*, 1970) and in atypical mast cells (Kawamura, 1986), but has been of no use for mapping histamine-containing cells and fibres in nervous tissue. Instructions for a simple OPT method (Enerback, 1969) were given in the first edition of this book, but are not repeated here.

Immunohistochemical methods (Chapter 19) are now preferred for examining the distribution of histamine. It is possible to immobilize the amine by fixation in a carbodiimide. Immunoreactivity is seen in all the cell-types known to contain histamine, and in other sites too (Panula *et al.*, 1988).

17.3. Sensitive methods for amines in neurons

17.3.1. Formaldehyde-induced fluorescence

The reactions for noradrenaline and serotonin discussed in Section 17.2 are not sensitive enough to demonstrate aminergic neurons. For this purpose it is necessary to make use of the formation of intensely fluorescent compounds by reaction of the amines with formaldehyde (or any of a few other reagents; see Section 17.3.2 below). These methods, which can detect as little as 5×10^{-4}pg of NA or DA, have contributed importantly to knowledge of the anatomy of the central and peripheral nervous systems. Theoretical and practical aspects of the techniques are comprehensively reviewed by Björklund (1983).

Formaldehyde reacts with NA, DA, and 5HT by condensation and cyclization to form non-fluorescent compounds. These then undergo either oxidation by atmospheric oxygen or reaction with more formaldehyde, to form brightly fluorescent products. While the first stage of the reaction can occur with formaldehyde in either the liquid or the gaseous phase, the second stage requires an almost dry proteinaceous matrix at 80–100°C. The overall reactions are:

The reactions are more complicated than is indicated by the equations below and the fluorescent products shown are not the only ones formed. Adrenaline yields only a weakly fluorescent product. The chemistry of formaldehyde-induced fluorescence is discussed in detail by Corrodi & Jonsson (1967) and Pearse (1985). A simpler account is given by Björklund et al. (1975).

It is usual to employ freeze-dried tissue in studies of the biogenic amines of the nervous system, though methods have been devised for cryostat sections and for spreads of membranous tissues such as the iris and mesentery. Glyoxylic acid or, alternatively, a formaldehyde–glutaraldehyde mixture (see below) is a more suitable reagent when facilities for freeze-drying are not available. In the simplest form of the method, freeze-dried blocks of tissue are treated with formaldehyde gas, derived from solid paraformaldehyde, under controlled conditions of humidity at about 80°C. The treated blocks are then embedded directly in paraffin wax. The sections must not be exposed to water or alcohols, which remove the fluorophores. There are several modified techniques (e.g. Laties et al., 1967; Watson & Ellison, 1976) in which a buffered solution of formaldehyde is perfused

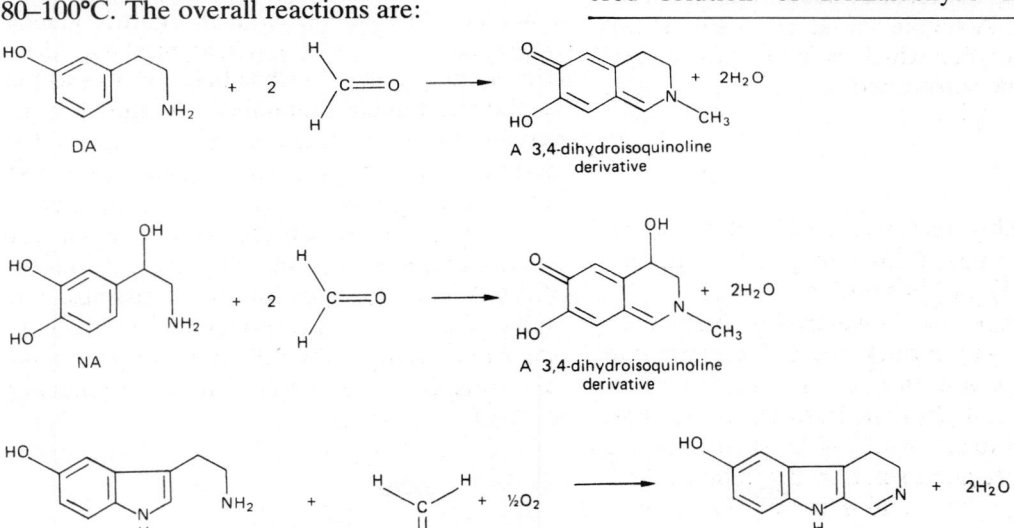

through the vascular system of the animal prior to either freeze-drying or the cutting of cryostat sections. Heating of the sections or of the freeze-dried blocks, with or without paraformaldehyde vapour, effects the second stage of the reaction in which the fluorophores are formed. The most sensitive of the formaldehyde-induced fluorescence techniques is the "ALFA" method, in which the animal is perfused with a solution of an aluminium salt before exposure of the tissue to formaldehyde. The metal ions catalyse the formation of fluorescent derivatives. The ALFA procedure can provide material for freeze-drying and sectioning in paraffin, or for cutting in a cryostat or a vibrating microtome (Loren *et al.*, 1980).

Formaldehyde-induced fluorescence is valuable not only for the identification and morphological study of cells containing biogenic amines, but also for investigating the pharmacological properties of such cells. For example, reserpine depletes cells of their monoamines, and drugs which inhibit monoamine oxidase produce visible increases in the intensity of fluorescence of aminergic neurons. A pharmacological method for distinguishing between DA and NA has been devised by Hess (1978). Animals are treated first with reserpine and then with DOPA, a metabolic precursor of the catecholamines. After a suitable interval of time, only the DA-containing cells exhibit fluorescence in formaldehyde-treated tissues. A much longer time is needed for replenishment of NA. Pearse (1968) has identified a series of cells that do not normally contain monoamines but which can take up the amino acids DOPA and 5-hydroxytryptophan and decarboxylate them to give DA and 5HT in histochemically demonstrable quantities. These cells, most of which have known or suspected endocrine functions, have been called APUD (amine precursor uptake and decarboxylation) cells. The synaptic varicosities of aminergic neurons normally reabsorb their transmitter substances and this property is occasionally exploited in order to enhance their formaldehyde-induced fluorescence.

17.3.2. *Other fluorescent methods for amines*

In addition to formaldehyde, several other carbonyl compounds are able to form fluorescent condensation products with biogenic amines. The most valuable is glyoxylic acid:

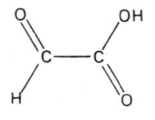

This substance was introduced as an alternative to formaldehyde by Axelsson *et al.* (1973). It reacts in a similar manner, the fluorescent products being closely similar though obtained in higher yield (see Björklund *et al.*, 1975).

A major advantage of methods making use of glyoxylic acid is that it is not necessary to prepare the tissue by freeze-drying. Usually the animal is perfused with a buffered aqueous solution of glyoxylic acid. Sections are cut with a vibrating microtome or a cryostat, dried onto slides and heated, either alone or in glyoxylic acid or formaldehyde vapour. It is also possible to use paraffin sections of freeze-dried specimens. Different variations of the method (e.g. Lindvall & Björklund, 1974; Furness & Costa, 1975; Bloom & Battenberg, 1976; Loren *et al.*, 1976; Watson & Barchas, 1977) have been developed to give optimum results with different tissues. The fluorescence of amine-containing nerve fibres is brighter after reaction with glyoxylic acid than after reaction with formaldehyde.

In another method (Furness *et al.*, 1977, 1978), tissues are treated with an aqueous solution of formaldehyde and glutaraldehyde by immersion or vascular perfusion. Fluorophores are formed with DA, NA, and 5HT, but the chemistry of the technique has not been studied. The fluorescence resists extraction by cold water, alcohols, several other organic solvents, and melted paraffin wax. Unfortunately, the non-specific "background" fluorescence associated with this technique is brighter than that seen with formaldehyde- or glyoxylic acid-induced fluorescence. For the demonstration of catecholamines, the FAGLU solution is used at pH 7. For serotonin the optimum pH is 10, and the sensitivity can be increased by adding an oxidizing agent, potassium ferricyanide, to the mixture (Wreford *et al.*, 1982).

17.4. Some fluorescence techniques for amines

The methods to be described are suitable, on account of their simplicity, for instructional purposes. In research work it is commonly necessary to use techniques of greater sensitivity; the ALFA method (Loren *et al.*, 1980) is probably the best of these. The use of fluorescence methods for amines has declined in recent years, and immunohistochemical methods for the enzymes involved in the synthesis of the amines are more popular. The reasons for this change in fashion include the commercial availability of reliable antisera, and the fact that the immunohistochemical methods do not require any special preparative procedures such as freeze-drying or cutting with a vibrating microtome.

17.4.1. Formaldehyde-induced fluorescence

The technique described below closely resembles the original Falck-Hillarp procedure (Carlsson *et al.*, 1962; Falck *et al.*, 1962).

Preparation of tissues

Alternative methods are prescribed for (a) thin whole-mounts (mesentery, iris, etc.), and (b) solid specimens, no more than 2mm thick.

(a) Stretch the thin tissue (removed immediately after killing the animal) onto glass slides. Place the slides in a glass rack in a desiccator containing a small beaker, one-third filled with phosphorus pentoxide. Partially evacuate the desiccator and wait for 1h. Remove the slides and treat with formaldehyde gas as described below.

(**Caution:** P_2O_5 is corrosive. It must not be allowed to come into contact with liquid water with which it reacts violently to form phosphoric acid. To discard used P_2O_5, place the vessel containing it in a safe place, open to the air for 12–24h. The white powder will then have deliquesced to a syrupy mass, which may be washed down the sink with copious running water.)

(b) The piece of tissue is rapidly frozen (by immersion in isopentane cooled by liquid nitrogen) and transferred to a freeze-drying apparatus. When freeze-drying is complete, the specimen is treated with gaseous formaldehyde as described below.

Treatment with formaldehyde vapour

Formaldehyde is generated by the action of heat on solid paraformaldehyde. The moisture content of the latter is a critical factor: the paraformaldehyde should be equilibrated with air of 50% relative humidity. This can be achieved by storing the powder for 7 days or longer in a desiccator which contains a mixture of concentrated sulphuric acid (specific gravity 1.84), 32ml, and water, 68ml. (**Caution:** add the acid, slowly with stirring, to the water.)

About 6g of the water-equilibrated paraformaldehyde is placed at the bottom of a wide-necked jar fitted with a tightly fitting screw cap. The slides or freeze-dried specimens are put into the jar, which is then closed and maintained at 80°C in an oven for 1h.

Subsequent processing

The jar containing formaldehyde vapour should be transferred to a fume cupboard before opening.

(a) Slides bearing spreads of tissue are covered, using liquid paraffin (mineral oil, heavy, U.S.P.) as mounting medium. Alternatively, rinse in xylene and mount in a non-fluorescent resinous mounting medium such as DPX.

(b) Freeze-dried specimens are vacuum-embedded in paraffin wax (time of infiltration, 15–30 min). Sections are cut and mounted onto slides without using water. The slides can be heated to 60°C to melt the wax and flatten the sections. The sections are then covered with liquid paraffin and coverslips applied. With further warming on the hotplate, the wax is dissolved by this mounting medium. Alternatively, the wax may be removed by xylene or petroleum ether and the coverslip applied with a non-fluorescent resinous mounting medium such as DPX. Clearing and mounting must be carried out very carefully to avoid loss of the sections.

Result

Green fluorescence from noradrenaline and dopamine; yellow fluorescence from serotonin. The optimum wavelength for excitation is 410nm (violet-blue).

If the sections are cleared and mounted in a resinous medium, there is slight suppression of the specific fluorescence, but the preparations are optically superior to those mounted in liquid paraffin.

Sections of freeze-dried blocks commonly contain numerous cracks and other deformities, though the preservation within small areas is satisfactory. In photomicrographs, the physical damage can be hidden by printing the pictures in such a way that the weakly autofluorescent background is black, and only the brighter, specific fluorescence is displayed.

17.4.2. Glyoxylic acid method

This technique (after Watson & Barchas, 1977) is applicable to cryostat sections of unfixed specimens of central nervous tissue. Each piece of fresh tissue is placed on a cryostat chuck, which is then stood in a slush of dry ice (solid CO_2) and acetone or in liquid nitrogen until the tissue has frozen.

Materials required

A. 0.1M phosphate buffer, pH 7.0
    (approximately) at 0–4°C: 45ml
    Glyoxylic acid
    (HOOC.CHO.H$_2$O): 1.0g
    Magnesium chloride
    (MgCl$_2$.6H$_2$O): 0.25g

1.0N (4%) sodium hydroxide
    (NaOH): add drops
    until the pH is 4.9–5.0
    Water (at 0–4°C): to make 50ml

Prepare immediately before use. Cool to 0°C by standing the coplin jar in iced water.

B. A hotplate, maintained at 45°C.

C. A dry screw-capped coplin jar containing approximately 2g of solid glyoxylic acid (HOOC.CHO.H$_2$O), heated to 100°C on a boiling water bath for 30 min prior to use. The lid should be only half-tightened during heating, to allow for expansion of air.

Procedure

1. Cut cryostat sections at any convenient thickness up to 20 μm. The temperature of the cryostat cabinet should be −17°C. Collect each section onto a warm (20–25°C) slide, where it will thaw and dry at once. Proceed **immediately** to stage 2, handling only 1–3 slides at a time.

2. Place slides in the glyoxylic acid solution (A) and leave in it at 0°C for 12 min (optimum for nerve terminals) **or** 4 min (optimum for cell bodies) **or** 45s (optimum for some axons).

3. Remove slides from solution A, blot with filter paper, and place on the hotplate at 45°C for about 5 min.

4. Place the slides in the coplin jar containing solid glyoxylic acid at 100°C for approximately 3 min.

5. Remove the slides and apply coverslips, using liquid paraffin (mineral oil, heavy, U.S.P.) as a mounting medium.

Result

Sites of NA and DA emit green fluorescence (excitation by violet-blue light).

Note

Glyoxylic acid is deliquescent and may deteriorate with repeated exposures to air. As soon as a bottle is received from the supplier, divide it into aliquots of 1.0g and store these in separate vials, **in a desiccator**, below 0°C. Do not allow solid glyoxylic acid to come into contact with the skin: it is irritant and corrosive.

17.4.3. A formaldehyde-glutaraldehyde (FAGLU) fluorescence method

This is a technique based on that of Furness *et al.* (1977). It differs from other aldehyde-condensation methods in that no heating of dry sections is needed. The fixative (solution A, below) is a satisfactory one for electron microscopy. See also *Note 1* below. For optimum demonstration of serotonin, use the modified FAGLU solution described in *Note 3*.

Solutions required

A. "FAGLU" reagent

Paraformaldehyde:	12g
0.1M phosphate buffer, pH7.0:	290ml

(Dissolve, with heating to 60°C. Cool to room temperature before adding next ingredient.)

25% aqueous glutaraldehyde:	6.0ml
Water:	to 300ml

This solution should be made on the day it is to be used.

Both the paraformaldehyde and the glutaraldehyde should be grades marketed as being suitable for electron microscopy.

B. Acidified DMP

2,2-dimethoxypropane (DMP):	25ml
Concentrated hydrochloric acid:	one drop (about 0.05ml)

This is prepared as required, but is stable for several hours or possibly for much longer. See Chapter 4 for discussion of DMP as a dehydrating agent.

C. Clearing agent

Chloroform:	50ml
Benzene:	50ml

Procedure

1. Remove small pieces of fresh tissue, no more than 2mm thick, and place them in the FAGLU reagent (solution A) for 3h. (See also *Note 2* below.)
2. Blot off excess FAGLU solution with filter paper and immerse specimen in acidified DMP (solution B) for 20min at 20–30°C, with occasional shaking.
3. Transfer to clearing agent (solution C). Shake at intervals of about 3min until the specimen has sunk to the bottom of the container, then leave for a further 10min.
4. Infiltrate with paraffin wax, three changes, each 10min, in a vacuum-embedding apparatus. Block out. Store blocks at 4°C until they are to be cut.
5. Cut sections at a convenient thickness: e.g. 15–25μm for peripheral tissues; thinner for CNS. Mount the sections onto dry slides **without using water**. Place the slides on a hotplate at about 60°C until the wax melts and the sections flatten, then proceed immediately to stage 6.
6. Remove slides from hotplate and put them into a staining rack. Immerse gently in a tank of previously unused xylene. Agitate very gently for 15–20s and leave for 5min.
7. Carefully remove slides individually from the xylene and apply coverslips, using a non-fluorescent mounting medium such as DPX. Place slides on a hotplate at about 60°C for 5min, then transfer them to a tray. If they are not to be examined immediately, keep them at 4°C for no more than 1 week.

Result

Sites of NA and DA green. Sites of 5HT yellow. Other components of the tissue also fluoresce but less intensely. Optimum excitation is by blue-violet light: the non-specific background fluorescence then appears as green when an orange-yellow barrier filter is used. If excitation is by broad-band ultraviolet, with a colourless or very pale yellow barrier filter, positively reacting elements appear yellow or greenish yellow against a blue background.

Notes

1. This technique is valuable on account of its technical simplicity, but the non-specific background fluorescence is more intense than with the two preceding methods. Furness *et al.* (1977) did not provide detailed practical instructions for the preparation of paraffin sections. The procedure described here is effective, but alternative schedules for dehydration and clearing may also be possible. It is important to avoid flotation of the paraffin sections on warm water, which results in loss of the specific fluorescence.

2. For the CNS, Furness *et al.* (1978) recommend perfusion of the animal with a buffered solution containing 4% formaldehyde and 1% glutaraldehyde, followed by sectioning with a vibrating microtome or a cryostat.

3. The following solution provides maximum fluorescence at sites containing serotonin (Wreford *et al.*, 1982). It should be perfused through the vascular system for 10min, and fixation continued by immersion at 4°C for 3h.

Water:	450ml
Sodium carbonate (Na_2CO_3):	5.85g
Sodium bicarbonate ($NaHCO_3$):	3.8g
Paraformaldehyde:	20g
Heat to dissolve (60°C), then cool to room temperature and add	
25% glutaraldehyde:	10ml
Water:	to make 500ml

The pH should be 10.0. Adjust if necessary. This solution is made on the day it is to be used, and kept at 4°C in a tightly capped bottle. **Immediately before use**, add 2.0g of potassium ferricyanide ($K_3Fe(CN)_6$).

17.4.4. *Other amines and free amino acids*

Acetylcholine. This quaternary ammonium ion can be precipitated by some metal complex anions, including reineckate, $[(NH_3)_2Cr(SCN)_4]^-$ (Gaddum, 1935), silicotungstate (Tsuji & Alameddine, 1981), tungstate and molybdate (Tsuji *et al.*, 1983). These can form electron-dense deposits in synaptic vesicles known to contain acetylcholine. Unfortunately, it is not possible to prove the chemical specificity of such reactions, which can be expected to give electron-dense deposits with many organic bases.

GABA (γ-aminobutyric acid, an inhibitory neurotransmitter) reacts with ninhydrin in octanol to form a fluorescent product. Pharmacological and histochemical studies indicate that this reaction demonstrates GABA in its expected sites in nervous tissue (Wolman, 1971; Gorne & Pfister, 1979), but the method has not often been used. The currently favoured approach to the localization of GABA is immunohistochemical. An antiserum is raised against a conjugate of GABA, glutaraldehyde and a protein such as albumin. The serum contains antibodies that can combine with GABA-glutaraldehyde-protein complexes in fixed tissue (Hodgson *et al.*, 1985). The same type of method has been used for the immunohistochemical staining of **glutamate** and **aspartate** ions (Hepler *et al.*, 1988) which are transmitters at excitatory synapses.

17.5. Ascorbic acid

Ascorbic acid is very soluble in water, so if it is to be localized histochemically it must be immobilized either by quick freezing or by instantaneous chemical precipitation. It is a strong reducing agent, becoming oxidized to dehydroascorbic acid:

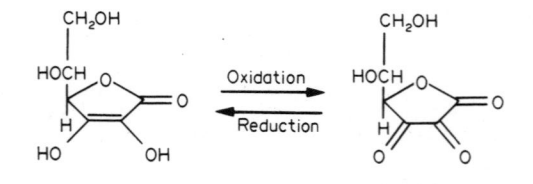

Both forms are likely to be present in tissues. The dehydroascorbic acid can be reduced *in situ* by treating the tissue with gaseous hydrogen sulphide, which is a strong reducing agent. The excess of H_2S can be removed by partially evacuating the container.

The only histochemical method for the reduced form of ascorbic acid is the acid silver nitrate method of Bourne (1933) and Giroud & Leblond (1934). This method is based on the biochemical observation that ascorbic acid is the only substance in living tissues that rapidly reduces silver nitrate in acid solution:

$$C_6H_8O_6 + Ag^+ \rightarrow C_6H_6O_6 + 2H^+ + 2Ag$$

The only common cause of a false-positive reaction is the presence of phosphate ions (as in bone or other calcified tissues), which might give a von Kossa reaction (see Chapter 14). This artifact can be avoided by treating the specimens, after immersion in $AgNO_3$, with ammonia or sodium thiosulphate, to remove insoluble salts of silver that might be reduced to the metal on exposure to light. The high solubility of ascorbic acid makes it unlikely that the silver deposits form in exactly the same subcellular sites, though the localization seen within tissues corresponds well with that determined by chemical assays (see Pearse, 1985).

The procedure described here is based on that of Barnett & Bourne (1941).

Tissue preparation

Freshly removed small organs (less than 2mm in least dimension) or similarly sized slices or pieces are usually immersed directly in the acidified silver nitrate solution. Bourne (1933) fixed in formaldehyde gas, generated from hot paraformaldehyde, and it is also possible to use paraffin sections of unfixed freeze-dried material. The reagent is applied to the sections before the wax is removed.

Solutions required

Acidified silver nitrate

Silver nitrate:	5g
Water:	500ml
Glacial acetic acid:	5ml

Stable, but it is convenient to make up as required, from an aqueous stock solution of silver nitrate.

Dilute ammonia

Ammonium hydroxide (S.G. 0.880):	5ml
Water:	95ml

This is made up immediately before using.

Procedure

1. Immerse specimens in the acidified silver nitrate, 1h, in a dark place. See *Note 1* below.
2. Rinse in water, then transfer to dilute ammonia. Leave for 10–60min.
3. Wash in 3 changes of water, each 10min.
4. *Either* prepare as a whole mount (thin specimens) *or* dehydrate, clear, embed in wax, cut sections, remove wax and mount in a resinous medium. See also *Note 2*.

Result

Tiny black granules indicate sites of reduction of silver ions by ascorbic acid. They are typically seen in cytoplasm, subjacent to the cell membrane and around the nucleus. Controls (see *Note 3*) should not show this appearance.

Notes

1. The time is not critical, but must be sufficient for penetration of the specimen. 15min may suffice for thin specimens to be examined as whole mounts. Paraffin sections are left overnight at 37°C, presumably to allow for slow penetration in the presence of wax.
2. Sections or whole mounts may be counterstained if desired. Suitable methods are given in Chapter 6.

3. Suitable controls include specimens soaked in water or methanol for 1h prior to immersion in the acidified silver nitrate, and specimens taken from experimental animals that have been fed a diet deficient in ascorbic acid.

17.6. Thiamine

Thiamine, which is present in many tissues of animals and plants, can be oxidized to **thiochrome**, a yellow substance that gives a blue fluorescence in response to ultraviolet irradiation.

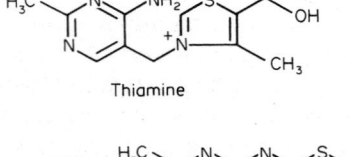

Thiamine

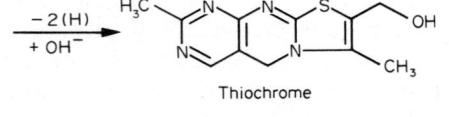

Thiochrome

This reaction is used in chemical assays for thiamine, and has been adapted as a histochemical method. The oxidizing agent may be alkaline potassium ferricyanide (Muralt, 1943) or cyanogen bromide vapour followed by ammonia (Tanaka *et al.*, 1973). Both procedures show an end-product in the myelin sheaths of peripheral nerves.

17.7. Exercises

Theoretical

1. A chromaffin granule in a cell of the adrenal medulla contains about 22,000 catecholamine molecules (see Winkler & Westhead, 1980). If these were all noradrenaline, could an individual granule be seen with the formaldehyde-induced fluorescence method? (For sensitivity of the technique, see p. 293. Avogadro's number, $N = 6.02 \times 10^{23}$ molecules per mole.)

2. What substances are responsible for the background staining in sections prepared by the azo-coupling method for argentaffin cells?

3. The mast cells of all mammals contain histamine and a sulphated proteoglycan. In rats and mice, the mast cells also contain serotonin. What histochemical methods would provide this information?

4. Would you expect amines precipitated by Reinecke salt (see Section 17.4.4) to exhibit:
(a) Formaldehyde-induced fluorescence?
(b) Positive azo-coupling reactions?
(c) Increased electron-density at their sites of storage?
Give reasons for your answers.

Practical

5. Carry out the azo-coupling reaction for argentaffin cells on paraffin sections of formaldehyde-fixed small intestine, tongue, and skin (from the rat). Identify cells which contain serotonin.

6. Process dried mesenteric spreads to demonstrate formaldehyde-induced fluorescence. Identify the sites of noradrenaline and serotonin in the preparations. If one of the spreads (not for fluorescence microscopy) is lightly stained with toluidine blue, this will assist in recognizing the cells that contain serotonin.

7. Observe the chromaffin reaction in paraffin sections of adrenal gland fixed in a chromate-dichromate mixture. It is advisable to embed both halves of the adrenal gland side by side in the same block; the small medulla may be in only one of the pieces. In making these preparations you may experience one of the adverse effects of fixation in potassium dichromate.

8. Cut frozen sections of a glutaraldehyde-fixed adrenal gland. Demonstrate the noradrenaline-containing cells in its medulla. What do you see in the adrenal cortex?

18

Neurohistological Techniques

It has long been recognized that the structure of the nervous system is not adequately revealed by the staining methods generally employed for other types of tissue. Special techniques are required and an enormous number of these exist, though many are merely slight variants of other methods. The procedures used in neurohistology are designed to display selectively certain elements characteristic of nervous tissues. These include normal components such as myelin sheaths and neuroglial cells, pathological features such as gliosis and viral inclusion bodies, and experimentally induced abnormalities such as axonal degeneration and substances artificially introduced as tracers.

The following account mentions most of the techniques used specially for nervous tissue, and a few are discussed in some detail. Unfortunately, the chemical or physical mechanisms of many neurological staining methods, especially those based on the reduction of silver and gold salts, are poorly understood. The procedures have been discovered and developed largely by trial and error.

For critical discussions of neurohistological methods and their uses and limitations, see Nauta & Ebbesson (1970), Ralis *et al.* (1973), Kiernan & Berry (1975), Santini (1975) and Jones & Hartman (1978).

18.1. Classification of techniques

This classification embraces most of the methods used in neuroanatomy and neuropathology. For techniques discussed in this or other chapters, the reader is referred to the appropriate numbered sections. For other methods, references are given to other publications.

Traditional techniques for neurons and nerve fibres

A. Nissl stains (Section 18.2.)
  (For neuronal cell-bodies)

B. Golgi methods (Section 18.3.)
(For occasional whole neurons, especially dendrites)

C. Methods for axons (Section 18.4.)
(a) Normal axons (silver methods)
(b) Normal presynaptic boutons
(c) Motor end-plates
(d) Osmium-iodide methods
(e) Vital staining methods
(f) Degenerating axons and boutons

D. Stains for myelin sheaths of axons
(a) Methods for normal myelin (Section 18.5.1.)
(i) Mordant dye methods
(ii) Luxol fast blue
(iii) Osmium tetroxide
(b) Degenerating myelin. (Section 18.5.2.)
(i) Solvent dyes
(ii) The Marchi method

Selective staining of neuroglial cells

E. Metallic impregnation methods (Section 18.6.1.)
(a) Cajal's gold-sublimate method (For astrocytes)
(b) Silver carbonate and similar methods (For oligodendrocytes and microglia)
(c) The Golgi methods (Section 18.3.) (Demonstrate occasional neuroglial cells, in addition to neurons)

F. Dye methods for a pathological neuroglia (Section 18.6.2.)

G. Immunocytochemical methods for glia (Section 18.6.3.)

Histochemical methods

H. Methods for amines (Chapter 17)

I. Methods for classical neurosecretory material (Section 18.7.)

J. Methods for peptides and enzymes of neurotransmitter metabolism. (Section 18.8: Chapter 19)

K. Miscellaneous histochemical methods (Section 18.9) (Specialized applications; often fortuitously discovered)

Modern neuroanatomical methods

L. Axonally transported tracers (Section 18.10)
(a) Radioactive amino acids
(b) Proteins, notably horseradish peroxidase and labelled lectins
(c) Fluorescent dyes

M. Filling methods (Section 18.11)

18.2. Nissl stains

The histochemical properties of the nucleic acids have already been discussed in Chapter 9. The nuclei of cells in the nervous system, together with the cytoplasmic Nissl bodies (RNA) of neurons, are easily stained with basic dyes. The procedure described in Chapter 11 for demonstrating metachromasia with toluidine blue gives excellent results when used as a Nissl stain. Many other cationic dyes, including cresyl violet acetate, neutral red, and thionine, are equally suitable for the purpose. Lillie's azure-eosin method (Chapter 6) provides preparations with blue nuclei and Nissl bodies, and other structures pink. Alum-haematoxylin does not stain nucleic acids specifically, but the cell-bodies of larger neurons are clearly visible in H. & E.-stained sections.

Chrome alum-gallocyanine (see Chapter 5) is a useful Nissl stain that resists extraction by alcohols or acidic counterstains to a greater extent than the simple cationic dyes mentioned in the preceding paragraph.

Preparation of staining solution

Gallocyanine:	300mg
Chrome alum (Chromium potassium sulphate; $K_2Cr(SO_4)_2.12H_2O$):	10g
Water:	200ml

Boil for 20min, and leave until cool. Filter, and

then make up the volume of the filtrate to 200ml by pouring water through the precipitate in the filter paper. The staining properties of this solution begin to weaken after a week, but it is usable for up to one month.

Procedure

1. De-wax and hydrate paraffin sections of tissue that has been fixed in any mixture that preserves RNA. Frozen sections of formalin-fixed material may also be used.
2. Stain *either* for 2h at 50–60°C, *or* for 24h at room temperature.
3. Wash in tap water, and counterstain (e.g. with eosin) if desired.
4. Dehydrate, clear, and mount in a resinous medium.

Result

Nissl substance and nuclear chromatin blue. Deposits of RNA in the cytoplasm of other cells are also blue. Proteoglycans (mast cells, cartilage) and some types of mucus are also stained.

18.3. The Golgi methods

Camillo Golgi (1843–1926) discovered his famous method when he observed the blackening of occasional whole neurons in dichromate-fixed blocks of tissue that had been subsequently immersed in a solution of silver nitrate (Corsi, 1987). The serendipity of this Italian histologist led to the elucidation of the morphology of the neuron and the development of the neuron theory. The theory was championed by a Spaniard, Santiago Ramon y Cajal (1852–1934), who is generally held to be the greatest of all neurohistologists. Although they disagreed bitterly about the structure of the nervous system, the two men shared the Nobel Prize for Physiology and Medicine in 1906.

There are three major variants of the Golgi procedure, known respectively as the Golgi, rapid Golgi, and Golgi-Cox methods. With all of these methods, whole neurons (dendrites, perikarya, and axons) and glial cells are coloured in shades of deep reddish-brown and black. Only about 1% of the cells in the tissues are stained, however, so it is possible to see the shapes of the individual cells very clearly. Examination with the electron microscope has revealed that the deposits are located mainly within the cytoplasm, and that within individual stained cells the impregnation is sometimes incomplete. The myelinated parts of the axon are rarely darkened by Golgi procedures, and the methods are used principally in the study of dendritic morphology and of the relationships between axonal terminals and the cells with which they synapse.

In the Golgi methods most closely related to the original one, pieces of central nervous tissue are treated with silver nitrate after prolonged fixation in potassium dichromate, while in the rapid Golgi methods a shorter fixation in a mixture containing potassium dichromate and osmium tetroxide precedes the immersion in silver nitrate. In the Golgi-Cox and related methods, tissues are fixed for several weeks in a solution containing $K_2Cr_2O_7$, K_2CrO_4 and a mercuric salt. Either whole blocks or sections of nitrocellulose-embedded tissue are then treated with an alkali, which precipitates black material in the randomly impregnated cells. Related techniques, often called Golgi-Kopsch methods, are available for tissue fixed in formaldehyde.

No satisfactory explanations are yet available for the peculiar specificity of staining achieved with any of the Golgi techniques. The chemical natures of the final deposits are not known with certainty, but there is evidence that the precipitate with the rapid Golgi method is silver chromate, Ag_2CrO_4 (Chan-Palay, 1973), and that with the Golgi-Cox technique it is mercuric oxide chromate, $Hg_3O_2CrO_4$ (Blackstad *et al.*, 1973). Braitenberg *et al.* (1967) noticed that with a Golgi method of the classical type, axonal staining was associated with the appearance of filamentous crystals on the surface of the tissue. They suggested that similar linear crystalline growth might occur within the sporadically impregnated cells. This suggestion receives support from electron microscopic observations of the early stages of impregnation (Spacek, 1989).

The precipitation of silver chromate began at randomly distributed sites, some of which were inside neurons. Only those neurons that contained these initial precipitates became filled by subsequent crystalline growth. The enlargement of the deposits was arrested by natural barriers, including the plasmalemma.

Before using a Golgi method for research purposes, the student is advised to consult the reviews by Ramon-Moliner (1988) and Kiernan & Berry (1975) in which the advantages and disadvantages of the different techniques are evaluated in relation to particular neurohistological applications. A Golgi-Cox method, described below, is suitable for demonstrating the dendritic architecture of most types of neurons in the mammalian CNS. Methods applicable to fixed tissues are given by Fairen *et al.* (1977), Peters (1981) and Gabbott & Somogyi (1984). These newer techniques can also be used in conjunction with electron microscopy.

Although it is slow, this **Golgi-Cox method** is possibly the most reliable of the Golgi techniques.

Solutions required

A. *Golgi-Cox fixative*

This is prepared from three stable stock solutions:

  (i) 5% aqueous mercuric chloride ($HgCl_2$).
  (ii) 5% aqueous potassium dichromate ($K_2Cr_2O_7$).
  (iii) 5% aqueous potassium chromate (K_2CrO_4).

Working solution. Mix in order:

  (i) 20 ml
  (ii) 20 ml ⎫
  Water: 40 ml ⎬ Mix before using
  (iii) 16 ml ⎭

B. *Alkaline developer*

Strong ammonia solution ⎫ Mix just
  (17% NH_3): 5 ml ⎬ before
Water: 95 ml ⎭ using

Procedure

1. Fix pieces of nervous tissue (not more than 10 mm thick) in the working solution of Golgi-Cox fixative for 6–8 weeks in a tightly capped container at 37°C. Pour off fixative and replace with fresh solution after the first day.
2. Wash in water, many changes, for 6–8 h, then dehydrate and embed in nitrocellulose. Cut sections 100 μm thick.
3. Treat the sections with the alkaline developer (solution B) for 2 or 3 min.
4. Wash in water, dehydrate, clear, and mount in a resinous medium. (See *Note* below.)

Result

Some neurons and neuroglial cells, including their cytoplasmic processes—black. Background—pale yellow.

Note

A mixture of chloroform (33 ml), xylene (33 ml), and absolute ethanol (33 ml) is preferred to absolute alcohol for the last stage of dehydration (following 95% alcohol). This can be followed by clearing for 10–15 min in creosote (beechwood) or in cedarwood oil, followed by a rinse in xylene.

The black colour fades with time if a coverslip is applied, so it is usual to mount the sections in **thick** Canada balsam, and allow to set at 40–45°C without a coverslip. Such preparations are stable for many years. If a synthetic mounting medium such as DPX is used, coverslips may be applied, but fading occurs after about 6 months. It is possible to modify the deposit chemically, to produce a more stable black substance (see Geisert & Updyke, 1977; Wouterlood *et al.*, 1983, for details).

18.4. Methods for axons

18.4.1. Silver methods for normal axons

The multitude of methods in this category is sufficient testimony to the fact that there is no entirely satisfactory way to stain every kind of axon. The argyrophilia of normal axons is associated with the neurofilaments. Chemical studies (Gambetti *et al.*, 1981; Phillips *et al.*, 1983) have revealed that the neurofilament proteins are responsible for stainability by at least one of the silver techniques, that of Bodian (1936). Although this technique has sometimes been regarded as histochemically specific for neurofilamentous material (e.g. Duyckaerts *et al.*, 1987), numerous other objects are stained, notably nuclei and the cytoplasm of many endocrine cells (Scopsi & Larsson, 1986). Although much is known of the chemical reactions that occur during silver staining, the reason for the specificity of the methods remains obscure. The subject has been reviewed by Kiernan & Berry (1975) and by Koski & Reyes (1986).

Silver methods that differ in their technical details from those for axons are used to stain neuroglial cells (see Section 18.6.1.), and also for the demonstration of nucleolar organizer regions of chromosomes, where staining is attributed to the presence of sulphydryl groups (de Capoa *et al.*, 1982; see also Moreno *et al.*, 1988).

18.4.1.1. Methodology and chemistry

The silver methods all have three features in common. These are:

1. Treatment of the tissue (blocks or frozen, paraffin or celloidin sections; nearly always after an aldehyde containing fixative) with a solution containing silver ions, whose concentration may be anything from about 10^{-5} to 1.0M.
2. Subsequent treatment of the specimen with a reducing agent capable of effecting the reaction: $Ag^+ + \varepsilon^- \rightarrow Ag \downarrow$.
3. The deposition of a darkly coloured material (consisting mainly or entirely of metallic silver) in the axons, which are said to be argyrophilic.

The chemical reactions involved in selective axonal staining with silver are fairly well understood as the result of the investigations of Holmes (1943), Samuel (1953a), and Peters (1955a, b). These studies were concerned principally with some of the technically simpler procedures applied to paraffin sections.

In the first step of such methods, sections are exposed for several hours to a mildly alkaline solution in which the concentration of free silver ions is low (10^{-5} to 10^{-3}M) but in which there is an adequate excess of the silver. The latter condition is attained by having a large volume of dilute aqueous $AgNO_3$ or by arranging for most of the metal to be present as a complex such as $[Ag(NH_3)_2]^+$, or a silver-protein compound, or by using a saturated solution of a sparingly soluble salt such as Ag_2O, Ag_2CO_3, or $AgOCN$ in equilibrium with the undissolved solid. Methods in which concentrated silver nitrate solutions are used have not been studied, but it is probable that the same chemical reactions occur. Silver is taken up by the section in two ways. The larger quantity is bound chemically by protein throughout the tissue. This chemically bound silver is not specifically related to axons and can easily be removed by treating the tissue with complexing agents such as ammonium hydroxide, citric acid, or sodium sulphite. A much smaller amount of silver is reduced at sites in the axons and precipitated as tiny "nuclei" of the metal. These nuclei are too small to be resolved with the electron microscope (Peters, 1955c). Similar nuclei, formed by the action of light upon crystals of silver halides, form the latent image in an exposed but undeveloped photographic emulsion. A latent image nucleus consists of 2–6 atoms of silver (see James, 1977). The nuclei formed in axons may be similar in size. The reduced silver cannot be extracted by NH_4OH or Na_2SO_3.

In the second stage of the method, the sections are transferred to a solution similar to a photographic developer. A commonly used one contains sodium sulphite and hydroquinone. This mixture is alkaline. The sulphite removes the chemically bound silver, introducing $[Ag(SO_3)_2]^{3-}$ ions into the solution. Hydroquinone reduces this complex ion to silver (metal) on the surfaces of the previously formed nuclei of metallic silver present in the axons. The nuclei are thereby enlarged until they are (in light microscopy) coalescent, and the axons appear as black or brown linear structures. In some techniques the chemically bound silver is removed by treatment with Na_2SO_3 alone and a developer containing a known concentration of silver is used. Such a mixture (known as a **physical developer**) must contain stabilizing agents to delay the spontaneous reduction of Ag^+ in the solution. Stability is obtained by including in the physical developer a substance which forms a soluble complex with silver, such as citric acid or sodium sulphite, so that the concentration of free silver ions is low. Addition of a macromolecular solute, commonly gelatin or gum acacia, provides additional stabilization.

If the processes of development in histological staining with silver are closely similar to the ones used in photography, it is likely that the silver precipitated as submicroscopic nuclei during the first stage of the procedure serves as a catalyst for reactions of the type

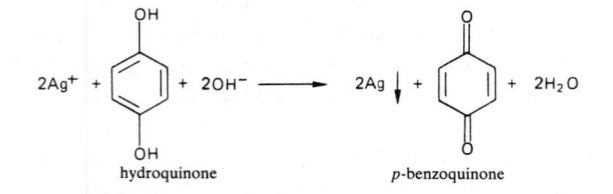

hydroquinone   p-benzoquinone

Metallic silver is insoluble, so it accumulates around the particles of the catalyst. Many substances other than silver can catalyse the above reaction, and physical developers have been used to produce visible end-products in a wide variety of histological and histochemical techniques (see Gallyas, 1971, 1979). The developing agent used in silver-staining methods should be one that is "weak" or "slow" by photographic standards. Hydroquinone, pyrogallol and p-hydroxyphenylglycine are suitable, but the more powerful

developers such as N-methyl-p-aminophenol (metol, elon) cause excessive deposition of silver which obscures the histological detail.

18.4.1.2. Gold toning

The coloration of axons impregnated with silver as described above does not always provide adequate contrast under the microscope. The contrast can be improved by adding a third stage of gold toning. To do this, the silver-stained sections are immersed in a solution of gold chloride. Sometimes this manoeuvre produces an adequate increase in contrast, but it is commonly necessary to add a further stage of reduction in oxalic acid. Finally, the sections are immersed in aqueous sodium thiosulphate to remove residual silver salts—a process analogous to the "fixing" of a developed photograph. The chemistry of gold toning was rather simply explained by Samuel (1953b) on the basis of the reaction

$$3Ag + AuCl_3 \rightarrow Au + 3AgCl$$

However, this explanation is unsatisfactory since it does not conform with the known chemistry of gold compounds and it assumes that one atom of gold would be optically more conspicuous than three of silver.

Histologists use "gold chloride" in two forms: sodium tetrachloroaurate ($NaAuCl_4.2H_2O$), which is an orange-yellow solid, and chloroauric acid which is $HAuCl_4$ with 3 or 4 molecules of water of crystallization. Chloroauric acid may be yellow or brown; the colour does not affect the properties. Both dissolve readily in water to give solutions containing chloroaurate anions, $[AuCl_4]^-$. Auric chloride, $AuCl_3$, which dissolves to form $H^+[AuCl_3OH]^-$, is not used in histology and neither is the sparingly soluble aurous chloride, $AuCl$.

The complex ion $[AuCl_4]^-$ is reduced by silver, with the formation of gold and silver chloride:

$$3Ag \quad + \quad [AuCl_4]^- \quad \longrightarrow$$
(Solid)   (Dissolved)

$$Au \quad + \quad 3AgCl \quad + \quad Cl^-$$
(Solid)   (Insoluble)   (in solution)

This reaction is essentially similar to the one proposed by Samuel (1953b). Silver chloride can be removed by treatment with sodium thiosulphate:

$$AgCl + 2S_2O_3^{2-}$$
(Solid)   ($Na_2S_2O_3$ in solution)

$$\longrightarrow [Ag(S_2O_3)_2]^{3-} + Cl^-$$
(Soluble)   (in solution)

In those techniques in which the toning procedure consists only of treatment with "gold chloride" followed by sodium thiosulphate, the increased contrast must be due to a reduction in the unwanted argyrophilia of the background rather than to an intensification of the metallic deposits in the axons.

The intensification of gold toning by treatment with oxalic acid is not so easily explained. Oxalic acid is not a photographic developer and does not reduce AgCl to Ag (which might have accounted for increased metallic deposition), but it does reduce chloroaurate anions to metallic gold. Now the particles of silver in stained axons are very small (Peters, 1955c; Sechrist, 1969), and they may well form, collectively, a colloidal suspension of the metal. After treatment with $HAuCl_4$, the silver is replaced by a mixture of Au and AgCl, presumably also in particles with colloidal dimensions. Since many colloids are able to adsorb ions and other molecules, it is not unreasonable to suppose that $[AuCl_4]^-$ ions are bound by physical forces at the sites of axonal staining. The bound chloroaurate ions would then be reduced on treatment with oxalic acid, possibly in the reaction

$$2[AuCl_4]^- + 3H_2C_2O_4$$
$$\rightarrow 2Au\downarrow + 6CO_2 + 6H^+ + 8Cl^-$$

Hence, substantial deposits of gold would be precipitated around each original particle of silver. The above speculations are supported by the observations of Zagon *et al.* (1970), who found by electron probe microanalysis that the metallic particles in silver stained and gold-toned protozoa contained considerably more gold than silver.

18.4.1.3. Choice of silver methods

Axons in most parts of the CNS are most easily demonstrated in paraffin sections by methods in which the primary impregnation is with a solution containing a very low concentration of free silver ions, applied for 24–48h at 37°C. The simplest technique is that of Holmes (1943). Another widely used technique is Bodian's (1936) protargol method. Protargol is a silver-protein complex, manufactured for use as an antiseptic, which slowly releases Ag^+ ions.† These procedures are also suitable for paraffin sections of peripheral nerves, but they often fail to demonstrate the innervation of tissues such as skin and viscera.

To stain peripheral nerve endings with silver, frozen sections of material thoroughly fixed in formaldehyde should be used. Primary impregnation is accomplished with a concentrated (10–20%) solution of silver nitrate. A simple and versatile technique is Mitchell's modification of the Hirano-Zimmermann procedure. This method can be applied to frozen, paraffin, and celloidin sections, and although it is rather capricious it is usually possible to obtain satisfactory results by varying the times of exposure of sections to the reagents. Cutaneous innervation is notoriously difficult to demonstrate by silver methods, but a simple technique devised by Winkelmann & Schmit (1957) is often satisfactory for this purpose. If it fails, the more troublesome Gros-Schultze method can be tried. The latter procedure is derived from earlier ones introduced by Bielschowsky, and makes use of an ammoniacal silver nitrate solution containing the silver diammine cation, $[Ag(NH_3)_2]^+$. The reduction of this complex ion by formaldehyde is probably catalysed by silver nuclei produced by

---

†Two types of silver proteinate, designated "strong" and "weak", have been used as antiseptics. Only the "strong" type, which is now obsolete as a pharmaceutical product, is suitable for Bodian's technique. Many unsatisfactory products are supplied by distributors of microscopical stains. The name "protargol-S" has been adopted by the Biological Stains Commission for silver proteinates tested and certified as satisfactory for the histological demonstration of axons.

axons during the primary impregnation. Thus, the ammoniacal solution appears to serve the same purpose as a physical developer (Horobin, 1988). Indeed, much of the untrustworthiness of Bielschowsky-type methods disappears when a controlled physical development replaces the alternation of treatments with formalin and silver diammine solutions.

There are many methods for staining with silver in the block before embedding and sectioning (see Jones, 1950). Excellent results can sometimes be obtained with these methods in both CNS and PNS, but the effects of uneven penetration of the reagents are usually apparent in the sectioned material.

18.4.1.4. Holmes's silver method for axons

The best results are obtained with tissue from the CNS fixed for 24h in neutral, buffered formaldehyde. Paraffin sections are used. The technique described below differs from the original (Holmes, 1943) only in the addition of a little pyridine to the silver solution, following Holmes (1947). The function of the pyridine is unknown, but if it is omitted there is a reduction in the contrast between the stained axons and their background. Addition of pyridine does not change the pH of the staining solution but it may reduce the concentration of free silver ions by forming complexes such as $[Ag(C_5H_5N)_2]^+$. A preliminary treatment with 20% $AgNO_3$ is often recommended (Holmes, 1947; Luna, 1968; Culling, 1974). This appears to be unnecessary.

Solutions required

A. 1% silver nitrate (AgNO₃) (stock solution)

Keep in a dark place. Discard if it becomes discoloured or contains a precipitate.

B. 1% (v/v) pyridine in water (stock solution)

Keeps for several months. (Many people find the odour of pyridine objectionable.)

C. Staining solution

Borate buffer, pH8.4:	100ml
1% $AgNO_3$ (solution A):	1.0ml
1% pyridine (solution B):	5.0ml
Water:	to 500ml

Mix just before using

D. Developer

Hydroquinone (= quinol):	2.0g
Sodium sulphite (Na_2SO_3):	10g
(**or** 20g of $Na_2SO_3.7H_2O$)	
Water: Dissolve and make up to	200ml

Mix on the day it is to be used

E. 0.2% gold chloride (stock solution)

May be $NaAuCl_4.2H_2O$ (yellow) or $HAuCl_4$ (yellow or brown). This solution can be used repeatedly but should be discarded (and recycled; see Kiernan, 1977b) when it has a greyish tinge or when there is an appreciable quantity of precipitated material in the bottle.

F. 1% oxalic acid ($H_2C_2O_4.2H_2O$)

This stock solution keeps indefinitely, but should be used only once.

G. 5% sodium thiosulphate ($Na_2S_2O_3.5H_2O$)

The remarks for solution F also apply here.

Procedure

1. De-wax and hydrate paraffin sections. Wash the sections in three changes of water (which **must** be distilled or de-ionized) for a total of 10min.
2. Place the slides, in a glass rack, in staining solution C for 24h at 37°C. The staining tank must be covered to reduce evaporation and there must be at least 50ml of the staining solution for each slide.
3. Remove the rack of slides from solution C, shake off excess liquid, but do not rinse in water. Transfer directly to the developer (solution D) for 3min. (See *Note 3* below.)
4. Rinse in distilled or de-ionized water, then

wash for 3 min in running tap water. Rinse again in distilled or de-ionized water.

5. Immerse in 0.2% gold chloride (solution E) for 3 min.
6. Wash in two changes of water.
7. Immerse in 1% oxalic acid (solution F) for 3–10 min until the sections are deep grey. Longer treatment than necessary produces a reddish colour and this is best avoided.
8. Wash in two changes (each 1 min) of tap water.
9. Immerse in sodium thiosulphate (solution G) for 5 min.
10. Wash in three changes of tap water, dehydrate, clear, and mount in a resinous medium.

Results

Axons black. Cell-bodies and nuclei are generally unstained and connective tissue fibres should be no darker than light grey. With over-treatment at stage 7, the grey and black tones are changed to pink and dark red.

Notes

1. The method as described is nearly always successful with formaldehyde-fixed CNS or peripheral nerve trunks. Peripheral nerve endings, however, are rarely demonstrable by this technique.
2. If poor results are obtained, variations in the composition of staining solution C should be tried. The volume of 1% silver nitrate may be halved or doubled: the pH may be varied between 7.0 and 9.0. All other aspects of the technique should be kept constant. The composition of solution C will usually need to be changed when fixatives other than neutral, buffered formaldehyde are used. Satisfactory results can be obtained with tissue fixed in most of the commonly used mixtures other than those containing potassium dichromate or osmium tetroxide.
3. The slides **must not be washed in water after stage 2.** It is necessary to carry some silver

ions into the developer. The axonal staining is invisible, or at best very faint, after the development at stage 2. The first part of the toning procedure (stage 5) causes the axons to disappear almost completely, but they reappear and become intensely coloured during stage 7 of the method.

18.4.1.5. Mitchell's rapid silver method

This technique (R. Mitchell, personal communication) is described because it is rapid. When it fails, minor variations can be introduced for subsequent sections without wasting too much time. The method is similar to one for celloidin sections described by Hirano & Zimmermann (1962), which can also be used on paraffin sections. Silver nuclei in axons are probably deposited during the impregnation with 20% $AgNO_3$. In the later stages of the method it is likely that a silver diammine complex is transiently formed and then immediately reduced to the metal by formaldehyde at the sites of the silver nuclei.

Material should be fixed in formaldehyde for at least 12 h and frozen sections cut (usually 15–25 μm thick). The technique is applicable to any tissue containing nerve fibres.

Solutions required

A. 20% aqueous silver nitrate

Keeps for several months in a dark place. It may be used repeatedly if not contaminated by other reagents.

B. Dilute ammonia

Ammonium hydroxide (= ammonia solution; S.G. 0.9, 28% NH_3): 20 drops	Freshly mixed
Tap water: 50 ml	

C. Formalin solution

Formalin (37–40% HCHO): 10 ml	Freshly mixed
Tap water: 40 ml	

Procedure

1. Rinse the frozen sections in three changes of water (which **must** be distilled or de-ionized). They can be left overnight in water if necessary. The sections must be handled with glass hooks. They are taken individually through the following steps of the method.
2. Place section in 20% $AgNO_3$ (solution A), for about 10s.
3. Transfer (without washing) to the dilute ammonia (solution B), for 10s.
4. Transfer (without washing) to formalin (solution C) for about 30s.
5. Rinse section in water and either mount in glycerin or dehydrate, clear, and mount in a resinous medium.
6. Examine the stained section. If it is unsatisfactory, proceed as directed in the *Notes* below.

Result

Axons black. Background (cells, collagen, etc.) in shades of yellow, gold, and brown.

Notes

1. If staining is **too light**, *either* rinse the section in water and repeat stages 2–5, *or* leave a subsequent section longer (up to 30min) in the 20% $AgNO_3$ at stage 2 and proceed as before.
2. If staining (especially the background) is **too dark**, try one of the following modifications on subsequent sections:
   (a) Use a stronger solution of ammonia at stage 3.
   (b) Leave for longer in the dilute ammonia at stage 3.
   (c) Rinse the section in water between stages 2 and 3.

18.4.1.6. Winkelmann and Schmit method for peripheral nerve endings

Material should be fixed in 4% formaldehyde for several days. Small pieces of tissue are then dehydrated in three changes of ethanol (each 1h), cleared in xylene (for 1h), rehydrated, and finally returned to the fixative until they are sectioned (Winkelmann & Schmit, 1957). (See *Note 1* below.)

This method was devised for skin, but axons in other tissues, such as muscle, are also stained by it.

Solutions required

A. Silver solution

20% aqueous silver nitrate. Keeps indefinitely; may be used repeatedly. Filter after use to remove particles derived from sections.

B. Developer

Hydroquinone:	0.2g	Make up
Sodium sulphite (Na_2SO_3):	1.0g	just before
Water:	100ml	using

C. 0.2% gold chloride

(See under solution E of Holmes's method, p. 308.)

D. 5% aqueous sodium thiosulphate ($Na_2S_2O_3.5H_2O$)

Keeps indefinitely, but use only once.

Procedure

1. Cut frozen sections 50μm thick and collect them into 4% aqueous formaldehyde. Take the free-floating sections through stages 2–10 of the method.
2. Rinse the sections in three changes of water (**must** be distilled or de-ionized) and leave for 30min in the last change.
3. Place sections in 20% $AgNO_3$ (solution A) for 20min. The sections should not be creased or folded.
4. Rinse sections in three changes of water, 3s in each.
5. Place sections (avoiding folds and creases) in the developer (solution B) for 10min.

6. Rinse in two changes of water.
7. Transfer sections to toning solution (C) for 2 min.
8. Wash in two changes of water.
9. Immerse sections in sodium thiosulphate (solution D) for 5 min.
10. Wash in two changes of water, mount onto slides, dehydrate, clear, and cover, using a resinous medium.

Result

Axons black. Other structures (cells, connective tissue, etc.) in shades of grey.

Notes

1. The preliminary treatment of the fixed blocks with alcohol and xylene is to remove soluble lipids which can interfere with the staining of axons. Other reagents (e.g. methanol, acetone, etc.) may be substituted.
2. Many beautiful photomicrographs of the results obtained with this method can be found in Winkelmann (1960). It is important that the sections be adequately thick, so that enough chemically bound silver will be carried through into stage 5 to allow solution B to function as a physical developer. Thick sections are also needed in order to reveal the full extent of axonal end-formations in the skin.

18.4.1.7. Gros-Schultze method for axons

This fairly reliable Bielschowsky-type method can be used with frozen or paraffin sections of formaldehyde-fixed material. The procedure described below (based on technical details given by Culling 1974) is also known as the "Gros-Bielschowsky" method.

Solutions required

A. 20% aqueous silver nitrate

Keeps indefinitely; may be used repeatedly. Filter after use to remove debris derived from sections.

B. Formalin solution

Prepare four dishes or staining tanks, each containing a freshly prepared mixture of 20 volumes of formalin (37–40% HCHO) and 80 volumes of tap water.

C. Ammoniacal silver solution (See Note 1 below.)

To 30 ml of 20% aqueous silver nitrate in a 100 ml conical flask add ammonium hydroxide (strong ammonia solution; S.G. 0.88–0.91) until the precipitate of brown silver oxide just redissolves. The ammonia should be added drop by drop, with swirling for a few seconds between each addition. About 2.5 ml of strong ammonia solution are needed. When the precipitate has dissolved, add a further 18 drops of the ammonium hydroxide.

D. Dilute ammonia solution

Strong ammonium hydroxide (as above): 5 ml } Freshly
Water: 95 ml } mixed

E. Dilute acetic acid

Glacial acetic acid: 1.0 ml } Freshly
Water: 99 ml } mixed

F. Toning solution

0.02% gold chloride. Conveniently made by ten-fold dilution of 0.2% gold chloride (see under Holmes's method, p. 308). This very dilute gold chloride should be used for only one or two batches of sections.

G. 5% aqueous sodium thiosulphate ($Na_2S_2O_3.5H_2O$)

Keeps indefinitely, but use only once.

Procedure

1. De-wax and hydrate paraffin sections. Collect frozen sections into water. Wash the sections (frozen or paraffin) in three

changes of water (**must** be distilled or de-ionized), for total time of 10 min.

2. Immerse in 20% silver nitrate (solution A) for 5 min (frozen sections) or 30 min (paraffin sections). (See also *Note 2* below.)

3. Without washing, transfer the sections directly to the first bath of formalin solution (B). Agitate and transfer sections successively through the other three formalin baths, allowing about 2 min in each. The last change of formalin solution must be free of turbidity. If it is not, move the sections into one or two more baths of solution B.

4. Without washing, transfer sections (or slides) directly into ammoniacal silver solution (C) for about 20s. (Slides bearing paraffin sections can be placed in a clean coplin jar and solution C poured in. Allow about 10ml for each slide.) The solution becomes grey and turbid and silver mirrors form on slides and on the inside of the staining vessel. The sections go dark brown. (See also *Notes 2* and *3* below.)

5. Transfer sections to dilute ammonia (solution D) for 1 min.

6. Transfer to dilute acetic acid (solution E) for 1 or 2 min.

7. Wash in water for about 30s.

8. Immerse in toning (solution F) for 10 min.

9. Wash in water for about 30s.

10. Immerse in sodium thiosulphate (solution G) for 3 to 5 min.

11. Wash in two changes of water, dehydrate, clear, and mount in a resinous medium.

Results

Axons black. In the CNS, fibrillary structures (neurofibrils) in the perikarya and dendrites of some large neurons are also coloured black. Other tissue components, grey.

Notes

1. The ammoniacal silver solution should be mixed just before using. The ammonium hydroxide should come from a recently opened stock bottle, because it loses NH_3 on standing. Handle strong ammonia carefully in a fume hood. Surplus and used solution should be discarded by washing down the sink. All ammoniacal silver solutions are potentially dangerous because if they evaporate to dryness the deposit may contain fulminating silver, which is probably a mixture of silver amide ($AgNH_2$) and nitride (Ag_3N). Fulminating silver explodes violently when touched.

2. The length of time in 20% $AgNO_3$ can be varied if unsatisfactory results are obtained with the times suggested. If no staining at all is obtained, it is probably that solution C contained too large an excess of ammonia.

3. Dirty precipitates on the sections can be due to too long a time in the ammoniacal silver solution. They are presumably due to the reaction

$$2Ag(NH_3)_2{}^+ + 2OH^- + HCHO \rightarrow$$
$$2Ag \downarrow + 4NH_3 + H_2O + HCOOH$$

18.4.1.8. A physical developer method for axons

This method, which was introduced in the first edition of this book, may be applied to frozen or paraffin sections of any formaldehyde-fixed tissue. Bouin's fluid is also a suitable fixative. This method differs from earlier techniques with physical developers in that (a) the initial impregnation is with a concentrated silver nitrate solution, so only a short time is needed for formation of nuclei within axons; (b) the action of the developer, which is modified from one used in photographic research (Berg & Ford, 1949), is more easily controlled than that of physical developers previously used for histological purposes. Some of the stable physical developers described by Gallyas (1979) can be substituted for the one prescribed below, but with no apparent advantage.

Solutions required

A. 20% aqueous silver nitrate

Keeps for several months and may be used repeatedly.

B. Sodium sulphite solution

Sodium sulphite (Na_2SO_3):	15.0g
Water:	to 500ml

Keeps indefinitely, but may be used only once.

C. Stock solution for physical developer

Hydroquinone (=quinol):	4.0g	
Citric acid ($C_6H_8O_7.H_2O$):	4.0g	
Gelatin (use a high quality bacteriological grade):	4.0g	Stable
Water:	400ml	for
Glacial acetic acid:	20ml	4–6
Dissolve all the above ingredients (30 min with occasional shaking) and add:		weeks
Water:	to make 500ml	

D. Physical developer (working solution)

Solution C:	100ml
Solution A:	8–10ml

This must be mixed immediately before use (see stage 6 of the procedure). The volume of solution A in the working solution is not critical, and may be as high as 20ml (though this is a waste of an expensive compound).

E. Gold chloride solution

This is a 0.2% aqueous solution of either sodium chloroaurate ($NaAuCl_4.2H_2O$) or chloroauric acid ($HAuCl_4$). It may be re-used many times.

F. 1% aqueous oxalic acid ($H_2C_2O_4. 2H_2O$)

Keeps indefinitely, but should be used only once.

G. 5% aqueous sodium thiosulphate ($Na_2S_2O_3.5H_2O$)

Keeps indefinitely but should be used only once.

Procedure

1. De-wax and hydrate paraffin sections. Attach frozen sections to slides. Either Mayer's albumen or chrome-gelatin may be used as adhesive. Wash in three changes of water.
2. Immerse slides in 20% $AgNO_3$ (solution A) for 15 min (time is not critical and may range from 10 min to 2h).
3. Wash in three changes of water.
4. Immerse in 3% Na_2SO_3 (solution B) for 5 min. (The slides may remain in this solution for 1h without ill effect.)
5. Rinse in three changes of water. Drain.
6. Mix the working solution of the physical developer (D) and pour it into the vessel containing the slides. The time for development depends on the ambient temperature: usually 4 min at 25°C or 6 min at 20°C. If the developer becomes grey and cloudy, proceed immediately to stage 7, whatever time has elapsed.
7. Pour off the developer and rinse slides with three changes of water. Examine under a microscope. If axonal staining is adequate, counterstain, dehydrate, clear, and mount. If more intense staining is required, proceed with the remaining stages of the method.
8. Tone in gold chloride (solution E) for 5 min. Rinse in three changes of water. If staining is now satisfactory, proceed to stage 11. If not, proceed to stage 9.
9. Immerse in oxalic acid (solution F) for 5–10 min. The time is not critical
10. Rinse in three changes of water.
11. Immerse in sodium thiosulphate (solution G) for 5 min.
12. Wash in running tap water for about 2 min.
13. Apply a counterstain if desired. Neutral red (see Chapter 6, p. 92) is suitable for the

CNS; the van Gieson method (see Chapter 8, p. 118) for peripheral tissues.

14. Wash, dehydrate, clear, and mount in a resinous medium.

Result

Axons black. Other structures are usually completely colourless if the procedure is stopped at stage 7 or stage 8. Treatment with oxalic acid (stage 9) produces a grey background and sometimes causes blackening of connective tissue fibres. If neutral red is used as a counterstain, nuclei, Nissl substance (RNA), and the proteoglycans are red. The van Gieson counterstain gives yellow cytoplasm and red collagen.

For some parts of the CNS, including the cerebral and cerebellar cortex, Holmes's method is superior. This method is superior to Holmes's technique for peripheral innervation and for the retina and optic nerve.

18.4.2. Methods for normal presynaptic terminals

Axonal endings are shown by some of the silver methods, but procedures are also available with which only the boutons terminaux are shown. Such methods involve pre-chromation of the tissue, followed by staining with haematoxylin or impregnation by a silver method (Armstrong et al., 1956; Rasmussen, 1957; Armstrong & Stevens, 1960; Abadia-Fenoll, 1968; Braak & Jacob, 1973; Desclin, 1973). Probably the phospholipids of mitochondria, which are concentrated within boutons, are immobilized by chromation (see Chapters 2, 12) and then stained in the second stage of the procedure.

These methods have been used in quantitative studies of synapses, but they cannot resolve the smaller presynaptic terminals, and they have now been largely supplanted by electron microscopy.

18.4.3. Motor end-plates

Neuromuscular junctions can be stained by various empirically derived techniques, reviewed thoroughly by Zacks (1973). The best known of these are Ranvier's lemon juice and gold chloride method and its more recent variants, which are applied to teased preparations (see Gray, 1954; Swash & Fox, 1972). The mechanisms of action of these methods are unknown.

The motor end-plate comprises an axon ending upon a specialized region of the sarcolemma (the subneural apparatus), so it is reasonable to combine different techniques for the two components. Nerve fibres in muscle are fairly reliably stained by silver methods, and the subneural apparatus is rich in acetylcholinesterase. Silver methods have been combined with thiocholine techniques for AChE (Gwyn & Heardman, 1965; Namba et al., 1967; Cunningham & FitzGerald, 1972), but when this is done both components of the end-plate are coloured black. By using an admittedly less specific indigogenic method for the enzyme of the subneural apparatus, the pre- and post-synaptic parts can be displayed in different colours (McIsaac & Kiernan, 1974).

The simplest procedure is to apply the indigogenic method for carboxylic esterases (Chapter 15, p. 247) to thick (60μm) frozen sections of muscle. The sections are then counterstained by any suitable silver method for axons, such as that of Winkelmann & Schmit given in this chapter. Optionally, van Gieson's picro-fuchsine (Chapter 6) may be applied as an additional counterstain, to show the muscle fibres and connective tissue.

18.4.4. Osmium-iodide methods

The innervation of some tissues is shown with striking clarity by using one of the osmium-iodide methods. Specimens are fixed in a solution containing iodide ions and osmium tetroxide, and subsequently are either examined as whole mounts or embedded and sectioned. The cation associated with the iodide was sodium or potassium in the earlier methods of this type (e.g. Champy et al., 1946), but more recently zinc iodide has been preferred (Maillet, 1963).

Although several ideas have been suggested and disproved (see Maillet, 1963), there is still no satisfactory explanation for the mechanism of axonal staining by iodide–osmium mixtures. The chemistry of the reagent has been partially elucidated by Gilloteaux & Naud (1979). When solutions of zinc iodide and osmium tetroxide are mixed, the latter compound is reduced by iodide ions to the osmate(VI) ion:

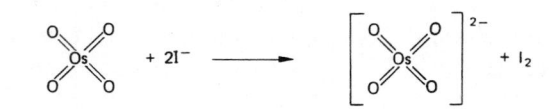

The solution rapidly becomes orange as iodine is liberated. After several hours it turns black; a black precipitate settles out during the next 1 to 2 weeks. This precipitate is zinc osmate, $ZnOsO_4$. It has been shown to be identical to the black, electron-dense substance formed within tissues fixed in zinc iodide–osmium tetroxide mixtures. Spectroscopic studies of solutions that initially contained osmium tetroxide and iodide ions or iodine indicate that there may also be oxidation of iodine (to iodate, in which the oxidation number of I is +5) by OsO_4, and deposition of metallic osmium:

$$I_2 + 2OsO_4 \rightarrow 2IO_3^- + 2Os \downarrow + O_2$$

The sensitivity of the compounds of osmium to reduction by organic material may be regulated by the concentration of elemental iodine, and by the oxidation states of this element in its ions in the staining solution (Carrapiço et al., 1984). Reduction of osmium-iodide mixtures occurs more rapidly in tissue than in the solution and may be catalysed at the sites of organic reducing groups such as —SH and —CH=CH—. Gilloteaux and Naud have also obtained evidence indicating that deposition of zinc osmate can occur at intracellular sites which, in life, were occupied by calcium ions.

Examination of zinc iodide–osmium-fixed central nervous tissues with the electron microscope has revealed electron-dense deposits within synaptic vesicles (Akert & Sandri, 1968). The reaction in these organelles has been shown by Reinecke & Walther (1978) to depend upon the presence of free sulphydryl groups in the tissue. It is not known whether similarly produced deposits

account for the optical blackness of stained peripheral axons.

Iodide–osmium methods have been found to be most valuable for the examination of nerve-endings in viscera. Both autonomic and sensory innervations are demonstrated. Central nervous tissue, which has a high content of osmiophilic membrane-bound lipids, is homogeneously blackened by fixation in iodide–osmium solutions and the preparations are uninformative when examined by light microscopy.

The following procedure is based on the method described by Rodrigo et al. (1970). Pieces of tissue must be taken from a freshly killed animal and should be less than 4mm thick, preferably only about 1.0mm. Hollow viscera are opened to make flat specimens, which can then be fixed to an improvised glass frame to facilitate penetration of the fixative.

Solutions required

A. 2% osmium tetroxide

1.0g of OsO_4 is dissolved in 50ml of water by breaking the cleaned ampoule under water in a clean, stoppered bottle. Do this in a fume hood. The solution keeps for several weeks at 4°C.

B. Zinc iodide

Put 6.0g of zinc powder in a 500ml conical flask. Add 10g of iodine (resublimed) and then 200ml of water. Swirl continuously. The elements combine with evolution of heat and the colour of the solution changes from reddish to grey. Allow the remaining solid material to settle and then filter. This solution should be prepared shortly before use and the excess discarded.

C. Working fixative

Add 25 ml of solution A to 75ml of solution B (or smaller quantities in the same proportions). The mixture is at first yellow, but soon darkens. Mix immediately before using.

D 2% potassium permanganate

Potassium permanganate (KMnO₄): 2.0g
Water: 100ml

Keeps for several months. Filter before using.

Alternatively, dilute a stock solution containing 6% KMnO₄.

E. 2% oxalic acid

Oxalic acid (H₂C₂O₄. 2H₂O):
                                          5.0g ⎫ Keeps
Water: 250ml ⎭ indefinitely

Procedure

1. Immerse the specimens in at least twenty times their own volumes of solution C, for 24h.
2. Wash in running tap water for 4–6h (if frozen sections are to be cut) or for 24h if the specimens are to be embedded in wax.
3. **Either** cut frozen sections **or** dehydrate, clear, embed in wax, and cut paraffin sections. Mount the sections onto slides and take to water. (See *Note* below.)
4. Examine the sections. If staining is satisfactory, dehydrate, clear, and mount. If the colour is too dark and nerve fibres cannot be distinguished from background, proceed as follows.
5. To differentiate the stain, pass the slides individually though (a) 2% KMnO₄ (solution D), (b) water, (c) 2% oxalic acid (solution E), (d) water, for about 5s in each. Re-examine the sections. Repeat the differentiating procedure until the desired appearance is obtained.
6. Wash in water, dehydrate through graded alcohols, clear in xylene, and mount in a resinous medium.

Result

Unmyelinated terminal branches of axons black. Myelin sheaths are also blackened. Other components of the tissue appear in shades of grey.

Note

Very thin specimens can be prepared as whole mounts. It is important to wash out all the osmium tetroxide before dehydrating, since alcohols reduce OsO₄ (See Chapter 2). Sections need to be fairly thick (20 to 100µm).

18.4.5. *Vital staining methods for axons*

The most important dye for staining axons in living (or freshly removed) tissue is **methylene blue**, which is considered below. **Quinacrine**, which is strongly fluorescent, is taken up by certain neurons of the autonomic and central nervous systems (Olson *et al.*, 1976; Alund *et al.*, 1979; Crowe & Burnstock, 1984) and also by the Merkel cells of the epidermis (Crowe & Whitear, 1978; Nurse *et al.*, 1983), and by cells that produce peptide hormones (Ekelund *et al.*, 1980). Quinacrine is sometimes used as a vital fluorochrome for the cells and fibres in which it accumulates. Fluorescent dyes of very low toxicity have recently been used to label individual neurons or axonal end-formations over long periods of time (Yoshikami & Okun, 1984; Lichtman *et al.*, 1985; Purves & Voyvodic, 1987) and for vital staining of cutaneous innervation (Nurse & Farraway, 1989).

Methylene blue has already been encountered as an ordinary cationic dye (see Chapters 5, 6, and 7), but its value in neurohistology depends on properties other than basicity. When living or freshly removed tissue is treated with methylene blue under suitable conditions, axons and their terminal branches are selectively stained. The biochemical mechanisms underlying this coloration are poorly understood despite several studies of the phenomenon (see Schabadasch, 1930; Richardson, 1969; Kiernan, 1974). The ability to stain nerve fibres is shared by some other N-methylated derivatives of thionine, but methylene blue itself is generally considered to be the best dye for the purpose.

For staining peripheral innervation, the most satisfactory results are obtained by immersing freshly removed tissue in a dilute oxygenated solution of methylene blue buffered to pH5.0–7.0, for about 30 min at 30°C. It is also possible to inject the dye locally or systemically into the living animal. Since the deposits of dye in the axons are soluble in water and in alcohol, the stained tissues are fixed in ammonium molybdate, which forms an insoluble salt with methylene blue. Ammonium picrate has also been used for the same purpose, but the picrate of methylene blue is more soluble in alcohol than the molybdate. The methylene blue-molybdate complex has been shown by electron microscopy to be associated with microtubules and also to occur as free crystals in the axoplasm of vitally stained nerve fibres (Chapman, 1982). Thin specimens are examined as whole mounts, but it is also possible to prepare frozen or paraffin sections of the vitally stained specimens. Ideally, the axons stand out as blue lines on a colourless background, but commonly there is also some blue staining of non-neuronal elements. Both somatic and autonomic fibres take up vital methylene blue, the former at pH 5.5–7.0, the latter at pH 5.0–5.5. Above pH 7.0, only the coarser, myelinated sensory axons are stained (FitzGerald & Fitter, 1971).

Vital methylene blue can also be used for the CNS. Usually an animal is perfused with a solution of the dye (commonly a higher concentration than is used for the immersion methods). Reducing agents in the anoxic tissues decolorize the methylene blue, forming a leuco-compound, which is thought to enter living cells more easily than the ionized dye (see Baker, 1958). Thin slices of the nervous tissue are removed and exposed to moist air to re-oxidize the dye, which is then fixed in ammonium molybdate. In a simpler method, methylene blue powder is sprinkled onto moist, freshly removed fragments of brain or spinal cord (see Gray, 1954). Vital staining with methylene blue is not a very reliable technique for the CNS, but when it works well, occasional whole neurons are coloured, as with the Golgi methods. Neuroglial cells are not stained. Cajal (1911) used the method to demonstrate the morphological characteristics of many types of central neuron.

The following method (from Richardson, 1969 and Kiernan, 1974) is for the demonstration of peripheral innervation in freshly removed tissue.

Solutions required

A. Buffered diluent. (Store at 4°C)

Sodium succinate, $(CH_2COONa)_2 . 6H_2O$:	14.0g
Sodium chloride (NaCl):	6.0g
Glucose (dextrose):	3.0g
Water:	800ml
Concentrated hydrochloric acid:	Add carefully until the pH is 5.5
Water:	to make final volume 1000ml

B. Methylene blue stock. (Keep at room temperature)

Methylene blue (C.I. 52015):	100mg
Water:	100ml

(It is advisable to use a grade of dye designated "for vital staining".)

C. Staining solution

(Usually mixed just before use, but can be kept for a few weeks at 4°C.)

Solution A:	100ml
Solution B:	1.5ml

Allow to warm to 37°C before using.

D. Ammonium molybdate fixative

Buffer (pH 5.5):

Sodium phosphate, dibasic (Na_2HPO_4):	2.0g	Keeps for several months at 4°C
Citric acid $(C_6H_8O_7 . H_2O)$:	1.2g	
Water:	500ml	

Working fixative solution:

Buffer (above):	50ml	Prepare before using
Ammonium molybdate, $(NH_4)_6Mo_7O_{24}.4H_2O$:	3g	

It takes about 15 min for the ammonium molybdate to dissolve (magnetic stirrer). Make up the fixative while the specimens are incubating in the stain. Cool to 0–4°C for use.

Procedure

1. Kill a small animal. Remove pieces of tissue, no more than 1.0mm thick. Collect into saline (0.9% NaCl) but do not leave them in it for more than 5 min.
2. Put the pieces in the warm staining solution (C) with continuous aeration at about 30°C. (These conditions are obtained by bubbling air through the solution and specimens while they are in a flask or beaker standing in a 37°C water bath. The air cools the staining solution to about 30°C). Leave for 30–40 min.
3. Remove specimens from the staining solution, rinse quickly (5–10s) in water and place in the working fixative solution (D) for 12–18h at 4°C.
4. Wash in water (4°C) for 10 min.
5. Transfer to absolute methanol at 4°C for 30 min.
6. Complete the dehydration in *n*-butanol at 4°C for 30 min.
7. Clear in benzene, 15 min (at room temperature).
8. Prepare as whole mounts in a resinous medium (See also *Notes 2* and *3* below).

Result

Axons blue. The background should be largely unstained.

Notes

1. If the method fails, try varying the pH between 5 and 7.5. In general, thicker axons are stained at higher pH.
2. As an alternative to whole mounts, the specimens may be (a) sectioned on a freezing microtome and collected into ice-cold water after stage 4, and then mounted onto slides, dehydrated in **cold** absolute methanol (2 quick changes), cleared, and covered, **or** (b) embedded in wax after stage 7 and the paraffin sections mounted onto slides, cleared, and covered. Sections should be thick: 50–100 μm.
3. Müller (1989) recommends post-fixation in a buffered formaldehyde-glutaraldeyde mixture after stage 3, followed by dehydration in *t*-butanol and paraffin embedding. The second fixation improves structural preservation.

18.4.6. *Degenerating axons and boutons*

When a nerve fibre is cut within the central nervous system, the fragmented remains of its axon remain *in situ* for about 2 weeks. They can be stained by special silver methods, which probably depend on the presence of unsaturated lipids (Evans & Hamlyn, 1956; Giolli, 1964). Methods for revealing the trajectories of degenerating axons were once the most important tool of the neuroanatomical researcher. The most useful procedures, introduced by Nauta & Gygax (1951), include a preliminary chemical treatment that prevents the deposition of silver in normal axons.

Degenerating presynaptic boutons, which persist for only a few days, can be recognized by careful examination of sections stained by a Bielschowsky-type method (Glees, 1946), which also stains normal axons and many normal boutons. The silver methods of Fink & Heimer (1967), derived from the Nauta-Gygax technique, largely suppress the staining of normal structures, enabling the sites of terminal degeneration to be seen more clearly.

These staining methods have now been largely supplanted by tracing methods based on axonal transport (Section 18.10). Methods for staining degenerating myelin sheaths (Section 18.5.2), though much less precise than the silver methods mentioned in the previous paragraphs, still find some use in human neuroanatomy and neuropathology, because the stainable debris derived from myelin is not removed by phagocytosis for several months following axonal transection caused by injury or disease.

18.5. Myelin

The myelin sheaths of axons are composed of concentric membranous lamellae derived from the ensheathing cells (Schwann cells in the peripheral nervous system; oligodendrocytes in the central nervous system). The membranes contain lipoproteins and proteolipids, and are stained histologically by virtue of the presence of these substances. The lipids of normal myelin are relatively hydrophilic compared with those of degenerating myelin and in both states unsaturated fatty acid residues are sufficiently abundant to permit their histochemical identification. The phospholipids, being for the most part protein-bound, are only partly extracted in the course of embedding in wax. Many of the methods used in lipid histochemistry will demonstrate myelin, but some procedures have been developed specially for the purpose. The latter techniques are valuable for the CNS and a few of them will now be discussed.

18.5.1. Normal myelin

18.5.1.1. Mordant-dye methods

A long-established staining method is the Weigert-Pal procedure in which formaldehyde-fixed blocks of nervous tissue are treated with a mixture of potassium dichromate and chromium fluoride. This renders all the phospholipid components insoluble in dehydrating and clearing agents and probably also acts as a mordant (see Chapter 12). Paraffin or celloidin sections are prepared, stained with a ripened haematoxylin solution, and then differentiated until only the myelin sheaths retain the blue–black chromium complex of haematein. There are many variations of the technique, and in some of them an iron mordant is used. The principal inconveniences of the methods using haematoxylin are the critical differentiation and the fact that some of the procedures take several days. The intensity

of staining, however, is higher than can be obtained with the simpler luxol fast blue method.

A more recent mordant-dye method is that of Page (1965) in which a ferric salt and chromoxane cyanine R are used. This method is applicable to paraffin sections, and prior insolubilization of phospholipids by chromation is not necessary. Presumably the fixed lipoproteins of myelin are retained in the sections in sufficient quantity to allow adequate staining by the dye–metal complex. The differentiation (by a solution of the ferric salt which is also used as the mordant) is technically simpler in Page's method than in the techniques employing haematoxylin.

The following method is that of Page (1965), modified by Kiernan (1984b). Use paraffin sections of material fixed in neutral, buffered formaldehyde. Zenker, Helly, and SUSA are also suitable fixatives, but when they are used it is not usually possible to eliminate nuclear staining by differentiation. Following alcoholic fixatives the myelin sheaths are damaged but staining is still possible. Precipitates formed by the action of mercuric chloride must be removed as appropriate.

Solutions required

A. Staining solution

This is the solution of ferric chloride and chromoxane cyanine R described in Chapter 6, Section 6.1.3.7 (page 99).

B. Differentiating solution

Any one of the following aqueous solutions may be used. See also *Note* below.

Iron alum, $NH_4Fe(SO_4)_2.12H_2O$:	10% w/v
Ferric chloride, $FeCl_3.6H_2O$:	5.6 w/v
Ferric nitrate, $Fe(NO_3)_3.6H_2O$:	7.3% w/v

The differentiating solution keeps indefinitely, but may be used only once.

C. Counterstain

0.5% aqueous neutral red or safranine is suitable.

Procedure

1. De-wax and hydrate paraffin sections.
2. Immerse in staining solution (A) for 15–20 min.
3. Wash in tap water (running, or several changes) until excess dye is removed (about 2 min).
4. Differentiate in the ferric salt solution (B) until only the myelin (white matter of CNS) retains the stain. This usually takes 5–10 min. It is sometimes impossible to decolorize the nuclei completely without losing some intensity in myelin.
5. Wash in tap water (running, or three or four changes) for about 5 min.
6. Apply counterstain if desired. See the luxol fast blue method (p. 92) for details.
7. Wash, dehydrate, clear in xylene, and cover, using a resinous mounting medium.

Result

Myelin and erythrocytes—deep blue. Nuclei and Nissl substance (with the counterstain suggested above)—red. Nuclei may be blue if perfect differentiation cannot be obtained.

Note

Clark (1979) recommends an alkaline differentiating solution: a freshly prepared 1% (v/v) dilution of ammonium hydroxide. This acts much more quickly than a ferric salt, and may remove too much of the blue dye-metal complex. The mechanisms of differentiation by Fe(III) or alkali have been discussed by Kiernan (1984b).

18.5.1.2. Luxol fast blue

As explained in Chapter 5 (p. 87), the luxol dyes are arylguanidinium salts of anionic chromogens. They are insoluble in water and are used as solutions in alcohol or other moderately polar organic liquids. Phospholipids are stained by these dyes, though probably not very specifically (Salthouse, 1963; Lycette et al., 1970). Luxol dyes probably also enter hydrophobic domains of protein molecules (Clasen et al.,

1973). It has been shown that the arylguanidinium cation is liberated into the solvent during the process of attachment of the dye to its substrate, leaving the coloured anion behind. The staining is then differentiated and made specific for myelin by treatment with dilute aqueous lithium carbonate, followed by 70% ethanol. The mechanism of differentiation is in need of investigation. A red or violet cationic dye subsequently applied as a counterstain binds not only to the nuclei and Nissl substance but also to the luxol fast blue anions present in the myelin sheaths, thereby increasing the intensity of coloration of the latter (Clasen et al., 1973).

Luxol fast blue MBS, G, and ARN, and methansol fast blue 2G are some of the dyes that can be used to stain myelin in the manner described above. The first such technique to be described was that of Klüver and Barrera (1953), which is still widely used. A somewhat simpler version is described below. Luxol fast blue is easy to use but slow in its action. It does not give such a strong colour to the myelin, especially in the PNS, as do the mordant-dye methods. The latter are therefore preferred for the demonstration of myelinated axons of narrow calibre. Fixation is as for the ferric salt–chromoxane cyanine R method (p. 319). Frozen sections of formaldehyde-fixed tissue are also satisfactorily stained by this procedure.

Solutions required

A. Staining solution

Luxol fast blue MBS:	0.25 g	} Keeps
95% ethanol:	250 ml	indefinitely

B. Differentiating solution (0.05% Li_2CO_3)

Lithium carbonate (Li_2CO_3):	0.25 g
Water:	500 ml

This keeps indefinitely and can be used repeatedly, but should be discarded when it is more than faintly blue.

C. Counterstain

0.5% aqueous neutral red.

Procedure

1. De-wax paraffin sections and take to absolute ethanol. Frozen sections should be dried onto slides (from water) and then equilibrated with absolute ethanol.
2. Stain in solution A in a screw-capped staining jar at 56–60°C (i.e. in wax oven) for 18–24h.
3. Rinse in 70% ethanol and take to water.
4. Immerse in the differentiating solution (B) until grey and white matter can be distinguished. This commonly takes 20–30s for thin paraffin sections, but thick frozen sections may require up to 30 min.
5. Transfer slides to 70% ethanol, two changes, each 1 min. More dye leaves the sections at this stage.
6. Rinse in water.
7. Counterstain for 1–5 min in 0.5% neutral red (solution C).
8. Wash in water and blot dry.
9. If necessary, differentiate the counterstain in 70% ethanol until only the nuclei and Nissl substance are red. The colour of the stained myelin is deepened by counterstaining.
10. Blot dry and dehydrate in two changes (each 3–5 min) of *n*-butanol. Clear in xylene and cover, using a resinous medium.

Result

Myelin—blue; nuclei and Nissl substance—red.

Notes

1. Some batches of luxol fast blue do not work very well. For other techniques, see Cook (1974). Suitable alternative dyes can be made in the laboratory by combining acid dyes with diphenylguanidine; for instructions, see Clasen *et al.* (1973).
2. An alternative counterstain is the PAS procedure (see Chapter 11). This provides a pink background without cellular staining.

18.5.1.3. Osmium tetroxide

The chemical reactions of OsO_4 with unsaturated lipids have already been described in connection with fixation (Chapter 2) and lipid histochemistry (Chapter 12). See also under "Degenerating myelin" (Section 18.5.2) below.

Osmium tetroxide is of little value for the staining of myelinated fibres in the CNS because it also causes blackening of all the other membranous components of this region. It is most valuable in morphometric studies of peripheral nerves. Small nerves may be teased in OsO_4 to display individual fibres. The nodes of Ranvier are clearly shown and internodal lengths can be measured. If nerves are fixed in OsO_4 or post-fixed in it after primary fixation by glutaraldehyde, the diameters and thicknesses of myelin sheaths can be measured in transverse sections.

18.5.2. Degenerating myelin

When an axon is severed, the part distal to the neuronal cell-body degenerates into a string of tiny fragments. The myelin sheath also undergoes fragmentation, and this process is associated with a change in the component lipids. Phospholipids, which are hydrophilic, predominate in the normal myelin sheath, but hydrophobic esters of cholesterol are the principal lipids of the degenerate fragments.

Occasionally, methods for hydrophobic and hydrophilic lipids are used sequentially, to stain degenerating and normal myelin in the same frozen section. Such methods, however, are not often used.

The **Marchi method** demonstrates the degenerated remains of myelin sheaths. There are several variations of the technique, but in all of them the tissue is treated simultaneously with OsO_4 (soluble in polar and non-polar substances) and an oxidizing agent (which must be one soluble only in polar substances). The oxidizing agent is most commonly potassium dichromate (Marchi, 1892) or potassium chlorate (Swank & Davenport, 1935). In degenerating myelin, OsO_4 is bound by the unsaturated fatty-acid residues of the hydrophobic cholesterol esters and then reduced to an insoluble black compound, probably $OsO_2.2H_2O$ (see Chapter 2 for chemistry). In normal myelin, which is hydrophilic, the oxidizing agent prevents the reduction of bound osmium.

Important differences exist between the early and late stages of degeneration of myelin. Strich (1968) and Fraser (1972) have shown that in the **early phase** (up to about 10 days after injury for PNS or about 100 days for CNS) the products are destroyed by freezing and thawing or by storage for more than 3 weeks in a formaldehyde-containing fixative. They can therefore be demonstrated only by treating blocks of tissue with OsO_4–oxidant mixtures. The products of the **late phase** of degeneration, however, persist much longer after axonal transection (up to 30 days in the PNS and at least 2 years in the CNS), and are not destroyed by freezing and thawing or removed by prolonged storage of tissues in solutions of formaldehyde. The degenerating myelin of the late phase can therefore be stained either in the block or in frozen sections. For staining in the block, the method of Swank & Davenport (1935) is recommended, and is described below.

The chief advantage of the Marchi method for neuroanatomical tracing is that the length of the time of survival after a destructive lesion is not very critical, so it is possible to apply the technique to human post-mortem material. The disadvantages are the impossibility of demonstrating degenerating unmyelinated axons (including all presynaptic branches) and the occasional blackening of normal myelin sheaths.

Solutions required

A. *Fixative*

4% formaldehyde (in water, saline or 0.1M phosphate buffer, pH 7.2–7.4).

B. *Staining solution*

Potassium chlorate ($KClO_3$):	1.5g	Keeps for a few months at 4°C in a stoppered bottle
Water:	200ml	
Osmium tetroxide:	0.5g	
Formalin (37–40% HCHO):	30 ml	
Glacial acetic acid:	2.5ml	

Procedure

1. Remove tissue from the central nervous system of a suitable experimental animal, or use human *post mortem* material from a patient in whom degenerated tracts are expected to be present. The specimens should be no more than 1cm thick. Fix in solution A for 2 to 4 days.

2. Trim into pieces no more than 3mm thick, and put these, without washing, into the staining solution (B). The volume of this solution should be approximately 15 times that of the specimens. Leave in a screw-capped jar for 7–10 days, with daily agitation to expose all surfaces of the tissue evenly to the reagent.

3. Wash in running tap water for 24 hours. Frozen sections may be prepared, if desired. Otherwise, proceed to steps 4 and 5.

4. Dehydrate in graded alcohols. *Either* embed in nitrocellulose *or* double-embed in paraffin wax. (See Chapter 4 for techniques.)
   Cut serial sections 20μm thick, and mount onto slides.

5. De-wax and clear the sections by passing through 3 changes of xylene. Cover, using a resinous mounting medium.

Result

Degenerating myelinated fibres appear as rows of black dots. Occasional normal nerve fibres are often stained, but can be recognized by their integrity. Fat in adipose tissue is also blackened.

Notes

1. For distinguishing between early and late products of degeneration, see the foregoing discussion of this method.
2. A counterstain may be applied to some sections, to facilitate topographical orientation. Alum-haematoxylin will stain nuclei, but cationic dyes do not work well after fixation in OsO_4.
3. Swank & Davenport (1935) recommended, for experimental animals, preliminary fixation by vascular perfusion of a solution consisting of 15g of $Mg(SO_4)_2.7H_2O$ and 5g of $K_2Cr_2O_7$ in 250ml of water.

18.6. Methods for neuroglia

18.6.1. Metallic impregnation methods

The traditional staining methods that were used by early histologists (notably Ramon y Cajal and Del Rio Hortega) to recognize and define the different types of neuroglial cell depend on impregnation with compounds of silver or gold. The chemical rationales of these methods are not understood. **Cajal's gold-sublimate method** for astrocytes is described below, as an example of one of these techniques.

Other methods for normal neuroglial cells include several in which sections are treated with an ammoniacal solution of silver carbonate. Such methods are used for the demonstration of oligodendrocytes and microglial cells. Instructions for these methods can be found in manuals of histological technique. Penfield & Cone (1950) is highly recommended. None of the metallic impregnations are completely selective for neuroglia; neuronal elements, such as cell-bodies and axons, are also stained to a variable extent, as are blood vessels. Like the silver methods for axons, neuroglial impregnation procedures are notoriously unreliable, and it is often difficult to determine why a method is not working properly.

Fixation is one important factor. The formal-ammonium bromide mixture (FAB) described below is suitable for several of the methods for neuroglia. Its properties are probably attribut-

able to its acidity (pH 1.5), rather than to any property of NH_4Br (Lascano, 1946). The acidity is due to reaction of the ammonium ion with formaldehyde. Hexamethylene tetramine is formed, and hydrogen ions are released, thus lowering the pH:

$$6HCHO + 4NH_4^+ \rightarrow C_6H_{12}N_4 + 6H_2O + 4H^+$$

Formalin acidified with sulphuric acid can replace the traditional FAB mixture, though it has been claimed (Polak, 1948) that this simpler formulation is inferior to FAB as a preparation for the silver carbonate methods.

Solutions required

A. FAB fixative

(Formal-ammonium bromide)

Formalin (37–40% HCHO): 15ml	Prepare
Ammonium bromide	before
$\quad$(NH₄Br):$\qquad\qquad\quad$2g	using
Water:$\qquad$to make 100ml	

B. Globus's reagents (see Note 1, below)

Prepare as required.

(i) *Ammonia–water:* Ammonium
$\qquad$hydroxide (S.G. 0.9; 27% NH_3): 5.0ml
$\qquad$Water:$\qquad\qquad\qquad\qquad\quad$45ml

(ii) *Globus's hydrobromic acid:*
$\qquad$Concentrated hydrobromic
$\qquad$acid (47% HBr):$\qquad\qquad$5.0ml
$\qquad$Water:$\qquad\qquad\qquad\qquad\quad$45ml

C. Gold-sublimate mixture

A *stock solution* of gold chloride (1% aqueous chloroauric acid) is needed. (This should not be a solution that has been previously used for other purposes.)

Working solution

Mercuric chloride ($HgCl_2$: the highest
$\qquad$available purity):$\qquad\qquad$0.5mg
Water:$\qquad\qquad\qquad\qquad\qquad$60ml

Dissolve the $HgCl_2$ (warm gently and stir with a

glass rod). To this solution, add 10ml of 1% gold chloride. This mixture should be made just before using.

D. 5% aqueous sodium thiosulphate ($Na_2S_2O_3.5H_2O$)

Procedure

1. Fix pieces of tissue up to 5mm thick in formal–ammonium–bromide for 24–48h. Cut frozen sections (about 20μm) and collect into water containing a few drops of formalin.
2. Place about twenty sections in the gold–sublimate mixture (solution C) for 4–8h in a petri dish in the dark. The sections must not be creased or folded. They should be manipulated with a glass hook. At hourly intervals remove one or two sections, take them through stages 3, 4 and 5, and examine under a microscope. Grossly, the sections in the gold–sublimate solution become purple. This colour deepens with time. The hourly checking of sections need not begin until the purple coloration appears.
3. When staining of astrocytes is adequate, pass all the remaining sections through two changes of water.
4. Immerse in 5% $Na_2S_2O_3.5H_2O$ (solution D) for 5 min.
5. Wash in two changes of water, mount onto slides, dehydrate, clear, and cover, using a resinous mounting medium.

Result

Astrocytes dark red, dark purple, and black. The cytoplasmic processes and pericapillary end-feet stain more intensely than the perinuclear parts of the cells. Other cells, notably neurons, are coloured in lighter shades of purple.

Notes

1. For material fixed in formaldehyde (without NH$_4$Br), cut frozen sections and wash them in three changes of water. Place the sections in ammonia water (solution B (i)) overnight in a tightly closed container, then rinse them quickly in two changes of water and put them into Globus's hydrobromic acid (solution B (ii)) for 1h at 37°C. Rinse quickly in two changes of water and place the sections into solution B as stage 2 of the procedure given above.
2. A modified technique for paraffin sections of formaldehyde-fixed tissue has been described by Naoumenko & Feigin (1961).

18.6.2. Dye methods for pathological neuroglia

The metallic impregnation methods and, more recently, electron microscopy show that the neuroglia consists entirely of cells. It had previously been thought that neuroglial cells were embedded in a fibrillary extracellular matrix. This false impression was gained from observations of nervous tissue stained with dyes, in which it could not be seen that the fibrous material consisted of the cytoplasmic processes of the cells.

Dyes are useful for showing abnormally dense aggregations of the cytoplasmic processes of astrocytes, which occur in and around sites of injury or disease in the central nervous system. The production of a tangled mass of astrocytes is known as **gliosis**, and the cytoplasmic processes are often called "glial fibres".

Holzer's crystal violet technique is the dye-based method most widely employed by neuropathologists. It also demonstrates normal fibrous astrocytes, but not very well. Another useful stain is **Mallory's phosphotungstic acid-haematoxylin** (PTAH), which imparts a variety of colours to the tissue, and permits the identification of sites of gliosis (blue), myelin (blue), and collagen fibres (reddish brown). Instructions for Holzer's technique and the PTAH method are given by Luna (1968), Clark (1981), and Bancroft & Cook (1984). Heteropolyacids (see Chapter 8) are used in both these staining methods, but the mechanisms are not understood. The sizes of the molecular aggregates of heteropolyacid and dye are probably involved in their penetration and retention in different components of the tissue, just as in the staining of connective tissue by the trichrome methods (see Horobin, 1982).

18.6.3. *Immunohistochemical methods for glia*

Immunohistochemical methods for antigens characteristic of the different cell-types are probably more reliable than the traditional staining methods for neuroglia. The principles of immunohistochemistry are reviewed in Chapter 19. Here, it may simply be stated that in these techniques the sites of attachment of antibodies to antigens can be made visible in either conventional or fluorescence microscopy.

Immunocytochemical methods are available for some human glial antigens, and some of the required antibodies are commercially available. Weller (1984) recommends staining for the following:

S100 protein	Astrocytes
	Oligodendrocytes
	Schwann cells
GFAP*	Astrocytes
Carbonic anhydrase, type C	Oligodendrocytes
α-1-antitrypsin or α-1-antichymotrypsin	Microglia and macrophages
Neuron-specific enolase	Neurons

These antigens do not necessarily provide completely specific staining of the indicated cell-types, but they remove much of the uncertainty associated with the older techniques. The chief disadvantages of immunohistochemistry are that the antisera are always expensive and not always easily available.

---

*Glial fibrillary acidic protein. The intermediate filaments in the cytoplasm of astrocytes are made of this protein. In gliosis, the filaments are particularly abundant.

18.7. Classical neurosecretory material

In vertebrate animals certain large neurons in the hypothalamus produce peptide hormones (oxytocin and vasopressin in mammals) that are elaborated in granules with characteristic staining properties, attributable to their content of neurosecretory material (NSM). The granules are present in the perikarya, the axons, and—in highest concentration—in the perivascular axonal terminals of the posterior lobe of the pituitary gland. The granules contain precursor or carrier proteins known as neurophysins, from which the peptide hormones are released by the action of proteolytic enzymes. In mammals these hypothalamic cells are the only neurons that contain NSM, but neurosecretory cells with the same tinctorial properties occur in various parts of the nervous systems of almost all invertebrates (see Gabe, 1966).

Neurophysins and their associated peptides contain numerous disulphide (cystine) cross-links, and the staining methods for NSM take advantage of this property. Indeed, any histochemical technique that displays sites of high cystine concentration (see Chapter 10 for chemistry) may be used as a stain for NSM. In practice, procedures are chosen that give distinctive colours and can easily be combined with other stains.

Paraffin sections are cut from tissue fixed in an aqueous fixative (ideally one that contains coagulant and non-coagulant ingredients, such as Bouin or SUSA; see Chapter 2). Alcoholic fixatives are unsuitable because they do not render the NSM insoluble in water. The hydrated sections are immersed for about 2 min in acidified potassium permanganate (0.25–0.3% w/v $KMnO_4$ in approximately 0.02M sulphuric acid). The brown manganese dioxide that forms on the sections is removed by rinsing in 0.5–1% aqueous oxalic acid. Oxidation by acidified $KMnO_4$ produces two residues of cysteic acid at the site of each disulphide bridge. In the second stage of the method, the sulphonate anions of cysteic acid are stained with a dye that has a suitably intense colour and does not bind to many other basophilic objects. Any cationic dye at pH <2 will do this, for reasons discussed in Chapters 5, 6, 10 and 11, but most ordinary basic dyes would be removed during subsequent counterstaining.

The basic dye with which oxidized NSM was first demonstrated (by Bargmann, 1949) was a chromium(III)-haematein complex. A more reliable reagent, however, is aldehyde-fuchsine, which was introduced for this purpose by Gabe (1953). This is used exactly as when staining elastin (Chapter 8, p. 129). Subsequent staining is carried out according to the requirements of the investigation, and following the general principles of microtechnique contained in Chapters 1, 6 and 8. Alum-haematein (for nuclei) and fast green FCF (for acidophilic material) provide a pleasing combination of colours.

It is also possible to stain neurosecretory cells immunohistochemically, using antibodies to either the neurophysins or the peptide hormones.

18.8 Histochemical methods for neurotransmitter-related substances

It is often useful, especially for research workers, to be able to stain certain functional types of neuron selectively. The traditional methods of neurohistology (e.g. Sections 18.2, 18.3 and 18.4) do not do this, but there are histochemical techniques that have greater specificity.

Several **amine neurotransmitters** can be demonstrated by the methods discussed in Chapter 17. It is also possible to use immunohistochemistry to localize some of the enzymes involved in the synthesis of the catecholamines and of serotonin.

Cholinergic neurons contain the synthetic enzyme choline acetyltransferase, which moves a

$$CH_3C{-}\!\!\!\!\underset{O}{\overset{\|}{}}$$ group from acetyl-coenzyme A to

choline. Conventional histochemical methods for this enzyme have been devised, but are not very satisfactory. Antibodies to some forms of the enzyme have been used for its immunohistochemical demonstration in some laboratory animals. The enzyme **acetylcholinesterase**, for which thoroughly reliable histochemical methods

are available, catalyses the destruction of acetylcholine. Although present at all cholinergic synapses, this enzyme is also present in many neurons that are known not to be cholinergic, so its presence does not identify any particular neurotransmitter.

The **amino acid transmitters** (GABA, glutamate, aspartate) can be demonstrated immunohistochemically despite their small molecular sizes (see Chapter 17), and antibodies are also available for some of their synthetic and catabolic enzymes. The many **peptides** secreted by neurons can also be demonstrated immunohistochemically without difficulty.

18.9. Miscellaneous histochemical methods

In addition to the empirically derived metallic methods discussed in Sections 18.3, 18.4 and 18.6, several histochemical techniques of known specificity have been found, usually by chance, to be useful for certain components of nervous tissue. A few of these are briefly considered below.

Lectin histochemistry (see Chapter 11) can display certain cells because of carbohydrates they contain. For example, microglial cells bind RCA_1, which attaches most strongly to galactosyl residues (Debbage *et al.*, 1981; Mannoji *et al.*, 1986). Other galactose-binding lectins demonstrate a population of small neurons in sensory ganglia (Streit *et al.*, 1985). SBA, which binds to β-galactosyl and *N*-acetylgalactosaminyl residues, has affinity for the subneural apparatus of the neuromuscular junction (Sanes & Cheney, 1982) and also for the primary neurons of the olfactory system and their axons (Key & Giorgi, 1986). In peripheral nerves, the affinity of lectins for certain axons is less conspicuous than the staining of endoneurial blood vessels and perineurial cells (Gulati *et al.*, 1986), which are usefully, though not selectively, stained by several lectins.

An enzyme that occurs in some populations of neurons but which is not related in any obvious way to synaptic transmission, is **carbonic an-**

hydrase. This enzyme has been found in primary olfactory neurons (Brown *et al.*, 1984) and in some large neurons in sensory ganglia (Kazimierczak *et al.*, 1986), including those that provide the sensory innervation of muscles (Peyronnard *et al.*, 1988).

Enzymes that catalyse the hydrolysis of adenosine triphosphate (**ATPases**) are ubiquitous in nervous tissue, but present at high concentration in microglial cells (Ibrahim, Khreis & Koshayan, 1974) and in unmyelinated cutaneous sensory nerve fibres (Ide & Sato, 1980). They are demonstrated by precipitation of the released phosphate ions as their insoluble lead or cobalt salts, which can then be converted to darkly coloured sulphides. (See also methods for phosphatases in Chapter 15.)

Among methods for the oxidoreductases, strong staining for **NADH diaphorase** (see Chapter 16) can be obtained in the neurons of autonomic ganglia (Gabella, 1968; Scheuermann & Stach, 1985), and has often been used in quantitative studies, especially with whole mounts (Gabella, 1971; Chiang & Gabella, 1986). Neurons in whole-mounts can also be stained with **cuprolinic blue** in the presence of a high concentration of magnesium ions (see Chapter 6). Used in this way, the dye stains the RNA of the Nissl substance and nucleolus (Heinicke *et al.*, 1987). The enzyme **NADPH diaphorase** occurs in some central neurons, and histochemical methods provide a Golgi-like staining of the somata and dendrites (Hope & Vincent, 1989; Mizukawa *et al.*, 1989). The functional significance of this neuronal enzyme is unknown.

Deposits of coloured formazan are formed in some nerve fibres and endings in tissue incubated in a **tetrazolium salt** without reduced NADH. The innervation of amphibian muscles can be demonstrated by this method (Letinsky & Decino, 1980; Letinsky, 1983), as can some degenerating boutons in the brain (Steward *et al.*, 1973). Presumably the formazan is produced by non-enzymatic reduction of the tetrazolium salt.

D-Amino acid oxidase is present in certain neuroglial cells of the cerebellum, which can be shown by demonstrating the enzyme histochemically (Horiike *et al.*, 1987). The hydrogen peroxide released when a D-amino acid is oxidized is made the substrate of horseradish peroxidase in the incubation medium. The peroxidase catalyses the oxidation of DAB at the substrate, in the presence of nickel ions (see Chapter 16).

Histochemical methods for substances other than enzymes can also be used as stains for nervous tissue. Thus, the solvent dye **Sudan black B** (see Chapters 5, 12) is useful for showing neural elements in whole-mounts (Filipski & Wilson, 1985; Nishikawa, 1987). This dye can stain some lipids other than those of myelin, because it works with the mammalian enteric nervous system (Goyal & Sengupta, 1986), which does not contain myelinated fibres. A sensitive histochemical test for metals with insoluble sulphides, **Timm's sulphide-silver method** detects the small amounts of zinc in certain presynaptic terminals, notably in the hippocampal formation of the mammalian brain (see Haug, 1973).

18.10. Axonally transported tracers

Proteins are synthesized principally in the cell-bodies of neurons, and those needed in the processes are transported within the cytoplasm. Cytoskeletal proteins are moved slowly (about 1mm per day) along the axon, but other substances are transported rapidly (1–2mm per hour). The rapidly transported materials include enzymes of neurotransmitter metabolism and components of the cell membrane. Rapid transport also occurs retrogradely, from synaptic boutons towards the cell-body.

The first neuroanatomical tracing method based on axonal transport (Goldberg & Kotani, 1967; Lasek *et al.*, 1968) made use of **radioactive amino acids** injected into a region containing neuronal cell-bodies. The tracer molecules were incorporated into protein and transported to the synaptic boutons, where they could be detected by autoradiography. This technique is still used, though anterograde tracing of exogenous protein tracers (see below) is usually preferred.

Kristensson *et al.* (1971) discovered that **exogenous proteins** injected into muscles were taken up by the axonal terminals and retro-gradely transported to the cell-bodies of motor neurons. The proteins they used were fluor-escently labelled albumin and horseradish per-oxidase. The latter was made visible by a simple histochemical method (see Chapter 16), and the "HRP method" of retrograde axonal tracing quickly became the most used method of experi-mental neuroanatomy. With the development of sensitive histochemical methods for peroxidase (see Chapter 16), it soon became apparent that small amounts of the exogenous enzyme were taken up by cell-bodies, and transported to the terminal parts of axons. Thus, anterograde and retrograde axoplasmic transport of the same tracer could be exploited.

Two significant improvements have been made since the mid-1970s. The first has been the use of **lectins** covalently coupled to HRP (Gonatas *et al.* 1979; Trojanowski, 1983). Sensitivity is increased, especially for anterograde tracing, and the amount of protein needed is much less than when HRP is used alone. Unlabelled lectins can also be used, and demonstrated immunohisto-chemically. In a variant of this method, known as **suicide transport**, a toxic lectin is used (Yama-moto *et al.*, 1983, 1984). This kills the neurons in whose cell-bodies it accumulates, and permits functional studies of the effects of ablation of populations of neurons with specific, known con-nections.

The other technical advance has been the use of **axonally transported fluorochromes** instead of proteins. The first to be used were Evans blue and primulin (Kuypers *et al.*, 1977), but these have been superseded by cyanine dyes that are more intensely fluorescent (see Aschoff & Hol-lander, 1982). Fluorescent dyes with differently coloured emissions can be injected into two sites in the nervous system of the same animal. Any neurons that send branches of their axons to both sites are then seen to contain both tracers. The most serious difficulties with fluorochromes and other axonally transported tracers are uptake by unintentionally injured fibres and diffusion from the site of injection. Both may give spurious labelling of neurons that do not project to the intended site of application.

18.11. Filling methods for neurons

Neurophysiologists frequently make electrical recordings from neurons that have been impaled by micropipettes. At the end of such an experi-ment, a visible tracer can be transferred from the pipette to the cytoplasm of the cell by injection or iontophoresis. Suitable intracellular tracers include fluorescent reactive dyes (Stretton & Kravitz, 1968), HRP, salts or complexes of cobalt or nickel (Kater & Nicholson, 1973; Hackney & Altman, 1982; Springer & Prokosch, 1982; Fred-man, 1987), and biocytin (Horikawa & Arm-strong, 1988). The metals are precipitated as their black sulphides, and the deposits can be intensified by physical development (see Section 18.4.1.1). Biocytin is a derivative of biotin, and is made visible by virtue of its specific affinity for avidin (see Chapter 19).

Intracellular tracers fill the cell-body and den-drites of the neuron, and can spread for several millimetres along the axon. They can pass through gap junctions, when these are present between adjacent cells. It is also possible to fill neurons retrogradely, by such manoeuvres as immersing the central stump of a cut nerve in a solution of the tracer and waiting for a few days. Neurons do not have to be alive to be filled. It is even possible to demonstrate short projections in the human brain by appropriate applications of HRP to *post mortem* material (Beach & McGeer, 1988; Haber, 1988).

Neuronal tracing *in vivo* and *post mortem* can also be done with fluorescent hydrophobic car-bocyanine dyes (Honig & Hume, 1986, 1989). These enter the lipid bilayer of the plasmalemma and spread laterally within the membrane, thus delineating axons and other neurites for distances up to about 3 mm.

18.12. Exercises

1. Which of the staining methods discussed in this chapter would be suitable for:

(a) Study of the sizes and shapes of neurons in the six layers of the cerebral cortex?

(b) Study of the extents of the dendritic trees of neurons?

(c) Examination of cell groups and major tracts in the brain stem, at low magnification?

(d) Comparison of the lengths and thicknesses of astrocytic processes in normal and pathological brains?

2. Why would Sudan IV (see Chapter 12) be unsatisfactory as a stain for myelinated nerve fibres in frozen sections?

3. Why does the developer used in Holmes's method for axons contain sodium sulphite?

4. If, in processing an exposed photographic film, you accidentally fixed it (in sodium thiosulphate) when you should have developed it, what could you do to obtain printable negatives? (If the fixer used is a simple aqueous solution of $Na_2S_2O_3$ it is possible to do this. An acid-fixing-hardening solution, which is normally used in photography, will dissolve the latent image and the negative is then irretrievable.)

5. What methods are available for the demonstration of unmyelinated autonomic axons in tissues such as the skin or viscera? Evaluate the advantages and disadvantages of each technique.

6. Show that the neurosecretory axons in the posterior lobe of the pituitary gland are unmyelinated.

19

Immunohistochemistry and Other Methods Based on Affinity

In the preceding chapters attempts have been made to show how the principles of chemistry and biochemistry are applied for the localization of substances within tissues. This chapter is concerned mainly with methods based on the precepts of immunology. Students totally unfamiliar with the science are advised to read an introductory text such as Barrett (1987), Roitt *et al.* (1989), or the relevant parts of a textbook of pathology. Fortunately the immunological principles applicable to microscopical techniques are fairly easy to grasp. Valuable works of reference in the field of immunocytochemistry are the books by Goldman (1968), Nairn (1976) and Sternberger (1979), and the multi-author treatise edited by Polak & Van Noorden (1986). Immunohistochemical methods are employed in many fields of biological research, and several are regularly used in diagnostic histopathology. Many of the required reagents, which once had to be made in the laboratory, are available commercially.

Immunohistochemical staining is based on **affinity** between antigens and antibodies. Affinity is the attractive force between molecules that causes them to join together and stay joined. In the case of immunological reactions, the forces of affinity are neither covalent nor electrostatic; they therefore resemble the binding of many dyes by their substrates (see Chapter 5, p. 55), but are stronger and much more specific. Comparable affinity exists between lectins and the sugars to which they bind (see Chapter 11). There are histochemical techniques that exploit biologically important affinities other than those of lectins and antibodies. Some of these methods, including the hybridization of nucleic acids, are discussed in the later sections of this chapter.

19.1. Antigens and antibodies

Vertebrate animals are able to defend themselves against the potentially harmful effects of macromolecules derived from other organisms. Any such foreign material that may enter an animal's body is called an **antigen**. When an antigen is introduced into the *milieu intérieur* of an animal, some of its molecules are carried, by lymph or blood-vessels, to lymph nodes. Here the antigen molecules come into contact with small lymphocytes of the B-type (bone-marrow derived), which react to the encounter by transforming themselves into plasma cells. The plasma cells synthesize and secrete **antibodies**. These are proteins of the γ-globulin class capable of combining specifically with the antigens that evoked their production. It will be noticed that the words "antigen" and "antibody" are difficult to define. An antibody is a substance produced in response to the presence of an antigen. The antibodies secreted by plasma cells in lymphoid tissues circulate in the blood plasma and so gain access to all parts of the body including the original sites of introduction of antigens. The combination of an antibody with its antigen commonly results in neutralization of the toxicity or pathogenicity of the latter.

When a molecule of antigen bumps into a molecule of its antibody, the two combine to form an **antigen-antibody complex**. This happens because part of the antibody molecule has been specially tailored to accommodate part of the antigen molecule. The reaction is reminiscent of the "lock-and-key" mechanism by which an enzyme combines with its substrate. The two components of the complex are held together by non-covalent forces such as ionic attraction, hydrogen bonding, and hydrophobic interaction.

The part of an antigen molecule that joins it to the antibody is known as an **antigenic determinant** or **epitope**. A large protein molecule, or an object such as a bacterium that consists of many different macromolecules, will have many antigenic determinants and will therefore evoke the synthesis of many different species of antibody molecule, all of which will be capable of combining with the same antigen. It is, however, a property of all antibodies that they form complexes only with the antigens that stimulated their production.† This specificity is fundamental to the techniques of immunohistochemistry: no purely chemical reactions can be used to identify individual macromolecular substances.

19.2. Antibody molecules

The antibodies circulating in the blood belong to the γ-globulin class of plasma proteins, and are known as **immunoglobulins**. The most abundant type is immunoglobulin G (IgG). This is not the only kind of immunoglobulin. The other types, however, are relatively unimportant to the immunohistochemist. The IgG molecule consists of two identical subunits joined by a disulphide (cystine) bridge. Each subunit comprises two polypeptide chains: a long one, the heavy (H) chain, and a short one, the light (L) chain. The H chain is joined to the L chain by a disulphide bridge. The structure of IgG is shown diagrammatically in Fig. 19.1. It can be seen that the molecule is Y-shaped. The stem of the Y and the proximal parts of the arms have the same amino-acid composition in all the IgG molecules of a given species of animal. Specificity for antigens resides in the distal (variable) part of the H and L chains constituting the two limbs of the Y. Each limb is capable of combining with an antigenic determinant and is therefore called an antibody fragment or F_{ab}. The stem of the Y, consisting of parts of both H chains, is called the constant fragment or F_c. The F_{ab} and F_c fragments can be isolated by collecting the products of digestion of IgG by papain, a proteolytic enzyme. Other

---

†This statement is not strictly true. Two closely similar proteins may react with an antibody raised by only one of them. Such cross-reactivity occurs when the antigenic determinant site is a sequence of amino acids common to both the proteins. Cross-reactivity is sometimes a source of confusion in the interpretation of immunohistochemically stained preparations because it cannot be detected by the usual controls for specificity. The subject is discussed by Swaab *et al*. (1977), Hutson *et al*. (1979), and Vandesande (1979), but will not be further considered in this introductory account.

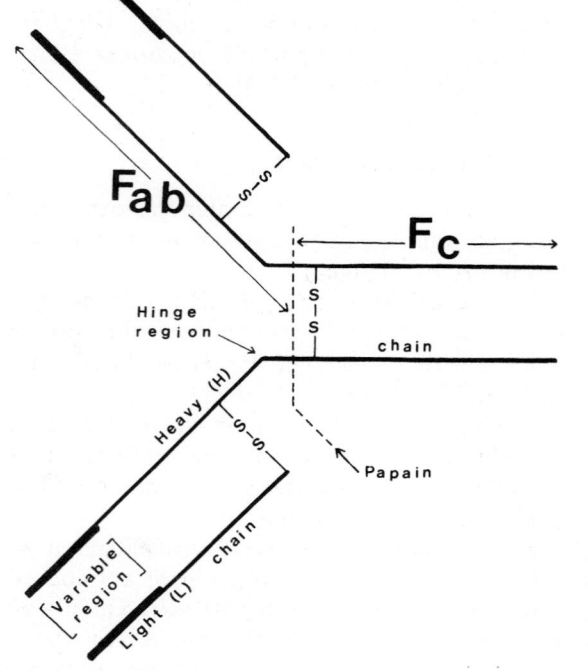

FIG. 19.1. Diagram showing the structure of one molecule of immunoglobulin G (IgG). The antibody-combining sites are the variable regions of the two F_{ab} segments. When IgG serves as an antigen in immunohistochemical techniques it is the F_c segment of the molecule that is the antigenic determinant. Each IgG molecule is potentially capable of combining with two specific antigenic determinant sites. The distance between these two sites may vary because there is some flexibility at the hinge region of the H chains. Note that papain releases one F_c and two F_{ab} fragments from each IgG molecule. Cleavage of the disulphide bridges yields two H chains and two L chains. Oligosaccharides are attached to the H chains in the F_c segment.

fragments have been isolated from the products of attack by other enzymes and by disulphide-splitting reagents.

From the point of view of the immunohistochemist, the IgG molecule has three important features:

(i) There are two sites, the ends of the F_{ab} segments, capable of binding to an antigenic determinant. Thus, the antibody molecule is **bivalent.**

(ii) Part of the IgG molecule, the F_c fragment, is common to all antibodies of the animal species concerned and is not involved in combining with antigens.

(iii) Immunoglobulins are themselves macromolecules and as such can behave as antigens when injected into different species of animals. Because the F_c fragment has a constant chemical structure, it is possible to raise in one species antibodies against all the possible immunoglobulins that might be produced by another species of animal. Such antibodies (contained in anti-γ-globulin antisera) are important immunohistochemical reagents. As an antigen, the IgG molecule may have several antigenic sites and be able to bind more than one molecule of anti-antibody.

Antibody molecules are large: the M.W. of IgG is 150,000. For this reason it is possible to conjugate some of their amino-acid side-chains with other compounds such as fluorochromes or enzymes. This process, known as **labelling**, usually involves the formation of covalent linkages with the ε-amino groups of lysine. The combining properties of an antibody will remain intact provided that the labelling molecule does not sterically obstruct the specific "keyholes" formed by the distal parts of the F_{ab} segments. In solution, the molecules of a labelling agent react randomly with different parts of the molecules of IgG, so some blocking of the combining sites is inevitable. Consequently the potency of an antibody-containing serum is always reduced by combination with a fluorescent or enzymatic label.

19.3. Antigen-antibody complexes

When a solution containing an antigen is mixed with a solution containing appropriate antibodies, the antigen-antibody complex often precipitates. Insoluble aggregates are formed because each antibody molecule is bivalent and each antigen molecule has multiple determinant sites. If either component of the mixture is present in excess, the aggregates will be smaller and will remain soluble.

If an antigen forms part of a solid structure, such as a cell in a section of a tissue, it is able to bind antibody molecules from an applied solution:

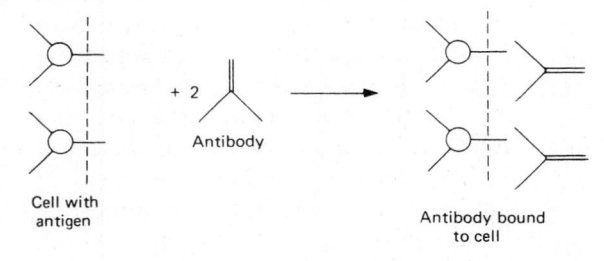

The antibody is quite firmly attached and is not removed when the excess solution is washed off. Similar attachment occurs when a dissolved antigen is brought into contact with an object such as a plasma cell which contains appropriate antibodies:

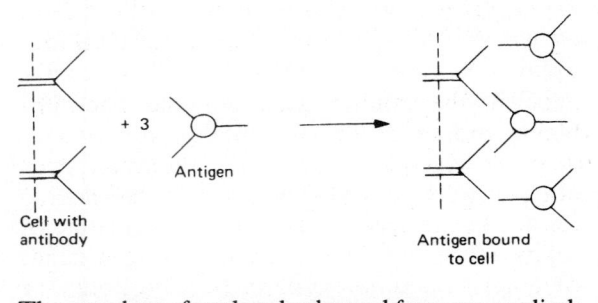

The number of molecules bound from an applied solution depends on the abundance of combining sites in the specimen. Steric factors are also important: if in the above example the two antibody molecules were further apart, they would be able to unite with four antigen molecules.

It is possible to label antibodies by conjugating them with fluorochromes or with histochemically demonstrable enzymes. Consequently, if a solution of a labelled antibody is applied to a section of tissue containing the appropriate antigen, the label will be detected at those sites at which antigen-antibody complexes have been formed. Since antigens and antibodies combine specifically with one another, it ought to be possible to localize individual macromolecular substances under the microscope. The techniques of immunohistochemistry are based on the use of labelled antibodies, but the methodology is, unfortunately, not quite as simple as might at first be thought. The more important techniques will be discussed in Sections 19.5–19.9.

19.4. Some definitions

The understanding of immunohistochemical methods will be easier if some commonly used terms are first defined.

Immunocytochemistry. Widely used as synonym for immunohistochemistry but strictly applicable only when cells or intracellular structures are the objects of interest.

Immunofluorescence. Secondary fluorescence introduced into tissues by the application of immunohistochemical methods in which fluorochromes are used as labels.

Immunization. Administration of antigen to an animal to evoke the production of antibodies. Originally the word was applied to the induction of immunity to infectious diseases. It now enjoys wider usage. The act of injecting antigenic material is often called **inoculation**.

Serum. Blood plasma from which the fibrinogen has been removed. Mammalian sera contain about 8% w/v protein, consisting of approximately equal proportions of albumin and globulin.

Antiserum. Serum containing antibodies to an antigen. Loosely used also for diluted sera and diluted solutions of the globulin fraction.

Globulin. The proteins remaining in serum after removal of the albumin by chromatographic or other separation methods. The γ-globulin fraction, comprising the immunoglobulins, constitutes approximately one-third of the total globulin. The concentration of γ-globulin in whole serum is 7–15mg per ml.

Monoclonal antibody or **MAB**. A single species of antibody globulin, produced by the cultured descendents of a single antibody-producing cell. A MAB recognizes a specific epitope, which may be quite a short sequence of amino acids.

Label. A molecule artificially attached to a protein. The usual labels are either fluorescent substances or enzymes for which simple, reliable histochemical methods are available. Horse-radish peroxidase (HRP) is the most popular label of the latter type. Ordinary dyes and radio-active isotopes are also used to label proteins, though mainly for applications other than immunohistochemistry.

Saline. Water is unphysiological and alcohol denatures proteins, so all immunological materials are dissolved in 0.9% sodium chloride, which is usually dissolved in 0.03M phosphate buffer, pH7.4. Phosphate-buffered saline is commonly called **PBS**. When used as a solvent in immunohistochemical procedures, saline may optionally contain a surfactant and a protein, as explained in Section 19.13.

Absorption. (i) Neutralization of specific anti-body in an antiserum by adding an excess of the appropriate antigen. (ii) Removal of unreacted fluorochrome from a solution containing conju-gated proteins by treatment with activated char-coal. (iii) Treatment of a labelled antiserum with powdered acetone-fixed tissue, such as liver or kidney, in order to remove unwanted labelled proteins that bind to tissues by non-immune mechanisms. The tissue powder must not, of course, contain the antigen to which the anti-serum was raised.

19.5. Direct fluorescent antibody methods

These are the simplest and oldest immuno-histochemical methods. A known antigen in a section or smear of tissue is localized by virtue of its combination with fluorescently labelled mol-ecules of its antibody. The technique has several shortcomings and is not often used, but it will be described in some detail in order to introduce principles that also apply to the more compli-cated modern methods.

19.5.1. The technique

It is first necessary to isolate the antigen and purify it as much as possible. An animal (of a species other than that from which the antigen was taken, if the antigen is of animal origin) is then immunized by injecting it, usually on several occasions, with the purified antigen. When the animal has a high circulating level of antibody, blood is collected and the serum, an antiserum, is separated. If possible, the globulin or, prefer-ably, the γ-globulin, is isolated from the anti-serum. The techniques for purification of antigen, immunization, and isolation of serum proteins require some expertise in practical bio-chemistry and immunology. Apparatus not usually available in a general histology laboratory is also needed. Very detailed descriptions of most of the practical procedures used in immunology are given by Garvey et al. (1977), but that book includes only a brief account of immunohisto-chemical methodology. Many pure antigenic sub-stances and antisera are now available commercially.

The antiserum must next be labelled by conju-gating its proteins with a fluorescent dye. The most widely used label is fluorescein isothiocyan-ate (FITC). In alkaline solution (pH9–10), FITC combines covalently with proteins, reacting prin-cipally with the ε-amino group of lysine:

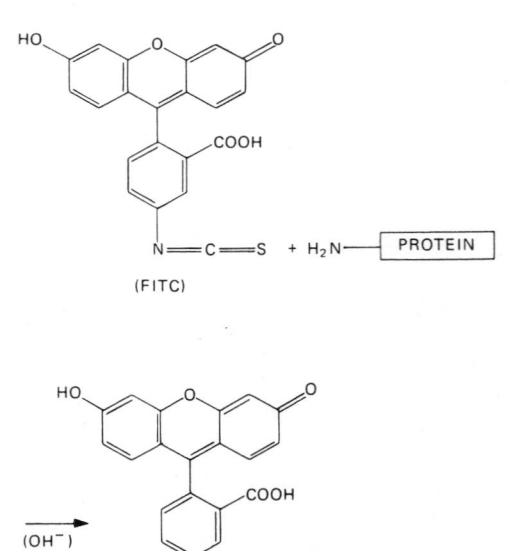

(FITC)

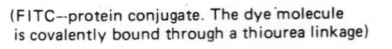

(FITC–protein conjugate. The dye molecule
is covalently bound through a thiourea linkage)

Other fluorochromes are also used. The reactive group is usually isothiocyanate, sulphonyl chloride (—SO$_2$Cl), or dichlorotriazinyl (see p. 51). The most popular alternatives to FITC are 1-dimethylaminonaphthalene-5-sulphonyl chloride (dansyl chloride) and various reactive derivatives of rhodamine B and related dyes.

The fluorochrome will, of course, conjugate with all the types of protein present in the serum, not just with the antibodies. Unbound fluorescent material will also be present in the conjugated serum because some hydrolysis of FITC occurs in alkaline solutions:

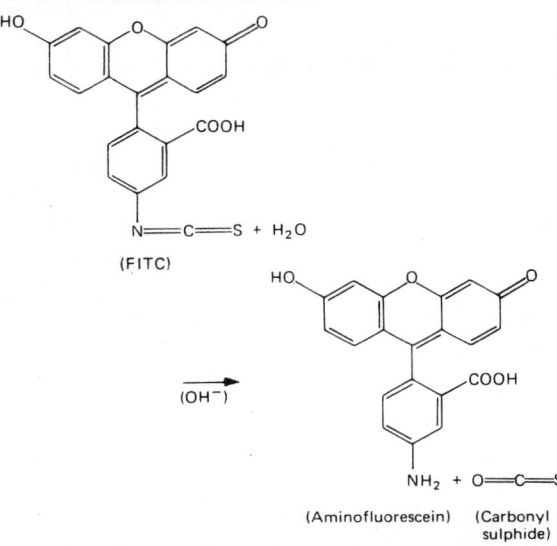

(FITC)

(OH⁻)

(Aminofluorescein) (Carbonyl sulphide)

The unconjugated fluorescent material must be removed from the serum since it would be capable of staining tissues in its own right. The serum is therefore dialysed to remove small molecules (M.W. < about 5000) and to replace the alkaline solvent with a saline solution of physiological pH. It may also be passed through a gel-fitration column, which retains small molecules, or it may be treated with activated charcoal, which does much the same thing. Sometimes all three methods are used. Finally, the serum is concentrated by dialysis against an inert synthetic polymer such as polyvinylpyrollidone. The labelled antiserum is now ready for use. Practical instructions for preparing labelled proteins are given by Pearse (1980) and Nairn (1976).

Sections of the antigen-containing specimen are cut and mounted onto slides or coverslips. The tissue must not have been fixed or otherwise altered by processing to an extent sufficient to destroy or distort the antibody-binding sites of the antigen. Some degree of fixation is desirable, however, to prevent disintegration of the sections during the staining procedure. Commonly fresh tissue is cut on a cryostat and the sections are fixed for a few minutes in alcohol, acetone, or neutral, buffered formaldehyde. Cross-linking fixatives such as formaldehyde and glutaraldehyde are more likely to destroy antigenicity than are simple coagulants. Salts of heavy metals are usually avoided because fluorescence can be quenched in the presence of elements with high atomic numbers. Many antigens survive embedding in wax but some do not. The optimum treatment for every tissue and antigen has to be found by trial and error.

Some soluble antigens, including serum proteins, may diffuse from their normal loci in the tissue during the course of freezing or cutting of unfixed material, giving rise to artifactual false-positive localizations in the stained preparations (Sparrow, 1980). Errors from this cause should be suspected when an antigen is detected in an unexpected part of a tissue, especially when one of the more sensitive immunohistochemical techniques, such as the unlabelled antibody–enzyme method, is used.

The sections are now rinsed with saline and a drop of labelled antiserum is placed on each. Various twofold dilutions of the serum in saline are tried. Dilutions weaker than 1 in 8 are unlikely to be needed for direct fluorescent antibody methods. The serum is left in contact with the sections, in a humid atmosphere, for approximately 1h. The serum is then washed off with saline and a coverslip is applied. Immunofluorescent preparations are usually mounted in a mixture of glycerol and phosphate buffer, pH7.4–7.6. Permanent mounting media are likely to weaken the intensity of the fluorescence.

The stained, mounted sections are examined by fluorescence microscopy. For **FITC conjugates** the **optimum wavelength** for excitation is

490nm (blue). The maximum emission occurs at 550nm (green-yellow). Exciting light of 320nm (in the ultraviolet) may also be used, but the emission will be less intense. However, with ultraviolet excitation and a colourless barrier filter it is easier to distinguish between the specific emission of FITC and the blue autofluorescence of the specimens. When blue exciting light is used, a yellow or orange barrier filter is necessary and the autofluorescence appears to be green. This colour is difficult to distinguish from the specific emission of FITC. **Rhodamine B derivatives** are optimally excited by green light (546nm) and have orange–red emissions (580–600nm). Dansylated proteins absorb in the ultraviolet and emit blue-green light.

19.5.2. Controls

The following control procedures must be carried out alongside the definitive staining technique:

1. Incubate in saline alone. The only fluorescence seen will be autofluorescence.
2. Incubate in a fluorescently labelled non-immune serum derived from the same species as that in which the antiserum was raised. Any fluorescence seen in this control but not in No. 1 will be due to non-specific attachment of fluorescent proteins to the section. Absorption of the labelled antiserum with a tissue powder (see p. 334) may help to eliminate this artifact.
3. Incubate with labelled antiserum an identically processed section known not to contain the antigen. This is only possible when the antigen is one foreign to the tissue (e.g. bacteria, foreign protein). The specimens for this control should be of the same tissue as that used for the definitive procedure.
4. Mix some labelled antiserum with a solution in saline of the purified antigen. A large excess of the latter should be present in the mixture. The added antigen combines with the antibody and little or no specific fluorescence should be seen in a section stained with this absorbed antiserum.

19.5.3. *Shortcomings of the direct method*

With the direct fluorescent antibody method it is not possible to localize antigens present at low concentrations in a tissue. There are two reasons for this lack of sensitivity:

1. The process of conjugation with FITC or any other reactive fluorochrome will block the combining sites of some of the antibody molecules, reducing the potency of the antiserum. The likelihood of attachment of label to the combining site will be reduced if the concentration of the labelling agent is low, but then the number of fluorescent groups usefully attached to antibody molecules will also be less. An ideally labelled antiserum has about ten fluorochrome molecules attached to each molecule of IgG.

2. The immunological reaction in the direct method is

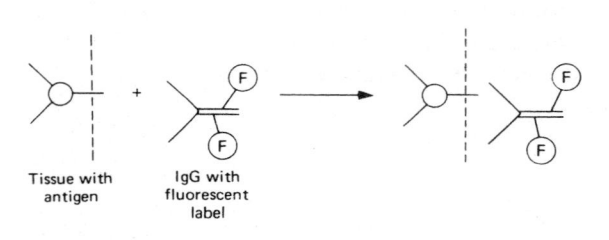

Tissue with antigen   IgG with fluorescent label

Thus only one molecule of antibody, with its attached fluorochrome, combines with each antigenic determinant site in the tissue. If the concentration of antigen is low, the density of bound fluorochrome molecules may not be high enough to permit their detection under the microscope. In other methods it is possible to accumulate greater numbers of molecules of visible substances at each antigenic site.

Another objection to this technique arises from the necessity to label a specific antiserum. Antisera are troublesome to prepare and often available only in small quantities. The conjugation also requires time and effort. An investigation of the localizations of several antigens would demand the preparation of as many labelled antisera, and this would entail an inordinate amount of work. In the methods to be described next, a single labelled antiserum will render visible the sites of localization of any number of different antigens.

19.6. Indirect fluorescence techniques

With indirect methods it is possible to identify either antigens or antibodies (see Section 19.9.1) in a tissue. The reagent common to the procedures of this type is a fluorescently labelled antiserum to immunoglobulin: an anti-antibody. The method for demonstration of a tissue antigen will illustrate the principle of the method.

19.6.1. Method for detection of an antigen

A section containing the antigen is first incubated for 30–60 min with a suitable dilution (commonly 1 in 20 or 1 in 50) of an appropriate antiserum. This antiserum, which is **not** labelled, is called the **primary antiserum:**

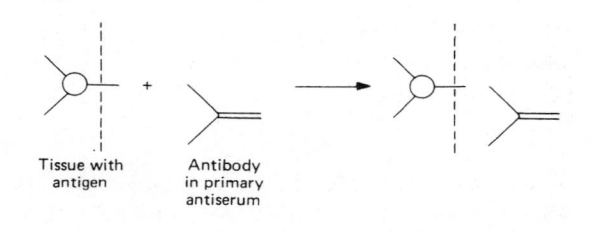

Tissue with antigen   Antibody in primary antiserum

The antibody molecules which attach to the antigenic sites in the tissue are immunoglobulins. The F_c segments of their molecules are the same as those of all other immunoglobulins of the same species and can themselves serve as antigens. It is possible to raise an anti-immunoglobulin antiserum by injecting into another species of animal some γ-globulin from the blood of the species that was the source of the primary antiserum. This **secondary antiserum** may be labelled by conjugating it with a fluorescent dye.

In the next stage of the staining method, the section is treated with fluorescently labelled secondary antiserum. The anti-IgG molecules in it will bind to the F_c segments of the already attached immunoglobulin molecules derived from the primary antiserum:

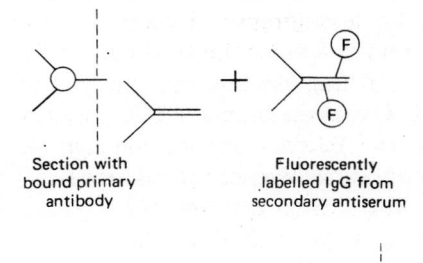

Section with bound primary antibody   Fluorescently labelled IgG from secondary antiserum

Thus the fluorochrome will be seen at the sites occupied by the antigen.

A specific example will further clarify the technique. Let us suppose that a guinea-pig has been given an intravenous injection of egg albumen. Several hours later the foreign protein will be present in the renal tubules. We have available an antiserum to egg albumen raised in a rabbit. This is **rabbit antiserum to egg albumen** (or "rabbit anti-egg albumen"). We also have a labelled antiserum to rabbit immunoglobulin. The latter antiserum was raised in a goat, so it is **goat antiserum to rabbit immunoglobulin** (often also known as "goat anti-rabbit γ-globulin"). Cryostat sections of one of the guinea-pig's kidneys are briefly fixed in alcohol and then treated as follows:

1. Rinse in saline.
2. Apply rabbit antiserum to egg albumen. Use serial twofold dilutions in saline, from 1 in 10 to 1 in 160. Leave in contact with the sections for 30–60 min at room temperature in a humid atmosphere.
3. Rinse sections with three changes of saline.
4. Apply labelled goat antiserum to rabbit immunoglobulin (try 1 in 10 and 1 in 40 dilutions) for 30–60 min.
5. Rinse sections with three changes of saline.
6. Mount in buffered glycerol (see Chapter 4) and examine by fluorescence microscopy.

The sera are diluted in order to minimize non-specific adherence of protein molecules to the sections. This could result in fluorescence at sites other than those containing the antigen. Generally the dilution of the primary antiserum is more critical than that of the labelled secondary antiserum. Diluted sera containing 0.1–1.0mg of protein per ml are often the most satisfactory for indirect immunofluorescence methods.

19.6.2. Controls

As with the direct fluorescent antibody method, control procedures are necessary. These are:

1. Omit treatment with both sera to reveal the autofluorescence of the tissue.
2. Omit treatment with the primary antiserum. Incubate with saline instead. Any fluorescence seen cannot be due to the antigen and is probably autofluorescence or a result of non-specific binding of fluorescently labelled proteins from the secondary antiserum.
3. In place of the primary antiserum, use non-immune serum from the same species as that in which the primary antiserum was raised. If fluorescence is seen that does not appear in controls 1 and 2, it is possible that the primary antiserum used in the definitive method is being bound to the tissue by a non-immune mechanism.
4. Mix the primary antiserum with an excess of the purified antigen. Absent or greatly diminished immunofluorescence should be obtained after application of this absorbed antiserum.

19.6.3. Critique of the method

This technique is superior to the direct method for two reasons:

1. It is not necessary to label and thereby reduce the potency of the valuable primary antiserum. Labelled secondary antisera can be obtained commercially and are potentially useful for the detection of an infinite number of antigens.

2. The determinant sites of antigens are multiple (see p. 333). This is true of the antigen in the tissue and of the F_c segments of the antibody molecules derived from the primary antiserum. Consequently, more than one molecule of fluorescent antibody from the secondary antiserum will be able to attach to each molecule of the primary antibody. For example, two suitably spaced antigenic determinant sites in a section will bind two antibody molecules from a primary antiserum:

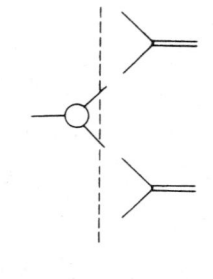

If each of these bound antibody molecules also has two antigenic determinant sites on its F_c segment, four molecules of fluorescent anti-antibody will be bound:

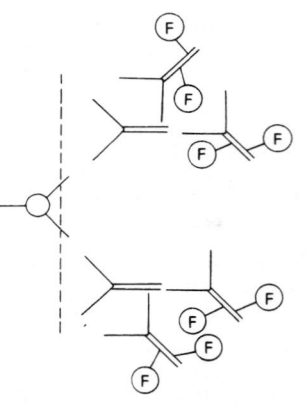

In this example, each antigen molecule has accreted twice as many fluorescent molecules as would have been possible had a direct immunofluorescent method been used. In reality the amplification factor may be much greater than 2, so indirect immunohistochemical techniques are always more sensitive than direct ones.

Even so, some antigens occur in tissues in quantities too small to be detected by the indirect fluorescent antibody method. Greater sensitivity is achieved when the label is a histochemically demonstrable enzyme rather than a fluorochrome.

The indirect immunofluorescence method is very versatile, and technically quicker than the enzymatic procedures discussed in the later sections of this chapter. By using primary antisera raised in different species, it is possible to localize as many as three antigens in the same section, by applying secondary antisera conjugated with different fluorochromes (Staines *et al.*, 1988).

19.7. Enzyme-labelled antibody methods

In these techniques antibodies are labelled by conjugation with enzymes. The enzyme currently most in favour is horseradish peroxidase (HRP). Simple and reliable histochemical methods are available for the localization of peroxidases and some of them yield products that can be detected by both light and electron microscopy (see Chapter 16).

Various methods of conjugation are feasible. A simple one consists of mixing solutions containing the antibody and HRP and then adding a little glutaraldehyde. Cross-linking of protein molecules occurs by the same chemical reactions as those involved in the fixation of tissues by glutaraldehyde (see Chapter 2). After reaction the solution can be expected to contain the following:

Unreacted glutaraldehyde
Unreacted HRP
Unreacted antibody
Unreacted other proteins
(Antibody)—$(CH_2)_5$—(Antibody)
(HRP)—$(CH_2)_5$—(HRP)
(Antibody)—$(CH_2)_5$—(HRP)
(Other proteins)—$(CH_2)_5$—(Other proteins)
(Other proteins)—$(CH_2)_5$—(HRP)

Larger aggregates derived from three or more protein molecules form a precipitate, which is removed by centrifugation. Any excess of glutaraldehyde is removed by dialysis or gelfiltration. Some of the double molecules (Antibody)—$(CH_2)_5$—(HRP) have the structure

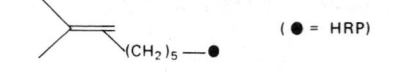

with the F_c segment labelled and the F_{ab} segments unhindered. This is the only useful component of the solution, but further purification is not usually necessary.

The solution containing HRP-labelled antibody is used in exactly the same way as a fluorochrome-labelled antiserum. The sites of attachment of HRP to the tissue are detected histochemically, usually by means of the DAB-hydrogen peroxide reaction (see Chapter 16). Enzyme-labelled antibodies may be used in direct or indirect techniques. The latter are generally preferred, for the reasons set out in Section 19.6.3 of this chapter. The bound HRP is capable of catalysing the oxidation of many molecules of DAB by hydrogen peroxide with consequent accumulation of substantial amounts of the insoluble brown product. The apparent size of each antigenic site in the tissue is therefore enlarged until such time as the enzyme is inhibited by the accumulated products of its activity. The "amplification factor" is greater than can be obtained in fluorescent antibody techniques.

As with the immunofluorescence methods, control procedures must accompany the use of enzyme-labelled antibodies. Indeed, false-positive results are more likely to be seen with the latter than with the former techniques, since many tissues contain endogenous peroxidases. With the indirect enzyme-labelled antibody methods the following controls are required:

1. Omit all immunological reagents. Carry out only the histochemical method for the enzymatic label. This will reveal sites of endogenous enzyme.
2. Omit treatment with the primary antiserum. This will reveal any non-specific attachment of enzyme-labelled proteins or of conjugated enzyme. HRP has a rather

strong tendency to bind to tissues in this way.

3. Substitute non-immune serum for the primary antiserum to detect non-immunological binding of γ-globulin.
4. Treat with primary antiserum that has been absorbed with an excess of its antigen. This should result in absent or greatly diminished intensity of staining.

The advantage of this technique over the indirect fluorescent antibody method lies in its higher sensitivity. The higher probability of obtaining false-positive results is a relative disadvantage. Both methods involve the use of covalently labelled antisera, whose potencies are reduced as a result of hindrance of the combining sites of many of their specific antibody molecules.

When antigenic sites are few and far between, the indirect fluorescent antibody technique is preferred. It is easier to see isolated fluorescent spots on a dark background than to find occasional small accumulations of enzymatic reaction product against a bright background. Histochemical methods yielding fluorescent end-products are available for a few enzymes, including peroxidases (Papadimitriou et al., 1976) but have not become popular in conjunction with immunohistochemical methods. Enzyme-labelled antibodies are valuable for ultra-structural immunohistochemistry because the end-products of many histochemical reactions for enzymes, including the DAB-H$_2$O$_2$ method, can be detected by electron microscopy.

The next methods to be described are much more sensitive than the indirect enzyme-labelled antibody procedure. Chemically labelled proteins are not required, and the likelihood of false-positive results is no higher than with other techniques.

Peroxidase is the most popular enzymatic label for antibodies, but it is not the only one used. Others include **alkaline phosphatase** (Ponder & Wilkinson, 1981; Cordell et al., 1984; Hohmann et al., 1988), **β-galactosidase** (Bondi et al., 1982) and glucose oxidase (Clark et al., 1982). These enzymes can also be used in unlabelled antibody-

enzyme methods (Section 19.8) and avidin-biotin methods (Section 19.9). One advantage of having a variety of enzymes available as immunohistochemical labels is the possibility of demonstrating more than one antigen in the same section in different colours (Appenteng et al., 1986; van der Loos et al., 1987; Sakanaka et al., 1988).

19.8. The unlabelled antibody-enzyme method

19.8.1. Principle of the method

Horseradish peroxidase (HRP) can serve as an antigen when it is injected into an animal (usually a rabbit). By careful mixing of the resultant immunoglobulin with HRP it has been possible to isolate a stable antigen-antibody complex. This is known as **peroxidase-antiperoxidase (PAP)** and has the structure

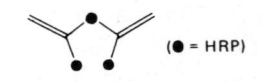

Two molecules of specific IgG are associated with three of HRP. The complex is soluble and is available commercially. Combination with antibody does not inhibit the activity of the enzyme. Each complex molecule of PAP has two F$_c$ segments, which are available for attachment to suitable anti-IgG molecules. For example, the F$_c$ segments of rabbit PAP will bind to the F$_{ab}$ segments of the specific IgG in a goat antiserum to rabbit γ-globulin. The PAP complex is thus a valuable and versatile immunohistochemical reagent for the detection of the sites of binding of anti-antibodies.

19.8.2. Procedure for detection of an antigen

Theoretical and practical aspects of the un-labelled antibody-enzyme method will be illustrated by considering the technique for localization of an antigen, X, in sections of a tissue. An antiserum to X is raised in a rabbit. The other reagents needed are goat antiserum to rabbit immunoglobulin (ideally an antiserum specific for the F_c segment of IgG), rabbit PAP, and the chemicals required for histochemical demonstration of peroxidase. Both the secondary antiserum and the PAP are available from several commercial suppliers. The procedure is as follows:

1. Apply rabbit antiserum to X (the primary antiserum) to the sections for 1h at room temperature. Several dilutions (in phosphate-buffered saline) are used, from 1 in 50 to 1 in 2000. A longer exposure (12–24h, at 4°C) is advantageous when a very dilute (1:2000–1:10,000) primary antiserum is necessary.

2. Rinse in three or four changes of phosphate-buffered saline.

3. Apply goat antiserum to rabbit immunoglobulin for 30min at room temperature. A 1 in 10 or 1 in 20 dilution of this serum is used. The concentration is not critical but it must be much higher than that of the primary antiserum or the PAP.

4. Rinse in three or four changes of phosphate-buffered saline.

5. Apply rabbit peroxidase-antiperoxidase (rabbit PAP) for 30min at room temperature. The stock solution (containing 1.0mg protein per ml) of the commercially available product is diluted 1 in 50 with phosphate-buffered saline.

6. Rinse in three or four changes of phosphate-buffered saline.

7. Incubate for demonstration of peroxidase activity (see Chapter 16). The DAB–H_2O_2 method (p. 285) is usually used with the incubation medium buffered to pH7.6.

8. Wash in water, dehydrate, clear, and mount in a resinous medium.

A three-layered "sandwich" is built up by the three reagents employed:

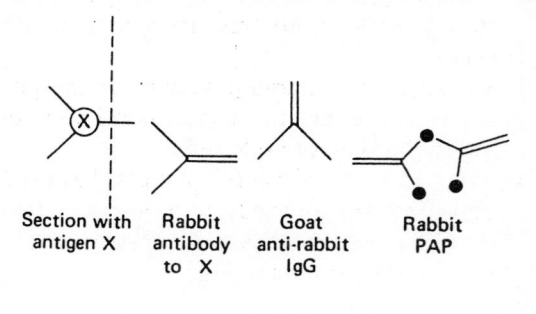

Section with   Rabbit   Goat   Rabbit
antigen X   antibody   anti-rabbit   PAP
                    to X   IgG

Because the goat antibody to rabbit IgG is bivalent, its molecules attach to the F_c segments of the primary rabbit anti-X antibodies **and** to those of the rabbit antiperoxidase antibodies in the PAP complex. The products of the histochemical reaction for HRP therefore accumulate at the sites of the antigen X.

The intensity of the colour of the end-product is influenced by the dilution of the primary antiserum. When the antigen is present at high concentration in the tissue, the density of the final product is higher after the application of more dilute solutions of the primary antiserum. The probable reason for this apparently paradoxical effect is given in Fig. 19.2.

19.8.3. Controls

The following control procedures should accompany any investigation in which the PAP method is used for the localization of an antigen in a tissue:

1. Omit treatment with both the antisera and the PAP. Incubate for peroxidase to demonstrate the endogenous enzymatic activity of the tissue.

2. Omit treatment with the primary antiserum. Any staining then observed cannot be due to the antigen.

3. Substitute a non-immune serum for the primary antiserum (e.g. normal rabbit serum, in the method for antigen X described above). This control will reveal any non-immunological attachment of γ-globulins to the tissue.

4. Omit treatment with both the primary and

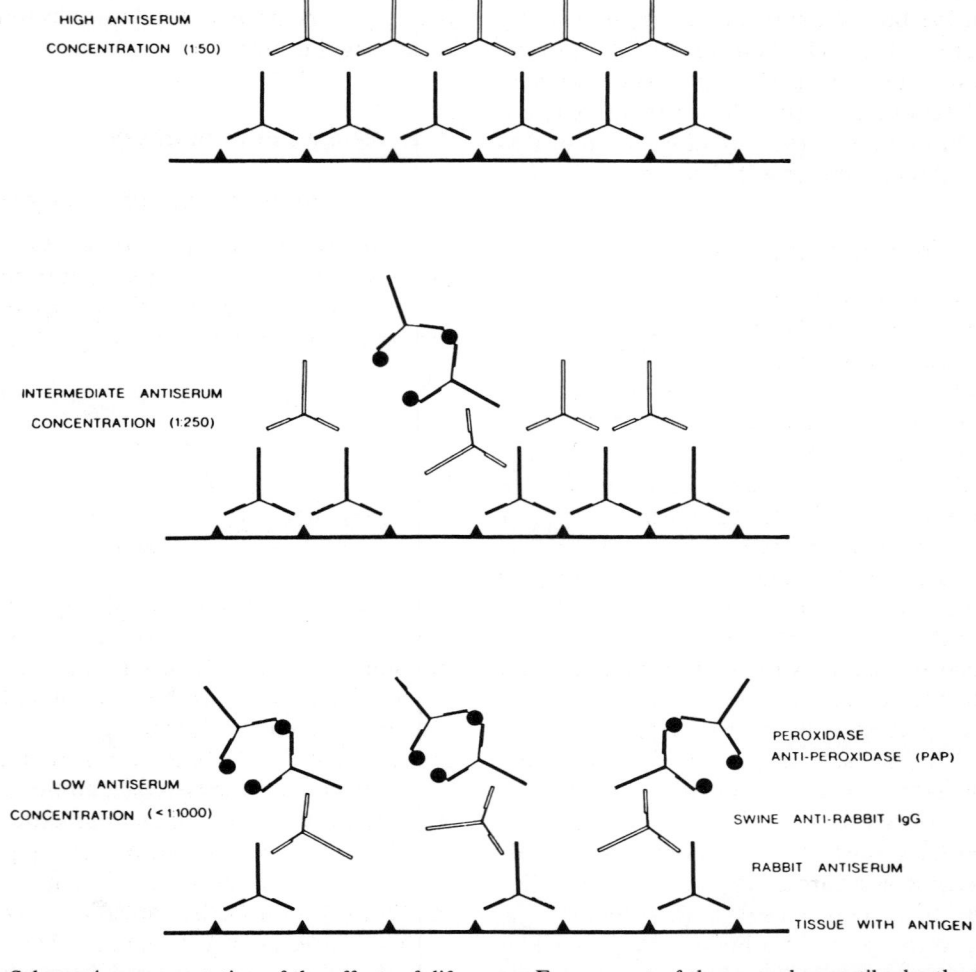

FIG. 19.2. Schematic representation of the effects of different primary antiserum concentrations on the staining of a tissue with high density of antigenic sites (unlabelled antibody-enzyme method). High concentration of primary antiserum (*top*) results in binding of the secondary antibody (in this case, swine anti-rabbit IgG) by both F_{ab} segments. With low concentrations of primary antiserum (*bottom*), the F_c segments of the primary antibody molecules will be further apart, so that each will bind only one F_{ab} segment of the secondary antibody: the free F_{ab} segments of the swine anti-rabbit IgG will therefore be available for combination with F_c segments of the PAP complex. Thus, when antigenic determinant sites in the tissue are close together, maximum staining is obtained with a dilute primary antiserum but no staining occurs after application of a concentrated primary antiserum. (After Bigbee *et al.*, 1977. Reproduced with permission.)

the secondary antisera. Non-specific binding of PAP, should it occur, will then be demonstrated. If non-specific binding is a function of the HRP moiety of PAP, it can be prevented by treating the sections (before stage 1 of the above procedure) with a solution of HRP, followed by 1% H_2O_2 in 30% methanol, which irreversibly inhibits the enzymatic activity of the bound HRP (Minard & Cawley, 1978).

5. Mix the primary antiserum with an excess of the purified antigen and apply this mixture to the sections instead of the primary antiserum. Greatly diminished staining should be observed. It is rarely possible to inhibit completely staining by the PAP method by the use of absorbed antisera.

19.8.4. Critique of the method

The main advantage of the unlabelled antibody-enzyme method is its great sensitivity. Positive results can often be obtained with sections of tissues fixed and processed by methods that denature most of the antigen present. It is even possible to use old H. & E.-stained slides from which the coverslips, mounting media, and dyes have been removed. The high sensitivity is due to the attachment of three molecules of peroxidase at each site of binding of the PAP complex. Amplification due to multiplicity of antigenic sites on the F_c segment of the primary antibody may also occur, as with indirect immunofluorescence techniques. The large size of the PAP complex molecule limits the amount of PAP binding to the secondary antibody. Although control procedures are necessary in order to exclude non-specific staining, this is a "cleaner" method than most others, probably because the PAP reagent is a pure antigen-antibody complex, unlikely to associate itself with proteins other than its own specific antibodies. With some tissues and some antibodies, there is even less non-specific background staining with the ABC method (see Section 19.9.2) than with PAP.

It has been pointed out that the intensity of staining is commonly increased when the primary antiserum is diluted. Consequently it is not possible when using the PAP method to draw even approximate conclusions concerning the quantity of antigen in a specimen or even to make meaningful comparisons between strongly and weakly stained regions within a single section. It can only be concluded that an antigen is either present or undetectable. With other immunohistochemical methods, especially those involving fluorescently labelled antibodies, it is widely assumed that the intensity of the observed staining has some quantitative significance.

19.9. Avidin-biotin methods

19.9.1. Useful properties of avidin and biotin

Biotin (M.W. 244.3), a metabolite of many kinds of microorganism, is a vitamin that forms the coenzyme or prosthetic group of several enzymes that transfer carboxyl groups.

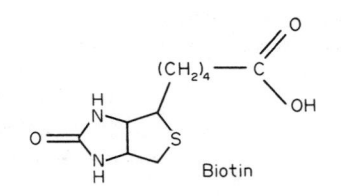

Avidin (M.W. 70,000) is a glycoprotein of egg-white. Each molecule consists of four identical subunits, each capable of binding a molecule of biotin. The affinity between biotin and avidin is very high, even though no covalent bonds are formed, so the binding occurs rapidly and is reversible only in extreme conditions, such as strong acidity or aqueous 6M guanidine. (Administration of large quantities of egg-white to experimental animals results in sequestration of the biotin produced by intestinal bacteria, and causes deficiency of the vitamin. The investigation of this apparent toxicity of egg-white led to the discovery of biotin and avidin.) **Streptavidin** (M.W. 60,000) is a microbial protein that contains no carbohydrate. Its properties are similar to those of avidin. The two proteins are used for the same purposes, but streptavidin is about 10 times as expensive as avidin.

It is possible to conjugate proteins and other large molecules with biotin. This reaction is known as **biotinylation**. A typical reactive derivative of biotin is the sodium salt of sulpho-*N*-succinimidobiotin, shown below. This combines with amino groups of protein, to which biotin becomes joined by an amide linkage:

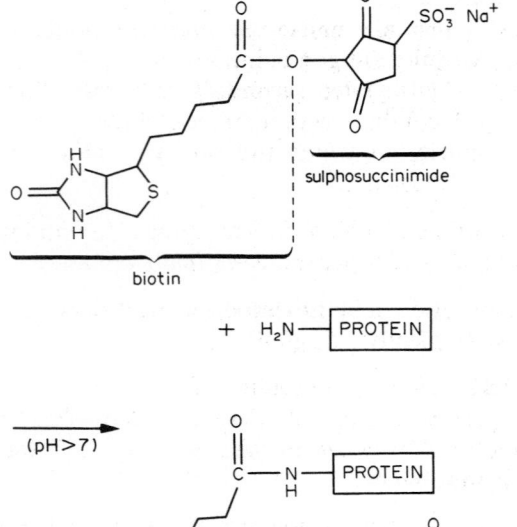

biotin

sulphosuccinimide

+ H₂N——| PROTEIN |

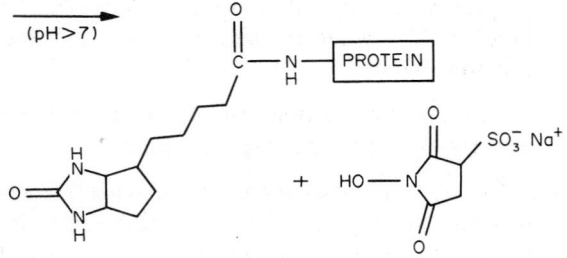

(pH>7)

biotinylated protein

sulpho –N– hydroxysuccinimide

The sulphonic acid group of the reagent makes it soluble in water. Several other reactive derivatives of biotin are available, for labelling proteins, carbohydrates and nucleic acids.

The biotin molecule is small enough not to interfere with the biological activity of the macromolecule that it labels, and its reactivity with avidin is unchanged. Many biotinylated antisera and enzymes are commercially available.

The first methods to make use of biotin and avidin for immunohistochemical detection of antigens (Guesdon, Ternynck & Avrameas, 1979) consisted of either three or four stages, separated by washes in PBS:

Labelled avidin-biotin procedure
1. Application of a primary antiserum (for example, rabbit antibody to the antigen in the tissue).
2. Application of a biotinylated secondary antiserum (for example, biotinylated goat anti-rabbit IgG).

3. Application of a solution of avidin that has been covalently conjugated with a fluorochrome or a histochemically demonstrable enzyme such as horseradish peroxidase. The enzyme is then localized histochemically in the usual way.

The final complex has the form:

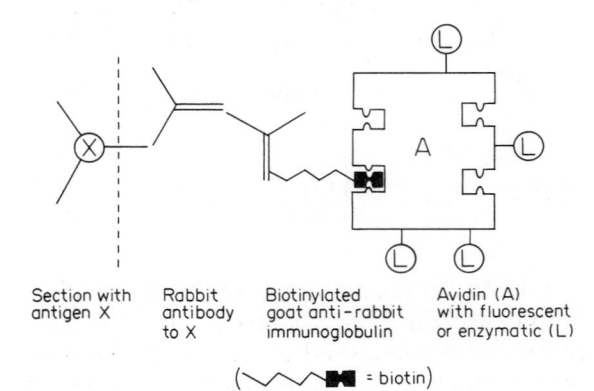

| Section with antigen X | Rabbit antibody to X | Biotinylated goat anti-rabbit immunoglobulin | Avidin (A) with fluorescent or enzymatic (L) |

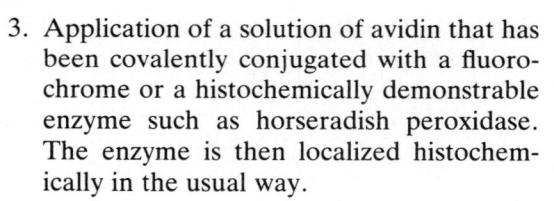

(‿‿‿■◖◗ = biotin)

Bridged avidin-biotin procedure
1 & 2. The first two steps are the same as those of the previous procedure.
3. Application of a solution of avidin to the section.
4. Application of a solution of a biotinylated enzyme, such as a conjugate of biotin with horseradish peroxidase. The bound enzyme is then demonstrated histochemically.

In this case, the final complex is:

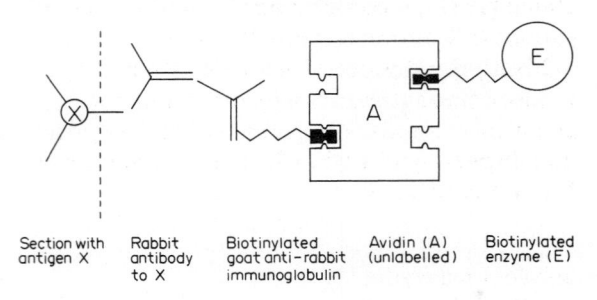

| Section with antigen X | Rabbit antibody to X | Biotinylated goat anti-rabbit immunoglobulin | Avidin (A) (unlabelled) | Biotinylated enzyme (E) |

Greater sensitivity is achieved by using as the detecting agent a freshly prepared **avidin-biotin complex ("ABC")**, made by mixing a solution of avidin with one of biotinylated HRP (Hsu *et al.*,

1981). The structure of the final product with the ABC method is:

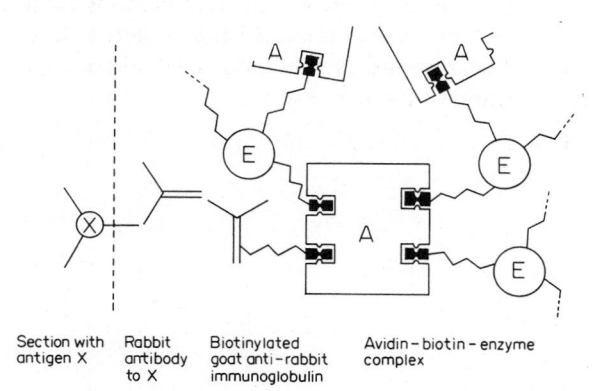

Section with   Rabbit   Biotinylated   Avidin–biotin–enzyme
antigen X   antibody   goat anti–rabbit   complex
                to X   immunoglobulin

This looks like the product of the bridged avidin-biotin procedure, but differs in that the pre-formed avidin-biotin-HRP complex is probably an enormous crosslinked molecule containing many more than three HRP molecules per free biotin-binding site. The size of this complex may restrain its diffusion into the section. The technical details of the ABC method are discussed in the next section.

19.9.2. Detection of antigens by the "ABC" method

The following procedure is that of Hsu & Raine (1981), incorporating the optional modifications of Cattoretti *et al.* (1988).

Fixation and processing are determined by the nature of the antigen, as with other immunohistochemical methods. The sections are hydrated and equilibrated with PBS, which is also the solvent for all the reagents.

Solutions required

A. *Phosphate-buffered saline (PBS)* as solvent for all immunological reagents, and for all aqueous washes. See Chapter 20, p. 361. This saline may optionally contain a surfactant and a protein, as explained in Section 19.13.

B. *Antisera:*
   (i) A **primary antiserum** that will combine with the antigen being sought.
   (ii) A **biotinylated secondary antiserum** that will combine with immunoglobulins of the species in which the primary antiserum was raised.

C. *An avidin solution.* This typically contains 10 μg of avidin per ml. See *Note 1* below.

D. A solution of **biotinylated horseradish peroxidase** (typically 2.5 μg/ml).

E. **ABC** made by combining solutions C and D (typically equal volumes, but see *Note 1* below). **Mix when needed: see step 7 of the method, below.**

F. **Reagents for demonstration of peroxidase** activity at neutral pH (see Chapter 17).

G. Any dye solutions needed to provide a **counterstain** appropriate to the tissue and the reason for demonstrating the antigen. Many of the methods in Chapters 6 and 7 can be used.

Procedure

1. (Optional) Inhibit endogenous peroxidase by treating sections for 30min with 0.3% hydrogen peroxide in PBS, followed by two washes in PBS.

2. (Optional) Block non-specific background staining by treating the sections for 30min with a 10% dilution of serum from a species of animal different from that in which the primary antiserum was raised. Wash in 2 changes of PBS.

3. Incubate in suitably diluted primary antiserum (e.g. rabbit antiserum to "antigen X"). The range of dilution may vary between 1:200 and 1:6400, and the optimum should be determined by trial, as with the PAP method. Long incubations (12–48h) at 4°C are preferred by most workers to shorter incubations at higher temperatures.

4. Wash in 3 changes of PBS.
5. Incubate in biotinylated secondary antiserum (e.g. goat antiserum to rabbit IgG). A 1:10 to 1:40 dilution, for 30–60m at room temperature is usual, but some workers use greatly diluted serum, at 4°C, for 12–48h.
6. Wash in 3 changes of PBS.
   15–20min before moving the sections out of the last wash, make the **working ABC solution** by mixing the recommended volumes of avidin and biotinylated HRP stock solutions.
7. Incubate sections in the working ABC solution for 60m at room temperature.
8. Wash in 3 changes of PBS.
   At this stage of the procedure, optionally carry out the steps given in *Note 2* below.
9. Demonstrate the peroxidase activity with one of the methods explained in Chapter 17. A DAB–H_2O_2 method is usually used.
10. Wash in water, counterstain as desired, and make a permanent preparation in a mounting medium appropriate to the counterstain.

Results

Sites of immunoreactivity recognized by the primary antiserum are brown (or black, if a Co^{2+}- or Ni^{2+}-enhanced DAB–H_2O_2 method is used). Other structures may be counterstained in contrasting colours.

Notes

1. Solutions of biotinylated secondary antiserum, avidin and biotinylated HRP are commercially available, often in proprietary kits such as the "Vectastain" system. For measurements and dilutions, the instructions that accompany commercial products or kits should be followed, even if they differ from the suggestions given here or in other publications.
2. To introduce more peroxidase into sites of immunoreactivity in the section, a biotin-rich peroxidase-avidin complex ("CBA") may be applied after stage 8, and the preparation can be treated with alternating appli-

cations of ABC and CBA. The CBA contains 5 times as much biotinylated HRP as the regular ABC reagent. Thus, with the "Vectastain" product ABC is made by mixing 4 volumes of the avidin solution with 1 volume of the biotinylated HRP solution. For CBA, the proportions are 4:5. Cattoretti *et al.* (1988) recommend three treatments with ABC and two with CBA: (following Step 7 above:

$$CBA \rightarrow ABC \rightarrow CBA \rightarrow ABC$$

with a wash between each application, then proceed with Step 8).

The amplification of staining intensity is due to the affinity between ABC and CBA, which results in the accumulation of increasing amounts of the labelling enzyme.

19.9.3. Controls

For the reader who has read the chapter up to this point, the controls for specificity of the ABC method should be obvious. The specimen should have no endogenous activity of the labelling enzyme (usually peroxidase). There should be no staining if the primary antiserum is omitted, and staining should be reduced by prior absorption of the primary antiserum with purified antigen. There should be no evidence of non-immune binding of the secondary antiserum or of the avidin-biotin-enzyme complex to the section.

19.9.4. Critique of the method

The ABC and PAP methods are probably of approximately equal sensitivity (Hsu & Raine, 1981). In both procedures, large complex molecules are used to introduce the detecting enzyme into the sites where antibodies are bound. It is therefore often necessary to treat the preparations, before staining, by one of the methods for increasing permeability (Section 19.13.1) of the tissue. According to Sternberger & Sternberger (1986), non-immune binding of reagents is more likely to occur with the ABC system than with the PAP method, at least with nervous tissue. However, the ABC technique is widely respected for giving "clean" preparations. Some workers claim that the results are even cleaner with streptavidin than with avidin.

When high sensitivity is not needed, as in the staining of cytoskeletal proteins, the indirect immunofluorescence techniques are often preferred to those in which the last step is the histochemical demonstration of an enzyme.

19.10. Other immunohistochemical methods

The techniques described so far are those most widely used for the demonstration of antigens in tissues. Other ingenious techniques also exist for the localization of both antigens and antibodies. Four of these will be described briefly to give the reader some idea of the scope and versatility of immunohistochemical methodology.

19.10.1. Sandwich technique for antibody in tissue

Antibodies in a tissue are demonstrated by applying the purified antigen followed by a fluorescently labelled antiserum to the antigen. The method is often used with impression smears of cells from lymph nodes and with cryostat sections. For example, a section of human tissue containing antibodies to an antigen Y could be treated with a solution of Y, followed by a fluorescently labelled rabbit antiserum to Y.

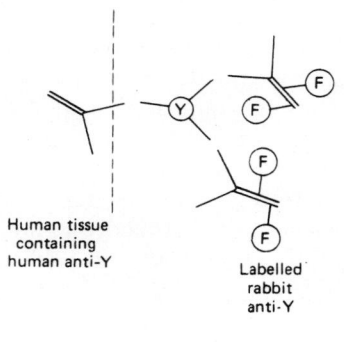

Human tissue
containing
human anti-Y

Labelled
rabbit
anti-Y

This is a method for demonstrating specific antibodies. Antibodies in general (i.e. immunoglobulins) are detected by considering them to be antigens. Human immunoglobulin G, for example, is localized in sections by means of a rabbit antiserum to human immunoglobulin G and a labelled goat antiserum to rabbit immunoglobulin.

19.10.2. Detection of antibody in serum

In autoimmune diseases the blood contains circulating antibodies to certain components of a patient's or animal's own tissues. The existence and specificity of such auto-antibodies can be detected immunohistochemically by applying the suspect serum (usually as a 1 in 10 dilution in saline) to unfixed cryostat sections of tissues known to contain appropriate auto-antigens. Different tissues serve as controls. The sites of binding of antibodies to the sections are made visible by the application of a fluorescently labelled antiserum to the immunoglobulins of the species from which the serum under test was obtained.

For example, to test human serum for antibodies against thyroglobulin one requires some sections of normal human thyroid gland and a fluorescently labelled rabbit antiserum to human γ-globulin. The stained preparation has the structure:

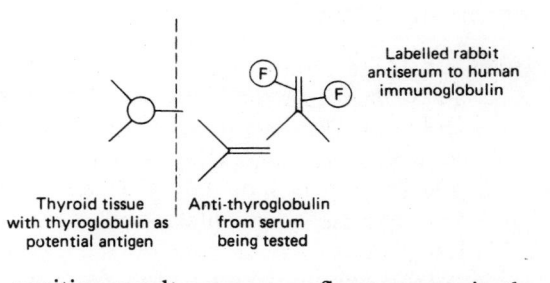

Thyroid tissue
with thyroglobulin as
potential antigen

Anti-thyroglobulin
from serum
being tested

Labelled rabbit
antiserum to human
immunoglobulin

A positive result appears as fluorescence in the follicles of the thyroid tissue. A section of another human organ, such as the liver or a kidney, serves as a control for specificity. Often the sections used are composite blocks containing several tissues, so that it is possible to seek the presence of several auto-antigens in a single drop of serum.

19.10.3. *Methods using staphylococcal protein A*

The cell-wall of *Staphylococcus aureus*, a common pathogen, contains an isolable component known as protein A which is able to bind to the F_c segments of most types of mammalian IgG molecule. The binding is non-immune but results from an affinity similar to that existing between antigens and antibodies. The molecule of protein A is bivalent and can therefore be made to serve as a bridge between the F_c portions of IgG molecules of different kinds. It is also possible to label protein A with fluorochromes or with HRP (Dubois-Dalcq *et al.*, 1977). Thus protein A may be used for the same purposes as an antiserum to IgG. Like the commoner anti-immunoglobulin sera, protein A is commercially available.

A conjugate of protein A with HRP is employed in a technique closely similar to an enzyme-labelled antibody method. The specific primary antiserum is applied to a section of tissue and followed by HRP-protein A conjugate:

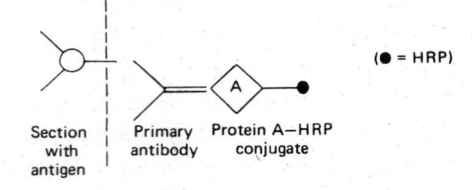

The sites of bound HRP are demonstrated histochemically in the usual way. This method may provide "cleaner" staining, with less non-immunological background coloration, than the equivalent enzyme-labelled antibody technique.

Protein A has also been employed instead of a secondary antiserum in the unlabelled antibody-enzyme method. Application of a specific primary antiserum is followed by protein A and then by PAP:

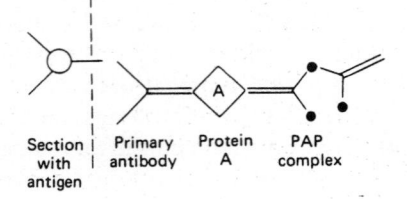

This technique, however, is somewhat less sensitive than those in which secondary anti-immunoglobulin sera are used (Celio *et al.*, 1979). Some other immunohistochemical applications of protein A are described by Notani *et al.* (1979).

The main advantage of protein A is that it is able to combine with IgG of all mammals, so that it has a greater number of potential uses than an antiserum to the immunoglobulin of a single species*. Obviously protein A is not suitable for use with any specimen that already contains IgG.

19.10.4. *Labelling with colloidal gold*

It is quite easy to make a colloidal suspension (sometimes called a "sol") of gold, containing particles of known diameter in the range 20–150nm (Frens, 1973). In order to prevent the suspended particles from clumping together and sinking, the sol contains a macromolecular stabilizing agent, which coats the surfaces of the particles. If this stabilizing agent is or includes a protein with specific affinity, such as a lectin or an immunoglobulin, the metal will serve as a label, in much the same manner as a fluorescent dye or a histochemically demonstrable enzyme (Geoghegan & Ackerman, 1977; Horisberger & Rosset, 1977). Typically either an anti-IgG serum or staphylococcal protein A is bound to gold particles in this way, and used to identify sites of attachment of a primary antibody (Roth, 1978; Bendayan, 1981). Avidin can also be used to coat colloidal gold particles (see van den Pol, 1984). These methods have two advantages over those that use HRP:

(a) The colloidal gold particles will stick to antigenic sites on the surface of an ultrathin section of plastic-embedded tissue. (If enzyme-labelled antibody methods are to be used in conjunction with electron microscopy, the immunostaining must be done on sections cut with a

---

*Before accepting a negative result, it is necessary to confirm that protein A does bind to the primary antibody, especially if this is a monoclonal antibody. Protein A binds only to some subtypes of IgG.

vibrating microtome, before embedding in plastic.)

(b) It is possible to have antibodies labelled with gold particles of different size, and thereby identify more than one antigen in the same section.

The colloidal gold methods are used mainly in conjunction with electron microscopy. The sols are strongly coloured, but the bound quantities are small, so they produce only faint staining for light microscopy (Roth, 1982). However, it is possible to enlarge the particles by treating the sections with a physical developer until dense, optically visible deposits are formed (Holgate *et al.*, 1983; Danscher & Norgaard, 1985). The principle of physical development is explained in Chapter 18 (p. 306). The protein that labels the gold particles can also be made visible by any appropriate immunohistochemical method.

Protein-gold sols can be made in the laboratory, or purchased. For details of their preparation and uses, see Polak & Varndell (1984), Bendayan & Duhr (1986) and Birrell *et al.* (1987).

19.11. *In situ* hybridization of nucleic acids

The affinities between the purine and pyrimidine bases of nucleic acids have been mentioned in Chapter 9. It is possible to synthesize polynucleotides in which the sequences of bases are complementary to those in fragments of naturally occurring DNA or RNA. These are known as **cDNA** or **cRNA probes**. (The "c" is for "complementary".) A cDNA or cRNA probe will adhere to a molecule of messenger RNA (mRNA) containing the complementary sequence of nucleotides. The specific adhesion between a natural nucleic acid and a complementary probe is known as base-pairing or **hybridization.** The term "*in situ*" is used when the nucleic acid probes are applied to whole cells or sections of tissue, rather than to solutions or cell-free suspensions.

A nucleic acid probe can be labelled in various ways. A radioactive tag, such as ^3H, ^{125}I or ^{35}S is most often used, and is eventually detected by autoradiography. Alternatively, the probe can be biotinylated, and the bound biotin made visible by virtue of its affinity for avidin (see Section 19.9). There are two reasons for wanting to identify the sites of hybridization of nucleic acid probes:

(1) A cDNA or cRNA probe identifies the mRNA for the protein encoded by a specific nucleotide sequence (or gene) of chromosal DNA. The presence of the mRNA is revealed in interphase nuclei and in cytoplasm. Thus, it is possible to determine whether a cell is actively transcribing the gene and thus presumably synthesizing the encoded protein. This can be important, because protein molecules are often degraded or expelled from cells (secreted) soon after they have been assembled. Immunohistochemical staining in these cases may be negative, but binding of cDNA can identify the cells that produce such proteins. Conversely, a positive immunohistochemical reaction may be obtained in a cell that has taken up a protein that was synthesized and secreted by another cell. *Nucleic acid hybridization reveals the site of origin, as distinct from the site of binding or storage of a protein.*

(2) A cRNA or cDNA probe may hybridize with a gene, which is the nucleotide sequence of DNA encoding the amino acid sequence of a particular protein. For this to occur, the two DNA strands of the chromosome at the site of the gene must be separated, either as part of the mitotic or meiotic cycle or by artificial denaturation. It is thus possible to detect the presumptive positions of some genes on some types of chromosome.

In situ hybridization is primarily a technique used by research workers, and only a few aspects of the subject are considered here. Detailed accounts, with practical instructions for the production, labelling, and application of nucleic acid probes, can be found in Hames & Higgins (1985), Uhl (1986) and Angerer *et al.* (1987).

Although the affinity between matched base-pairs of nucleic acids is high, the conditions required for *in situ* hybridization are far from physiological. Thus, the proteins and nucleic acids of the tissue must be permeable to the large molecules of the probe. This may be achieved by chemical fixation. According to Tournier *et al.* (1987), buffered formaldehyde is preferable to coagulant mixtures. Aldehydes do not react chemically with nucleic acids (see Chapter 2), so mRNA is probably immobilized by being trapped among cross-linked protein molecules, and DNA by the insolubilization of chromosomal proteins. The section or smear is often treated with a proteolytic enzyme after fixation but before application of the probe. This is reputed to enhance penetration and it also destroys any nucleases in the preparation that were not inactivated by the fixative. RNase, for example, might destroy endogenous mRNA or an applied cRNA probe. Sections may also be treated with 0.2M HCl or 0.07M NaOH to enhance penetration and to separate the paired strands of nuclear DNA. An acid treatment must not be severe enough to hydrolyse nucleic acids (see Chapter 9). Before hybridization, the preparation is incubated in solutions containing single-stranded nucleic acids that would not be expected to hybridize with the mRNA being sought by the specific probe. DNA from salmon sperm or tRNA from yeast are used with mammalian tissues. These substances will exclude the probe from any sites that bind nucleotides non-specifically and they may also be added to the hybridization solution. Their function is comparable to that of non-immune serum included in immunohistochemical incubation media.

The conditions for hybridization include the presence of a fairly high concentration of salt (typically 0.3M Na$^+$) in a solvent consisting of 50% formamide and 50% water. The formamide reduces hydrogen-bonding of nucleic acids to water, thereby increasing the probability of union between complementary base-pairs. It probably also destroys the activity of any RNases that have survived fixation and the other pre-treatments. Dextran sulphate can also be included in the medium, to displace water and effectively increase the concentration of the probe. The temperature is critical, and should be some 10°C below the melting temperature of the nucleic acid hybrid that is to be formed. It may vary between 20°C and 70°C, according to the hybrid and the other ingredients of the medium. The melting temperature is that at which the complementary nucleic acid strands are half-dissociated or denatured. After several hours, the preparations are washed. It is necessary to remove all unbound or non-specifically bound probe. In the case of cRNA, the removal can be accelerated by treatment with ribonuclease A. The molecules of probe that have hybridized with nucleic acids in the specimen are not attacked by this enzyme, which catalyses the hydrolysis only of single-stranded RNA.

Finally, sites at which the probe has been bound must be made visible. Autoradiography is the method most often chosen, because grains can be counted and quantitative data obtained. Superior resolution can be obtained by using a biotinylated probe. After hybridization, the preparation is treated with avidin, and the sites of avidin binding are detected immunohistochemically, using an anti-avidin serum. Monoclonal antibodies that bind to biotin have also been used (Pringle *et al.*, 1987).

From the foregoing summary, it will be appreciated that several potentially destructive steps are involved in techniques of *in situ* hybridization. Sections of tissue that have been subjected to all the procedures mentioned above are rarely in a condition that would be considered satisfactory with other histochemical procedures. Hybridization is less damaging to chromosomes than to histological sections, and the appearances are similar to those of conventional cytological preparations. It seems probable that severe pre-treatment of sections with acids and proteolytic enzymes is often unnecessary, and that *in situ* hybridization will provide more satisfactory morphological images in the future. Recent developments include procedures for combined immunohistochemistry and *in situ* hybridization on the same section (Mullink *et al.*, 1989).

19.12. Non-immunological affinity techniques

Specific attachments between molecules are by no means confined to the reactions of antigens with antibodies. The affinities of other biologically significant substances for one another have been exploited in a variety of histochemical methods. The following types of technique are worthy of mention.

1. The use of **lectins** in carbohydrate histochemistry (Chapter 11).

2. The use of labelled or fluorescent inhibitors for the localization of **enzymes** (Chapter 14). Another application of enzyme-substrate affinity is the use of polymers of an enzyme to bind to its substrate in sections of tissue. Thus, each molecule of glucose oxidase has two substrate-binding sites. Aggregates (made by mixing the enzyme with glutaraldehyde) will attach to α-D-glucosyl groups in sections of fixed tissue. The free active sites of the enzyme aggregates are then demonstrable histochemically (Dermietzel et al., 1985).

3. The use of **labelled drugs** to bind to physiologically or pharmacologically defined receptor sites in tissues. For example, fluorescently labelled α-bungarotoxin (from a snake venom) has been used to demonstrate receptors for acetylcholine, including those at neuromuscular junctions (Anderson & Cohen, 1974). With some fluorescent derivatives of drugs, however, the microscopically observed binding has been shown not to coincide with the sites of pharmacological action (Corréa et al., 1980).

4. The filamentous protein actin, which is important in the contraction and locomotion of cells, selectively binds heavy meromyosin, which is isolated from the actomyosin of striated muscle. The binding results in thickening of the actin filaments, as observed in the electron microscope. For light microscopy, heavy meromyosin can be fluorescently labelled (Sanger, 1975) or detected immunohistochemically. Phalloidin, a fungal toxin that binds to actin, can be biotinylated and used for affinity staining in light and electron microscopy (Faulstich et al., 1989).

5. **Receptors for hormones** can be demonstrated in various ways. The oldest is autoradiography, following administration of radioactively labelled hormone to an experimental animal. More recent techniques include the use of oestrogen or progesterone conjugated to fluorescently labelled albumin (Bergqvist et al., 1984), and exposure of tissue to biotinylated parathyroid hormone (Niendorf et al., 1988). The bound biotinylated hormone was made visible by means of an avidin-peroxidase conjugate.

6. Detection of GM_1 ganglioside, a lipid component (see Chapter 12) of cell membranes, by virtue of its specific affinity for **choleragenoid**, which is the non-toxic (cell-binding) portion of the cholera toxin molecule. The bound choleragenoid is detected with an immunohistochemical procedure (Willinger & Schachner, 1980). Alternatively, biotinylated choleragen is used, and is detected with an avidin-HRP conjugate (Asou et al., 1983).

7. Connective tissue matrix contains proteoglycans and glycoproteins that bind specifically to **hyaluronic acid**. This affinity has been exploited for the histochemical detection of hyaluronic acid (Ripellino et al., 1985; Girard et al., 1986).

8. If cells that are about to divide are exposed to **5-bromo-2'-deoxyuridine** (BrDU), this synthetic compound is incorporated into their DNA. Monoclonal antibodies are available that recognize BrDU, and these can be used for the identification of cells that contain the tracer (Morstyn et al., 1986; de Fazio et al., 1987). The method gives results similar to those obtained with [^{3}H]thymidine, but autoradiography is not needed. BrDU is therefore useful in studies of cell kinetics.

19.13. Miscellaneous practical considerations

This chapter concludes with a few technical tips of a general nature.

19.13.1. Preservation of structure and antigenicity

There is no ideal way to fix and section tissues for immunohistochemistry. Buffered formaldehyde is suitable for many purposes, though a coagulant agent is advantageous for some antigens, in which distortion is necessary in order to expose the epitopes. Some other fixatives often used for immunohistochemical work are described in Chapter 2 (p. 29). The most popular of these is that of Stefanini *et al.* (1967) containing formaldehyde and neutral picrate ions. Glutaraldehyde is sometimes added to such mixtures (Newman *et al.*, 1982; Somogyi & Takagi, 1982), especially if ultrastructural preservation is required.

A fixative mixture containing 4% $HgCl_2$ and 8% HCHO (from formalin) was recommended by Hickey *et al.* (1983). After vascular perfusion or immersion for 24h, sections were cut with a vibrating microtome and collected into TRIS buffer. TRIS is a primary amine, and combines rapidly with aldehydes, so it terminates the fixation and ensures that the primary antiserum will not react with formaldehyde that leaches out of the tissue.

19.13.2. Reduction of non-specific binding of immunoglobulins

Tissue-bound aldehyde groups, which bind all proteins strongly and non-specifically, are a major source of trouble when the fixative contains **glutaraldehyde**. Although glutaraldehyde is best avoided for immunohistochemical work, it is necessary for satisfactory preservation of ultrastructure, and also for the immobilization of some soluble antigens (see Chapter 17, p. 298). The aldehyde groups should be converted to unreactive hydroxyls by **reduction with sodium borohydride** before attempting any immunohistochemical procedure (Willingham, 1983; see also Chapter 10, p. 164).

The other way to reduce non-specific sticking of protein molecules to sections is to **treat with a solution of a protein** that will not interfere in any way with the staining method to be used. It is usual to use a 1:100 to 1:10 dilution in saline (the concentration is not critical) of a normal (i.e. non-immune) serum from the species in which the secondary antiserum was raised. Thus, if the secondary antibody is goat anti-rabbit IgG, all the solutions used could contain 1–10% serum from a goat that has not been immunized against any rabbit proteins. **The use of non-immune serum in this way is desirable as a routine measure, whatever fixative has been used,** especially if sensitive methods like the PAP or ABC procedures are to be used.

19.13.3. Helping large molecules to penetrate the tissue

Failure to demonstrate the presence of an antigen in a tissue does not necessarily mean the antigen is not there. It may be masked by other molecules that obstruct the access of the antibody molecules in the primary antiserum. The most obvious barriers to penetration are lipoprotein membranes (fixed or unfixed), and matrices of cytoplasmic or extracellular material that have been tightly cross-linked by an aldehyde or a chromium-containing fixative. There are several ways to increase the permeability of a tissue to large molecules; some were mentioned in connection with nucleic acid hybridization. The methods most often used are pre-treatment of the sections with either a proteolytic enzyme or a surfactant. These procedures are critically discussed by Horobin (1982) and Feldmann *et al.* (1983).

A suitable **proteolytic enzyme** is trypsin. A convenient solution is one containing 0.01–0.1% (w/v) enzyme in $0.1M$ $CaCl_2$, adjusted to pH7.4 with a little $0.1M$ NaOH. Sections are incubated for 10–60min, at room temperature or 37°C. The optimum conditions can be determined only by trial. Most antigens are proteins or peptides, so it is important not to overdo a proteolytic digestion. A pH well removed from the optimum for the enzyme (as in the trypsin solution above) helps to moderate the destructive action. Proteolytic digestion is damaging to structure, so it is used only when necessary, as when demonstrating types I, III and IV collagen (see Chapter 8) in paraffin sections (Bedossa *et al.*, 1987).

Surfactants (detergents) probably emulsify the lipid components of membranes, making holes through which large molecules can pass. Non-ionic surfactants are used because they do not impart electrical charge, which might cause non-specific protein binding to the specimen. Digitonin, saponin, and the synthetic compound known as Triton X-100 are all suitable. The surfactant is applied to the section as a pre-treatment or, more commonly, it is added to the antibody and saline washing solutions, in a concentration of 0.05 to 1.0% (w/v). Saponin may even be added to fixative mixtures (Pignal *et al.*, 1982).

19.13.4. Economizing with expensive reagents

Many of the reagents necessary for histochemical methods based on affinity are available only in small quantities. It is not usually possible to use large excesses of solutions, as is the custom when carrying out ordinary staining techniques. Primary antisera and monoclonal antibodies for immunohistochemistry and nucleic acid probes for hybridization procedures are the reagents available in the smallest amounts. Often they are made in the laboratory with great expenditure of effort, or if obtained from commercial sources they are very expensive. Reagents used in many methods, such as FITC-conjugated goat anti-rabbit γ-globulin or rabbit PAP, are less expensive, but still costly enough to justify parsimony in their use.

When about 0.05 ml of a reagent is available for each section, the slides are placed horizontally in an atmosphere saturated with water vapour, and a drop of the solution is deposited on each section. If the available volume is less, a tiny drop is applied and the smallest possible coverslip (or piece broken from a coverslip) is applied. A large drop of mineral oil is then put on top of and around the small coverslip, and covered again with a large coverslip. Thus, the expensive reagent is kept in contact with the slide, and evaporation is prevented. This procedure is often used with nucleic acid probes (Angerer *et al.*, 1987). Another method is to smear Vaseline thinly on both surfaces of a piece of Parafilm, and then make a hole, slightly larger than the section, in the middle. Two slides, each bearing a section, are applied to the greased surfaces to make the roof and floor of the incubation chamber. A drop of antiserum or other expensive reagent (about 20 μl per cm^2 of area) is applied to one of the sections before the chamber is assembled (Abbuhl & Velasco, 1985). Slides with a 75 μm layer of paint at the corners are commercially available. Capillary action draws a drop of reagent into the 150 μm space between an opposing pair of such slides (Kumar, 1989).

The value of prolonged incubation in very dilute antiserum has already been mentioned. This manoeuvre increases the sensitivity of immunohistochemical methods (Brandtzaeg, 1981), and is also economical. Sofroniew & Schrell (1982) have found that diluted antisera can be kept in coplin jars for several weeks at 4°C, and used repeatedly. They recommend a solvent with this composition:

Triton X-100:	5.0 g
λ-Carrageenan:	7.0 g
TRIS (See Chapter 20, p. 358):	5.0 g
Na$_2$HPO$_4$:	12 g
NaH$_2$O$_4$.2H$_2$O:	0.25 g
NaCl:	7.0 g
Sodium azide (NaN$_3$):	2.0 g
Water:	to make 1000 ml

The function of the Triton X-100 is explained in Section 19.13.3. The carrageenan (a sulphated galactose polysaccharide) is said to suppress non-specific staining, and the sodium azide inhibits growth of bacteria and fungi. Thiomersal (= thimerosal), 10^{-3}M (405 mg of the sodium salt per litre), is a safe substitute for sodium azide.

19.14. Exercises

1. Why are antisera labelled by conjugating them with fluorochromes rather than with ordinary dyes?

2. Suggest reasons why alcohol and picric acid are more suitable fixatives for antigens than formaldehyde and glutaraldehyde.

3. Devise an immunohistochemical method suitable for localizing the large amounts of HRP present in the root of the horseradish plant.

4. It has occasionally been possible to localize antibodies in tissues by a direct method using fluorescently labelled antigens. Why is this simple technique usually unsuccessful?

5. What are the advantages of the indirect over the direct fluorescent antibody techniques?

6. Devise an immunohistochemical method for the localization of an antigen A making use of the following:

(a) purified A;
(b) antiserum to A;
(c) covalent conjugate of A with an enzyme such as HRP.

Discuss the value and limitations of the method (described by Mason & Sammons, 1979).

7. When a tissue has been fixed in glutaraldehyde, widely distributed non-specific staining by the unlabelled antibody-enzyme (PAP) method is observed in addition to the specific staining of antigenic sites in the sections. This non-specific staining is suppressed if the sections are incubated with non-immune goat serum before applying the primary antiserum. Explain.

8. In the unlabelled antibody-enzyme method it is necessary to use a high concentration of the secondary antiserum but a low concentration of primary antiserum. Why is a high concentration of secondary antiserum necessary if adequate quantities of PAP are to be bound in the third stage of the technique?

9. In an attempt to stain a hormone in sections of an endocrine gland, the following reagents were applied sequentially to sections:

(a) dog antiserum to the hormone;
(b) goat antiserum to dog γ-globulin;
(c) rabbit peroxidase-antiperoxidase (PAP);
(d) histochemical detection of peroxidase.

Why did this procedure fail to demonstrate the hormone?

Using the same primary antiserum, what could be done to obtain a successful immunohistochemical result?

10. Immunohistochemical staining reveals the presence of immunoglobulins in the lesions of many diseases. What could you do to find out whether the immunoglobulin is synthesized in or bound by the cells of the affected organs?

20

Miscellaneous Data

20.1. Buffer solutions

A buffer is chosen for (a) its efficacy in the required pH range, and (b) the absence of components which would react undesirably with other substances in the solution to be buffered. Prescriptions for the commonly used buffers follow. The pH values are given to the nearest 0.1pH unit; greater accuracy is rarely required in histochemistry. For more extensive tables of buffers, see Perrin & Dempsey (1974). Some useful practical information concerning the preparation of buffers is given by Kalimo & Pelliniemi

(1977). Variations in temperature affect the pH of any solution, but most of the buffers given below should be accurate to ± 0.1 pH unit over the range 15–25°C.

The pH of a mixture containing a buffer should be checked with a pH meter. It is usual to standardize the meter against buffer solutions obtained from a chemical supply house. However, the following standard solutions are easily made in the laboratory:

0.05 M *potassium hydrogen phthalate*

$KHC_8H_4O_4$: 10.21g; water to 1000ml.
The pH is 4.0 from 0°C to 40°C.

(Although it was recommended by Lillie & Fullmer (1976) **this solution should not be used as a solvent for dyes** because it inhibits staining, even when 4.0 is the optimum pH, and it sometimes modifies the colour. These undesirable effects might be due to a combination of ionic and non-ionic attraction between the aromatic phthalate ion and the dye molecules.)

0.01M *borax*

$Na_2B_4O_7.10H_2O$: 3.81g; water to 1000ml
The pH is 9.3 from 10°C to 15°C; 9.2 from 20°C to 25°C; 9.1 from 30°C to 35°C.

20.1.1. pH0.7–5.2: acetate-hydrochloric acid buffer

This buffer is compatible with all reagents other than those metals whose chlorides are insoluble (e.g. Ag, Pb).

Stock solutions

A. 1.0M sodium acetate

    Either CH_3COONa: 82.04g; water to
           1000ml;
    or $CH_3COONa.3H_2O$: 136.09g;
        water to 1000ml.

B. 1.0M hydrochloric acid (see Section 20.3, p. 362)

pH	A ml1.0M csodium acetate	B ml1.0N HCl	Water: to make total volume (ml)
0.7	50	95	250
0.9	50	80	250
1.1	50	70	250
1.2	50	65	250
1.4	50	60	250
1.7	50	55	250
1.9	50	53	250
2.3	50	51	250
3.2	50	48	250
3.6	50	45	250
3.8	50	42.5	250
4.0	50	40	250
4.2	50	35	250
4.4	50	30	250
4.6	50	25	250
4.8	50	18	250
4.9	50	15	250
5.2	50	10	250

The final volume of 250ml must include all other ingredients of a buffered mixture. It is important to check the final pH with a meter. The buffering capacity is poor between pH1.8 and 3.5.

20.1.2. pH3.6–5.6: acetate-acetic acid buffer

This buffer is compatible with all commonly used reagents.

The instructions here are for preparation of a 0.1M buffer. Some techniques require 0.2M acetate buffer, so it is convenient to keep 0.2M stock solutions, with double the strengths of those described below. (See also *Note* following the table.)

Stock solutions

A. 0.1M sodium acetate

    Either CH₃COONa: 8.20g; water to 1000ml;
    or CH₃COONa.3H₂O: 13.61g; water to 1000ml.

B. 0.1M acetic acid (see Section 20.3, p. 362)

pH	A ml 0.1M sodium acetate	B ml 0.1M acetic acid
3.6	15	185
3.8	24	176
4.0	36	164
4.2	53	147
4.4	74	126
4.6	98	102
4.8	120	80
5.0	141	59
5.2	158	42
5.4	171	29
5.6	181	19

Note

This buffer may also be made up with 0.2M, 0.05M, or 0.01M reagents, but the pH values for the proportions given in the table are not accurate for concentrations other than 0.1M. Check with a meter and adjust the pH when making acetate buffer at strengths other than 0.1M.

20.1.3. pH 2.7–7.7: phosphate-citrate buffer

This buffer is useful because it is effective over a wide range. It should not be used in the presence of any metal ions that form insoluble phosphates or complexes with citric acid (e.g. Ag, Al, Ba, Ca, Co, Cu, Fe, Mg, Mn, Ni, Pb, Zn).

Stock solutions

A. 0.2M disodium hydrogen phosphate (= sodium phosphate, dibasic)

    Either Na₂HPO₄: 28.39g; water to 1000ml;
    or Na₂HPO₄.2H₂O: 35.60g; water to 1000ml;
    or Na₂HPO₄.7H₂O: 53.61g; water to 1000ml.

(The dihydrate is the most stable form, since it is neither hygroscopic nor efflorescent.)

B. 0.1M *citric acid*

> **Either** $C_6H_8O_7$ (anhydrous): 19.21g; water to 1000ml;
>
> **or** $C_6H_8O_7.H_2O$: 21.01g; water to 1000ml.

(The monohydrate is the form of citric acid most commonly used.)

pH	A ml 0.2M Na_2HPO_4	B ml 0.1M citric acid
2.6	22	178
2.8	32	168
3.0	41	159
3.2	49	151
3.4	57	143
3.6	64	136
3.8	71	129
4.0	77	123
4.2	83	117
4.4	88	112
4.6	93.5	106.5
4.8	99	101
5.0	103	97
5.2	107	93
5.4	111	89
5.6	116	84
5.8	121	79
6.0	126	74
6.2	132	68
6.4	138.5	61.5
6.6	145.5	54.5
6.8	154.5	45.5
7.0	165	35
7.2	174	26
7.4	182	18
7.6	187	13
7.8	191.5	8.5

20.1.4. *pH5.3–8.0: phosphate buffer*

The table gives quantities for preparing 0.1M sodium phosphate buffer solutions. These may be diluted with water to obtain weaker solutions (e.g. 0.06M, 0.05M, etc.). Twofold dilution increases the pH by approximately 0.05. Phosphate buffers cannot be used in the presence of those metal ions which would be precipitated as insoluble phosphates (i.e. all common cations other than Na^+, K^+, and NH_4^+). Equimolar quantities of potassium salts may be substituted for the sodium phosphates prescribed here.

Stock solutions

A. 0.1M sodium dihydrogen phosphate (=sodium phosphate, monobasic; sodium acid phosphate)

> **Either** NaH_2PO_4: 12.40g; water to 1000ml;
>
> **or** $NaH_2PO_4.H_2O$: 13.80g; water to 1000ml;
>
> **or** $NaH_2PO_4.2H_2O$: 15.60g; water to 1000ml.

(The dihydrate is preferred, since it is neither hygroscopic nor efflorescent.)

B. 1M disodium hydrogen phosphate (=sodium phosphate, dibasic)

> **Either** Na_2HPO_4: 14.20g; water to 1000ml;
>
> **or** $Na_2HPO_4.2H_2O$: 17.80g; water to 1000ml;
>
> **or** $Na_2HPO_4.7H_2O$: 26.81g; water to 1000ml.

(The dihydrate is preferred, since it is neither hygroscopic nor efflorescent.)

pH	A ml 0.1M NaH_2PO_4	B ml 0.1M Na_2HPO_4
5.3	192	8
5.5	188	12
5.7	184	16
5.8	180	20
5.9	174	26
6.0	168	32
6.1	162	38
6.2	154	46
6.3	146	54
6.4	136	64
6.5	128	72
6.6	112	88
6.7	104	96
6.8	96	104
6.9	82	118
7.0	68	132
7.1	56	144
7.2	48	152
7.3	40	160
7.4	32	168
7.5	28	172
7.6	23	177
7.7	17	183
7.8	12	188
7.9	8	192

20.1.5. pH6.6–7.5: HEPES buffer

HEPES (*N*-[2-hydroxyethyl]piperazine-*N*'-2-ethanesulphonic acid) is a zwitterionic amino acid that does not complex with the physiologically important ions Ca^{2+}, Mg^{2+} and Mn^{2+}. It is used in tissue culture media and in a few histochemical procedures.

Stock solutions

A. 0.05 M HEPES

(See above for full name. Use the acid, not its sodium salt.)

HEPES: 11.9 g
Water: to make 1000 ml

B. 1.0 M Sodium chloride (NaCl, 58.5 g per litre)

C. 1.0 M Sodium hydroxide (NaOH, 40.0 g per litre)

pH	A ml	B ml	C ml
6.8	100	9.25	0.75
7.0	100	8.90	1.10
7.2	100	8.46	1.54
7.4	100	7.93	2.07
7.6	100	7.36	2.64
7.8	100	6.80	3.20
8.0	100	6.31	3.69
8.2	100	5.91	4.09

It is important to check the pH with an accurately calibrated meter, because slight inaccuracy with ingredient C will give a buffer with the wrong pH.

20.1.6. pH7.2–9.0: TRIS buffer

This is a 0.05 M TRIS buffer. It is compatible with salts of all heavy metals other than those with insoluble chlorides (e.g. Ag, Pb). The amino group of TRIS reacts with aldehydes, so these buffer solutions are not used in fixatives. The effective concentrations of some metal ions may be reduced by complex formulation with TRIS.

Stock solutions

A. 0.2 M TRIS

Tris (hydroxymethyl) aminomethane: 24.2g
Water: to 1000ml

B. 0.1 M hydrochloric acid (see p. 362)

pH (see *Note*, below)	A ml 0.2 M TRIS	B ml 0.1 M HCl	Water; to make total volume (ml)
7.2	50	89.5	200
7.4	50	84	200
7.6	50	77	200
7.8	50	69	200
8.0	50	58.5	200
8.2	50	46	200
8.4	50	35	200
8.6	50	25	200
8.8	50	17	200
9.0	50	11.5	200

Note

TRIS buffers are more strongly affected by temperature than most others. The pH values in this table are correct at 25°C. For each degree C below 25°C the pH will be higher by 0.03. Thus the mixture in the eighth line of the table (pH8.6 at 25°C) will have pH 8.75 at 20°C and pH 8.39 at 32°C.

20.1.7. pH5.0–7.4: cacodylate-hydrochloric acid buffer

This buffer (Plumel, 1948) is much used in fixatives for electron microscopy, but it is doubtful whether it has any advantage over cheaper, less toxic buffers. Sodium cacodylate is poisonous. Its solution must not be pipetted by mouth.

Stock solutions

A. 0.2 M sodium cacodylate

$Na(CH_3)_2AsO_2.3H_2O$:   42.8g;   water to 1000ml.

This reagent is expensive, so use only as required (e.g. 2.16g for 50ml solution).

B. 0.2M hydrochloric acid (See Section 20.3, p. 362)

pH	A ml 0.2M sodium cacodylate	B ml 0.2M HCl	Water: to make total volume (ml)
5.0	50	47	200
5.2	50	45	200
5.4	50	43	200
5.6	50	39	200
5.8	50	35	200
6.0	50	29.5	200
6.2	50	24	200
6.4	50	18.5	200
6.6	50	13.5	200
6.8	50	9.5	200
7.0	50	6.5	200
7.2	50	4.2	200
7.4	50	2.5	200

20.1.8. pH7.4–9.0: borate buffer

This buffer (Holmes, 1943), which is compatible with low concentrations of silver nitrate, should not be confused with other borax-containing buffers.

Stock solutions

A. 0.2M boric acid

H_3BO_3: 12.4g; water to 1000ml.

Takes 1–2h to dissolve with vigorous magnetic stirring.

B. 0.05M borax

$Na_2B_4O_7.10H_2O$: 19.0g; water to 1000ml.

Dissolves easily.

pH	A ml 0.2M boric acid	B ml 0.5M borax
7.4	180	20
7.6	170	30
7.8	160	40
8.0	140	60
8.2	130	70
8.4	110	90
8.7	80	120
9.0	40	160

20.1.9. pH7.0–9.6: barbitone buffer

This is compatible with most commonly used reagents. Barbitone sodium (also known as barbital sodium, sodium diethylbarbiturate, veronal, and medinal) is subject to dangerous drug control legislation in most countries. It is a toxic substance and should not be pipetted by mouth. It is probably rarely necessary to use barbitone buffer, because several other buffer systems cover the same pH range.

Stock solutions

A. 0.1M barbitone sodium

Barbitone sodium ($C_6H_{11}O_3N_2Na$): 20.62g; water to 1000ml.

B. 0.1M hydrochloric acid (See Section 20.3, p. 362)

pH	A ml 0.1M barbitone sodium	B ml 0.1M HCl
7.0	107	93
7.2	111	89
7.4	116	84
7.6	123	77
7.8	132.5	67.5
8.0	143	57
8.2	154	46
8.4	164.5	35.5
8.6	174	26
8.8	181.5	18.5
9.0	187	13
9.2	190.5	9.5
9.4	195	5
9.6	197	3

20.1.10. pH8.5–12.8: glycine-sodium hydroxide buffer

Stock solutions

A. 0.1M glycine with 0.1M sodium chloride

Glycine (aminoacetic acid; H_2NCH_2COOH):	7.51g
Sodium chloride (NaCl):	5.84g
Water:	to make 1000ml

B. 0.1M sodium hydroxide (See Section 20.3, p. 362)

pH	A ml 0.1M glycine + NaCl	B ml 0.1M NaOH
8.5	190	10
8.8	180	20
9.2	160	40
9.6	140	60
10.0	120	80
10.3	110	90
10.9	102	98
11.1	100	100
11.4	98	102
11.9	90	110
12.2	80	120
12.5	60	140
12.7	40	160
12.8	20	180

20.1.11. pH9.2–10.6: carbonate-bicarbonate buffer

Stock solutions

A. 0.2M sodium carbonate

Sodium carbonate (Na_2CO_3): 21.2g; water to 1000ml.

B. 0.2M sodium bicarbonate

Sodium bicarbonate ($NaHCO_3$): 16.8g; water to 1000ml.

pH	A ml 0.2M Na_2CO_3	B ml 0.2M $NaHCO_3$	Water: to make total volume (ml)
9.2	4	46	200
9.4	9.5	40.5	200
9.6	16	34	200
9.8	22	28	200
10.0	27.5	22.5	200
10.2	33	17	200
10.4	38.5	11.5	200
10.6	42.5	7.5	200

20.2. Physiological solutions

A physiological solution is one in which the structural and functional integrity of a living tissue is sustained for several hours after removal from the body. Physiological solutions mimic the ionic compositions of extracellular fluids and have similar pH and osmotic pressure. They may also contain glucose and other simple nutrients such as glutamate, lactate and pyruvate. They are not tissue culture media, however, and they do not accommodate prolonged life or growth *in vitro*.

The following mixtures are derived largely from the data of Dawson *et al.* (1969). *Each ingredient should be weighed and dissolved in 50–100ml of the water. These solutions should then be added, **in the order given**, to the remaining water.* Precipitation of insoluble salts will thus be avoided. The solutions can be kept until signs of infection are seen; cloudiness or mould usually appears after 2 days at 20°C or 2 weeks at 4°C.

20.2.1. Mammalian Ringer-Locke

Sodium chloride (NaCl):	9.0g
Potassium chloride (KCl):	0.25g
Calcium chloride ($CaCl_2$):	0.30g
Sodium bicarbonate ($NaHCO_3$):	0.5g
Glucose ($C_6H_{12}O_6$):	1.0g
Water:	to make 1000ml

20.2.2. Krebs bicarbonate-phosphate Ringer

Sodium chloride (NaCl):	6.9g
Potassium chloride (KCl):	0.4g
Calcium chloride ($CaCl_2$):	0.3g
Potassium phosphate, monobasic (KH_2PO_4):	0.2g

Magnesium sulphate
(MgSO$_4$.7H$_2$O): 0.3g
Sodium bicarbonate
(NaHCO$_3$): 2.1g
Water: to make 1000ml

This solution is isotonic with mammalian tissues. It can also be made by mixing 0.154M aqueous solutions of the reagents in the following proportions, by volume: NaCl 100; KCl 4; CaCl$_2$ 3; KH$_2$PO$_4$ 1; NaHCO$_3$ 21 (Total = 130 volumes). If the salts are to be kept as stock solutions, they should be stored in concentrated form (1.0M or higher) to inhibit microorganisms, and diluted to 0.154M before using.

20.2.3. Dulbecco's balanced salt solution

Sodium chloride
(NaCl): 8.0g
Potassium chloride
(KCl): 0.2g
Sodium phosphate,
dibasic (Na$_2$HPO$_4$): 1.15g
Potassium phosphate,
monobasic (KH$_2$PO$_4$): 0.2g
Calcium chloride
(CaCl$_2$): 0.1g
Magnesium chloride
(MgCl$_2$.6H$_2$O): 0.1g

This solution was devised for washing tissue cultures (Dulbecco & Vogt, 1954), but it is also suitable for vascular perfusion and many other purposes requiring a buffered, isotonic solution. It contains no bicarbonate, so it is not suitable for prolonged maintenance of isolated organs.

20.2.4. Amphibian Ringer

This, which is the fluid developed by Ringer (1893), differs from the mammalian solution (20.2.1, above) in containing only 6.5g NaCl per litre, and no glucose. The addition of glucose was found by Locke (1895) to prolong the active life of the isolated frog's heart from 5h to 24h.

20.2.5. Phosphate-buffered saline (PBS)

This is not a truly physiological solution, but it is widely used for rinsing unfixed tissue, and as a solvent for antibodies and other reagents used in immunohistochemistry.

0.1M phosphate
buffer, pH7.4: 1000ml
Sodium chloride: 9g

(Keeps indefinitely at room temperature. Discard when infected.)

20.2.6. Normal saline

Sodium chloride
(NaCl): 9.0g
Water: to make 1000ml

This is the simplest isotonic solution. It will not cause osmotic damage to mammalian or avian cells. For use with unfixed material, a buffered solution containing potassium and calcium ions is preferable for all but the briefest rinses.

For tissues of cold-blooded animals, the concentration of NaCl should be reduced to 0.6–0.65%. (See also Section 20.2.4.)

20.3. Dilution of acids and alkalis

The simplest way to express the concentration of a substance is as **molarity** (number of moles per litre of the solution; not to be confused with the much less often used "molality", which is the number of moles per kilogram of solution). However, concentrations of acids and bases are still quite commonly expressed as **normalities**, following an older convention.

A normal solution of an acid or alkali contains the equivalent weight of the substance in a volume of 1 litre.

$$\text{Equivalent weight} = \frac{\text{Molecular weight}}{\text{Number of available H}^+ \text{ or OH}^- \text{ ions per molecule}}$$

Thus the equivalent weight of NaOH (M.W. 40) is $40 \div 1 = 40$, so that a 1.0N solution contains 40g of NaOH per litre. For sulphuric acid (H_2SO_4, M.W. 98.1) the equivalent weight is $98.1 \div 2 = 49.05$, so a 1.0N solution will contain 49.05g of 100% H_2SO_4 per litre.

The common laboratory acids are liquids, so it is convenient to dispense them by volume rather than by weight. Furthermore, these reagents are rarely 100% pure. The bottle in which a concentrated acid is supplied bears a label on which will be found the molecular weight, the specific gravity, and the w/w percentage assay (g of 100% acid per 100g of the liquid in the bottle). From these data, a solution of any normality may be prepared by calculating the volume of the acid needed to make 1 litre of the solution.

$$V = \frac{100MN}{BPD}$$

where V is the required volume of concentrated

acid (**in ml**) to make 1l; N is the desired normality of the solution; M is the molecular weight; P is the percentage assay of concentrated acid (**w/w**); D is the specific gravity of concentrated acid; B is the basicity (i.e. number of available H^+ per molecule). The only common mineral acid whose basicity is greater than 1 is H_2SO_4 ($B = 2$). For all organic acids and for weak inorganic acids the concentration should always be expressed as molarity, not as normality.

Table 20.1 is useful when preparing 1.0N or 1.0M† solutions of common acids and alkalis. When the specific gravity and percentage assay stated on the label differ appreciably from the data in the table, it will be necessary to use the formula explained above.

---

†Do not confuse "molar" (moles per litre of solution, abbreviated to M) with "molal" (moles per kilogram of solution). In histochemical practice, concentrations are hardly ever expressed as molalities.

TABLE 20.1. *Data for preparation of 1 l of 1.0 N or 1.0 M solutions of some acids and alkalis.* Add the "quantity required" to 800 ml water, mix well, then add water to 1000 ml

Name	Concentrated product			Quantity required	Strength of dilute solution
	Assay (w/w) (%)	S.G.	Normality or molarity		
Hydrochloric acid	36	1.18	12 N (= 12 M)	83 ml	1.0 N (= 1.0 M)
Hydrobromic acid	40	1.38	6.8 N (= 6.8 M)	147 ml	1.0 N (= 1.0 M)
Nitric acid	71	1.42	16 N (= 16 M)	63 ml	1.0 N (= 1.0 M)
Perchloric acid	60	1.54	9.2 N (= 9.2 M)	109 ml	1.0 N (= 1.0 M)
Sulphuric acid	96	1.84	36 N (= 18 M)	28 ml	1.0 N (= 0.5 M)
Acetic acid	99.5	1.05	17.4 M (= 17.4 N)	57 ml	1.0 M (= 1.0 N)
Formic acid	90	1.20	23.4 M (= 23.4 N)	42.5 ml	1.0 M (= 1.0 N)
Sodium hydroxide	100 (solid)	2.13	—	40 g	1.0 N (= 1.0 M)
Potassium hydroxide	100 (solid)	2.04	—	56 g	1.0 N (= 1.0 M)
Ammonium hydroxide (= ammonia water)	27% NH_3 35% NH_3	0.901 0.880	14.3 M 18.2 M	70 ml 55 ml	1.0 M 1.0 M

TABLE 20.2. *Approximate atomic weights of the commoner elements, radicals and ions.*

Element Name	Symbol	Atomic weight
Aluminium	Al	27.0
Antimony	Sb	121.8
Arsenic	As	74.9
Barium	Ba	137.3
Beryllium	Be	9.0
Bismuth	Bi	209.0
Boron	B	10.8
Bromine	Br	79.9
Cadmium	Cd	112.4
Calcium	Ca	40.1
Carbon	C	12.0
Cerium	Ce	140.1
Chlorine	Cl	35.5
Chromium	Cr	52.0
Cobalt	Co	58.9
Copper	Cu	63.5
Fluorine	F	19.0
Gold	Au	197.0
Hydrogen	H	1.0
Iodine	I	126.9
Iron	Fe	55.8
Lanthanum	La	138.9
Lead	Pb	207.2
Lithium	Li	6.9
Magnesium	Mg	24.3
Manganese	Mn	54.9
Mercury	Hg	200.6
Molybdenum	Mo	95.9
Nickel	Ni	58.7
Nitrogen	N	14.0
Osmium	Os	190.2
Oxygen	O	16.0
Palladium	Pd	106.4
Phosphorus	P	31.0
Platinum	Pt	195.1
Potassium	K	39.1
Ruthenium	Ru	101.1
Selenium	Se	79.0
Silicon	Si	28.1
Silver	Ag	107.9
Sodium	Na	23.0
Strontium	Sr	87.6
Sulphur	S	32.1
Tellurium	Te	127.6
Thallium	Tl	204.4
Thorium	Th	232.0
Tin	Sn	118.7
Titanium	Ti	47.9
Tungsten	W	183.9
Uranium	U	238.0
Vanadium	V	50.9
Zinc	Zn	65.4
Zirconium	Zr	91.2

Radical or ion	Molecular weight
CH_2	14.0
CH_3	15.0
C_2H_5	29.1
C_6H_5	77.1
CN	26.0
H_2O	18.0
NH	15.0
NH_2	16.0
NH_3	17.0
NH_4	18.0
NO_2	46.0
NO_3	62.0
OH	17.0
PO_4	95.0
SO_3	80.1
SO_4	96.1

Notes

1. **Caution**. Always add the concentrated acid slowly to the larger volume of water, stirring thoroughly to avoid overheating. Sodium or potassium hydroxide should also be added to water in the same way.

   Concentrated HCl, HBr, HNO_3, CH_3COOH, HCOOH, and NH_4OH have pungent vapours. They should be poured in a fume cupboard.

   All the concentrated acids and alkalis in the table are caustic. Avoid contact with skin. If you get concentrated acid on your skin it must be washed off with copious running tap water within 5–15 s if burning is to be prevented.

   Never pipette concentrated acids or alkalis by mouth.

2. **Accuracy**. Volatile substances (HCl, HBr, and especially, NH_4OH) lose potency after the bottles have been opened. When a reagent from an old, two-thirds-empty bottle is used in the preparation of a buffer solution, it is important to check the pH with a meter. Be sure that the meter has been standardized against a reliable buffer solution.

20.4. Atomic weights

Table 20.2 may be used in calculating molecular weights of compounds. In histology and histochemistry, molecular weights accurate to the nearest whole number are adequate. The values of atomic weights in this table are approximated to the first decimal place.

20.5. Suitable tissues for histochemical techniques

This list includes some mammalian tissues with which positive histochemical reactions may easily be obtained. For a longer list, including invertebrate material, see Gabe (1976). For botanical materials, see Jensen (1962) and Klein & Klein (1970).

NUCLEIC ACIDS

DNA	Any cellular tissue (nuclei)
RNA	Brain, spinal cord, ganglia (Nissl substance of neurons); glands (e.g. salivary glands, pancreas, stomach, intestine, pituitary); active lymph nodes (plasma cells)

PROTEINS AND FUNCTIONAL GROUPS

Protein (general, amino and carboxyl groups, tyrosine)	Any tissue (cytoplasm, collagen)
Arginine	Intestine (Paneth cells); lymphoid tissue; any tissue with many nuclei
Tryptophan	Pancreas (exocrine cells, α-cells of islets); amyloid; fibrin
Cysteine	Skin (hair follicles)
Cystine	Skin (stratum corneum, hair shafts); pituitary (neurosecretory material); pancreas (β-cells of islets)

Aldehyde groups	Arterial elastic laminae in young rodents; any tissue fixed in glutaraldehyde (especially cytoplasm and collagen)

CARBOHYDRATES

Glycogen	Liver (hepatocytes)
Proteoglycans:	
Hyaluronic acid	Umbilical cord (Wharton's jelly); eye (vitreous); joints (synovial fluid)
Chondroitin sulphates	Cartilage matrix
Dermatan sulphate	Skin (dermis); tendon; lung (connective tissue)
Keratan sulphates	Cornea
Heparin	Mast cells (e.g. in skin, tongue, mesentery); blood basophils
Neutral glycoproteins	Stomach (surface mucus); thyroid (follicular cells); salivary glands (serous cells); collagen, reticulin
Acid glycoproteins:	
Sulphated	Rat or mouse tongue (mucous glands); rat or mouse duodenum (Brunner's glands, goblet cells); colon (goblet cells)
With sialic acids, labile to neuraminidase	Rat or mouse rectum (goblet cells); mouse sublingual salivary gland (mucous cells)
With sialic acids, labile to neuraminidase only after saponification	Rat sublingual salivary gland (mucous cells)

LIPIDS

Neutral fats	Adipose connective tissue
Phospholipids	Brain, peripheral nerve (myelin); heart, kidney (mitochondria); erythrocytes
Cholesterol esters	Degenerating myelin (2-4 weeks after transection of axons or destruction of neuronal somata in CNS); atherosclerotic lesions in human arteries
Steroids	Adrenal cortex; testis (Leydig cells)
Cholesterol	Brain (myelin)

INORGANIC IONS

Calcium (phosphate and carbonate)	Sites of pathological or senile calcification (kidney, tendons, human pineal gland); incompletely decalcified bones or teeth
Calcium (soluble salts)	Kidney, muscle, nervous tissue
Iron	Liver, spleen, bone marrow (phagocytic cells); any tissue at site of old injury or haemorrhage
Zinc	Blood, haemopoietic tissue (granular leukocytes); prostate gland (cells of ducts); pancreas (β-cells of islets); brain (neuropil of various regions, especially hippocampus)

ENZYMES

Acid phosphatase	Kidney (proximal tubule epithelium); liver (hepatocytes, phagocytic cells); prostate; intestine (cytoplasm of epithelial cells). Nuclear staining is a common artifact
Alkaline phosphatase	Kidney (brush-border of proximal tubules); intestine (brush-border of epithelium); rat brain (endothelium)
Esterases ("non-specific")	Liver (hepatocytes); kidney (tubules); brain (neuroglia, pericytes)
Acetylcholin-esterase	Muscle (motor endplates); brain (some neurons, neuropil, and axons)
Cholinesterase ("pseudo-cholinesterase")	Brain (some neurons, capillary endothelium in rat)
Dehydrogenases	Liver, kidney, heart, intestine, etc. (cytoplasm, mitochondria)
Cytochrome oxidase	Liver, kidney, heart, intestine, etc. (mitochondria)
Monophenol oxygenase	Skin (melanocytes in epidermis and dermis); eye (retina, choroid, iris). Do not use albino animal

Peroxidase	Blood, haemopoietic tissue (granular leukocytes). Exogenous HRP in motor neurons 24–48h after injection into muscle. Erythrocytes exhibit peroxidase-like activity due to haemoglobin	Noradrenaline	Adrenal medulla (large amounts in some chromaffin cells); brain (small amounts in some axons); ductus deferens (sympathetic axons)
		Adrenaline	Adrenal medulla (large amounts in some chromaffin cells)
AMINES		Dopamine	Mast cells in lungs of ruminants (large amounts); stomach of rat (small amounts in endocrine cells in mucosa of fundus)
Serotonin	Intestine (large amounts in argentaffin cells of epithelium); rat or mouse mast cells (large amounts in granules). Brain (medulla), spinal cord (small amounts in some neurons and axons)		

Glossary

Many terms are defined in the text. For these, consult the index. The following list includes a variety of chemical and histological terms with which some readers may be unfamiliar.

Acetal. Compound formed by condensation of one molecule of an aldehyde with two molecules of an alcohol to give the structure:

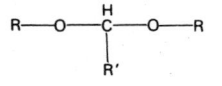

Adduct. Compound formed by combination of two others without the loss of any atoms. Often used when the precise structure of the addition compound is uncertain, or when one of the components is a reagent used for analytical purposes.

Albumen. The principal protein of egg-white. The penultimate *e* distinguishes it from the **albumins**, which are proteins soluble in water and precipitated by high concentrations of salts (e.g. saturation with $(NH_4)_2SO_4$), such as serum albumin.

Aprotic solvent. A polar solvent which does not contain an ionizable hydrogen atom. Its molecules cluster around (solvate) cations, but leave anions relatively unimpeded, so that the latter will be more reactive than when dissolved in an ordinary (protic) polar solvent. Examples are dimethylsulphoxide and *N*-dimethylformamide.

Astrocyte. A neuroglial cell with numerous cytoplasmic processes, some of which form end-feet on capillary blood vessels.

Axon. That cytoplasmic process of a neuron which is specialized for the conduction of trains of impulses, usually away from the soma.

Canonical forms. The different structures which may exist, at instants in time, of an organic compound in which resonance occurs. In writing canonical structures, only bonds and sites of electrical charge may be varied; the positions of the atoms may not be changed.

Caudal. Towards the tail of an animal. Used mainly to refer to relative positions along the axis of the central nervous system.

Chromatin. The material in the nucleus of a cell (excluding the nucleolus) which is stained by cationic dyes and by some dye-metal complexes such as aluminium-haematein. Consists of the DNA and nucleoprotein of the chromosomes.

Common ion effect. The tendency of a salt to become less soluble when the concentration of one of its ions in a solution greatly exceeds that of the other ion.

Condensation. Combination of two molecules with elimination of a compound of low molecular weight such as water.

Delocalized π-electrons. Electrons, associated with double or triple bonds between atoms or with resonant structures such as aromatic rings, which are shared by more than two atoms and cannot therefore be said to form part of any individual covalent bond.

Dendrites. Processes of a neuron specialized for receiving synaptic connections and conducting impulses towards the soma. Dendrites are usually multiple and shorter than the axon. They also differ ultrastructurally and electrophysiologically from axons.

Dialysis. Passage of small, but not large, molecules through a semi-permeable membrane. A technique for the purification or concentration of solutions of proteins or other macromolecular substances.

Dimer. A molecule formed by the union of two molecules of the same compound.

Enantiomers. Isomers whose three-dimensional structures are mirror-images of one another.

Furanose. A sugar whose ring structure consists of four carbon atoms and one oxygen atom, so that it could be thought of as a derivative of **furan**:

Gel. A colloidal solution with a semi-solid consistency due to extensive hydrogen bonding between the suspended macromolecules and the "solvent", which is usually water.

Gel filtration. A technique whereby molecules of different size are separated by virtue of their entry or non-entry into the pores contained in beads of a suitably designed polymer. Usually, the polymer is packed in a column and a solution containing the substances to be separated is applied. When the column is eluted with a suitable solvent, the larger molecules are released first and the smaller molecules later.

Glycocalyx. The carbohydrate-containing material present on the outer surfaces of all plasmalemmae.

Haem. The non-protein portion of the haemoglobin molecule. Often used more generally for iron-porphyrin prosthetic groups of proteins, such as occur in many enzymes and cytochromes.

Haematin. A pigment formed when haemoglobin is degraded in acid conditions. Also known as acid haematin. Do not confuse with haematein.

Haemopoietic tissue. Tissue such as red bone marrow in which the cells of the blood are produced.

Hemiacetal. Compound formed by condensation of one molecule of an aldehyde with one molecule of an alcohol to give the structure:

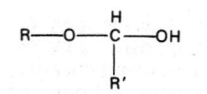

Hydrophilic. Describes substances which attract water: water molecules are able to come into intimate contact with a hydrophilic compound because the latter contains oxygen or nitrogen atoms with which hydrogen bonds can be formed.

Hydrophobic. Describes substances which repel water: a hydrophobic compound has few or no atoms capable of forming hydrogen bonds.

Hydroxyalkylation. Addition of an aldehyde or ketone to an aromatic ring.

Hypertonic. Having a higher osmotic pressure than blood or extracellular fluid.

Hypotonic. Having a lower osmotic pressure than blood or extracellular fluid.

Imide. A compound in which the two bonds of the $>$NH radical are joined to acyl groups, to give the structure:

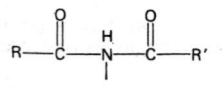

Imine. A compound containing the configuration $-\overset{H}{\underset{}{C}}=N-$. Such compounds are also known as azomethines, anils, or Schiff's bases. The term "imino" is sometimes applied (though not in this book) to the $>$NH radical of secondary amines.

Isotonic. Having the same osmotic pressure as blood or extracellular fluid.

Ketal. A compound formed by condensation of a ketone with an alcohol, to give the structure:

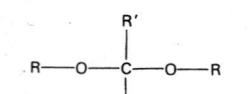

Le Chatelier's principle. When a constraint is applied to any system in equilibrium, the system will always react in a direction which will tend to counteract the applied constraint. For chemical equilibria, the "constraint" may be a change in concentration of a reactant, or a change of temperature, etc. The common ion effect is a consequence of this principle.

Lipofuscin. A pigment containing lipids and proteins, found as granules within the cytoplasm of cells, especially in old animals. Thought to be the indigestible remains of phagocytosed material.

Metabolite. Any substance participating in a chemical reaction in a living organism.

Milieu intérieur. The "internal environment" of the cells of an organism. In higher animals, this is the extracellular fluid.

Monoamine oxidase. An enzyme which catalyses the oxidative degradation of biogenic monoamines such as noradrenaline and serotonin.

Myelin. The sheath surrounding many of the axons of the central and peripheral nervous systems of vertebrate animals. It contains numerous proteins and lipids and is formed from the plasmalemmae of the non-neuronal cells that ensheath the axon. The myelin is trophically dependent upon the axon; it disintegrates and is phagocytosed if the axon dies.

Neuroglia. The cells other than neurons that occur in the central nervous system. Often the term is used more generally to include also the non-neuronal cells of peripheral nerves (Schwann cells) and ganglia (satellite cells) and the non-neuronal cells in the nervous systems of invertebrate animals.

Neurohypophysis. The portion of the pituitary gland derived from the central nervous system. Comprises the median eminence of the ventral surface of the brain, the stalk of the gland and the posterior lobe of the gland.

Neuropil. Tissue within the nervous system consisting of axons and dendrites, with numerous synapses, but without neuronal somata or tracts of myelinated axons.

Neurosecretion. The secretion of a substance by a neuron for release into the blood. The word is sometimes also applied to neurons whose axons terminate upon endocrine cells. Neurons of the latter type do not usually contain products stainable by the traditional methods for neurosecretory material.

Non-polar solvent. A hydrophobic liquid, not miscible with water, such as benzene or carbon tetrachloride. A molecule is non-polar because its electrons are symmetrically distributed.

Oligodendrocyte. A neuroglial cell with few cytoplasmic processes; responsible for formation of myelin sheaths in the central nervous system.

Periodontal membrane. The connective tissue which anchors a tooth into its bony socket.

Plasmalemma. The membrane forming the outside surface of a cell. Also called the cell membrane. Not to be confused with the cell-wall in plants, which is external to the plasmalemma.

Polar solvent. A liquid miscible with water and capable of dissolving ionized substances. A molecule is polar because its electrons are unevenly distributed, so that one end is relatively electropositive and the other end relatively electronegative. Such a molecule is known as a dipole.

Pyranose. A sugar whose ring structure consists of five carbon atoms and one oxygen atom, so that it can be thought of as a derivative of the hypothetical substance **pyran:**

Quinaldine. An aromatic heterocyclic compound, also known as 2-methylquinoline:

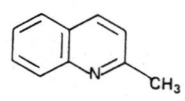

Quinhydrone. The darkly coloured substance formed when hydroquinone is part oxidized to quinone. Formed by hydrogen bonding of hydroquinone to *p*-quinone:

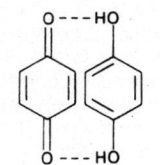

Reserpine. A drug which causes biogenic monoamines to be released from the cells in which they are stored. The cells are thereby depleted of amines.

Rostral. Towards the beak or nose of an animal. Mainly applied to levels of the axis of the central nervous system.

Salting out. Precipitation of an ionic compound due to addition of an excess of one of its ions to the solution. Also applied to precipitation of protein by addition of an inorganic salt to its solution.

Sol. A colloidal dispersion of an inorganic substance, such as gold, ferric hydroxide or sulphur, in a "solvent", which is usually water. Unlike a gel, a sol is a mobile liquid. The particles suspended in a sol are charged; the balancing opposite charge is carried by the solvent molecules surrounding each particle.

Soma (plural: **somata**). A body. Applied to the cell-body of a neuron, in which the nucleus is contained.

Sulphoamino. The radical $-\underset{\underset{\text{N}}{|}}{\overset{\overset{\text{H}}{|}}{}}-SO_3H.$

Tunicates. A subphylum of the Chordata, also known as Urochordata, including the ascidians or sea-squirts. Only the larval form has a notochord. The adult animal is tubular and is covered externally by a "test" composed of cellulose.

Vital staining. The application of dyes to living cells. With **intravital staining**, the dye is administered to the whole animal or plant; with **supravital** (or supervital) staining, freshly excised tissue is treated with a dye solution. Effects due to vital staining cannot be obtained when the tissue is dead.

Wallerian degeneration. The fragmentation and eventual disappearance of axons and their myelin sheaths following severance from or destruction of neuronal somata.

Zwitterionic. A zwitterion is a single molecule bearing both positively and negatively charged groups, such as an amino acid.

Bibliography

ABADIA-FENOLL, F. (1968) Staining pericellular boutons of glutaraldehyde-fixed nervous tissue: a chromation-silvering technique for frozen sections. *Stain Technology* **43**, 190–195.

ABBUHL, M. F. and VELASCO, M. E. (1985) An economical minichamber for immunohistochemical incubation. *Journal of Histochemistry and Cytochemistry* **33**, 162–164.

ABRAHART, E. N. (1968) *Dyes and their Intermediates.* Oxford: Pergamon Press.

ACCINI, L., HSU, K. C., SPIELE, H. and DE MARTINO, C. (1974) Picric acid-formaldehyde fixation for immunoferritin studies. *Histochemistry* **42**, 257–264.

ADAMS, C. W. M. (1965) Histochemistry of lipids. Ch. 2 in *Neurohistochemistry*, ed. C. W. M. Adams. pp. 6–66. Amsterdam: Elsevier.

ADAMS, C. W. M. and BAYLISS, O. B. (1962) The release of protein, lipid and polysaccharide components of the arterial elastica by proteolytic enzymes and lipid solvents. *Journal of Histochemistry and Cytochemistry* **10**, 222–226.

ADAMS, C. W. M. and BAYLISS, O. B. (1968) Reappraisal of osmium tetroxide and OTAN histochemical reactions. *Histochemie* **16**, 162–166.

ADAMS, C. W. M. and SLOPER, J. C. (1955) Technique for demonstrating neurosecretory material in the human hypothalamus. *Lancet* **1955–I**, 651–652.

ADAMS, C. W. M. and TUQAN, N. A. (1961) The histochemical demonstration of protease by a gelatin-silver film substrate. *Journal of Histochemistry and Cytochemistry* **9**, 469–472.

ADAMS, C. W. M., ABDULLAH, Y. H. and BAYLISS, O. B. (1967) Osmium tetroxide as a histochemical and histological reagent. *Histochemie* **9**, 68–77.

ADAMS, J. C. (1981) Heavy metal intensification of DAB-based HRP reaction product. *Journal of Histochemistry and Cytochemistry* **29**, 775.

AKERT, K. and SANDRI, C. (1968) An electron microscope study of zinc-iodide-osmium impregnation of neurones. I. Staining of synaptic vesicles and cholinergic junctions. *Brain Research* **7**, 286–295.

ALLEN, R. D. (1987) The microtubule as an intracellular engine. *Scientific American* **256 (No. 2)**, 42–49.

ALLEN, R. L. M. (1971) *Colour Chemistry.* London: Nelson.

ALLISON, R. T. (1987) The effects of various fixatives on subsequent lectin binding to tissue sections. *Histochemical Journal* **19**, 65–74.

ALUND, M. and OLSON, L. (1979) Quinacrine affinity

of endocrine cell systems containing dense core vesicles as visualized by fluorescence microscopy. *Cell and Tissue Research* **204**, 171–186.

ANDERSON, M. J. and COHEN, M. W. (1974) Fluorescent staining of acetylcholine receptors in vertebrate skeletal muscle. *Journal of Physiology* **237**, 385–400.

ANGERER, L. M., COX, K. H. and ANGERER, R. C. (1987) Demonstration of tissue-specific gene expression by *in situ* hybridization. Ch. 68 in *Guide to Molecular Cloning Techniques*, ed. S. L. Berger and A. R. Kimmel. (*Methods in Enzymology*, Vol. **152**, pp. 649–661.) Orlando, Florida: Academic Press.

ANGERMULLER, S. and FAHIMI, H. D. (1987) Electron microscopic cytochemical localization of oxidases with the cerium technique. *Journal of Anatomy* **155**, 222–223.

APPENTENG, K., BATTEN, T. F. C. and OAKLEY, B. A. (1986) An alternative method for visualization of immunohistochemical reactions in HRP-labelled neurones. *Journal of Physiology* **238**, 179P.

ARCHIMBAUD, E., ISLAM, A. and PREISLER, H. D. (1986) Alcian blue method for attaching glycol methacrylate sections to glass slides. *Stain Technology* **61**, 121–123.

ARMSTRONG, J. and STEPHENS, P. R. (1960) A modified chrome-silver paraffin wax technique for staining neural end-feet. *Stain Technology* **35**, 71–75.

ARMSTRONG, J., RICHARDSON, K. C. and YOUNG, J. Z. (1956) Staining neural end-feet mitochondria after postchroming and Carbowax embedding. *Stain Technology* **31**, 263–270.

ASCHOFF, A. and HOLLANDER, H. (1982) Fluorescent compounds as retrograde tracers compared with horseradish peroxidase (HRP). I. A parametric study in the central nervous system of the albino rat. *Journal of Neuroscience Methods* **6**, 179–197.

ASHWOOD-SMITH, M. J. (1971) Radioprotective and cryoprotective properties of DMSO. Ch. 5 in *Dimethyl Sulfoxide*, ed. S. W. Jacob, E. E. Rosenbaum and D. C. Wood. Vol. 1, pp. 147–187. New York: Marcel Dekker.

ASOU, H., BRUNNGRABER, E. G. and JENG, I. (1983) Cellular localization of GM-1 ganglioside with biotinylated choleragen and avidin peroxidase in primary cultured cells from rat brain. *Journal of Histochemistry and Cytochemistry* **31**, 1375–1379.

AXELSSON, S., BJORKLUND, A., FALCK, A., LINDVALL, O. and SVENSSON, L-A. (1973) Glyoxylic acid condensation: a new fluorescence method for the histochemical demonstration of biogenic monoamines. *Acta Physiologica Scandinavica* **87**, 57–62.

BACKSTROM, G., HALLEN, A., HOOK, M., JANSSON, L. and LINDAHL, U. (1975) Biosynthesis of heparin. In *Heparin*, ed. R. A. Bradshaw & S. Wessler (*Advances in Experimental Medicine and Biology*, Vol. **52**). pp. 61–72. New York: Plenum Press.

BAHR, G. F. (1954) Osmium tetroxide and ruthenium tetroxide and their reactions with biologically important substances. *Experimental Cell Research* **7**, 457–479.

BAKER, J. R. (1946) The histochemical recognition of lipine. *Quarterly Journal of Microscopical Science* **87**, 441–470.

BAKER, J. R. (1947) The histochemical recognition of certain guanidine derivatives. *Quarterly Journal of Microscopical Science* **88**, 115–121.

BAKER, J. R. (1956) The histochemical recognition of phenols especially tyrosine. *Quarterly Journal of Microscopical Science* **97**, 161–164.

BAKER, J. R. (1958) *Principles of Biological Microtechnique*. (Reprinted 1970, with corrections) London: Methuen.

BAKER, J. R. (1960) Experiments on the action of mordants. I. "Single-bath" mordant dyeing. *Quarterly Journal of Microscopical Science* **101**, 255–272.

BAKER, J. R. (1962) Experiments on the action of mordants. 2. Aluminium-haematein. *Quarterly Journal of Microscopical Science* **103**, 493–517.

BAKER, J. R. and WILLIAMS, E. G. M. (1965) The use of methyl green as a histochemical reagent. *Quarterly Journal of Microscopical Science* **106**, 3–13.

BALD, W. B. (1983) Optimizing the cooling block for the quick freeze method. *Journal of Microscopy* **131**, 11–23.

BANCROFT, J. D. and COOK, H. C. (1984) *Manual of Histological Techniques*. Edinburgh & London: Churchill-Livingstone.

BANCROFT, J. D. and STEVENS, A. (ed) (1982) *Theory and Practice of Histological Techniques*. 2nd ed. Edinburgh: Churchill-Livingstone.

BANGLE, R. (1954) Gomori's paraldehyde-fuchsin stain. I. Physico-chemical and staining properties of the dye. *Journal of Histochemistry and Cytochemistry* **2**, 291–299.

BANGLE, R. and ALFORD, W. C. (1954) The chemical basis of the periodic acid Schiff reaction of collagen fibres with reference to periodate consumption by collagen and insulin. *Journal of Histochemistry and Cytochemistry* **2**, 62–76.

BARBER, P. C. (1982) Differentiation of neurons containing olfactory marker protein in adult rat olfactory epithelium transplanted to the anterior chamber of the eye. *Neuroscience* **7**, 2687–2695.

BARGMANN, W. (1949) Über die neurosekretorische Verknupfung von Hypothalamus und Neurohypophyse. *Zeitschrift für Zellforschung* **34**, 610–634.

BARKA, T. and ANDERSON, P. J. (1963) *Histochemistry. Theory, Practice and Bibliography*. New York: Harper & Row.

BARNETT, R. J. and SELIGMAN, A. M. (1958) Histochemical demonstration of protein-bound alpha-acylamido carboxyl groups. *Journal of Biophysical and Biochemical Cytology* **4**, 169–176.

BARNETT, S. J. and BOURNE, G. (1941) Use of silver nitrate for the histochemical demonstration of ascorbic acid. *Nature* **147**, 542–543.

BARRETT, A. J. (1971) The biochemistry and function of mucosubstances. *Histochemical Journal* **3**, 213–221.

BARRETT, J. T. (1987) *Textbook of Immunology: An Introduction to Immunochemistry and Immunobiology*. 5th ed. St. Louis: Mosby.

BAYLISS-HIGH, O. (1977) Lipids. Ch. 12 in *Theory and Practice of Histological Techniques*, ed. J. D. Bancroft and A. Stevens. pp. 168–185. Edinburgh: Churchill-Livingstone.

BAYLISS-HIGH, O. (1984) *Lipid Histochemistry* (Royal Microscopical Society; Microscopy Handbooks 06). Oxford: Oxford University Press.

BAYLISS, O. B. and ADAMS, C. W. M. (1972) Bromine-Sudan black: a general stain for lipids including free cholesterol. *Histochemical Journal* **4**, 505–515.

BAYLISS, O. B. and ADAMS, C. W. M. (1979) The pH dependence of borohydride as an aldehyde reductant. *Histochemical Journal* **11**, 111-116.

BEACH, T. G. and McGEER, E. G. (1988) Retrograde filling of pyramidal neurons in postmortem human cerebral cortex using horseradish peroxidase. *Journal of Neuroscience Methods* **23**, 187–93.

BECKMANN, H-J. and DIERICHS, R. (1982) Lipid extracting properties of 2,2-dimethoxypropane as revealed by electron microscopy and thin layer chromatography. *Histochemistry* **76**, 407–412.

BEDOSSA, P., BACCI, J., LEMAIGRE, G. and MARTIN, E. (1987) Effects of fixation and processing on the immunohistochemical visualization of type-I, III and IV collagen in paraffin-embedded tissue. *Histochemistry* **88**, 85–89.

BELL, M. A. and SCARROW, W. G. (1984) Staining for microvascular alkaline phosphatase in thick celloidin sections of nervous tissue: morphometric and pathological applications. *Microvascular Research* **27**, 189-203.

BENDAYAN, M. (1981) Ultrastructural localization of actin in insulin-containing granules. *Biology of the Cell* **41**, 157–160.

BENDAYAN, M. and DUHR, M.-A. (1986) Modification of the protein A-gold immunocytochemical technique for the enhancement of its efficiency. *Journal of Histochemistry and Cytochemistry* **34**, 569–575.

BENHAMOU, N., GILBOA-GARBER, N., TRUDEL, J. and ASSELIN, A. (1988) A new lectin-gold complex for ultrastructural localization of galacturonic acids. *Journal of Histochemistry and Cytochemistry* **36**, 1403–1411.

BENKOEL, L., CHAMLIAN, A., BARRAT, E. and LAFFARGUE, P. (1976) The use of ferricyanide for the electron microscopic demonstration of dehydrogenases in human steroidogenic cells. *Journal of Histochemistry and Cytochemistry* **24**, 1194–1203.

BENNETT, H. S., WYRICK, A. D., LEES, S. W. and McNEIL, J. H. (1976) Science and art in preparing tissues embedded in plastic for light microscopy, with special reference to glycol methacrylate, glass knives and simple stains. *Stain Technology* **51**, 71–79.

BENNION, P. J., HOROBIN, R. W. and MURGATROYD, L. B. (1975) The use of a basic dye (azure A or toluidine blue) plus a cationic surfactant for selective staining of RNA: a technical and mechanistic study. *Stain Technology* **50**, 307–313.

BENSON, R. C., MEYER, R. A., ZARUBA, M. E. and McKHANN, G. M. (1979) Cellular autofluorescence—is it due to flavins? *Journal of Histochemistry and Cytochemistry* **27**, 44–48.

BERG, W. F. and FORD, D. G. (1949) Latent image distribution as shown by physical development. *Photographic Journal* **89B**, 31–36. Cited from *Abridged Scientific Publications from the Kodak Research Laboratories* **31**, 194–199 (1951). Rochester, NY: Eastman-Kodak.

BERGERON, J. A. and SINGER, M. (1958) Metachromasy: an experimental and theoretical reevaluation. *Journal of Biophysical and Biochemical Cytology* **4**, 433–457.

BERGQVIST, A., CARLSTROM, K. and LJUNGBERG, O. (1984) Histochemical localization of estrogen and progesterone receptors: evaluation of a method. *Journal of Histochemistry and Cytochemistry* **32**, 493–500.

BERNHARD, G. R. (1974) Microwave irradiation as a generator of heat for histological fixation. *Stain Technology* **49**, 215–224.

BERRY, J. P., HOURDRY, J., STERNBERG, M. and GALLE, P. (1982) Aluminium phosphate visualization of acid phosphatase activity: a biochemical and X-ray microanalysis study. *Journal of Histochemistry and Cytochemistry* **30**, 86–90.

BERTALANFFY, L. von and BICKIS, I. (1956) Identification of cytoplasmic basophilia (ribonucleic acid) by fluorescence microscopy. *Journal of Histochemistry and Cytochemistry* **4**, 481–493.

BERUBE, G. R., POWERS, M. W., KERKAY, J. and CLARK, G. (1966) The gallocyanin-chrome alum stain: influence of methods of preparation on its activity and separation of the active staining compound. *Stain Technology* **41**, 73–81.

BIGBEE, J. W., KOSEK, J. C. and ENG, L. F. (1977) Effects of primary antiserum dilution on staining of "antigen-rich" tissues with the peroxidase-antiperoxidase technique. *Journal of Histochemistry and Cytochemistry* **25**, 443–447.

BILLMAN, J. H. and DIESING, A. C. (1957) Reduction of Schiff bases with sodium borohydride. *Journal of Organic Chemistry* **22**, 1068–1070.

BIRD, C. L. and BOSTON, W. S. (ed.) (1975) *The Theory of Coloration of Textiles*. Bradford: Dyer's Company Publications Trust.

BIRRELL, G. B., HEDBERG, K. K. and GRIFFITH, O. H. (1987) Pitfalls of immunogold labelling: analysis by light microscopy, transmission electron microscopy, and photoelectron microscopy. *Journal of Histochemistry and Cytochemistry* **35**, 843–853.

BJORKLUND, A. (1983) Fluorescence histochemistry of biogenic amines. Ch. 2 in *Handbook of Chemical Neuroanatomy*, ed. A. Bjorklund & T. Hokfelt, Vol. 1, pp. 50–121. Amsterdam: Elsevier.

BJORKLUND, A., FALCK, B. and LINDVALL, O. (1975) Microspectrofluorometric analysis of cellular monoamines after formaldehyde or glyoxylic acid condensation. Ch. 5 in *Methods in Brain Research*, ed. P. B. Bradley, pp. 249–294. London: Wiley.

BLACKSTAD, T. W., FREGERSLEV, S., LAURBERG, S. and ROKKEDAL, K. (1973) Golgi impregnation with potassium dichromate and mercurous or mercuric nitrate: identification of the precipitate by X-ray diffraction methods. *Histochemie* **36**, 247–268.

BLOOM, F. E. and BATTENBERG, E. L. F. (1976) A rapid, simple and sensitive method for the demonstration of central catecholamine-containing axons. II. A detailed description of methodology. *Journal of Histochemistry and Cytochemistry* **24**, 561–571.

BODIAN, D. (1936) A new method for staining nerve fibres and nerve endings in mounted paraffin sections. *Anatomical Record* **65**, 89–97.

BONDI, A., CHIEREGATTI, G., EUSEBI, G., FULCHERI, E. and BUSSOLATI, G. (1982) The use of β-galactosidase as a tracer in immunocytochemistry. *Histochemistry* **76**, 153–158.

BONILLA, E. and PRELLE, A. (1987) Application of Nile blue and Nile red, two fluorescent probes, for detection of lipid droplets in human skeletal muscle. *Journal of Histochemistry and Cytochemistry* **35**, 619–621.

BOON, M. E. and DRIJVER, J. S. (1986) *Routine Cytological Staining Techniques. Theoretical Background and Practice*. New York: Elsevier.

BOTTCHER, C. J. F. and BOELSMA-VAN HOUTE, E. (1964) Method for the histochemical identification of choline-containing compounds. *Journal of Atherosclerosis Research* **4**, 109–112.

BOURNE, G. (1933) Vitamin C in the adrenal gland. *Nature* **131**, 874.

BOYD, W. C. (1970) Lectins. *Annals of the New York Academy of Sciences* **169**, 168–190.

BRAAK, H. and JACOB, K. (1973) Dye staining of neural end-feet for light microscopy after glutaraldehyde fixation. *Stain Technology* **48**, 181–183.

BRADBURY, S. (1973) *Peacock's Elementary Microtechnique*. 4th ed. London: Edward Arnold.

BRAITENBERG, V., GUGLIELMOTTI, V. and SADA, E. (1967) Correlation of crystal growth with the staining of axons by the Golgi procedure. *Stain Technology* **42**, 277–283.

BRANDTZAEG, P. (1981) Prolonged incubation time in immunocytochemistry: effects on fluorescence staining of immunoglobulins and epithelial components in ethanol- and formaldehyde-fixed paraffin-embedded tissues. *Journal of Histochemistry and Cytochemistry* **29**, 1302–1315.

BREUER, A. C., EAGLES, P. A., LYNN, M. P., ATKINSON, M. B., GILBERT, S. P., WEBER, L., LEATHERMAN, J. and HOPKINS, J. M. (1988) Long-term analysis of organelle translocation in isolated axoplasm of *Myxicola infundibulum*. *Cell Motility and the Cytoskeleton* **10**, 391–399.

BRIDGES, J. W. (1968) Fluorescence of organic compounds. Ch. 2 in *Luminescence in Chemistry*, ed. E. J. Bowen. London: van Nostrand. pp. 77–115.

BRIGGS, R. T., DRATH, D. B., KARNOVSKY, M. L. and KARNOVSKY, M. J. (1975) Localization of NADH oxidase on the surface of human polymorphonuclear leukocytes by a new cytochemical method. *Journal of Cell Biology* **67**, 566–586.

BRIMACOMBE, J. S. and WEBBER, J. M. (1964) *Mucopolysaccharides*, Amsterdam: Elsevier.

BRITTON, H. T. S. (1956) *Hydrogen Ions*. 4th ed. (2 vols). London: van Nostrand.

BROOKER, L. G. S. (1966) Sensitizing and desensitizing dyes. Ch. 11 in *The Theory of the Photographic Process*, ed. T. H. James. New York: Macmillan. pp. 198–232.

BROWN, D., GARCIA-SEGURA, L-M. and ORCI, L. (1984) Carbonic anhydrase is present in olfactory receptor cells. *Histochemistry* **80**, 307–309.

BUEHNER, T. S., NETTLETON, G. S. and LONG-LEY, J. B. (1979) Staining properties of aldehyde fuchsin analogs. *Journal of Histochemistry and Cytochemistry* 27, 782–787.

BULLOCK, G. R. (1984) The current status of fixation for electron microscopy: a review. *Journal of Microscopy* 133, 1–15.

BULMER, D. (1962) Observations on histological methods involving the use of phosphotungstic and phosphomolybdic acids, with particular reference to staining with phosphotungstic acid/haematoxylin. *Quarterly Journal of Microscopical Science* 103, 311–323.

BURNESS, D. M. and POURADIER, J. (1977) The hardening of gelatin and emulsions. Ch. 2 in *The Theory of the Photographic Process*, 4th ed., ed. T. H. James. pp. 77–87. New York: Macmillan.

BURNS, J. (1982) The unlabelled antibody peroxidase-anti-peroxidase method (PAP). *In Techniques in Immunocytochemistry*, ed. G. R. Bullock & P. Petrusz, Vol. 1, pp. 91–105. London: Academic Press.

BURNS, V. W. F. (1972) Location and molecular characteristics of fluorescent complexes of ethidium bromide in the cell. *Experimental Cell Research* 75, 200–206.

BURSTONE, M. S. (1959) New histochemical techniques for the demonstration of tissue oxidase (cytochrome oxidase). *Journal of Histochemistry and Cytochemistry* 7, 112–122.

BURSTONE, M. S. (1962) *Enzyme Histochemistry*. New York: Academic Press.

BUTCHER, L. L. (1983) Acetylcholinesterase histochemistry. Ch. 1 in *Handbook of Chemical Neuroanatomy*, ed. A. Bjorklund & T. Hokfelt, Vol. 1, Methods in Chemical Neuroanatomy, pp. 1–49. Amsterdam: Elsevier.

CAIN, A. J. (1947) Use of Nile blue in the examination of lipoids. *Quarterly Journal of Microscopical Science* 88, 383–392.

CAINELLI, G. and CARDILLO, G. (1984) *Chromium Oxidations in Organic Chemistry*. Berlin: Springer-Verlag.

CAJAL, S. RAMON Y. (1911) *Histologie du Système Nerveux de l'homme et des vertébrés*. (2 vols; transl. L. Azoulay). Paris: Maloine.

CANNON, H. G. (1937) A new biological stain for general purposes. *Nature* 139, 549.

CARES, R. (1945) A note on stored formaldehyde and its easy reconditioning. *Journal of Technical Methods and Bulletin of the International Association of Medical Museums* 25, 67–70.

CARLQUIST, S. (1982) The use of ethylenediamine in softening hard plant structures for paraffin sectioning. *Stain Technology* 57, 311–317.

CARLSSON, A., FALCK, B. and HILLARP, N-A. (1962) Cellular localization of brain monoamines. *Acta Physiologica Scandinavica* 54, Suppl. 196, 1–28.

CARRAPIÇO, F., MADALENA-COSTA, F. and PAIS, M. S. S. (1984) Impregnation of biological material by ZnI_2-OsO_4, KI-OsO_4 and NaI-OsO_4 mixtures for electron microscopic observations: chemical

interpretation of the reaction. *Journal of Microscopy* 134, 193–202.

CARTER, H. E., GLICK, F. J., NORRIS, W. P. and PHILLIPS, G. E. (1947) Biochemistry of the sphingolipids. III. The structure of sphingosine. *Journal of Biological Chemistry* 170, 285–294.

CASON, J. E. (1950) A rapid one-step Mallory-Heidenhain stain for connective tissue. *Stain Technology* 25, 225–226.

CASSELMANN, W. G. B. (1955) Cytological fixation by chromic acid and dichromates. *Quarterly Journal of Microscopical Science* 96, 203–222.

CAWOOD, A. H., POTTER, U. and DICKINSON, H. G. (1978) An evaluation of coomassie brilliant blue as a stain for quantitative microdensitometry of protein in section. *Journal of Histochemistry and Cytochemistry* 26, 645–650.

CELIO, M. R., LUTZ, H., BINZ, H. and FEY, H. (1979) Protein A in immunoperoxidase techniques. *Journal of Histochemistry and Cytochemistry* 27, 691–698.

CHABEREK, S. and MARTELL, A. E. (1959) *Organic Sequestering Agents*. New York: Wiley.

CHALMERS, G. R. and EDGERTON, V. R. (1989) Marked and variable inhibition by chemical fixation of cytochrome oxidase and succinate dehydrogenase in single motoneurons. *Journal of Histochemistry and Cytochemistry* 37, 899–901.

CHAMPY, C., COUJARD, R., et COUJARD-CHAMPY, C. (1946) L'innervation sympathétique des glandes. *Acta anatomica* 1, 233–283.

CHAN-PALAY, V. (1973) A brief note on the chemical nature of the precipitate within nerve fibres after the rapid Golgi reaction: selected area diffraction in high voltage electron microscopy. *Zeitschrift für Anatomie und Entwicklungs-Geschichte* 139, 115–117.

CHAPLIN, A. J. (1985) Tannic acid in histology: an historical perspective. *Stain Technology* 60, 219–231.

CHAPMAN, D. M. (1977) Eriochrome cyanin as a substitute for haematoxylin and eosin. *Canadian Journal of Medical Technology* 39, 65–66.

CHAPMAN, D. M. (1982) Localization of methylene blue paramolybdate in vitally stained nerves. *Tissue and Cell* 14, 475–487.

CHAYEN, J., BITENSKY, L. and BUTCHER, R. G. (1973) *Practical Histochemistry*. London: John Wiley.

CHEW, E. C., RICHES, D. J., LAM, T. K. and HOU-CHAN, H. J. (1984) Microwave fixation as a substitute for chemical fixation of tissues for light and electron microscopy. *Journal of Anatomy* 138, 586.

CHIANG, C.-H. and GABELLA, G. (1986) Quantitative study of the ganglion neurons of the mouse trachea. *Cell and Tissue Research* 246, 243–252.

CHIECO, P., NORMANNI, P. and BOOR, P. J. (1984) Improvement in soluble dehydrogenase histochemistry by nitroblue tetrazolium preuptake in sections: a qualitative and quantitative study. *Stain Technology* 59, 201–211.

CHRISTIE, K. N. and STOWARD, P. J. (1982) Distribution of endogenous hydrogen peroxide in cardiac muscle. *Journal of Anatomy* 135, 837–838.

CLARK, C. A., DOWNS, E. C. and PRIMUS, J. F.

(1982) An unlabelled antibody method using glucose oxidase-antiglucose oxidase complexes (GAG). *Journal of Histochemistry and Cytochemistry* **30**, 27–34.

CLARK, G. (1979) Staining with chromoxane cyanine R. *Stain Technology* **54**, 337–344.

CLARK, G. (ed.) (1981) *Staining Procedures used by the Biological Stains Commission*. 4th ed. Baltimore: Williams & Wilkins.

CLARK, P. G. (1954) A comparison of decalcifying methods. *American Journal of Clinical Pathology* **24**, 1113–1116.

CLARK, W. M. (1972) *Oxidation-reduction Potentials of Organic Systems*. Huntington, NY: R. E. Krieger.

CLASEN, R. A., SIMON, G., SCOTT, R. V., PANDOLFI, S. and LESAK, A. (1973) The staining of the myelin sheath by Luxol dye techniques. *Journal of Neuropathology and Experimental Neurology* **32**, 271–283.

CLELAND, W. W. (1964) Dithiothreitol, a new protective reagent for SH groups. *Biochemistry* **3**, 480–482.

CLONIS, Y. D., ATKINSON, T., BRUTON, C. J. and LOWE, C. R. (edit.) (1987) *Reactive Dyes and Enzyme Technology*, Basingstoke & London: Macmillan.

COLLINS, J. S. and GOLDSMITH, T. H. (1981) Spectral properties of fluorescence induced by glutaraldehyde fixation. *Journal of Histochemistry and Cytochemistry* **29**, 411–414.

COMBS, J. W., LAGUNOFF, D. and BENDITT, E. P. (1965) Differentiation and proliferation of embryonic mast cells of the rat. *Journal of Cell Biology* **25**, 577–592.

CONGER, K. A., GARCIA, J. H., LOSSINSKY, A. S. and KAUFMANN, F. C. (1978) The effect of aldehyde fixation on selected substrates for energy metabolism and amino acids in mouse brain. *Journal of Histochemistry and Cytochemistry* **26**, 423–433.

CONTESTABILE, A. and ANDERSEN, H. (1978) Methodological aspects of the histochemical localization of some cerebellar dehydrogenases. *Histochemistry* **56**, 117–132.

COOK, G. M. W. and STODDART, R. W. (1973) *Surface Carbohydrates of the Eukaryotic Cell*. London: Academic Press.

COOK, H. C. (1974) *Manual of Histological Demonstration Methods*. London: Butterworths.

CORBETT, J. F. (1971) Hair dyes. Ch. VII in *The Chemistry of Synthetic Dyes*, ed. K. Venkataraman, Vol. V, pp. 475–534. New York: Academic Press.

CORDELL, J. L., FALINI, B., ERBER, W. N., GHOSH, A. K., ABDULAZIZ, Z., MacDONALD, S., PULFORD, K. A. F., STEIN, H. and MASON, D. Y. (1984) Immunoenzymatic labelling of monoclonal antibodies using immune complexes of alkaline phosphatase and monoclonal anti-alkaline phosphatase (APAAP complexes). *Journal of Histochemistry and Cytochemistry* **32**, 219–229.

CORREA, F. M. A., INNIS, R. B., ROUOT, B., PASTERNAK, G. W. and SNYDER, S. H. (1980) Fluorescent probes of α- and β-adrenergic and opiate receptors: biochemical and histochemical evaluation. *Neuroscience Letters* **16**, 47–53.

CORRODI, H. and JONSSON, G. (1967) The formaldehyde fluorescence method for the histochemical demonstration of biogenic monoamines. *Journal of Histochemistry and Cytochemistry* **15**, 65–78.

CORSI, P. (1987) Camillo Golgi's morphological approach to neuroanatomy. In *Neuroplasticity: a New Therapeutic Tool in the CNS Pathology*, ed. R. L. Masland, A. Portera-Sanchez and G. Toffano. Berlin: Springer. pp. 1–7.

COTTON, F. A. and WILKINSON, G. (1980) *Advanced Inorganic Chemistry*. 4th ed. New York: Interscience Publisher.

COUPLAND, R. E., KOBAYASHI, S. and CROWE, J. (1976) On the fixation of catecholamines including adrenaline in tissue sections. *Journal of Anatomy* **122**, 403–413.

COUPLAND, R. E., PYPER, A. S. and HOPWOOD, D. (1964) A method for differentiating between nor-adrenaline- and adrenaline-storing cells in the light and electron microscope. *Nature* **201**, 1240–1242.

COWDEN, R. R. and CURTIS, S. K. (1970) Demonstration of protein-bound sulphydryl and disulphide groups with fluorescent mercurials. *Histochemie* **22**, 247–255.

COWDEN, R. R. and CURTIS, S. K. (1974) Use of a fluorescent probe for hydrophobic groups, anilino-naphthalene sulphonic acid, in the supravital study of unusual slug oocyte nuclei. *Histochemical Journal* **6**, 447–450.

COX, B. A., SHACKLEFORD, J. M. and YIELDING, L. W. (1982) Histochemical application of two phenanthridium compounds. *Stain Technology* **57**, 211–218.

CROWE, R. and BURNSTOCK, G. (1984) Quinacrine-positive neurones in some regions of the guinea-pig brain. *Brain Research Bulletin* **12**, 387—391.

CROWE, R. and WHITEAR, M. (1978) Quinacrine fluorescence of Merkel cells in *Xenopus laevis*. *Cell and Tissue Research* **190**, 273–283.

CULLING, C. F. A. (1974) *Handbook of Histopathological and Histochemical Techniques*. 3rd ed. London: Butterworths.

CULLING, C. F. A. and REID, P. E. (1977) The apparent failure of sodium borohydride reduction to block further PAS reactivity in rat epithelial mucins. *Histochemical Journal* **9**, 781–785.

CULLING, C. F. A., REID, P. E. and DUNN, W. L. (1976) A new histochemical method for the identification and visualization of both side chain acylated and nonacylated sialic acids. *Journal of Histochemistry and Cytochemistry* **24**, 1225–1230.

CUNNINGHAM, F. O. and FITZGERALD, M. J. T. (1972) Encapsulated nerve endings in hairy skin. *Journal of Anatomy* **112**, 93–97.

CUNNINGHAM, L. (1967) Histochemical observations of the enzymatic hydrolysis of gelatin films. *Journal of Histochemistry and Cytochemistry* **15**, 292–298.

DANIELLI, J. F. (1947) A study of techniques for the cytochemical demonstration of nucleic acids and some components of protein. *Symposia of the Society for Experimental Biology* **1**, 11–113.

DANSCHER, G. and NORGAARD, J. O. R. (1983) Light microscopic visualization of colloidal gold on resin-embedded tissue. *Journal of Histochemistry and Cytochemistry* **31**, 1394–1398.

DANSCHER, G. and NORGAARD, J. O. R. (1985)

Ultrastructural autometallography: a method for silver amplification of catalytic metals. *Journal of Histochemistry and Cytochemistry* 33, 706–710.

DANSCHER, G. and ZIMMER, J. (1978) An improved Timm sulphide-silver method for light and electron microscopic localization of heavy metals in biological tissues. *Histochemistry* 55, 27–40.

DAVIS, R. P. and JANIS, R. (1966) Free aldehydic groups in collagen and other tissue components. *Nature* 210, 318–319.

DAWSON, R. M. C., ELLIOTT, D. C., ELLIOTT, W. H. and JONES, K. M. (1969) *Data for Biochemical Research* 2nd ed. New York: Oxford University Press.

DE, A. K. KHOPKAR, S. M. and CHALMERS, R. A. (1970) *Solvent Extraction of Metals*. London: van Nostrand Reinhold.

DE BRUIJN, W. C., MEMELINK, A. A. and RIEMERSMA, J. C. (1984) Cellular membrane contrast and contrast differentiation with osmium triazole and tetrazole complexes. *Histochemical Journal* 16, 37–50.

DE CAPOA, A., FERRARO, M., AVIA, P., PELLICCIA, F. and FINAZZI-AGRO, A. (1982) Silver staining of the nucleolus organizer regions (NOR) requires clusters of sulfhydryl groups. *Journal of Histochemistry and Cytochemistry* 30, 908–911.

DE FAZIO, A., LEARY, J. A., HEDLEY, D. W. and TATTERSALL, M. H. N. (1987) Immunohistochemical detection of proliferating cells *in vivo*. *Journal of Histochemistry and Cytochemistry* 35, 571–577.

DEBBAGE, P. L., O'DELL, D. S. and JAMES, D. W. (1981) The identification of microglia in tissue culture by use of a lectin. *Journal of Anatomy* 132, 459.

DEIERKAUF, F. A. and HESLINGA, F. J. M. (1962) The action of formaldehyde on rat brain lipids. *Journal of Histochemistry and Cytochemistry* 10, 79–82.

DEMALSY, P. and CALLEBAUT, M. (1967) Plain water as a rinsing agent preferable to sulfurous acid after the Feulgen method. *Stain Technology* 42, 133–136.

DERMIETZEL, R., LEIBSTEIN, A., SIFFERT, W., ZAMBOGLOU, N. and GROS, G. (1985) A fast screening method for histochemical detection of carbonic anhydrase. *Journal of Histochemistry and Cytochemistry* 33, 93–98.

DESCLIN, J. C. (1973) A simplified silver impregnation of neural end-feet in paraffin sections. *Stain Technology* 48, 327–331.

DINSDALE, D. (1984) Ultrastructural localization of zinc and calcium within the granules of rat Paneth cells. *Journal of Histochemistry and Cytochemistry* 32, 139–145.

DIXON, M. and WEBB, E. C. (1979) *Enzymes*. 3rd ed. New York: Academic Press; London: Longmans.

DRURY, R. A. B. and WALLINGTON, E. A. (1980) *Carleton's Histological Technique*. 5th ed. Oxford: Oxford University Press.

DRZENIEK, R. (1973) Substrate specificity of neuraminidases. *Histochemical Journal* 5, 271–290.

DUBOIS-DALCQ, M., McFARLAND, H. and McFARLIN, D. (1977) Protein A-peroxidase: a valuable tool for the localization of antigens. *Journal of Histochemistry and Cytochemistry* 25, 1201–1206.

DUJINDAM, W. A. L. and VAN DUIJN, P. (1975) The interaction of apurinic aldehyde groups with pararosaniline in the Feulgen-Schiff and related staining procedures. *Histochemistry* 44, 67–85.

DULBECCO, R. and VOGT, M. (1954) Plaque formation and isolation of pure lines with poliomyelitis viruses. *Journal of Experimental Medicine* 99, 167–182.

DUYCKAERTS, C., BRION, J. P., HAUW, J-J. and FLAMENT-DURAND, J. (1987) Quantitative assessment of the density of neurofibrillary tangles and senile plaques in senile dementia of the Alzheimer type. Comparison of immunocytochemistry with a specific antibody and Bodian's protargol method. *Acta neuropathologica (Berlin)* 73, 167–170.

EDWARDS, R. and PRICE, R. (1982) Butvar B-98 resin as a section adhesive. *Stain Technology* 57, 50.

EGGERT, F. M. and GERMAIN, J. P. (1979) Rapid demineralization in acidic buffers. *Histochemistry* 59, 215–224.

EGGERT, F. M., LINDER, J. E. and JUBB, R. W. (1981) Staining of demineralized cartilage. 1. Alcoholic versus aqueous demineralization at neutral and acidic pH. *Histochemistry* 73, 385–390.

EKELUND, M., AHREN, B., HAKANSON, R., LUNDQUIST, I. and SUNDLER, F. (1980) Quinacrine accumulates in certain peptide hormone-producing cells. *Histochemistry* 66, 1–9.

ELFTMAN, H. (1954) Controlled chromation. *Journal of Histochemistry and Cytochemistry* 2, 1–8.

ELGHETANY, M. T. and SALEEM, A. (1988) Methods for staining amyloid in tissues: a review. *Stain Technology* 63, 201–212.

ELIAS, J. (1982) *Principles and Techniques in Diagnostic Histopathology*. Park Ridge, New Jersey: Noyes Publications.

ELLEDER, M. and LOJDA, Z. (1970) Studies in lipid histochemistry. III. Reaction of Schiff's reagent with plasmalogens. *Histochemie* 24, 328–335.

ELLEDER, M. and LOJDA, Z. (1971) Studies in lipid histochemistry. VI. Problems of extraction with acetone in lipid histochemistry. *Histochemie* 28, 68–87.

ELLEDER, M. and LOJDA, Z. (1972) Studies in lipid histochemistry. IX. The specificity of Holczinger's reaction for fatty acids. *Histochemie* 32, 301–305.

ELMES, M. E. and JONES, J. G. (1981) Paneth cell zinc: a comparison of histochemical and microanalytical techniques. *Histochemical Journal* 13, 335–337.

EMERMAN, M. and BEHRMAN, E. J. (1982) Cleavage and cross-linking of proteins with osmium(VIII) reagents. *Journal of Histochemistry and Cytochemistry*. 30, 395–397.

EMMEL, V. M. and STOTZ, E. H. (1986) Certified biological stains: a stability study. *Stain Technology* 61, 385–387.

ENERBACK, L. (1969) Detection of histamine in mast cells by *o*-phthalaldehyde reaction after liquid fixation. *Journal of Histochemistry and Cytochemistry* 17, 757–759.

ENGEN, P. C. and WHEELER, R. (1978) *n*-butyl methacrylate and paraffin as an embedding medium for light microscopy. *Stain Technology* 53, 17–22.

EPSTEIN, E. H., MUNDERLOH, N. H. and FUKUYAMA, K. (1979) Dithiothreitol separation of newborn rodent dermis and epidermis. *Journal of Investigative Dermatology* **73**, 207–210.

ERLEY, D. S. (1957) 2,2-dimethoxypropane as a drying agent for preparation of infrared samples. *Analytical Chemistry* **29**, 1564.

EVANS, D. H. L. and HAMLYN, L. H. (1956) A study of silver degeneration methods in the central nervous system. *Journal of Anatomy* **90**, 193–203.

EVERETT, M. M. and MILLER, W. A. (1974) The role of phosphotungstic and phosphomolybdic acids in connective tissue staining. *Histochemical Journal* **6**, 25–34.

FAIREN, A., PETERS, A. and SALDANHA, J. (1977) A new procedure for examining Golgi impregnated neurons by light and electron microscopy. *Journal of Neurocytology* **6**, 311–337.

FALCK, B., HILLARP, N.-A., THIEME, G. and TORP, A. (1962) Fluorescence of catechol amines and related compounds condensed with formaldehyde. *Journal of Histochemistry and Cytochemistry* **10**, 348–354.

FAULSTICH, H., ZOBELEY, S., BENTRUP, U. and JOCKUSCH, B. M. (1989) Biotinylphallotoxins: preparation and use as actin probes. *Journal of Histochemistry and Cytochemistry* **37**, 1035–1045.

FEIGL, F. (1960) *Spot Tests in Organic Analysis*. 6th edn. Transl. R. E. Oesper. Amsterdam: Elsevier.

FEIGL, F. and ANGER, V. (1972) *Spot Tests in Inorganic Analysis*. 6th English edn, Transl. R. E. Oesper. Amsterdam: Elsevier.

FELDMANN, G., MAURICE, M., BERNUAU, D., ROGIER, E. and DURAND, A-M. (1983) Penetration of enzyme-labelled antibodies into tissues and cells: a review of the difficulties. In *Immunoenzymatic Techniques*, ed. S. Avrameas, P. Druet, R. Masseyeff and G. Feldmann, pp. 3–15. Amsterdam: Elsevier.

FERRARI, F. A., MACCARIO, R., MARCONI, M., VITIELLO, M. A., UGAZIO, A. G., BURGIO, V. and SICCARDI, A. G. (1980) Reliability of alpha-naphthyl acetate esterase staining of blood smears for the enumeration of circulating human T lymphocytes. *Clinical and Experimental Immunology* **41**, 358–362.

FILIPSKI, G. T. and WILSON, M. H. (1985) Staining nerves in whole cleared amphibians and reptiles using Sudan black B. *Copeia* **1985** (2): 500–502.

FINE, A., AMOS, W. B., DURBIN, R. M. and McNAUGHTON, P. A. (1988) Confocal microscopy: applications in neurobiology. *Trends in Neurosciences* **11**, 346-351.

FINK, S. (1987a) Some new methods for affixing sections to glass slides. I. Aqueous adhesives. *Stain Technology* **62**, 27-33.

FINK, S. (1987b) Some new methods for affixing sections to glass slides. II. Organic solvent-based adhesives. *Stain Technology* **62**, 93-99.

FISCHER, J. M. C., PETERSON, C. A. and BOLS, N. C. (1985) A new fluorescent test for cell vitality using calcofluor white M2R. *Stain Technology* **60**, 69–79.

FITZGERALD, M. J. T. and FITTER, W. F. (1971) Significance of pH in staining cutaneous nerves with methylene blue. *Laboratory Practice* **20**, 783–800.

FOWLER, S. D. and GREENSPAN, P. (1964) Application of Nile red, a fluorescent hydrophobic probe, for the detection of neutral lipid deposits in tissue sections: comparison with oil red O. *Journal of Histochemistry and Cytochemistry* **33**, 833–836.

FOX, C. H., JOHNSON, F. B., WHITING, J. and ROLLER, R. P. (1985) Formaldehyde fixation. *Journal of Histochemistry and Cytochemistry* **33**, 845–853.

FRANGIONI, G. and BORGIOLI, G. (1979) Polystyrene embedding: a new method for light and electron microscopy. *Stain Technology* **54**, 167–172.

FRANKLIN, A. L. and FILION, W. G. (1985) A new technique for retarding fading of fluorescence: DPX-BME. *Stain Technology* **60**, 125–135.

FRANKLIN, W. A. and LOCKER, J. D. (1981) Ethidium bromide: a nucleic acid stain for tissue sections. *Journal of Histochemistry and Cytochemistry* **29**, 572–576.

FRANZBLAU, C. and FARIS, B. (1981) Elastin. Ch. 3 in E. D. Hay (ed.): *Cell Biology of the Extracellular Matrix*. pp. 65—93. New York: Plenum Press.

FRASER, F. J. (1972) Degenerating myelin: comparative histochemical studies using classical myelin stains and an improved Marchi technique minimizing artifacts. *Stain Technology* **47**, 147–154.

FREDERIKS, W. M. (1977) Some aspects of the value of Sudan Black B in lipid histochemistry. *Histochemistry* **54**, 27–37.

FREDMAN, S. M. (1987) Intracellular staining of neurons with nickel-lysine. *Journal of Neuroscience Methods* **20**, 181–194.

FRENS, G. (1973) Controlled nucleation for the regulation of the particle size in monodisperse gold suspensions. *Nature Physical Science* **241**, 20–22.

FRIED, B., GILBERT, J. J. and FEESE, R. C. (1976) Mercuric bromophenol blue to reveal gelatin substrates for protease. *Stain Technology* **51**, 140–141.

FRIEDBERG, S. H. and GOLDSTEIN, D. J. (1969) Thermodynamics of orcein staining of elastic fibres. *Histochemical Journal* **1**, 261–376.

FURNESS, J. B. and COSTA, M. (1975) The use of glyoxylic acid for the fluorescence histochemical demonstration of peripheral stores of noradrenaline and 5-hydroxytryptamine in whole-mounts. *Histochemistry* **41**, 335–352.

FURNESS, J. B., COSTA, M. and WILSON, A. J. (1977) Water-stable fluorophores, produced by reaction with aldehyde solutions, for the histochemical localization of catechol- and indolethylamines. *Histochemistry* **52**, 159–170.

FURNESS, J. B., HEATH, J. W. and COSTA, M. (1978) Aqueous aldehyde (Faglu) methods for the fluorescence histochemical localization of catecholamines and for ultrastructural studies of central nervous tissue. *Histochemistry* **57**, 285–295.

GABBOTT, P. L. and SOMOGYI, J. (1984) The "single" section Golgi-impregnation procedure: methodological description. *Journal of Neuroscience Methods* **11**, 221–230.

GABE, L. (1953) Sur quelques applications de la coloration par la fuchsine-paraldéhyde. *Bulletin de Microscopie Appliqué* (2è série) **3**, 153–162.

GABE, M. (1966) *Neurosecretion* (transl. R. Crawford). Oxford: Pergamon Press.

GABE, M. (1976) *Histological Techniques* (English ed., transl. E. Blackith and A. Kavoor). Paris: Masson.

GABELLA, G. (1968) Detection of nerve cells by a histochemical method. *Experientia* **25**, 218–219.

GABELLA, G. (1971) Neuron size and number in the myenteric plexus of the newborn and adult rat. *Journal of Anatomy* **109**, 81–95.

GABELLA, G. (1987) The number of neurons in the small intestine of mice, guinea-pigs and sheep. *Neuroscience* **22**, 737–752.

GADDUM, J. H. (1935) Choline and allied substances. *Annual Review of Biochemistry* **4**, 311–330.

GALLYAS, F. (1971) A principle for silver staining of tissue elements by physical development. *Acta Morphologica Academiae Scientiarum Hungaricae* **19**, 57–71.

GALLYAS, F. (1979) Light insensitive physical developers. *Stain Technology* **54**, 173–176.

GALLYAS, F. and WOLFF, J. R. (1986) Metal-catalysed oxidation renders silver intensification selective. Applications for the histochemistry of diaminobenzidine and neurofibrillary changes. *Journal of Histochemistry and Cytochemistry* **34**, 1667–1672.

GALLYAS, F., GORCS, T. and MERCHENTHALER, I. (1982) High-grade intensification of the end-product of the diaminobenzidine reaction for peroxidase. *Journal of Histochemistry and Cytochemistry* **30**, 183–184.

GAMBETTI, P., AUTILIO-GAMBETTI, L. and PAPASOZOMENOS, S. C. (1981) Bodian's silver method stains neurofilament polypeptides. *Science* **213**, 1521–1522.

GANTER, P. et JOLLES, G. (1969, 1970) *Histochimie Normale et Pathologique* (2 vols). Paris: Gauthier-Villars.

GARVEY, J. S., CREMER, N. E. and SUSSDORF, D. H. (1977) *Methods in Immunology*, 3rd ed. Reading, Mass.: W. A. Benjamin.

GATENBY, J. B. and BEAMS, H. W. (ed.) (1950) *The Microtomist's Vade-mecum (Bolles Lee)*. 11th ed. London: Churchill.

GEISERT, E. E. and UPDYKE, B. V. (1977) Chemical stabilization of Golgi silver chromate impregnations. *Stain Technology* **52**, 137–141.

GEOGHEGAN, W. D. and ACKERMAN, G. A. (1977) Adsorption of horseradish peroxidase, ovomucoid and anti-immunoglobulin to colloidal gold for the indirect detection of concanavalin A, wheat germ agglutinin and goat anti-human immunoglobulin G on cell surfaces at the electron microscopic level: a new method, theory and application. *Journal of Histochemistry and Cytochemistry* **25**, 1187–1200.

GILAD, G. M. and GILAD, V. H. (1981) Cytochemical localization of ornithine decarboxylase with rhodamine or biotin-labelled α-difluoromethylornithine. An example for the use of labelled irreversible enzyme inhibitors as cytochemical markers. *Journal of Histochemistry and Cytochemistry* **29**, 687–692.

GILES, C. H. (1975) Dye-fibre bonds and their investigation. Ch. 2 in *The Theory of Coloration of Textiles*, ed. C. L. Bird and W. S. Boston, pp. 41–110. Bradford: Dyer's Company Publications Trust.

GILL, J. E. and JOTZ, M. M. (1976) Further observations on the chemistry of pararosaniline-Feulgen staining. *Histochemistry* **46**, 147–160.

GILLOTEAUX, J. and NAUD, J. (1979) The zinc iodide-osmium tetroxide staining-fixative of Maillet. Nature of the precipitate studied by X-ray microanalysis and detection of Ca^{2+}-affinity subcellular rites in a tonic smooth muscle. *Histochemistry* **63**, 227–243.

GIOLLI, R. A. (1964) A note on the chemical mechanism of the Nauta-Gygax technique. *Journal of Histochemistry and Cytochemistry* **13**, 206–210.

GIRARD, N., DELPECH, A. and DELPECH, B. (1986) Characterization of hyaluronic acid on tissue sections with hyaluronectin. *Journal of Histochemistry and Cytochemistry* **34**, 539–541.

GIROUD, A. and LEBLOND, C-P. (1934) Étude histochimique de la vitamine C dans la glande surrénale. *Archives d'Anatomie Microscopique* **30**, 105–129.

GLAZER, A. N. (1976) The chemical modification of proteins by group-specific reagents. In *The Proteins*, 3rd ed., ed. H. Neurath and R. L. Hill, Vol. **2**, pp. 1–103. New York: Academic Press.

GLEES, P. (1946) Terminal degeneration within the central nervous system as studied by a new silver method. *Journal of Neuropathology and Experimental Neurology* **5**, 54–59.

GLEGG, R. E., CLERMONT, Y and LEBLOND, C. P. (1952) The use of lead tetraacetate, benzidine, *o*-anisidine and a "film test" in investigating the periodic acid-Schiff technique. *Stain Technology* **27**, 277–305.

GLENNER, G. G. (1957) The histochemical demonstration of indole derivatives by the rosindole reaction of E. Fischer. *Journal of Histochemistry and Cytochemistry* **5**, 297–304.

GLENNER, G. G., PINHO E COSTA, P. and FALCAO DE FREITAS, A. (1980) *Amyloid and Amyloidosis* (Proceedings of the 3rd International Symposium on Amyloidosis). Amsterdam: Excerpta Medica.

GOLAND, P., GRAND, N. G. and KATELE, K. V. (1967) Cyanuric chloride and *N*-methylmorpholine in methanol as a fixative for polysaccharides. *Stain Technology* **42**, 41–51.

GOLD, H. (1971) Fluorescent brightening agents. Ch. VIII in *The Chemistry of Synthetic Dyes*, ed. K. Venkataraman. Vol. **V**, pp. 535–679. New York: Academic Press.

GOLDBERG, S. and KOTANI, M. (1967) The projection of optic nerve fibres in the frog Rana catesbeiana as studied by radioautography. *Anatomical Record* **158**, 325–332.

GOLDFISCHER, S., KRESS, Y., COLTOFF-SCHILLER, B. and BERMAN, J. (1981) Primary fixation in osmium-potassium ferrocyanide: the staining of glycogen, glycoproteins, elastin, an intranuclear reticular structure, and intercisternal trabeculae. *Journal of Histochemistry and Cytochemistry* **29**, 1105–1111.

GOLDMAN, M. (1968) *Fluorescent Antibody Methods*. London: Academic Press.

GOLDSTEIN, D. J. (1962) Ionic and non-ionic bonds in staining, with special reference to the action of urea and sodium chloride on the staining of elastic fibres and glycogen. *Quarterly Journal of Microscopical Science* **103**, 477–492.

GOLDSTEIN, D. J. (1963) An approach to the thermodynamics of histological dyeing, illustrated by experiments with azure A. *Quarterly Journal of Microscopical Science* **104**, 413–439.

GOLDSTEIN, D. J. (1964) Relation of effective thickness and refractive index to permeability of tissue components in fixed sections. *Journal of the Royal Microscopical Society* **84**, 43–54.

GOLDSTEIN, I. J. and PORETZ, R. D. (1986) Isolation, physicochemical characterization, and carbohydrate-binding specificity of lectins. Ch. 2 in Liener, I. E., Sharon, N. and Goldstein, I. J. (ed.): *The Lectins. Properties, Functions and Applications in Biology and Medicine*, pp. 35–248. Orlando, Florida: Academic Press.

GONATAS, N. K., HARPER, C., MIZUTAINI, T. and GONATAS, J. O. (1979) Superior sensitivity of conjugates of horseradish peroxidase with wheat germ agglutinin for studies of retrograde axonal transport. *Journal of Histochemistry and Cytochemistry* **27**, 728–734.

GOODRICH, E. S. (1942) A new method of dissociating cells. *Quarterly Journal of Microscopical Science* **83**, 245–258.

GORBSKY, G. and BORISY, G. G. (1986) Reversible embedment cytochemistry (REC): a versatile method for the ultrastructural analysis and affinity labelling of tissue sections. *Journal of Histochemistry and Cytochemistry* **34**, 177–188.

GORDON, H. and SWEETS, H. H. (1936) A simple method for the silver impregnation of reticulum. *American Journal of Pathology* **12**, 545–552.

GORDON, P. F. and GREGORY, P. (1983) *Organic Chemistry in Colour*. Berlin: Springer-Verlag.

GORDON, S. R. (1988) Use of selected excitation filters for enhancement of diaminobenzidine photomicroscopy. *Journal of Histochemistry and Cytochemistry* **36**, 701–704.

GORNE, R. C. und PFISTER, C. (1979) Pharmakologische Beeinflussung der γ-Aminobuttersaure-(GABA)-Fluoreszenz in Hirnstrukturen der Ratte, mit Bemerkungen zum Chemismus der Nachweisreaktion. *Acta histochemica* **65**, 168–183.

GOTTSCHALK, A. (1972) *Glycoproteins*, 2nd ed. (2 vols). Amsterdam: Elsevier.

GOYAL, R. K. and SENGUPTA, A. (1986) A rapid silver-free method for staining the myenteric plexus. *Stain Technology* **61**, 127–134.

GOYER, R. A. and CHERIAN, M. G. (1977) Tissue and cellular toxicology of metals. In *Clinical Chemistry and Chemical Toxicology of Metals*, ed. S. S. Brown, pp. 89–103. Amsterdam: Elsevier-North Holland.

GRAHAM, R. C. and KARNOVSKY, M. J. (1966) The early stages of absorption of injected horseradish peroxidase in the proximal tubules of mouse kidney, ultrastructural cytochemistry by a new technique. *Journal of Histochemistry and Cytochemistry* **14**, 291–302.

GRAY, P. (1954) *The Microtomist's Formulary and Guide*. New York: Blakiston. Reprinted 1975 by R. E. Krieger, Huntington, N.Y., U.S.A.

GREEN, M. R. and PASTEWKA, J. V. (1974a) Simultaneous and differential staining by a cationic carbocyanine dye of nucleic acids, proteins and conjugated proteins. I. Phosphoproteins. *Journal of Histochemistry and Cytochemistry* **22**, 767–773.

GREEN, M. R. and PASTEWKA, J. V. (1974b) Simultaneous and differential staining by a cationic carbocyanine dye of nucleic acids, proteins and conjugated proteins. II. Carbohydrate and sulfated carbohydrate-containing proteins. *Journal of Histochemistry and Cytochemistry* **22**, 774–781.

GREEN, M. R. and PASTEWKA, J. V. (1979) The cationic carbocyanine dyes stains-all, DBTC, and ethyl stains-all, DBTC-3,3′,9 triethyl. *Journal of Histochemistry and Cytochemistry* **27**, 797–799.

GREENHALGH, C. W. (1976) Aspects of anthraquinone dyestuff chemistry. *Endeavour* **35**, 134–140.

GREENSPAN, P. and FOWLER, S. D. (1985) Spectrofluorometric studies of the lipid probe, Nile red. *Journal of Lipid Research* **26**, 781–789.

GREENSPAN, P., MAYER, E. P. and FOWLER, S. D. (1985) Nile red: a selective fluorescent stain for intracellular lipid droplets. *Journal of Cell Biology* **100**, 965–973.

GREGORY, R. E. (1980) Alcoholic Bouin fixation of insect nervous systems for Bodian silver staining. II. Modified solutions. *Stain Technology* **55**, 151–160.

GREGORY, R. E., GREENWAY, A. R. and LORD, K. A. (1980) Alcoholic Bouin fixation of insect nervous systems for Bodian silver staining. I. Composition of "aged" fixative. *Stain Technology* **55**, 143–149.

GUESDON, J.-L., TERNYNCK, T. and AVRAMEAS, S. (1979) The use of avidin-biotin in immunoenzymatic techniques. *Journal of Histochemistry and Cytochemistry* **27**, 1131–1139.

GULATI, A. K., ZALEWSKI, A. A., SHARMA, K. B., OGROWSKY, D. and SOHAL, G.S. (1986) A comparison of lectin binding in rat and human peripheral nerve. *Journal of Histochemistry and Cytochemistry* **34**, 1487–1493.

GURR, E. (1971) *Synthetic Dyes in Biology, Medicine and Chemistry*. London: Academic Press.

GURR, E., ANAND, N., UNNI, M. K. and AYYANGAR, N. R. (1974) Applications of synthetic dyes to biological problems. Ch. 5 in *The Chemistry of Synthetic Dyes*, ed. K. Venkataraman, Vol. **VII**, pp. 277–351. New York: Academic Press.

GUSTAVSON, K. H. (1956) *The Chemistry of Tanning Processes*. New York: Academic Press.

GWYN, D. G. and HEARDMAN, V. (1965) A cholinesterase-Bielschowsky staining method for mammalian motor end plates. *Stain Technology* **40**, 15–18.

HAASE, P. (1975) The development of nephrocalcinosis in the rat following injections of neutral sodium phosphate. *Journal of Anatomy* **119**, 19–37.

HABER, S. (1988) Tracing intrinsic fibre connections in postmortem human brain with WGA-HRP. *Journal of Neuroscience Methods* **23**, 15–22.

HACKNEY, C. M. and ALTMAN, J. S. (1982) Cobalt mapping of the nervous system: how to avoid artifacts. *Journal of Neurobiology* **13**, 403–411.

HADLER, W. A. and SILVEIRA, S. R. (1978) Histochemical technique to detect choline-containing lipids. *Acta histochemica* **63**, 265–270.

HADLER, W. A., LUCCA, O. de, ZITI, L. M. and PATELLI, A. S. (1969) An analysis of the effect of some fixatives on the histochemical distribution of nonhaem ferric iron in spleen sections. *Revista Brasileira de Pesquisas Medicas e Biologicas* **2**, 378–383.

HAHN von DORSCH, H., KRAUSE, R., FEHR-MANN, P. und SULZMANN, R. (1975) Histochemische Nachweismethoden für biogene Amine. *Acta histochemica* **52**, 281–302.

HAKANSON, R., OWMAN, C. and SUNDLER, E. (1972). *o*-Phthalaldehyde (OPT). A sensitive detection reagent for glucagon, secretin and vasoactive intestinal peptide. *Journal of Histochemistry and Cytochemistry* **20**, 138–140.

HAKANSON, R., OWMAN, C. H., SJOBERG, N. O. and SPORRONG, B. (1970) Amine mechanisms in enterochromaffin-like cells of gastric mucosa in various mammals. *Histochemie* **26**, 189–220.

HALBHUBER, K.-J., ZIMMERMAN, N. and LINSS, W. (1988) New, improved lanthanide-based methods for the ultrastructural localization of acid and alkaline phosphatase activity. *Histochemistry* **88**, 375–381.

HAMES, B. D. and HIGGINS, S. J. (1985) *Nucleic Acid Hybridization, A Practical Approach*. Oxford: IRL Press.

HANKER, J. S., THORNBURG, L. P. and YATES, P. E. (1973) The demonstration of cholinesterases by the formation of osmium blacks at the sites of Hatchett's brown. *Histochemie* **37**, 223–242.

HANKER, J. S., YATES, P. E., METZ, C. B. and RUSTIONI, A. (1977) A new specific, sensitive and non-carcinogenic reagent for the demonstration of horseradish peroxidase. *Histochemical Journal* **9**, 789–792.

HARRIS, C. M and LIVINGSTONE, S. E. (1964) Bidentate chelates. Ch. 3 in *Chelating Agents and Metal Chelators*, ed. F. P. Dwyer and D. P. Mellor, pp. 95–141. New York: Academic Press.

HARRISON, F., VAN HOOF, J. and VANROELEN, C. (1986) On the presence of proteolytic activity in glycosaminoglycan-degrading enzymes. *Journal of Histochemistry and Cytochemistry* **34**, 1231–1235.

HASCALL, V. C. and HASCALL, G. K. (1981) Proteoglycans. Ch. 2 in *Cell Biology of the Extracellular Matrix*, ed. E. D. Hay, pp. 39–63. New York: Plenum Press.

HASEGAWA, J. and HASEGAWA, J. (1977) Substrate limitations of the colour film technique for the localization of proteases. *Journal of Histochemistry and Cytochemistry* **25**, 234.

HAUG, F-M. S. (1973) Heavy metals in the brain. A light microscopic study of the rat with Timm's sulphide-silver method. Methodological considerations and cytological and regional staining patterns. *Advances in Anatomy, Embryology and Cell Biology* **47**, 1–71.

HAYAT, M. A. (edit.) (1973–1977) *Electron Microscopy of Enzymes. Principles and Methods*. Vols **1–5**. New York: van Nostrand-Reinhold.

HAYAT, M. A. (1975) *Positive Staining for Electron Microscopy*. New York: van Nostrand-Reinhold.

HAYAT, M. A. (1981) *Principles and Techniques of Electron Microscopy. Biological Applications*. 2nd edn, Vol. **1**, Baltimore: University Park Press.

HAYHOE, F. G. J. and QUAGLINO, D. (1988) *Haematological Cytochemistry*. 2nd ed. Edinburgh: Churchill-Livingstone.

HEATH, I. D. (1962) Observations on a highly specific method for the histochemical detection of sulphated mucopolysaccharides, and its possible mechanisms. *Quarterly Journal of Microscopical Science* **103**, 457–475.

HEIDEMANN, E. (1988). The chemistry of tanning. Ch. 2 in *Collagen*, Vol. III. (ed. M. E. Nimni), pp. 39–61. Boca Raton, Fla: CRC Press.

HEINICKE, E. A., KIERNAN, J. A. and WIJSMAN, J. (1987) Specific, selective, and complete staining of neurons of the myenteric plexus, using cuprolinic blue. *Journal of Neuroscience Methods* **21**, 45–54.

HENDRICKSON, J. B., CRAM, D. J. and HAMMOND, G. S. (1970) *Organic Chemistry* (3rd ed.). New York: McGraw-Hill.

HEPLER, J. R., TOOMIM, C. S., McCARTHY, C., CONTI, F., BATTAGLIA, G., RUSTIONI, A. and PETRUSZ, P. (1988) Characterization of antisera to glutamate and aspartate. *Journal of Histochemistry and Cytochemistry* **36**, 13–22.

HESLINGA, J. M. and DEIERKAUF, F. A. (1961) The action of histological fixatives on tissue lipids. Comparison of the action of several fixatives using paper chromatography. *Journal of Histochemistry and Cytochemistry* **9**, 572–577.

HESS, A. (1978) A simple procedure for distinguishing dopamine from noradrenaline in peripheral nervous structures in the fluorescence microscope. *Journal of Histochemistry and Cytochemistry* **26**, 141–144.

HIGHMAN, B. (1946) Improved methods for demonstrating amyloid in paraffin sections. *Archives of Pathology* **41**, 559–562.

HILWIG, I. and GROPP, A. (1972) Staining of constitutive heterochromatin in mammalian chromosomes with a new fluorochrome. *Experimental Cell Research* **75**, 122–126.

HILWIG, I. and GROPP, A. (1975) pH-Dependent fluorescence of DNA and RNA in cytologic staining with 33258 Hoechst. *Experimental Cell Research* **91**, 457–460.

HIRANO, A. and ZIMMERMANN, H. M. (1962) Silver impregnation of nerve cells and fibres in celloidin sections. *Archives of Neurology* **6**, 114–122.

HIXSON, D. C., YEP, J. M., GLENNEY, J. R., HAYES, T. and WALBERG, E. F. (1981) Evaluation of periodate/lysine/paraformaldehyde fixation as a method for cross-linking plasma membrane glycoproteins. *Journal of Histochemistry and Cytochemistry* **29**, 561–566.

HODGSON, A. J., PENKE, B., ERDEI, A., CHUBB, I. W. and SOMOGYI, P. (1985) Antisera to γ-aminobutyric acid. I. Production and characterization using a new model system. *Journal of Histochemistry and Cytochemistry* **33**, 229–239.

HOGG, R. M. and SIMPSON, R. (1975) An evaluation of solochrome cyanine R.S. as a nuclear stain similar to haematoxylin. *Medical Laboratory Technology* **32**, 301–306.

HOHMANN, A., HODGSON, A. J., SKINNER, J. M., BRADLEY, J. and ZOLA, H. (1988) Monoclonal alkaline phosphatase-anti-alkaline phosphatase (APAAP) complex: production of antibody, optimization of activity, and use in immunostaining. *Journal of Histochemistry and Cytochemistry* **36**, 137–143.

HOLGATE, C. S., JACKSON, P., COWEN, P. N. and

BIRD, C. C. (1983) Immunogold-silver staining: a new method of immunostaining with enhanced sensitivity. *Journal of Histochemistry and Cytochemistry* **31**, 938–944.

HOLMES, W. (1943) Silver staining of nerve axons in paraffin sections. *Anatomical Record* **86**, 157–187.

HOLMES, W. (1947). The peripheral nerve biopsy. Ch. XXXVIII in *Recent Advances in Clinical Pathology*, ed. S. C. Dyke. pp. 402–417. London: Churchill.

HOLT, S. J. and WITHERS, R. F. J. (1952) Cytochemical localization of esterases using indoxyl derivatives. *Nature* **170**, 1012–1014.

HONIG, M. G. and HUME, R. I. (1986) Fluorescent carbocyanine dyes allow living neurons of identified origin to be studied in long-term cultures. *Journal of Cell Biology* **103**, 171–187.

HONIG, M. G. and HUME, R. I. (1989) DiI & DiO: versatile fluorescent dyes for neuronal labelling and pathway tracing. *Trends in Neurosciences* **12**, 333–341.

HOOGHWINKEL, G. J. M. and SMITS, G. (1957) The specificity of the periodic acid-Schiff technique studied by a quantitative test-tube method. *Journal of Histochemistry and Cytochemistry* **5**, 120–126.

HOPE, B. T. and VINCENT, S. R. (1989) Histochemical characterization of neuronal NADPH-diaphorase. *Journal of Histochemistry and Cytochemistry* **37**, 653–661.

HOPWOOD, D. (1969) Fixation of proteins by osmium tetroxide, potassium dichromate and potassium permanganate. *Histochemie* **18**, 250–260.

HOPWOOD, D. (1977) Fixation and fixatives. Ch. 2 in *Theory and Practice of Histological Techniques*, ed. J. D. Bancroft and A. Stevens. pp. 16–28. Edinburgh: Churchill-Livingstone.

HOPWOOD, D. and BOON, M. E. (edit.) (1988) Special issue: application of microwaves. *Histochemical Journal* **20** (**6 & 7**): 311–404.

HOPWOOD, D. and SLIDDERS, W. (1989) Tissue fixation with phenol-formaldehyde for routine histopathology. *Royal Microscopical Society Proceedings* **24**, A55.

HOPWOOD, D., SLIDDERS, W. and YEAMAN, G. R. (1989) Tissue fixation with phenol-formaldehyde for routine histopathology. *Histochemical Journal* **21**, 228–234.

HOPWOOD, D., YEAMAN, G. and MILNE, G. (1988) Differentiating the effects of microwave and heat on tissue proteins and their cross linking by formaldehyde. *Histochemical Journal* **20**, 341–346.

HOPWOOD, D., COGHILL, G., RAMSAY, J., MILNE, G. and KERR, M. (1984) Microwave fixation: its potential for routine techniques, histochemistry, immunocytochemistry and electron microscopy. *Histochemical Journal* **16**, 1171–1191.

HORIIKE, K., TOJO, H., ARAI, R., YAMONO, T., NOZAKI, M. and MAEDA, T. (1987) Localization of D-amino acid oxidase in Bergmann glial cells and astrocytes of rat cerebellum. *Brain Research Bulletin* **19**, 587–596.

HORIKAWA, K. and ARMSTRONG, W. E. (1988) A versatile means of intracellular labelling: injection of biocytin and its detection with avidin conjugates. *Journal of Neuroscience Methods* **25**, 1–11.

HORISBERGER, M. and ROSSET, J. (1977) Colloidal gold, a useful marker for transmission and scanning electron microscopy. *Journal of Histochemistry and Cytochemistry* **25**, 295–305.

HOROBIN, R. W. (1980) Structure-staining relationships in histochemistry and biological staining. I. Theoretical background and a general account of correlation of histochemical staining with the chemical structure of the reagents used. *Journal of Microscopy* **119**, 345–355.

HOROBIN, R. W. (1982). *Histochemistry: An Explanatory Outline of Histochemistry and Biophysical Staining*. Stuttgart: Gustav Fischer.

HOROBIN, R. W. (1983). What the textile-dyeing literature has to offer histochemists. *Histochemical Journal* **15**, 1151–1154.

HOROBIN, R. W. (1988) *Understanding Histochemistry: Selection, Evaluation and Design of Biological Stains*. Chichester: Ellis Horwood.

HOROBIN, R. W. and FLEMMING, L. (1980) Structure-staining relationships in histochemistry and biological staining. II. Mechanistic and practical aspects of the staining of elastic fibres. *Journal of Microscopy* **119**, 357–372.

HOROBIN, R. W. and FLEMMING, L. (1988) One-bath trichrome staining: investigation of a general mechanism based on a structure-staining correlation analysis. *Histochemical Journal* **20**, 23–34.

HOROBIN, R. W. and GOLDSTEIN, D. J. (1974) The influence of salt on the staining of tissue sections with basic dyes: an investigation into general applicability of the critical electrolyte concentration theory. *Histochemical Journal* **6**, 599–609.

HOROBIN, R. W. and JAMES, N. T. (1970) The staining of elastic fibres with direct blue 152. A general hypothesis for the staining of elastic fibres. *Histochemie* **22**, 324–336.

HOROBIN, R. W. and KEVILL-DAVIES, I. M. (1971a) Basic fuchsin in acid alcohol: a simplified alternative to Schiff reagent. *Stain Technology* **46**, 53–58.

HOROBIN, R. W. and KEVILL-DAVIES, I. M. (1971b) A mechanistic study of the histochemical reactions between aldehydes and basic fuchsin in acid alcohol used as a simplified Schiff's reagent. *Histochemical Journal* **3**, 371–378.

HOROBIN, R. W. and WALTER, K. J. (1987) Understanding Romanowsky staining. I. The Romanowsky-Giemsa effect in blood smears. *Histochemistry* **86**, 331–336.

HOWARD, W. B., WILLHITE, C. C. and SMART, R. A. (1989). Fixative evaluation and histologic appearance of embryonic rodent tissue. *Stain Technology* **64**, 1–8.

HOYER, P. E., LYON, H., JAKOBSEN, P. and ANDERSEN, A. P. (1986) Standardized methyl green-pyronin Y procedures using pure dyes. *Histochemical Journal* **18**, 91–94.

HSU, S.-M. and RAINE, L. (1981) Protein A, avidin and biotin in immunohistochemistry. *Journal of Histochemistry and Cytochemistry* **29**, 1349–1353.

HSU, S.-M. and SOBAN, E. (1982) Colour modification of diaminobenzidine (DAB) precipitation by metallic ions and its application for double immunocytochemistry. *Journal of Histochemistry and Cytochemistry* **30**, 1079–1082.

HSU, S.-M., RAINE, L. and FANGER, H. (1981) Use of avidin-biotin-peroxidase complex (ABC) in immunoperoxidase techniques. A comparison between ABC and unlabelled antibody (PAP) procedures. *Journal of Histochemistry and Cytochemistry* **29**, 577–580.

HUGHES, R. C. (1976) *Membrane Glycoproteins*. London: Butterworths.

HULSTAERT, C. E., KALICHARAN, D. and HARDONK, M. J. (1983) Cytochemical demonstration of phosphatases in the rat liver by a cerium-based method in combination with osmium tetroxide and potassium ferrocyanide postfixation. *Histochemistry* **78**, 71–79.

HUMASON, G. L. (1979) *Animal Tissue Techniques*. 4th ed. San Francisco: Freeman.

HUTSON, J. C., CHILDS, G. V. and GARDNER, P. J. (1979) Considerations for establishing the validity of immunocytological studies. *Journal of Histochemistry and Cytochemistry* **27**, 1201–1202.

HYMAN, J. M. and POULDING, R. H. (1961) Solochrome cyanin-iron alum for rapid staining of frozen sections. *Journal of Medical Laboratory Technology* **18**, 107.

IBRAHIM, M. Z. M., KHREIS, Y. and KOSHAYAN, D. S. (1974) The histochemical identification of microglia. *Journal of the Neurological Sciences* **22**, 211–233.

IDE, C. and SATO, T. (1980) Adenosine triphosphatase activity of cutaneous nerve fibres. *Histochemistry* **65**, 83–92.

IPPOLITO, E., LAVELLE, S. and PEDRINI, V. (1981) The effect of various decalcifying agents on cartilage proteoglycans. *Stain Technology* **56**, 367–372.

IRONS, R. D., SCHENK, E. A. and LEE, C. K. (1977) Cytochemical methods for copper. *Archives of Pathology and Laboratory Medicine* **101**, 298–301.

JACOBS, G. F. and LIGGETT, S. J. (1971) An oxidation-distillation procedure for reclaiming osmium tetroxide from used fixative solutions. *Stain Technology* **46**, 207–208.

JAIN, M. K. (1982) *Handbook of Enzyme Inhibitors (1965–1977)*. New York: John Wiley.

JAMBOR, B. (1954) Reduction of tetrazolium salt. *Nature* **173**, 774–775.

JAMES, J. (1976) *Light Microscopic Techniques in Biology and Medicine*. The Hague: Martinus Nijhoff.

JAMES, J. and TAS, J. (1984) *Histochemical Protein Staining Methods*. (Royal Microscopical Society Microscopy Handbooks 94). Oxford: Oxford University Press.

JAMES, T. H. (ed.) (1977) *The Theory of the Photographic Process*. 4th ed. New York: Macmillan.

JANCSO, G., FERENCSIK, M., SUCH, G., KIRALY, E., NAGY, A. and BUJDOSO, M. (1985) Morphological effects of capsaicin and its analogues in newborn and adult mammals. In *Tachykinin Antagonists*, ed. R. Hakanson and F. Sundler. pp. 35–44. Amsterdam: Elsevier.

JANSSON, L., OGREN, S. and LINDAHL, U. (1975) Macromolecular properties and end-group analysis of heparin isolated from bovine liver capsule. *Biochemical Journal* **145**, 53–62.

JAQUES, L. B. (1978) The nature of mucopolysaccharides. *Medical Hypotheses* **4**, 123–135.

JAQUES, L. B., MAHADOO, J. and RILEY, J. F. (1977) The mast cell/heparin paradox. *Lancet* **1977-I**, 411–413.

JARVINEN, M. and RINNE, A. (1983) The use of polyvinyl acetate glue to prevent detachment of tissue sections in immunohistochemistry. *Acta histochemica* **72**, 751–752.

JASMIN, G. et BOIS, P. (1961) Coloration differentielle des mastocytes chez le rat. *Revue Canadienne de Biologie* **20**, 773–774.

JEANLOZ, R. W. (1975) The chemistry of heparin. In *Heparin: Structure, Function and Clinical Implications* ed. R. A. Bradshaw and S. Wessler. pp. 3–15. New York: Plenum Press.

JENSEN, W. A. (1962) *Botanical Histochemistry*. San Francisco: Freeman.

JHAVERI, S., CARMAN, L. and HAHM, J-O. (1988) Visualizing anterogradely transported HRP by use of TMB histochemistry: comparison of the TMB-SNF and TMB-AHM methods. *Journal of Histochemistry and Cytochemistry* **36**, 103–105.

JOHANNES, M.-L. and KLESSEN, C. (1984) Alcian blue/PAS or PAS/alcian blue? Remarks on a classical technique used in carbohydrate histochemistry. *Histochemistry* **80**, 129–132.

JONES, D. (1972) Reactions of aldehydes with unsaturated fatty acids during histological fixation. *Histochemical Journal* **4**, 421–465.

JONES, E. G. and HARTMAN, B. K. (1978) Recent advances in neuroanatomical methodology. *Annual Review of Neuroscience* **1**, 215–296.

JONES, K. H. and KNISS, D. A. (1987) Propidium iodide as a nuclear counterstain for immunofluorescence studies on cells in culture. *Journal of Histochemistry and Cytochemistry* **35**, 123–125.

JONES, R. M. (ed.) (1950) *McClung's Handbook of Microscopical Technique*. 3rd ed. New York: Hoebner.

JUARRANZ, A., HOROBIN, R. W. and PROCTOR, G. B. (1986) Prediction of *in situ* fluorescence of histochemical reagents using a structure-staining correlation procedure. *Histochemistry* **84**, 426–431.

KALIMO, H. and PELLINIEMI, L. J. (1977) Pitfalls in the preparation of buffers for electron microscopy. *Histochemical Journal* **9**, 241–246.

KARNOVSKY, M. J. (1965) A formaldehyde-glutaraldehyde fixative of high osmolality for use in electron microscopy. *Journal of Cell Biology* **27**, 137a-138a.

KARNOVSKY, M. J. and FASMAN, G. D. (1960) A histochemical method for distinguishing between side-chain and terminal (α-acylamido) carboxyl groups of proteins. *Journal of Biophysical and Biochemical Cytology* **8**, 319–325.

KARNOVSKY, M. J. and MANN, M. S. (1961) The significance of the histochemical reaction for carboxyl groups of proteins in cartilage matrix. *Histochemie* **2**, 234–243.

KARNOVSKY, M. J. and ROOTS. L. (1964) A "direct colouring" thiocholine method for cholinesterases. *Journal of Histochemistry and Cytochemistry* **12**, 219–221.

KASHIWA, H. K. and ATKINSON, W. B. (1963) The

applicability of a new Schiff base, glyoxal bis(2-hydroxyanil), for the cytochemical localization of ionic calcium. *Journal of Histochemistry and Cytochemistry* **11**, 258–264.

KASHIWA, H. K. and HOUSE, C. M. (1964). The glyoxal bis(2-hydroxyanil) method modifed for localizing insoluble calcium salts. *Stain Technology* **39**, 359–367.

KASTEN, F. H. and LALA, R. (1975) The Feulgen reaction after glutaraldehyde fixation. *Stain Technology* **50**, 197–201.

KATER, S. B. and NICHOLSON, C. (ed.) (1973) *Intracellular Staining in Neurobiology*. New York: Springer.

KATSUYAMA, T. and SPICER, S. S. (1978) Histochemical differentiation of complex carbohydrates with variants of the concanavalin A-horseradish peroxidase method. *Journal of Histochemistry and Cytochemistry* **26**, 233-250.

KAWAMURA, K. (1986) Occurrence and release of histamine-containing granules in summer cells in adrenal glands of the frog *Rana catesbeiana*. *Journal of Anatomy* **148**, 111–119.

KAZIMIERCZAK, J., SOMMER, E. W., PHILIPPE, E. and DROZ, B. (1986) Carbonic anhydrase activity in primary sensory neurons. I. Requirements for the cytochemical localization in the dorsal root ganglion of chicken and mouse by light and electron microscopy. *Cell and Tissue Research* **245**, 487–495.

KEILIN, D. and HARTREE, E. F. (1938). Cytochrome oxidase. *Proceedings of the Royal Society of London B* **125**, 171–186.

KEY, B. and GIORGI, P. P. (1986) Selective binding of soybean agglutinin to the olfactory system of *Xenopus*. *Neuroscience* **18**, 507–515.

KIERNAN, J. A. (1964) Carboxylic esterases of the hypothalamus and neurohypophysis of the hedgehog. *Journal of the Royal Microscopical Society* **83**, 297–306.

KIERNAN, J. A. (1974) Effects of metabolic inhibitors on vital staining with methylene blue. *Histochemistry* **40**, 51–57.

KIERNAN, J. A. (1977a) Histochemical demonstration of unsaturated hydrophilic lipids with palladium chloride. *Journal of Histochemistry and Cytochemistry* **25**, 200–205.

KIERNAN, J. A. (1977b) Recycling procedure for gold chloride used in neurohistology. *Stain Technology* **52**, 245-248.

KIERNAN, J. A. (1978) Recovery of osmium tetroxide from used fixative solutions. *Journal of Microscopy* **113**, 77–82.

KIERNAN, J. A. (1981) *Histological and Histochemical Methods: Theory and Practice*. (1st ed.) Oxford: Pergamon Press.

KIERNAN, J. A. (1984a) Chromoxane cyanine R. I. Physical and chemical properties of the dye and of some of its iron complexes. *Journal of Microscopy* **134**, 13–23.

KIERNAN, J. A. (1984b) Chromoxane cyanine R. II. Staining of animal tissues by the dye and its iron complexes. *Journal of Microscopy* **134**, 25–39.

KIERNAN, J. A. (1985) The action of chromium (III) in fixation of animal tissues. *Histochemical Journal* **17**, 1131–1146.

KIERNAN, J. A. and BERRY, M. (1975) Neuroanatom-ical Methods. Ch. 1 in P. B. Bradley ed.: *Methods in Brain Research*. pp. 1–77. London: Wiley.

KIERNAN, J. A. and STODDART, R. W. (1973) Fluorescent-labelled aprotinin: a new reagent for the histochemical detection of acid mucosubstances. *Histochemie* **34**, 77–84.

KIRKENBY, S. and MOE, D. (1986) Studies on the actions of glutaraldehyde, formaldehyde, and mixtures of glutaraldehyde and formaldehyde on tissue proteins. *Acta histochemica* **79**, 115–121.

KJELLSTRAND, P. T. T. (1977) Temperature and acid concentration in the search for optimum Feulgen hydrolysis conditions. *Journal of Histochemistry and Cytochemistry* **25**, 129–134.

KLAUSHOFER, K. and VON MAYERSBACH, H. (1979) Freeze-substituted tissue in 5'-nucleotidase histochemistry. Comparative histochemical and biochemical investigations. *Journal of Histochemistry and Cytochemistry* **27**, 1582–1587.

KLEENE, S. J. and GESTELAND, R. C. (1983) Dissociation of frog olfactory epithelium. *Journal of Neuroscience Methods* **9**, 173–183.

KLEIN, R. M. and KLEIN, D. T. (1970) *Research Methods in Plant Science*. Garden City, N.Y.: American Museum of Natural History.

KLESSEN, C. (1974) Histochemical demonstration of thyrotropic and gonadotropic cells in the pituitary gland of the rat by the use of a lead-tetraacetate sodium bisulphite technique. *Histochemical Journal* **6**, 311–318.

KLUVER, H. and BARRERA, E. (1953) A method for the combined staining of cells and fibres in the central nervous system. *Journal of Neuropathology and Experimental Neurology* **12**, 400–403.

KORN, E. D. (1967) A chromatographic and spectrophotometric study of the products of the reaction of osmium tetroxide with unsaturated lipids. *Journal of Cell Biology* **34**, 627–638.

KORNBLIHTT, A. R. and GUTMAN, A. (1988) Molecular biology of the extracellular matrix proteins. *Biological Reviews* **63**, 465–507.

KOSKI, J. P. and REYES, P. F. (1986). Silver impregnation techniques for neuropathologic studies. I. Variations on Bodian's technic with a review of the theory and methods. *Journal of Histotechnology* **9**, 265–272.

KRAJIAN, A. A. and GRADWOHL, R. B. H. (1952) *Histopathological Technic*. 2nd ed. St Louis: Mosby.

KRAMER, H. and WINDRUM, G. M. (1955) The metachromatic staining reaction. *Journal of Histochemistry and Cytochemistry* **3**, 226–237.

KRIKELIS, H. and SMITH, A. (1988). Palladium toning of silver-impregnated reticular fibers. *Stain Technology* **63**, 97–100.

KRISTENSSON, K., OLSSON, Y. and SJOSTRAND, J. (1971) Axonal uptake and retrograde transport of exogenous proteins in the hypoglossal nerve. *Brain Research* **32**, 399–406.

KUGLER, P. and WROBEL, K-H. (1978) Meldola blue: a new electron carrier for the histochemical demonstration of dehydrogenases (SDH, LDH, G-6-PDH). *Histochemistry* **59**, 97–109.

KUHN, K. (1987) The classical collagens: Types I, II and III. In R. Mayne & R. E. Burgeson ed: *Structure and*

Function of Collagen Types. Orlando, Florida: Academic Press.

KUMAR, R. K. (1989) Immunogold-silver cytochemistry using a capillary action staining. *Journal of Histochemistry and Cytochemistry* **37**, 913–917.

KUYPERS, H. G. J. M., CATSMAN-BERREVOETS, C. E. and PADT, R. E. (1977) Retrograde axonal transport of fluorescent substances in the rat's forebrain. *Neuroscience Letters* **6**, 127–135.

LAI, M., LAMPERT, I. A. and LEWIS, P. D. (1975) The influence of fixation on staining of glycosaminoglycans in glial cells. *Histochemistry* **41**, 275–279.

LASCANO, E. F. (1946) Importancia del pH en la fijacion del tejido nervioso. Creacion artificial de fijadores tipo formol-bromuro y formol-nitrato de urano de Cajal. *Archivos de la Sociedad Argentina de Anatomia Normal y Patologica* **8**, 185–194.

LASCANO, E. F. and BERRIA, M. I. (1988) PAP labelling enhancement by osmium tetroxide-potassium ferrocyanide treatment. *Journal of Histochemistry and Cytochemistry* **36**, 697–699.

LASEK, R., JOSEPH, B. S. and WHITLOCK, D. G. (1968) Evaluation of a radioautographic neuroanatomical tracing method. *Brain Research* **8**, 319–336.

LATGE, J.-P., MONSIGNY, M. and PREVOST, M-C. (1988) Visualization of exocellular lectins in the entomopathogenic fungus *Conidiobolus obscurus. Journal of Histochemistry and Cytochemistry* **36**, 1419–1424.

LATIES, A. M., LUND, R. and JACOBOWITZ, D. (1967) A simplified method for the histochemical localization of cardiac catecholamine-containing nerve fibres. *Journal of Histochemistry and Cytochemistry* **15**, 535–541.

LATIMER, W. M. (1952) *The Oxidation States of the Elements and their Potentials in Aqueous Solutions.* 2nd ed. Englewood Cliffs, N. J.: Prentice-Hall.

LEBLOND, C. P., GLEGG, R. E. and EIDINGER, D. (1957) Presence of carbohydrates with free 1,2-glycol groups in sites stained by the periodic acid-Schiff technique. *Journal of Histochemistry and Cytochemistry* **5**, 445–458.

LETINSKY, M. S. (1983) Staining normal and experimental motor nerve terminals with tetrazolium salts. *Stain Technology* **58**, 21–27.

LETINSKY, M. S. and DECINO, P. A. (1980) Histological staining of pre- and postsynaptic components of amphibian neuromuscular junctions. *Journal of Neurocytology* **9**, 305–320.

LEVER, J. D., SANTER, R. M., LU, K.-S. and PRESLEY, R. (1977) Electron probe X-ray microanalysis of small granulated cells in rat sympathetic ganglia after sequential aldehyde and dichromate treatment. *Journal of Histochemistry and Cytochemistry* **25**, 275–279.

LEVINSON, J. W., RETZEL, S. and McCORMICK, J. J. (1977) An improved acriflavine-Feulgen method. *Journal of Histochemistry and Cytochemistry* **25**, 355-358.

LEWIS, P. R. (1987) The mechanisms of positive staining. *Proceedings of the Royal Microscopical Society* **22**, 359–363.

LHOTKA, J. F. (1952) Histochemical use of sodium bismuthate. *Stain Technology* **27**, 259–262.

LHOTKA, J. F. (1956) On tissue argyrophilia. *Stain Technology* **31**, 185–188.

LI, C-Y., ZIESMER, S. C. and LAZCANO-VILLAREAL, O. (1986) Use of azide and hydrogen peroxide as an inhibitor for endogenous peroxidase in the immunoperoxidase method. *Journal of Histochemistry and Cytochemistry* **35**, 1457–1460.

LIAO, J. C., PONZO, J. L. and PATEL, C. (1981) Improved stability of methanolic Wright's stain with additive reagents. *Stain Technology* **56**, 251–263.

LICHTENSTEIN, S. J. and NETTLETON, G. S. (1980) Effects of fuchsin variants in aldehyde fuchsin staining. *Journal of Histochemistry and Cytochemistry* **28**, 683–688.

LICHTMAN, J. W., WILKINSON, R. S. and RICH, M. M. (1985) Multiple innervation of tonic endplates revealed by activity-dependent uptake of fluorescent probes. *Nature* **314**, 357–359.

LIENER, I. E., SHARON, N. and GOLDSTEIN, I. J. (ed.) (1986) *The Lectins. Properties, Functions and Applications in Biology and Medicine.* Orlando, Florida: Academic Press.

LIEVREMONT, M., POTUS, J. and GUILLOU, B. (1982) Use of alizarin red S for histochemical staining of Ca^{2+} in the mouse; some parameters of the chemical reaction *in vitro. Acta anatomica* **114**, 268–280.

LILLIE, R. D. (1962) The histochemical reaction of aryl amines with tissue aldehydes produced by periodic and chromic acids. *Journal of Histochemistry and Cytochemistry* **19**, 303–314.

LILLIE, R. D. (1964) Histochemical acylation of hydroxyl and amino groups. Effect on the periodic acid-Schiff reaction, anionic and cationic dye and van Gieson collagen stains. *Journal of Histochemistry and Cytochemistry* **12**, 821–841.

LILLIE, R. D. (1969) Mechanisms of chromation haematoxylin stains. *Histochemie* **20**, 338–354.

LILLIE, R. D. (1977) *H. J. Conn's Biological Stains.* 9th ed. Baltimore: Williams & Wilkins.

LILLIE, R. D. and BURTNER, H. J. (1953) The ferric ferricyanide reduction test in histochemistry. *Journal of Histochemistry and Cytochemistry* **1**, 87–92.

LILLIE, R. D. and DONALDSON, P. T. (1974) The mechanism of the ferric ferricyanide reduction reaction. *Histochemical Journal* **6**, 679–684.

LILLIE, R. D. and FULLMER, H. M. (1976) *Histopathologic Technic and Practical Histochemistry* (4th ed). New York: McGraw-Hill.

LILLIE, R. D., HENDERSON, R. and GUTIERREZ, A. (1968) The diazosafranin method: control of nitrite concentration and refinements in specificity. *Stain Technology* **43**, 311–313.

LILLIE, R. D., PIZZOLATO, P. and DONALDSON, P. T. (1976a) Nuclear stains with soluble metachrome metal mordant lake dyes. The effect of chemical endgroup blocking reactions and the artificial introduction of acid groups into tissues. *Histochemistry* **49**, 23–35.

LILLIE, R. D., PIZZOLATO, P. and DONALDSON, P. T. (1976b) Haematoxylin substitutes: a survey of mordant dyes tested and consideration of the relation of their structure to performance as nuclear stains. *Stain Technology* **51**, 25–41.

LILLIE, R. D., PIZZOLATO, P., DESSAUER, H. C.

and DONALDSON, P. T. (1971) Histochemical reactions at tissue arginine sites with alkaline solutions of β-naphthoquinone-4-sodium sulfonate and other *o*-quinones and oxidized *o*-diphenols. *Journal of Histochemistry and Cytochemistry* **19**, 487–497.

LINDVALL, O. and BJORKLUND, A. (1974) The glyoxylic fluorescence histochemical method: a detailed account of the methodology for the visualization of central catecholamine neurons. *Histochemistry* **39**, 97–127.

LISON, L. (1955) Staining differences in cell nuclei. *Quarterly Journal of Microscopical Science* **96**, 227–237.

LIU, C-C., SHERRARD, D. J., MALONEY, N. A. and HOWARD, G. A. (1987). Reactivation of bone acid phosphatase and its significance in bone histomorphometry. *Journal of Histochemistry and Cytochemistry* **35**, 1355–1363.

LLEWELLYN, B. D. (1974) Mordant blue 3: a readily available substitute for haematoxylin in the routine hematoxylin and eosin stain. *Stain Technology* **49**, 347–349.

LLEWELLYN, B. D. (1978) Improved nuclear staining with mordant blue 3 as a haematoxylin substitute. *Stain Technology* **53**, 73–77.

LOACH, P. A. (1976) Oxidation-reduction potentials, absorbance bands and molar absorbance of compounds used in biochemical studies. *Handbook of Biochemistry and Molecular Biology (3rd ed.): Physical and Chemical Data*, ed. G. D. Fasman, Vol. 1, pp. 122–130. Cleveland, Ohio: CRC Press.

LOCKE, F. S. (1958) Towards the ideal circulating fluid for the isolated frog's heart. *Journal of Physiology* **18**, 232–233.

LOJDA, Z. and MALIS, F. (1972) Histochemical demonstration of enterokinase. *Histochemie* **32**, 23–39.

LOJDA, Z., GOSSRAU, R. and SCHIEBLER, T. H. (1979) *Enzyme Histochemistry. A Laboratory Manual.* Berlin: Springer-Verlag.

LOREN, I., BJORKLUND, A., FALCK, B. and LINDVALL, O. (1976) An improved histofluorescence procedure for freeze-dried paraffin-embedded tissue based on combined formaldehyde-glyoxylic acid perfusion with high magnesium content and acid pH. *Histochemistry* **49**, 177–192.

LOREN, I., BJORKLUND, A., FALCK, B. and LINDVALL, O, (1980) The aluminium-formaldehyde (ALFA) histofluorescence method for improved visualization of catecholamines and indoleamines. I. A detailed account of the methodology for central nervous tissue using paraffin, cryostat or vibratome sections. *Journal of Neuroscience Methods* **2**, 277–300.

LUNA, L. G. (1968) *Manual of Histologic Staining Methods of the Armed Forces Institute of Pathology.* 3rd ed. New York: McGraw-Hill.

LUPPA, H. and ANDRA, J. (1983) The histochemistry of carboxyl ester hydrolases: problems and possibilities. *Histochemical Journal* **15**, 111–137.

LUTHER, P. W. and BLOCH, R. J. (1989) Formaldehyde-amine fixatives for immunocytochemistry of cultured *Xenopus* myocytes. *Journal of Histochemistry and Cytochemistry* **37**, 75–82.

LYCETTE, R. M., DANFORTH, W. F., KOPPEL, J. L. and OLWIN, J. H. (1970) The binding of luxol fast blue ARN by various biological lipids. *Stain Technology* **45**, 155–160.

LYMAN, W. J., REEHL, W. F. and ROSENBLATT, D. H. (ed.) (1982) *Handbook of Chemical Property Estimation Methods.* New York: McGraw-Hill.

LYON, H., JAKOBSEN, P., HOYER, P. and ANDERSEN, A. P. (1987) An investigation of new commercial samples of methyl green and pyronine Y. *Histochemical Journal* **19**, 381–384.

MacCULLUM. D. K. (1973) Positive Schiff reactivity of aortic elastin without prior HIO_4 oxidation: influence of maturity and a suggested source of the aldehyde. *Stain Technology* **48**, 117–122.

MAEDA, I., IMAI, H., ARAI, R., TAGO, M., NAGAI, T., SAKAMOTO, T., KITAHAMA, K., ONTENIENTE, B. and KIMURA, H. (1987) An improved coupled peroxidatic oxidation method of MAO histochemistry for neuroanatomical research at light and electron microscopic levels. *Cellular and Molecular Biology* **33**, 1–11.

MAILLET, M. (1963) Le réactif au tetraoxyde d'osmium-iodure du zinc. *Zeitschrift für mikroskopische-anatomische Forschung* **76**, 397–425.

MAINWARING, W. I. P., PARISH, J. H., PICKERING, J. D. and MANN, N. H. (1982) *Nucleic Acid Biochemistry and Molecular Biology.* Oxford: Blackwell.

MAKELA, O. (1957) Studies in haemagglutinins of Leguminosae seeds. *Acta Medicinae experimentalis et Biologiae Fenniae* **35**, Suppl. **11**, 1–133.

MALININ, G. (1977) Stable sudanophilia of "bound" lipids in tissue culture cells is a staining artifact. *Journal of Histochemistry and Cytochemistry* **25**, 155–156.

MALININ, G. (1980) The *in situ* determination of melting-solidification points of lipid inclusions in fixed cultured cells. *Journal of Histochemistry and Cytochemistry* **28**, 708–709.

MALM, M. (1962) *p*-Toluenesulphonic acid as a fixative. *Quarterly Journal of Microscopical Science* **103**, 163–171.

MALMGREN, H. and SYLVEN, B. (1955) On the chemistry of the thiocholine method of Koelle. *Journal of Histochemistry and Cytochemistry* **3**, 441–448.

MANN, G. (1902) *Physiological Histology. Methods and Theory.* Oxford: Clarendon Press.

MANNOJI, H., YEGER, H. and BECKER, L. E. (1986) A specific histochemical marker (lectin *Ricinus communis* agglutinin-I) for normal human microglia, and application to routine histopathology. *Acta neuropathologica* **71**, 341–343.

MARCH, J. (1977) *Advanced Organic Chemistry* (2nd edn). New York: McGraw-Hill.

MARCHI, V. (1892) Sur l'origine et le cours des pédoncles cérébellaux et sur leurs rapports avec les autres centres nerveux. *Archives Italiennes de Biologie* **17**, 190–201.

MARSHALL, P. N. (1977) Thin layer chromatography of Sudan dyes. *Journal of Chromatography* **136**, 353–357.

MARSHALL, P. N. (1978) Romanowsky-type stains in haematology. *Histochemical Journal* **10**, 1–29.

MARSHALL, P. N. and HOROBIN, R. W. (1972a) The chemical nature of the gallocyanin-chrome alum staining complex. *Stain Technology* **47**, 155–161.

MARSHALL, P. N. and HOROBIN, R. W. (1972b) The oxidation products of haematoxylin and their role in biological staining. *Histochemical Journal* **4**, 493–503.

MARSHALL, P. N. and HOROBIN, R. W. (1973) The mechanism of action of "mordant" dyes—a study using preformed metal complexes. *Histochemie* **35**, 361–371.

MARSHALL, P. N. and HOROBIN, R. W. (1974) A simple assay procedure for mixtures of haematoxylin and haematein. *Stain Technology* **49**, 137–142.

MASON, D. Y. and SAMMONS, R. E. (1979) The labelled antigen method of immunoenzymatic staining. *Journal of Histochemistry and Cytochemistry* **27**, 832–840.

MAURO, A., GERMANO, I., GIACCONE, G., GIORDANA, M. T. and SCHIFFER, D. (1985) 1-Naphthol basic dye (1-NBD). An alternative to diaminobenzidine (DAB) in immunoperoxidase techniques. *Histochemistry* **83**, 97–102.

MAXWELL, M. H. (1988) Osmium tetroxide: an historical appreciation. *Proceedings of the Royal Microscopical Society* **23**, 229–232.

MAYALL, B. H. and GLEDHILL, B. L. (ed.) (1977) The fifth Engineering Foundation Conference on Automatic Cytology. *Journal of Histochemistry and Cytochemistry* **25** (7), 479–952.

MAYNE, R. and BURGESON, R. E. (1987) *Structure and Function of Collagen Types*. Orlando, Florida: Academic Press.

MAYS, E. T., FELDHOFF, R. C. and NETTLETON, G. S. (1984) Determination of protein loss during aqueous and phase partition fixation using formalin and glutaraldehyde. *Journal of Histochemistry and Cytochemistry* **32**, 1107–1112.

MAZURKIEWICZ, J. E., HOSSLER, F. E. and BARRNETT, R. J. (1978) Cytochemical demonstration of sodium, potassium, adenosine triphosphatase by a haemepeptide derivative of ouabain. *Journal of Histochemistry and Cytochemistry* **26**, 1042–1052.

McAULIFFE, C. A. (ed.) (1977) *The Chemistry of Mercury*. London: Macmillan.

McAULIFFE, W. G. and NETTLETON, G. S. (1984) Phase partition fixation for electron microscopy. *Journal of Histochemistry and Cytochemistry* **32**, 913.

McISAAC, G. and KIERNAN, J. A. (1974) Complete staining of neuromuscular innervation with bromo-indigo and silver. *Stain Technology* **49**, 211–214.

McKAY, R. B. (1962) An investigation of the anomalous staining of chromatin by the acid dyes, methyl blue and aniline blue. *Quarterly Journal of Microscopical Science* **103**, 519–530.

McMILLAN, P. J., ENGEN, P. C., DALGLEISH, A. and McMILLAN, J. (1983) Improvement of the butyl methacrylate-paraffin embedment. *Stain Technology* **58**, 125–130.

MELOAN, S. N., VALENTINE, L. S. and PUCHTLER, H. (1971) On the structure of carminic acid and carmine. *Histochemie* **27**, 87–95.

MENDELSON, D., TAS, J. and JAMES, J. (1983) Cuprolinic Blue: a specific dye for single-stranded RNA in the presence of magnesium chloride. II. Practical applications for light microscopy. *Histochemical Journal* **15**, 1113–1121.

MERA, S. L. and DAVIES, J. D. (1984) Differential Congo red staining: the effects of pH, non-aqueous solvents and the substrate. *Histochemical Journal* **16**, 195–210.

MESULAM, M-M. (1978) Tetramethyl benzidine for horseradish peroxidase neurohistochemistry: a non-carcinogenic blue reaction-product with superior sensitivity for visualizing neural afferents and efferents. *Journal of Histochemistry and Cytochemistry* **26**, 106–117.

MESULAM, M-M. and ROSENE, D. L. (1979) Sensitivity in horseradish peroxidase neurohistochemistry: a comparative and quantitative study of nine methods. *Journal of Histochemistry and Cytochemistry* **27**, 763–773.

MILLER, M. W. and NOWAKOWSKI, R. S. (1988) Use of bromodeoxyuridine immunohistochemistry to examine the proliferation, migration and time of origin of cells in the central nervous system. *Brain Research* **457**, 44–52.

MINARD, B. J. and CAWLEY, L. P. (1978) Use of horseradish peroxidase to block nonspecific enzyme uptake in immunoperoxidase microscopy. *Journal of Histochemistry and Cytochemistry* **26**, 685–687.

MIZUKAWA, K., VINCENT, S. R., McGEER, P. L. and McGEER, E. G. (1989). Distribution of reduced-nicotinamide-adenine-dinucleotide-phosphate diaphorase-positive cells and fibers in the cat central nervous system. *Journal of Comparative Neurology* **279**, 281–311.

MOLNAR, J. (1952) The use of rhodizonate in enzymatic histochemistry. *Stain Technology* **27**, 221–222.

MONSAN, P., PUZO, G. et MARZARGUIL, H. (1975) Étude du mécanisme d'établissement des liaisons glutaraldéhyde-protéins. *Biochimie* **57**, 1281–1292.

MORENO, F. J., VILLAMARIN, A., GARCIA-HERDUGO, G. and LOPEZ-CAMPOS, J. L. (1988) Silver staining of the nucleolar organizer regions (NORs) in semithin lowicryl sections. *Stain Technology* **63**, 27–31.

MORETON, R. B. (1981) Electron-probe X-ray microanalysis: techniques and applications in biology. *Biological Reviews* **56**, 409–461.

MORRIS, S. M., STONE, P. J., ROSENKRANS, W. A., CALORE, J. D., ALBRIGHT, J. T. and FRANZENBLAU, C. (1978) Palladium chloride as a stain for elastin at the ultrastructural level. *Journal of Histochemistry and Cytochemistry* **26**, 635–644.

MORRISON, R. T. and BOYD, R. N. (1973) *Organic Chemistry*, 3rd ed. Boston: Allyn and Bacon.

MORSTYN, G., PYKE, K., GARDNER, J., ASHCROFT, R., de FAZIO, A. and BHATHAL, P. (1986) Immunohistochemical identification of proliferating cells in organ culture using bromodeoxyuridine and a monoclonal antibody. *Journal of Histochemistry and Cytochemistry* **34**, 697–701.

MORTHLAND, F. W., de BRUYN, P. P. H. and SMITH, N. H. (1954) Spectrophotometric studies on the interaction of nucleic acids with aminoacridines and other basic dyes. *Experimental Cell Research* **7**, 201–214.

MORTON, D. (1978) A comparison of iron histochemical methods for use on glycol methacrylate embedded tissues. *Stain Technology* **53**, 217–223.

MOWRY, R. W. (1978) Aldehyde fuchsin staining, direct or after oxidation: problems and remedies, with special reference to human pancreatic B cells, pituitaries and elastic fibres. *Stain Technology* **53**, 141–154.

MOWRY, R. W. and EMMEL, V. M. (1977) The production of aldehyde fuchsin depends on the pararosaniline (C.I. No. 42500) content of basic fuchsins which is sometimes negligible and is sometimes mislabelled. *Journal of Histochemistry and Cytochemistry* **25**, 239.

MULLER-WALZ, R. und ZIMMERMANN, H. W. (1987) Über Romanowsky-Farbstoffe und den Romanowsky-Giemsa-Effekt. 4. Mitteilung: Bindung von Azur B an DNA. *Histochemistry* **87**, 157–172.

MULLINK, H., WALBOOMERS, J. M. M., TADEMA, T. M., JANSEN, D. J. and MEIJER, C. J. L. M. (1989) Combined immuno- and non-radioactive hybridocytochemistry on cells and tissue sections: influence of fixation, enzyme pretreatment, and choice of chromogen on detection of antigen and DNA sequences. *Journal of Histochemistry and Cytochemistry* **37**, 603–609.

MURALT, A. von (1943) Die sekundäre Thiochromfluorescenz des peripheren Nerven und ihre Beziehung zu Bethes Polarizationsbild. *Pflügers Archiv für gesamte Physiologie* **247**, 1–10.

MURGATROYD, L. B. (1976) The preparation of thin sections from glycol methacrylate embedded tissue using a standard rotary microtome. *Medical Laboratory Sciences* **33**, 67–71.

MURRAY, G. I., BURKE, M. D. and EWEN, S. W. B. (1986) Glutathione localization by a novel *o*-phthalaldehyde histofluorescence method. *Histochemical Journal* **18**, 434–440.

MURRAY, G. I., BURKE, M. D. and EWEN, S. W. B. (1989) Enzyme histochemistry on freeze-dried, resin-embedded tissue. *Journal of Histochemistry and Cytochemistry* **37**, 643–652.

MUSTO, L. (1981) Improved iron-haematoxylin stain for elastic fibres. *Stain Technology* **56**, 185–187.

NAIRN, R. C. (1976) *Fluorescent Protein Tracing*. 4th ed. Edinburgh: Churchill-Livingstone.

NAKAO, K. and ANGRIST, A. A. (1968) A histochemical demonstration of aldehyde in elastin. *American Journal of Clinical Pathology* **49**, 65–67.

NAMBA, T., NAKAMURA, T. and GROB, D. (1967) Staining for nerve fibre and cholinesterase activity in fresh frozen sections. *American Journal of Clinical Pathology* **47**, 74–77.

NAOUMENKO, J. and FEIGIN, I. (1961) A modification for paraffin sections of the Cajal gold-sublimate stain for astrocytes. *Journal of Neuropathology and Experimental Neurology* **20**, 602–604.

NAUTA, W. J. H. and EBBESSON, S. O. B. (ed.) (1970) *Contemporary Research Methods in Neuroanatomy*. Berlin: Springer.

NAUTA, W. J. H. and GYGAX, P. A. (1951) Silver impregnation of degenerating axon terminals in the central nervous system: (1) Technic. (2) Chemical notes. *Stain Technology* **26**, 5–11.

NEDZEL, G. A. (1951) Intranuclear birefringent inclusions, an artifact occurring in paraffin sections. *Quarterly Journal of Microscopical Science* **92**, 343–346.

NEISS, W. F. (1984) Electron staining of the cell surface coat by osmium-low ferrocyanide. *Histochemistry* **80**, 231–242.

NETTLETON, G. S. (1982) The role of paraldehyde in the rapid preparation of aldehyde fuchsin. *Journal of Histochemistry and Cytochemistry* **30**, 175–178.

NETTLETON, G. S. and CARPENTER, A. M. (1977) Studies on the mechanism of the periodic acid-Schiff histochemical reaction for glycogen using infrared spectroscopy and model chemical compounds. *Stain Technology* **52**, 63–77.

NETTLETON, G. S. and McAULIFFE, W. G. (1986) A histological comparison of phase-partition fixation with fixation in aqueous solutions. *Journal of Histochemistry and Cytochemistry* **34**, 795–800.

NEWMAN, G. R., JASANI, B. and WILLIAMS, E. D. (1982) The preservation of ultrastructure and antigenicity. *Journal of Microscopy* **127**, RP5-RP6.

NEWMAN, S. B., BORYSKO, E. and SWERDLOW, M. (1949) New sectioning techniques for light and electron microscopy. *Science* **110**, 66–68.

NGUYEN-LEGROS, J., BIZOT, J., BOLESSE, M. et PULICANI, J-P. (1980) "Noir de diaminobenzidine": une nouvelle methode histochimique de revelation du fer exogène. *Histochemistry* **66**, 239–244.

NICHOLSON, M. L. and MONKHOUSE, W. S. (1985) Cholesterol retention in tissue sections: an assessment of different techniques. *Journal of Anatomy* **140**, 538.

NICOLET, B. H. and SHINN, L. A. (1939) The action of periodic acid on α-amino alcohols. *Journal of the American Chemical Society* **61**, 1615.

NICOLSON, G. L. (1974) The interactions of lectins with animal cell surfaces. *International Review of Cytology* **39**, 89–190.

NIELSON, A. J. and GRIFFITH, W. P. (1978) Tissue fixation and staining with osmium tetroxide: the role of phenolic compounds. *Journal of Histochemistry and Cytochemistry* **26**, 138–140.

NIELSON, A. J. and GRIFFITH, W. P. (1979) Tissue fixation by osmium tetroxide. A possible role for proteins. *Journal of Histochemistry and Cytochemistry* **27**, 997–999.

NIENDORF, A., DIETEL, M., ARPS, H. and CHILDS, G. V. (1988) A novel method to demonstrate parathyroid hormone binding on unfixed living target cells in culture. *Journal of Histochemistry and Cytochemistry* **36**, 307–309.

NISHIKAWA, K. C. (1987) Staining amphibian peripheral nerves with Sudan black B: progressive vs regressive methods. *Copeia* **1987 (2)**: 489–491.

NOLLER, C. R. (1965) *Chemistry of Organic Compounds*, 3rd ed. Philadelphia: Saunders.

NORTON, W. T., KOREY, S. R. and BROTZ, M. (1962) Histochemical demonstration of unsaturated lipids by a bromine-silver method. *Journal of Histochemistry and Cytochemistry* **10**, 83–88.

NOTANI, G. W., PARSONS, J. A. and ERLANDSEN, S. L. (1979) Versatility of *Staphylococcus aureus* protein A in immunocytochemistry. Use in unlabelled antibody enzyme systems and fluorescent methods. *Journal of Histochemistry and Cytochemistry* **27**, 1438–1444.

NURSE, C. A. and FARRAWAY, L. (1989) Characterization of Merkel cells and mechanosensory axons of the rat by styryl pyridinium dyes. *Cells and Tissue Research* **255**, 125–128.

NURSE, C. A., MEAROW, K. M., HOLMES, M., VISHEAU, B. and DIAMOND, J. (1983) Merkel cell distribution in the epidermis as determined by quinacrine fluorescence. *Cell and Tissue Research* **228**, 511–524.

OHNISHI, T., AMAMOTO, K. and TERAYAMA, H. (1973) Polysaccharides associated with chromosomes and their behaviour in the cell cycle. *Histochemie* **35**, 1–10.

OLSON, L., ALUND, M. and NORBERG, K-A. (1976) Fluorescence-microscopical demonstration of a population of gastro-intestinal nerve fibres with a selective affinity for quinacrine. *Cell and Tissue Research* **171**, 407–423.

ORNSTEIN, L., MAUTNER, W., DAVIS, B. J. and TAMURA, R. (1957) New horizons in fluorescence microscopy. *Journal of the Mount Sinai Hospital* **24**, 1066–1078.

OSTERBERG, R. (1974) Metal ion-protein interactions in solution. Ch. 2 in *Metal Ions in Biological Systems*, ed. H. Sigel. Vol. 3, pp. 45–88. New York: Marcel Dekker.

OVEREND, W. G. and STACEY, M. (1949) Mechanism of the Feulgen nucleal reaction. *Nature* **163**, 538-540.

PAGE, K. M. (1965) A stain for myelin using solochrome cyanin. *Journal of Medical Laboratory Technology* **22**, 224-225.

PALADINO, G. (1890) D'un nouveau procédé pour les recherches microscopiques due système nerveux central. *Archives Italiennes de Biologie* **13**, 484-486.

PALJARVI, L., GARCIA, J. H. and KALIMO, H. (1979) The efficiency of aldehyde fixation for electron microscopy: stabilization of rat brain tissue to withstand osmotic stress. *Histochemical Journal* **11**, 267–276.

PANULA, P., HAPPOLA, O., AIRAKSINEN, M. S., AUVINEN, S. and VIRKAMAKI, A. (1988) Carbodiimide as a tissue fixative in histamine immunohistochemistry and its application in developmental neurobiology. *Journal of Histochemistry and Cytochemistry* **36**, 259–269.

PAPADIMITRIOU, J. M., VAN DUIJN, P., BREDEROO, P. and STREEFKERK, J. G. (1976) A new method for the cytochemical demonstration of peroxidase for light, fluorescence and electron microscopy. *Journal of Histochemistry and Cytochemistry* **24**, 82–90.

PARK, C. M., REID, P. E., OWEN, D. A., DUNN, W. L. and VOLZ, D. (1987) Histochemical procedures for the simultaneous visualization of neutral sugars and either sialic acid and its *O*-acyl variants or *O*-sulphate ester. II. Methods based upon the periodic acid-phenylhydrazine-Schiff reaction. *Histochemical Journal* **19**, 257–263.

PARMLEY, R. T., SPICER, S. S. and ALVAREZ, C. J. (1978) Ultrastructural localization of nonhaeme cellular iron with ferrocyanide. *Journal of Histochemistry and Cytochemistry* **26**, 729–741.

PASTEELS, J. L. et HERLANT, M. (1962) Notions nouvelles sur la cytologie de l'antéhypophyse chez le rat. *Zeitschrift für zellforschung* **56**, 20–39.

PEARSE, A. G. E. (1968, 1972) *Histochemistry, Theoretical and Applied*, 3rd edn. (2 vols). Edinburgh: Churchill-Livingstone.

PEARSE, A. G. E. (1980, 1985) *Histochemistry, Theoretical and Applied*, 4th edn. Vol. 1. Preparative and Optical Technology. Vol. 2. Analytical Technique. Edinburgh: Churchill-Livingstone.

PEARSE, A. G. E. and POLAK, J. (1975) Bifunctional reagents as vapour- and liquid-phase fixatives for immunohistochemistry. *Histochemical Journal* **7**, 179–186.

PENFIELD, W. and CONE, W. V. (1950) Neuroglia and microglia (the metallic methods). In *McClung's Handbook of Microscopical Technique*, 3rd ed., ed. R. M. Jones, pp. 399–431. New York: Hoeber.

PEPLER, W. J. and PEARSE, A. G. E. (1957) The histochemistry of the esterases of the rat brain, with special reference to those of the hypothalamic nuclei. *Journal of Neurochemistry* **1**, 193–202.

PERRIN, D. D. and DEMPSEY, B. (1974) *Buffers for pH and Metal Ion Control*. London: Chapman & Hall.

PETERS, A. (1955a) Experiments on the mechanism of silver staining. I. Impregnation. *Quarterly Journal of Microscopical Science* **96**, 84–102.

PETERS, A. (1955b) Experiments on the mechanism of silver staining. II. Development. *Quarterly Journal of Microscopical Science* **96**, 103–115.

PETERS, A. (1955c) Experiments on the mechanism of silver staining. III. Electron microscope studies. *Quarterly Journal of Microscopical Science* **96**, 317–322.

PETERS, A. (1981) The Golgi-electron microscope technique. Ch. 5 in *Current Trends in Morphological Techniques*, ed. J. E. Johnson, Vol. 1, pp. 187–212. Boca Raton, Florida: CRC Press.

PEYRONNARD, J. M., CHARRON, L., LAVOIE, J., MESSIER, P. and DUBREUIL, M. (1988) Carbonic anhydrase and horseradish peroxidase: double labelling of rat dorsal root ganglion neurons innervating motor and sensory peripheral nerves. *Anatomy and Embryology* **177**, 353–359.

PFULLER, U., FRANZ, H. and PREISS, A. (1977) Sudan black B: chemical studies and histochemistry of the blue main component. *Histochemistry* **54**, 237–250.

PHILLIPS, L. L., AUTILIO-GAMBETTI, L. and LASEK, R. J. (1983) Bodian's silver method reveals molecular variation in the evolution of neurofilament proteins. *Brain Research* **278**, 219–223.

PIGNAL, F., MAURICE, M. and FELDMANN, G. (1982) Immunoperoxidase localization of albumin and fibrinogen in rat liver fixed by perfusion of immersion:

effect of saponin on the intracellular penetration of labelled antibodies. *Journal of Histochemistry and Cytochemistry* 30, 1004–1014.

PLATT, J. L. and MICHAEL, A. F. (1983) Retardation of fading and enhancement of intensity of immunofluorescence by *p*-phenylenediamine. *Journal of Histochemistry and Cytochemistry* 31, 840–842.

PLUMEL, M. (1948) Tampon au cacodylate de sodium. *Bulletin de la Société de Chimie Biologique* 30, 129–130.

POCHHAMMER, C., DIETSCH, P. and SIEGMUND, P. R. (1979) Histochemical detection of carbonic anhydrase with dimethylaminonaphthalene-5-sulfonamide. *Journal of Histochemistry and Cytochemistry* 27, 1103–1107.

PODOLSKY, D. K. (1985) Oligosaccharide structures of human colonic mucin. *Journal of Biological Chemistry* 260, 8262–8271.

POLAK, J. M. and VAN NOORDEN, S. (ed.) (1986) *Immunocytochemistry. Modern Methods and Applications.* 2nd ed. Bristol: Wright.

POLAK, J. M. and VARNDELL, I. M. (ed.) (1984) *Immunolabelling for Electron Microscopy.* Amsterdam: Elsevier.

POLAK, M. (1948) Sobre la importancia del bromuro de amonio de la solucion fijadora de Cajal en la impregnacion argentica del tejido nervioso. *Archivos de la Sociedad Argentina de Anatomia Normal y Patologica* 10, 224–234.

POLICARD, A., BESSIS, M. et BRICKA, M. (1952) La fixation des cellules isolées observée au contraste de phase et au microscope électronique. I. Action des différents fixateurs. *Bulletin de Microscopie Appliquée (2è Ser.)* 2, 29–42.

PONDER, B. A. and WILKINSON, M. M. (1981) Inhibition of endogenous tissue alkaline phosphatase with the use of alkaline phosphatase conjugates in immunohistochemistry. *Journal of Histochemistry and Cytochemistry* 29, 981–984.

POURADIER, J. (1977) Properties of gelatin in relation to its use in the preparation of photographic emulsions. Ch. 2, Sec. II in *The Theory of the Photographic Process*, 4th ed., ed. T. H. James. pp. 67–76. New York; Macmillan.

PRENTØ, P. (1978) Rapid dehydration-clearing with 2,2-dimethoxypropane for paraffin embedding. *Journal of Histochemistry and Cytochemistry* 26, 865–867.

PRENTØ, P. (1980) The effect of histochemical methylation on the phosphate groups of nucleic acids: interpretation of the absence of nuclear basophilia. *Histochemical Journal* 12, 661–668.

PRINGLE, J. H., HOMER, C. E., WARFORD, A., KENDALL, C. H. and LAUDER, I. (1987) *In situ* hybridization: alkaline phosphatase visualization of biotinylated probes in cryostat sections. *Histochemical Journal* 19, 488–496.

PUCHTLER, H. and ISLER, H. (1958) The effect of phosphomolybdic acid on the stainability of connective tissues by various dyes. *Journal of Histochemistry and Cytochemistry* 6, 265–270.

PUCHTLER, H. and SWEAT, F. (1964a) Histochemical specificity of staining methods for connective tissue fibres: resorcin-fuchsin and van Gieson's picro-fuchsin. *Histochemie* 4, 24–34.

PUCHTLER, H. and SWEAT, F. (1964b) Effect of phosphomolybdic acid on the binding of Sudan black B. *Histochemie* 4, 20–23.

PUCHTLER, H. and WALDROP, F. S. (1978) Silver impregnation methods for reticular fibres and reticulin: a reinvestigation of their origins and specificity. *Histochemistry* 57, 177–187.

PUCHTLER, H. and WALDROP, F. S. (1979) On the mechanism of Verhoeff's elastica stain: a convenient stain for myelin sheaths. *Histochemistry* 62, 233–247.

PUCHTLER, H., MELOAN, S. N. and WALDROP, F. S. (1988) Are picro-dye reactions for collagen quantitative? Chemical and histochemical considerations. *Histochemistry* 88, 243–256.

PUCHTLER, H., SWEAT, F. and KUHNS, J. G. (1964) On the binding of direct cotton dyes by amyloid. *Journal of Histochemistry and Cytochemistry* 12, 900–907.

PURVES, D. and VOYVODIC, J. T. (1987) Imaging mammalian nerve cells and their connections over time in living animals. *Trends in Neurosciences* 10, 398–404.

QUINTARELLI, G., SCOTT, J. E. and DELLOVO, M. C. (1964a). The chemical and histochemical properties of alcian blue. II. Dye binding by tissue polyanions. *Histochemie* 4, 86–98.

QUINTARELLI, G., SCOTT, J. E. and DELLOVO, M. C. (1964b). The chemical and histochemical properties of alcian blue. III. Chemical blocking and unblocking. *Histochemie* 4, 99–112.

RAAP, A. K. (1983) Studies on the phenazine methosulphate-tetrazolium salt capture reaction in $NAD(P)^+$-dependent dehydrogenase cytochemistry. III. The role of superoxide in tetrazolium reduction. *Histochemical Journal* 15, 977–986.

RAAP, A. K. and VAN DIUJN, P. (1983) Studies on the phenazine methosulphate-tetrazolium salt capture reaction in $NAD(P)^+$-dependent dehydrogenase cytochemistry. II. A novel hypothesis for the mode of action of PMS and a study of the properties of reduced PMS. *Histochemical Journal* 15, 881–893.

RAAP, A. K., VAN HOOF, G. R. M. and VAN DUIJN, P. (1983) Studies on the phenazine methosulphate-tetrazolium salt capture reaction in $NAD(P)^+$-dependent dehydrogenase cytochemistry. I. Localization artefacts caused by the escape of reduced coenzyme during cytochemical reactions for $NAD(P)^+$-dependent dehydrogenases. *Histochemical Journal* 15, 861–879.

RALIS, H. M., BEESLEY, R. A. and RALIS, Z. A. (1973) *Techniques in Neurohistology*. London: Butterworths.

RAMON-MOLINER, E. (1988) The Golgi-Cox technique. In *Contemporary Research Methods in Neuroanatomy*, edit. W. J. H. Nauta and S. O. B. Ebbesson, pp. 32–55. Berlin: Springer.

RANKI, A., REITAMO, S., KONITTINEN, Y. T. and HAYRY, P. (1980) Histochemical identification of human T lymphocytes from paraffin sections. *Journal of Histochemistry and Cytochemistry* 28, 704–707.

RASMUSSEN, G. L. (1957) Selective silver impregnation of synaptic endings. In *New Research Techniques in Neuroanatomy*, ed. W. F. Windle, pp. 27–39. Springfield, Illinois: C. C. Thomas.

REID, L. and CLAMP, J. R. (1978) The biochemical and

histochemical nomenclature of mucus. *British Medical Bulletin* **34**, 5–8.

REID, P. E. and CULLING, C. F. A. (1983) Apparent failure of 2-hydroxy-3-naphthoic acid hydrazide to block the periodic acid-Schiff reactivity of sialic acids. *Journal of Histochemistry and Cytochemistry* **31**, 1142–1144.

REID, P. E. and OWEN, D. A. (1988) Some comments on the mechanism of the periodic acid-Schiff-Alcian blue method. *Histochemical Journal* **20**, 651–654.

REID, P. E., VOLZ, D., CHO, K. Y. and OWEN, D. A. (1988) A new method for the histochemical demonstration of *O*-acyl sugars in human colonic epithelial glycoproteins. *Histochemical Journal* **20**, 510–518.

REID, P. E., CULLING C. F. A., DUNN, W. L., CLAY, M. G. and RAMEY, C. W. (1978) A correlative chemical and histochemical study of the *O*-acetylated sialic acids of human colonic epithelial glycoproteins in formalin fixed paraffin embedded tissues. *Journal of Histochemistry and Cytochemistry* **26**, 1033–1041.

REID, P. E., VOLZ, D., PARK, C. M., OWEN, D. A. and DUNN. W. L. (1987) Methods for the identification of side chain *O*-acyl substituted sialic acids and for the simultaneous visualization of sialic acid, its side chain *O*-acyl variants and *O*-sulphate ester. *Histochemical Journal* **19**, 396–398.

REID, P. E., DUNN, W. L., RAMEY, C. W., CORET, E., TRUEMAN, L. and CLAY, M. G. (1984a) Histochemical identification of side chain substituted *O*-acylated sialic acids: the PAT-KOH-Bh-PAS and the PAPT-KOH-Bh-PAS procedures. *Histochemical Journal* **16**, 623–639.

REID, P. E., DUNN, W. L., RAMEY, C. W., CORET, E., TRUEMAN, L. and CLAY, M. G. (1984b) Histochemical studies of the mechanism of the periodic acid-phenylhydrazine-Schiff (PAPS) procedure. *Histochemical Journal* **16**, 641–649.

REINER, A. and GAMLIN, P. (1980) On noncarcinogenic chromogens for horseradish peroxidase. *Journal of Histochemistry and Cytochemistry* **28**, 187–189.

RICHARDSON, K. C. (1969) The fine structure of autonomic nerves after vital staining with methylene blue. *Anatomical Record* **164**, 359–378.

RIECK, G. D. (1967) *Tungsten and its Compounds*. Oxford: Pergamon.

RINGER, S. (1893) The influence of carbonic acid dissolved in saline solutions on the ventricle of the frog's heart. *Journal of Physiology* **14**, 125–130.

RIPELLINO, J. A., KLINGER, M. M., MARGOLIS, R. U. and MARGOLIS, R. K. (1985) The hyaluronic acid binding region as a specific probe for the localization of hyaluronic acid in tissue sections. *Journal of Histochemistry and Cytochemistry* **33**, 1060–1066.

RITTMAN, B. R. and MACKENZIE, I. C. (1983) Effects of histological processing on lectin binding patterns in oral mucosa and skin. *Histochemical Journal* **15**, 467–474.

ROBERTS, G. P. (1977) Histochemical detection of sialic acid residues using periodate oxidation. *Histochemical Journal* **9**, 97–102.

ROBINSON, M. M. (1987) Fixation and immunofluorescent analysis of creatine kinase isozymes in embryonic skeletal muscle. *Journal of Histochemistry and Cytochemistry* **35**, 717–722.

RODRIGO, J., NAVA, B. E. and PEDROSA, J. (1970) Study of vegetative innervation in the oesophagus. I. Perivascular endings. *Trabajos del Instituto Cajal de Investigaciones Biologicas* **62**, 39–65.

ROE, R., CORFIELD, A. P. and WILLIAMSON, R. C. N. (1989) Sialic acid in colonic mucin: an evaluation of modified PAS reactions in single and combination histochemical procedures. *Histochemical Journal* **21**, 216–222.

ROITT, I., BROSTOFF, J. and MALE, D. (1989) *Immunology*. 2nd ed. London: Churchill Livingstone.

ROMANYI, G., DEAK, G. and FISCHER, J. (1975) Aldehyde-bisulphite-toluidine blue (ABT) staining as a topo-optical reaction for demonstration of linear order of vicinal OH groups in biological structures. *Histochemistry* **43**, 333–348.

ROQUE, A. L., JAFERAY, N. A. and COULTER, P. (1965) A stain for the histochemical demonstration of nucleic acids. *Experimental and Molecular Pathology* **4**, 266–274.

ROS BARCELO, A., MUNOZ, R. and SABATER, F. (1989) Activated charcoal as an adsorbent of oxidized 3,3′-diaminobenzidine in peroxidase histochemistry. *Stain Technology* **64**, 97–98.

ROSEN, A. D. (1981) End-point determination in EDTA decalcification using ammonium oxalate. *Stain Technology* **56**, 48–49.

ROSENE, D. L. and MESULAM, M-M. (1978) Fixation variables in horseradish peroxidase neurohistochemistry. I. The effects of fixation time and perfusion procedures upon enzyme activity. *Journal of Histochemistry and Cytochemistry* **26**, 28–39.

ROTH, J. (1982) Applications of immunocolloids in light microscopy. Preparation of protein A-silver and protein A-gold complexes and their application for localization of single and multiple antigens in paraffin sections. *Journal of Histochemistry and Cytochemistry* **30**, 691–696.

ROTH, J., BENDAYAN, M. and ORCI, L. (1978) Ultrastructural localization of intracellular antigens by the use of protein A-gold complex. *Journal of Histochemistry and Cytochemistry* **26**, 1074–1081.

RUTENBERG, A. M., ROSALES, C. L. and BENNETT, J. M. (1965) An improved method for the demonstration of leukocyte alkaline phosphatase activity in clinical application. *Journal of Laboratory and Clinical Medicine* **65**, 698–705.

RYE, D. B., SAPER, C. B. and WAINER, B. H. (1984) Stabilization of the tetramethylbenzidine (TMB) reaction product: Application for retrograde and anterograde tracing, and combination with immunohistochemistry. *Journal of Histochemistry and Cytochemistry* **32**, 1145–1153.

SABATINI, D. D., BENSCH, K. and BARNETT, R. J. (1963) Cytochemistry and electron microscopy. The preservation of cellular ultrastructure and enzymatic activity by aldehyde fixation. *Journal of Cell Biology* **17**, 19–58.

SAKANAKA, M., MAGARI, S., SHIBASAKI, T., SHINODA, K. and KOHNO, J. (1988) A reliable method combining horseradish peroxidase with immuno-β-galactosidase staining. *Journal of Histochemistry and Cytochemistry* **36**, 1091–1096.

SALTHOUSE, T. N. (1962) Luxol fast blue ARN: a new solvent azo dye with improved staining qualities for myelin and phospholipids. *Stain Technology* **37**, 313–316.

SALTHOUSE, T. N. (1963) Reversal of solubility characteristics of "Luxol" dye-phospholipid complexes. *Nature* **199**, 821.

SAMUEL, E. P. (1953a) The mechanism of silver staining. *Journal of Anatomy* **87**, 278–287.

SAMUEL, E. P. (1953b) Gold toning. *Stain Technology* **28**, 225–229.

SANES, J. R. and CHENEY, J. M. (1982) Lectin binding reveals a synapse-specific carbohydrate in skeletal muscle. *Nature* **300**, 646–647.

SANGER, J. W. (1975) Intracellular localization of actin with fluorescently labelled heavy meromyosin. *Cell and Tissue Research* **161**, 431–444.

SANTINI, M. (ed.) (1975) *Golgi Centennial Symposium: Perspectives in Neurobiology*. New York: Raven Press.

SANWICKI, E., HAUSER, T. R., STANLEY, T. W. and ELBERT, W. (1961) The 3-methyl-benzothiazolone hydrazone test. *Analytical Chemistry* **33**, 93–96.

SAYK, J. (1954) Ergebnisse neuer liquor-cytologischer Untersuchungen mit den Sedimentierkammer-Verfahren. *Ärztliche Wochenschrift* **9**, 1042–1046.

SCARSELLI, V. (1961) Histochemical demonstration of aldehydes by *p*-phenylenediamine. *Nature* **190**, 1206–1207.

SCHABADASCH, A. (1930) Untersuchungen zur Methodik der Methylenblaufarbung des vegetativen Nervensystems. *Zeitschrift für Zellforschung* **10**, 221–243.

SCHEUERMANN, D. W. and STACH, W. (1985) NADH-dehydrogenase reaction in combination with immunoperoxidase (PAP) staining for light microscopic observation on the interneuronal relations of the enteric nervous system of the pig. *Acta anatomica* **124**, 31–34.

SCHMID, K. W., HITTMAIR, A., SCHMID-HAMMER, H. and JASANI, B. (1989) Non-deleterious inhibition of endogenous peroxidase activity (EPA) by cyclopropanone hydrate: a definitive approach. *Journal of Histochemistry and Cytochemistry* **37**, 473–477.

SCHMIDT, R. and MOENKE-BLANKENBURG, L. (1986) Modern physical methods for analysing elements and structures in histochemistry. *Acta histochemica* **80**, 205–213.

SCHOOK, P. (1980) The effective osmotic pressure of the fixative for transmission and scanning electron microscopy. *Acta Morphologica Neerlando-Scandinavica* **18**, 31–45.

SCHRODER, M. (1980) Osmium tetraoxide cis hydroxylation of unsaturated substrates. *Chemical Reviews* **80**, 187–213.

SCHUBERT, M. and HAMERMAN, D. (1956) Metachromasia: chemical theory and histochemical use. *Journal of Histochemistry and Cytochemistry* **4**, 159–189.

SCHULZ-HARDER, B. and GRAF v. KEYSER-

LINGK, D. (1988) Comparison of brain ribonucleases of rabbit, guinea-pig, rat, mouse and gerbil. *Histochemistry* **88**, 587–594.

SCOPSI, L. and LARSSON, L-I. (1986) Bodian's silver impregnation of endocrine cells. A tentative explanation to the staining mechanism. *Histochemistry* **86**, 59–62.

SCOTT, J. E. (1967) On the mechanism of the methyl green-pyronin stain for nucleic acids. *Histochemie* **9**, 30–47.

SCOTT, J. E. (1972a) Histochemistry of alcian blue. II. The structure of alcian blue 8GX. *Histochemie* **30**, 215–234.

SCOTT, J. E. (1972b) Histochemistry of alcian blue. III. The molecular biological basis of staining by alcian blue 8GX and analogous phthalocyanins. *Histochemie* **32**, 191–212.

SCOTT, J. E. and DORLING, J. (1969) Periodate oxidation of acid polysaccharides. III. A PAS method for chondroitin sulphates and other glycosamino-glycuronans. *Histochemie* **19**, 295–301.

SCOTT, J. E. and HARBINSON, R. J. (1969) Periodate oxidation of acid polysaccharides. II. Rates of oxidation of uronic acids in polyuronides and acid mucopolysaccharides. *Histochemie* **19**, 155–161.

SCOTT, J. E., QUINTARELLI, G. and DELLOVO, M. C. (1964) The chemical and histochemical properties of alcian blue. I. The mechanism of alcian blue staining. *Histochemie* **4**, 73–85.

SECHRIST, J. W. (1969) Neurocytogenesis. I. Neurofibrils, neurofilaments and the terminal mitotic cycle. *American Journal of Anatomy* **124**, 117–134.

SEGADE, L. A. G. (1987) Pyrocatechol as a stabilizing agent for *o*-toluidine and *o*-dianisidine: a sensitive new method for HRP. *Journal für Hirnforschung* **28**, 331–340.

SEIDLER, E. (1979) Zum Mechanismus der Tetrazoliumsalzreduktion und Wirkungsweise des Phenazinmethosulphates. *Acta histochemica* **65**, 209–218.

SEIDLER, E. (1980) New nitro-monotetrazolium salts and their use in histochemistry. *Histochemical Journal* **12**, 619–630.

SHACKLEFORD, J. M. (1963) Histochemical comparison of mucous secretions in rodent, carnivore, ungulate and primate major salivary glands. *Annals of the New York Academy of Sciences* **106**, 572–582.

SHANNON, W. A. (1981) Light and electron microscopy cytochemistry of monoamine oxidase and other amine oxidative enzymes. Ch. 10 in *Current Trends in Morphological Techniques*, ed. J. E. Johnson; Vol. 3, pp. 193–242. Boca Raton, Florida: CRC Press.

SHOTTON, D. (1988) The current renaissance in light microscopy. II. Blur-free optical sectioning of biological specimens by confocal scanning fluorescence microscopy. *Proceedings of the Royal Microscopical Society* **23**, 289–297.

SILBERT, J. E., KLEINMAN, H. K. and SILBERT, C. K. (1975) Heparins and Heparin-like substances of cells. In *Heparin*. ed. R. A. Bradshaw and S. Wessler (*Advances in Experimental Medicine and Biology*, Vol. **52**). pp. 51–60. New York: Plenum Press.

SILVEIRA, S. R. and HADLER, W. A. (1978) Catalases and peroxidases detection techniques suitable to discriminate these enzymes. *Acta histochemica* **63**, 1–10.

SILVERMAN, M. S. and TOOTELL, R. B. H. (1987) Modified technique for cytochrome oxidase histochemistry: increased staining intensity and compatibility with 2-deoxyglucose autoradiography. *Journal of Neuroscience Methods* **19**, 1–10.

SMITH, I. C., CARSON, B. L. and FERGUSON, T. L. (1978) *Trace Metals in the Environment*. Vol. **4**. *Palladium and Osmium*. Ann Arbor, Michigan: Ann Arbor Science Publishers.

SMITH, J. L. (1908) On the simultaneous staining of neutral fat and fatty acids by oxazine dyes. *Journal of Pathology and Bacteriology* **12**, 1–4.

SNEED, M. C. and BRASTED, R. C. (1955) Ch. 6: The Lanthanide Series. In *Comprehensive Inorganic Chemistry*, Vol. **4**, pp. 151–187. Princeton, NJ: van Nostrand.

SOBELL, H. M., AI, C-C., JAIN, S. C. and GILBERT, S. G. (1977) Visualization of drug-nucleic acid interactions at atomic resolution. III. Unifying structural concepts in understanding drug-DNA interactions and their broader implications in understanding protein-DNA interactions. *Journal of Molecular Biology* **144**, 333–365.

SOFRONIEW, M. V. and SCHRELL, U. (1982) Long-term storage and repeated use of diluted antisera in glass staining jars for increased sensitivity, reproducibility, and convenience of single- and two-colour light microscopic immunocytochemistry. *Journal of Histochemistry and Cytochemistry* **30**, 504–511.

SOMOGYI, P. and TAKAGI, H. (1982) A note on the use of picric acid-paraformaldehyde-glutaraldehyde fixative for correlated light and electron microscopic immunocytochemistry. *Neuroscience* **7**, 1779–1783.

SORVARI, T. E. and LAUREN, P. A. (1973) The effect of various fixation procedures on the digestibility of sialomucins with neuraminidase. *Histochemical Journal* **5**, 405–512.

SPACEK, J. (1989) Dynamics of the Golgi method: a time-lapse study of the early stages of impregnation in single sections. *Journal of Neurocytology* **18**, 27–38.

SPARROW, J. R. (1980) Immunohistochemical study of the blood-brain barrier. Production of an artifact. *Journal of Histochemistry and Cytochemistry* **26**, 570–572.

SPICER, S. S. (1960) Siderosis associated with increased lipofuscins and mast cells in aging mice. *American Journal of Pathology* **37**, 457–475.

SPICER. S. S. and LILLIE, R. D. (1961) Histochemical identification of basic proteins with Biebrich scarlet at alkaline pH. *Stain Technology* **36**, 365–370.

SPRINGER, A. D. and PROKOSCH, J. H. (1982) Surgical and intensification procedures for defining visual pathways with cobaltous-lysine. *Journal of Histochemistry and Cytochemistry* **30**, 1235–1242.

STAINES, W. A., MEISTER, B., MELANDER, T., NAGY, J. I. and HOKFELT, T. (1988) Three-colour immunofluorescence histochemistry allowing triple labelling within a single section. *Journal of Histochemistry and Cytochemistry* **36**, 145–151.

STEEDMAN, H. F. (1960) *Section Cutting in Microscopy*. Oxford: Blackwell.

STEFANINI, M., de MARTINO, C. and ZAMBONI, L. (1967) Fixation of ejaculated spermatozoa for electron microscopy. *Nature* **216**, 173–174.

STERNBERGER, L. A. (1979) *Immunocytochemistry*. 2nd ed. New York: Wiley.

STERNBERGER, L. A. and PETRALI, J. P. (1976) The unlabelled antibody enzyme method: immunocytochemistry of hormone receptors at target cells. In *Immunoenzymatic Techniques*, ed. G. Feldmann, P. Druet, J. Bignon and S. Avrameas. pp. 43–58. Amsterdam: North-Holland.

STERNBERGER, L. A. and STERNBERGER, N. H. (1986) The unlabelled antibody method: comparison of peroxidase-antiperoxidase with avidin-biotin complex by a new method of quantification. *Journal of Histochemistry and Cytochemistry* **34**, 599–605.

STEWARD, O., COTMAN, C. and LYNCH, G. (1973) The nature of increased histochemical deposition of INT formazan in fields of degenerating synaptic terminals. *Brain Research* **63**, 182–193.

STODDART, R. W. (1984) *The Biosynthesis of Polysaccharides*. London: Croom Helm.

STODDART, R. W. and KIERNAN, J. A. (1973) Aprotinin, a carbohydrate-binding protein. *Histochemie* **34**, 275–280.

STOWARD, P. J. (1967) Studies in fluorescence histochemistry. III. The demonstration with salicylhydrazide of the aldehydes present in periodate oxidized mucosubstances. *Journal of the Royal Microscopical Society* **87**, 247–257.

STOWARD, P. J. and BURNS, J. (1971) Studies in fluorescence histochemistry. VII. The mechanism of the complex reactions that may take place between protein carboxyl groups and hot mixtures of acetic anhydride and pyridine in the acetic anhydride-salicylhydrazide-zinc (or fluorescent ketone) method for localizing protein C-terminal carboxyl groups. *Histochemical Journal* **3**, 127–141.

STRANGEWAYS, T. S. P. and CANTI, R. G. (1927) The living cell *in vitro* as shown by dark-ground illumination and the changes induced in cells by fixing reagents. *Quarterly Journal of Microscopical Science* **71**, 1–14 (& plates 1–5).

STRAUS, W. (1964) Factors affecting the cytochemical reaction of peroxidase with benzidine and the stability of the blue reaction product. *Journal of Histochemistry and Cytochemistry* **12**, 462–469.

STREEFKERK, J. G. and VAN DER PLOEG, M. (1974) The effect of methanol on granulocyte and horseradish peroxidase quantitatively studied in a film model system. *Histochemistry* **40**, 105–111.

STREIT, P. and REUBI, C. (1977) A new and sensitive staining method for axonally transported horseradish peroxidase (HRP) in the pigeon visual system. *Brain Research* **126**, 530–537.

STREIT, W. J., SCHULTE, B. A., BALENTINE, J. D. and SPICER, S. S. (1985) Histochemical localization of galactose-containing glycoconjugates in sensory neurons and their processes in the central and peripheral nervous system of the rat. *Journal of Histochemistry and Cytochemistry* **33**, 1042–1052.

STRETTON, A. O. W. and KRAVITZ, E. A. (1968)

Neuronal geometry: determination with a technique of intracellular dye injection. *Science* **162**, 132–134.

STRICH, S. J. (1968) Notes on the Marchi method of staining degenerating myelin in the peripheral and central nervous system. *Journal of Neurology, Neurosurgery and Psychiatry* **31**, 110–114.

SUMI, Y., MURAKI, T. and SUZUKI, T. (1983) The choice of a masking agent in the histochemical staining of metals. *Histochemical Journal* **15**, 231–238.

SUMI, Y., INOUE, T., MURAKI, T. and SUZUKI, T. (1985) A highly sensitive chelator for metal staining, bromopyridylazo-diethylaminophenol. *Stain Technology* **58**, 325–328.

SUMNER, B. E. H. (1965) A histochemical study of aldehyde-fuchsin staining. *Journal of the Royal Microscopical Society* **84**, 329–338.

SWAAB, D. F., POOL, C. W. and VAN LEEUWEN, F. W. (1977) Can specificity ever be proved in immunocytochemical staining? *Journal of Histochemistry and Cytochemistry* **25**, 388–390.

SWANK, L. and DAVENPORT, H. A. (1935) Chlorate-osmic-formalin method for degenerating myelin. *Stain Technology* **10**, 87–90.

SWASH, M. and FOX, K. P. (1972) Techniques for the demonstration of human muscle spindle innervation in neuromuscular disease. *Journal of the Neurological Sciences* **15**, 291–302.

SZERDAHELYI, P. and KASA, P. (1986a) A highly sensitive method for the histochemical demonstration of copper in normal rat tissues. *Histochemistry* **85**, 349–352.

SZERDAHELYI, P. and KASA, P. (1986b) Histochemical demonstration of copper in normal rat brain and spinal cord. *Histochemistry* **85**, 341–347.

TAATJES, D. J., ROTH, J., PEUMANS, W. and GOLDSTEIN, I. J. (1988) Elderberry bark lectin-gold techniques for the detection of Neu5Ac(α2,6)Gal/Gal-NAc sequences: applications and limitations. *Histochemical Journal* **20**, 478–490.

TAGO, H., KIMURA, H. and MAEDA, T. (1986) Visualization of detailed acetylcholinesterase fibre and neuron staining in rat brain by a sensitive histochemical procedure. *Journal of Histochemistry and Cytochemistry* **34**, 1431–1438.

TANAKA, C., ITOKAWA, Y. and TANAKA, S. (1973) The axoplasmic transport of thiamine in rat sciatic nerve. *Journal of Histochemistry and Cytochemistry* **21**, 81–86.

TANDLER, C. J. (1980) Dithiocarbamylation in histochemistry: carbon disulfide as a reagent for the visualization of primary amino groups with the light and electron microscope. *Journal of Histochemistry and Cytochemistry* **28**, 499–506.

TAS, J. (1977) The alcian blue and combined alcian blue-safranin O staining of glycosaminoglycans studied in a model system and in mast cells. *Histochemical Journal* **9**, 205–230.

TAYLOR, K. B. (1961) The influence of molecular structure of thiazine and oxazine dyes on their metachromatic properties. *Stain Technology* **36**, 73–83.

TERNER, J. Y. and HAYES, E. R. (1961) Histochemistry of plasmalogens. *Stain Technology* **36**, 265–278.

TERRACIO, L. and SCHWABE, K. G. (1981) Freezing and drying of biological tissues for electron microscopy. *Journal of Histochemistry and Cytochemistry* **29**, 1021–1028.

TEWARI, J. P., SEHGAL, S. S. and MALHOTRA, S. K. (1982) Microanalysis of the reaction product in Karnovsky and Roots histochemical localization of acetylcholinesterase. *Journal of Histochemistry and Cytochemistry* **30**, 436–440.

THORBALL, N. and TRANUM-JENSEN, J. (1983) Vascular reactions to perfusion fixation. *Journal of Microscopy* **129**, 123–139.

THORSTENSEN, T. C. (1969) *Practical Leather Technology*. New York: van Nostrand Reinhold.

TOMS, G. C. and WESTERN, A. (1971) Phytohaemagglutinins. Ch. 10 in *Chemotaxonomy of the Leguminosae*, ed. J. B. Harborne, D. Boulter & B. L. Turner. pp. 367–462. London: Academic Press.

TOURNIER, I., BERNUAU, D., POLIARD, A., SCHOEVAERT, D. and FELDMANN, G. (1987) Detection of albumin mRNAs in rat liver by *in situ* hybridization: usefulness of paraffin embedding and comparison of various fixation procedures. *Journal of Histochemistry and Cytochemistry* **35**, 453–459.

TRAMEZZANI, J. H., CHIOCCHIO, S. and WASSERMANN, G. F. (1964) A technique for light and electron microscopic identification of adrenalin- and noradrenalin-storing cells. *Journal of Histochemistry and Cytochemistry* **12**, 890–899.

TRANZER, J. P. and RICHARDS, J. G. (1976) Ultrastructural cytochemistry of biogenic amines in nervous tissue: methodologic improvements. *Journal of Histochemistry and Cytochemistry* **24**, 1178–1193.

TROJANOWSKI, J. Q. (1983) Native and derivatized lectins for *in vivo* studies of neuronal connectivity and neuronal cell biology. *Journal of Neuroscience Methods* **9**, 185–204.

TSUJI, S. (1974) On the chemical basis of the thiocholine methods for demonstration of acetylcholinesterase activities. *Histochemistry* **42**, 99–110.

TSUJI, S. and ALAMEDDINE, H. S. (1981) Silicotungstic acid for cytochemical localization of water soluble substance(s) of cholinergic motor nerve terminal. *Histochemistry* **73**, 33–37.

TSUJI, S. and LARABI, Y. (1983). A modification of the thiocholine-ferricyanide method of Karnovsky and Roots for localization of acetylcholinesterase activity without interference by Koelle's copper thiocholine iodide precipitate. *Histochemistry* **78**, 317–323.

TSUJI, S., ALAMEDDINE, H. S., NAKANISHI, S. and OHOKA, T. (1983) Molybdic and tungstic heteropolyanions for "ionic fixation" of acetylcholine in cholinergic motor nerve terminals. *Histochemistry* **77**, 51–66.

TULL, A. G. (1972) Hardening of gelatin by direct oxidation. In *Photographic Gelatin*, ed. R. J. Cox. pp. 127–134. London: Academic Press.

UHL, G. R. (ed.) (1986) *In Situ Hybridization in Brain*. New York: Plenum Press.

VALNES, K. and BRANDTZAEG, P. (1985) Retardation of immunofluorescence fading during microscopy. *Journal of Histochemistry and Cytochemistry* **33**, 755–761.

VAN DEN POL, A. N. (1984) Colloidal gold and biotin-avidin conjugates as ultrastructural markers for neural antigens. *Quarterly Journal of Experimental Physiology* **69**, 1–33.

VAN DER LOOS, C., DAS, P. K. and HOUTHOFF, H-J. (1987) An immunoenzyme triple staining method using both polyclonal and monoclonal antibodies from the same species. Application of combined direct, indirect, and avidin-biotin complex (ABC) technique. *Journal of Histochemistry and Cytochemistry* **35**, 1199–1204.

VAN DUIJN, P. (1956). A histochemical specific thionine-SO$_2$ reagent and its use in a bi-color method for deoxyribonucleic acid and periodic acid-Schiff positive substances. *Journal of Histochemistry and Cytochemistry* **4**, 55–63.

VAN NOORDEN, C. J. F. and BUTCHER, R. W. (1984) Histochemical localization of NADP-dependent dehydrogenase activity with four different tetrazolium salts. *Journal of Histochemistry and Cytochemistry* **32**, 998–1004.

VAN NOORDEN, C. J. F. and TAS, J. (1982) The role of exogenous electron carriers in NAD(P)-dependent dehydrogenase cytochemistry studied *in vitro* and with a model system of polyacrylamide films. *Journal of Histochemistry and Cytochemistry* **30**, 12–20.

VANDESANDE, F. (1979) A critical review of immuno-cytochemical methods for light microscopy. *Journal of Neuroscience Methods* **1**, 3–23.

VAUGHN, K. C. (edit.) (1987) *CRC Handbook of Plant Cytochemistry*. Boca Raton, Florida: CRC Press.

VELICAN, C. and VELICAN, D. (1970) Structural heterogeneity of basement membranes and reticular fibres. *Acta anatomica* **77**, 540–559.

VELICAN, C. and VELICAN, D. (1972) Silver impregnation techniques for the histochemical analysis of basement membranes and reticular fibre networks. In *Techniques of Biochemical and Biophysical Morphology*, Vol. **1**, ed. D. Glick and R. M. Rosenblaum. New York: Wiley. pp. 143–190.

VENKATARAMAN, K. (1952–1978) *The Chemistry of Synthetic Dyes*. Vols **I-VIII**. New York: Academic Press.

VERMEER, B. J., VAN GENT, C. M., DE BRUIJN, W. C. and BOONDERS, T. (1978) The effect of digitonin-containing fixatives on the retention of free cholesterol and cholesterol esters. *Histochemical Journal* **10**, 287–298.

VIAL, J. and PORTER, K. R. (1975) Scanning microscopy of dissociated cells. *Journal of Cell Biology* **67**, 345–360.

VIDAL, B. de C. (1978) The use of the fluorescent probe 8-anilinonaphthalene sulfate (ANS) for collagen and elastin histochemistry. *Journal of Histochemistry and Cytochemistry* **26**, 196–201.

VOLLMANN, H. (1971) Phthalogen dyestuffs. Ch. 5 in *The Chemistry of Synthetic Dyes*, ed. K. Venkataraman. Vol. **V**, pp. 283–311. New York: Academic Press.

VOLZ, D., REID, P. E., PARK, C. M., OWEN, D. A. and DUNN, W. L. (1987) A new method for the selective periodate oxidation of total tissue sialic acids. *Histochemical Journal* **19**, 311–318.

VOLZ, D., REID, P. E., PARK, C. M., OWEN, D. A.,

DUNN, W. L. and RAMEY, C. W. (1986) Can "mild" periodate oxidation be used for the specific histochemical identification of sialic acid residues? *Histochemical Journal* **18**, 579–582.

WALKER, J. F. (1964) *Formaldehyde*. 3rd ed. New York: Reinhold.

WATERS, S. E. and BUTCHER, R. G. (1980) Studies on the Gomori acid phosphatase reaction: the preparation of the incubation medium. *Histochemical Journal* **12**, 191–200.

WATERS, W. A. (1958) Mechanisms of oxidation by compounds of chromium and manganese. *Quarterly Reviews* **12**, 277–300.

WATSON, J. D. and CRICK, F. H. C. (1953) Molecular structure of nucleic acids. A structure for deoxyribose nucleic acid. *Nature* **171**, 737–738.

WATSON, S. J. and BARCHAS, J. D. (1977) Catecholamine histofluorescence using cryostat sectioning and glyoxylic acid in unperfused frozen brain: a detailed description of the technique. *Histochemical Journal* **9**, 183–195.

WATSON, S. J. and ELLISON, J. P. (1976) Cryostat technique for central nervous system histofluorescence. *Histochemistry* **50**, 119–127.

WATTENBERG, L. W. and LEONG, J. L. (1960) Effects of coenzyme Q10 and menadione on succinic dehydrogenase activity as measured by tetrazolium salt reduction. *Journal of Histochemistry and Cytochemistry* **8**, 296–303.

WEBER, P., HARRISON, F.W. and HOF, L. (1975) The histochemical application of dansylhydrazine as a fluorescent labeling reagent for sialic acids in glycoconjugates. *Histochemistry* **45**, 271–277.

WEIR, E. E., PRETLOW, T. G., PITTS, A. and WILLIAMS, E. E. (1974) Destruction of endogenous peroxidase activity in order to locate antigens by peroxidase-labelled antibodies. *Journal of Histochemistry and Cytochemistry* **22**, 51–54.

WEISBLUM, B. and de HASETH, P. L. (1972) Quinacrine: a chromosome stain specific for deoxyadenylate-deoxythymidylate-rich regions in DNA. *Proceedings of the National Academy of Sciences of the United States of America* **69**, 629–632.

WEISS, J. B. and JAYSON, M. I. V. (ed.) (1982) *Collagen in Health and Disease*. Edinburgh: Churchill-Livingstone.

WEISSBERGER, A. (1966) Principles and chemistry of colour photography. Ch. 17 in *The Theory of the Photographic Process*, ed. T. H. James. New York: Macmillan. pp. 382–396.

WELLER, R. O. (1984) *Colour Atlas of Neuropathology*. London: Harvey Miller.

WHITE, D. L., MAZURKIEWICZ, J. E. and BARNETT, R. J. (1979) A chemical mechanism for staining by osmium tetroxide-ferrocyanide mixtures. *Journal of Histochemistry and Cytochemistry* **27**, 1084–1091.

WHITE, E. H. and WOODCOCK, D. J. (1968) Cleavage of the carbon nitrogen bond. Ch. 8 in *The Chemistry of the Amino Group*, ed. S. Patai, pp. 407–497. London: Wiley-Interscience.

WHITE, R., HU, F. and ROMAN, N. A. (1983) False dopa reaction in studies of mammalian tyrosinase:

some characteristics and precautions. *Stain Technology* **58**, 13–19.

WHITMORE, F. C. (1921) *Organic Compounds of Mercury*. New York: Chemical Catalog Company.

WHYTE, A., LOKE, Y. W. and STODDART, R. W. (1978) Saccharide distribution in human trophoblast demonstrated using fluorescein-labelled lectins. *Histochemical Journal* **10**, 417–423.

WIGGLESWORTH, V. B. (1957) The use of osmium tetroxide in the fixation and staining of tissue. *Proceedings of the Royal Society B* **147**, 185–199.

WIGGLESWORTH, V. B. (1988) Histological staining of lipids for the light and electron microscope. *Biological Reviews* **63**, 417–431.

WILLIAMS, G. and JACKSON, D. S. (1956) Two organic fixatives for acid mucopolysaccharides. *Stain Technology* **31**, 189–191.

WILLIAMS, R. M. and ATALLA, R. H. (1981) Interactions of Group II cations and borate anions with nonionic saccharides. Studies on model glycols. Ch. 22 in *Solution Properties of Polysaccharides*, ed. D. A. Brant (A.C.S. Symposium Series, No. 150). Vol. **2**, pp. 317–330. Washington: American Chemical Society.

WILLINGER, M. and SCHACHNER, M. (1980) GM1 ganglioside as a marker for neuronal differentiation in mouse cerebellum. *Developmental Biology* **74**, 101–117.

WINKELMANN, R. K. (1960) *Nerve Endings in Normal and Pathologic Skin*. Springfield, Ill.: Thomas.

WINKELMANN, R. K. and SCHMIT, R. W. (1957) A simple silver method for nerve axoplasm. *Proceedings of Staff Meetings of the Mayo Clinic* **32**, 217–222.

WINKLER, H. and WESTHEAD, E. (1980) The molecular organization of adrenal chromaffin granules. *Neuroscience* **5**, 1803–1823.

WITTEKIND, D. H. (1983) On the nature of Romanowsky-Giemsa staining and its significance for cytochemistry and histochemistry: an overall review. *Histochemical Journal* **15**, 1029–1047.

WITTEKIND, D. H. and KRETSCHMER, V. (1987) On the nature of Romanowsky-Giemsa staining and the Romanowsky-Giemsa effect. II. A revised Romanowsky-Giemsa staining procedure. *Histochemical Journal* **19**, 399–401.

WOHLRAB, F., SEIDLER, E. und KUNZE, D. K. (1979) *Histo- und Zytochemie dehydrierender Enzyme. Grundlagen und Problematik*. Leipzig: Barth.

WOLLIN, A. and JAQUES, L. B. (1973) Metachromasia: an explanation of the colour change produced in dyes by heparin and other substances. *Thrombosis Research* **2**, 377–382.

WOLMAN, M. (1971) A fluorescent histochemical technique for gamma-aminobutyric acid. *Histochemie* **28**, 118–130.

WOLMAN, M. and BUBIS, J. J. (1965) The cause of the green polarization colour of amyloid stained with Congo red. *Histochemie* **4**, 351–356.

WOLTERS, G. H. J., PASMA, A., KONIJNENDIJK, W. and BOUMAN, P. R. (1979) Evaluation of the glyoxal-bis-(2-hydroxyanil)-method for staining of calcium in model gelatin films and pancreatic islets. *Histochemistry* **62**, 137–151.

WOUTERLOOD, F. G., NEDERLOF, J. and PANIRY, S. (1983) Chemical reduction of silver chromate: a procedure for electron microscopical analysis of Golgi-impregnated neurons. *Journal of Neuroscience Methods* **7**, 235–308.

WREFORD, N. G. M., SINGHANIYAM, W. and SMITH, G. C. (1982) Microspectrofluorometric characterization of the fluorescent derivatives of biogenic amines produced by aqueous aldehyde (Faglu) fixation. *Histochemical Journal* **14**, 491–505.

YAMABAYASHI, S. (1987) Periodic acid-Schiff-alcian blue: a method for the differential staining of glycoproteins. *Histochemical Journal* **19**, 565–571.

YAMAMOTO, T., IWASAKI, Y. and KONNO, H. (1983) Retrograde axoplasmic transport of toxic lectins is useful for transganglionic tracings of the peripheral nerve. *Brain Research* **274**, 325–328.

YAMAMOTO, T., IWASAKI, Y. and KONNO, H. (1984) Experimental sensory ganglionectomy by way of suicide axoplasmic transport. *Journal of Neurosurgery* **60**, 108–114.

YOSHIKAMI, D. and OKUN, L. M. (1984) Staining of living presynaptic nerve terminals with selective fluorescent dyes. *Nature* **310**, 53–56.

ZACKS, S. I. (1973) *The Motor Endplate*. Huntington, N.Y.: Kreiger.

ZAGON, I. S., VAVRA, J. and STEELE, I. (1970) Microprobe analysis of protargol stain deposition in two protozoa. *Journal of Histochemistry and Cytochemistry* **18**, 559–564.

ZIRKLE, C. (1928) The effect of hydrogen-ion concentration upon the fixation image of various salts of chromium. *Protoplasma* **4**, 201–227.

ZIRKLE, C. (1933) Cytological fixation with the lower fatty acids their compounds and derivatives. *Protoplasma* **18**, 90–111.

ZOLLINGER, H. (1961) *Azo and Diazo Chemistry. Aliphatic and Aromatic Compounds*. (Transl. H. E. Nursten). London: Interscience Publishers.

Answers to Exercises

Chapter 1

1. (a) A whole mount, with the mesentery stretched and spread over a slide. (b) Wax- or nitrocellulose-embedded sections of a fixed, decalcified specimen, or sections of plastic-embedded undecalcified material.

2. Before and after staining, avoid solvents that are more hydrophobic than that in which the dye is dissolved (typically 70% alcohol). An aqueous mounting medium (see Chapter 4) must be used.

3. The used alcohol contains fatty material dissolved from the specimen. The fatty substances become insoluble when the concentration of water in the solvent is increased.

4. The outside of the block is likely to be overstained and the middle understained. The effect can be reduced by using small pieces of tissue. If possible, the staining solution should be one that deposits colour very slowly, so that the penetration of the reagents occurs more rapidly than the coloration of the tissue.

5. Immerse the slide in xylene or toluene until the mounting medium is dissolved. Remove the coverslip. Treat with the solvent until all mounting medium is removed (no turbidity when slide is immersed in alcohol). Pass through graded alcohols to water. Removal of the coverslip and resin can take several days.

8. The volume of macerating fluid is small so that the concentration of released cells will be as high as possible.

Chapter 2

1. If fixation is brief (less than a few hours), the freshly diluted unbuffered formalin will not depolymerize, so penetration and chemical fixation will be inadequate. In the centre of the block there will be ruptured cells, osmotically damaged by the most rapidly penetrating ingredient (water). Shrinkage spaces will be present around cells throughout the specimen. These develop because the proteins have not been fixed (toughened) enough to resist deformation in the dehydrating alcohols. With prolonged fixation (12h or more), there is much less physical damage. Buffered formalin is fully active as soon as it is made. Its isotonicity prevents osmotic damage in the centre of the specimen, and the duration of fixation can be brief, though some days are needed for full structural stabilization of the tissue.

2. See Table 2.1. Picric acid alone has several undesirable effects on the tissue. Formaldehyde is needed to cross-link the precipitated proteins, and acetic acid to prevent shrinkage and to enhance preservation of chromatin.

3. Slow penetration, blackening, and adverse effect on cutting properties make OsO_4 useful only for tiny specimens that are to be cut into very thin sections. In electron microscopy, this is what is done. In peripheral nerves, OsO_4 blackens myelin sheaths, which are composed entirely of membranes. Central nervous tissue contains high concentrations of membrane everywhere, so it is uniformly blackened by OsO_4, and structural details are not easily seen by light microscopy. Fixation in osmium tetroxide also precludes the use of many stains.

4. The amount of cross-linking is so great that the specimens become brittle. Unless the specimen is tiny, fractures develop during dehydration and paraffin embedding. Frozen sections can be cut without difficulty, however, and cellular structure is well preserved. The free aldehyde groups remaining in the tissue have to be taken into account when staining (see Section 2.4.8 and Chapter 10, Section 10.10.6).

5. (a) Modified Carnoy and AFA probably best; Neutral buffered formaldehyde equally good for RNA though structure is somewhat damaged in paraffin sections. Strongly acid fixatives should be avoided or used only for a few hours.

(b) Aqueous formaldehyde or formal-calcium.

(c) Aqueous formaldehyde, formal-calcium, Helly and Zenker are all satisfactory. Non-aqueous fixatives cause structural distortion and extraction of much lipid. Osmium tetroxide is useful for small peripheral nerves. Much of the lipid of myelin is protein-bound, so it is not extracted by organic solvents (See Chapters 12 and 18).

6. Large anions, which form insoluble ion-pairs with the CPC cation. Glycosaminoglycans (Chapter 11) are the most important substances in this class. As a surfactant, CPC damages membranes and other lipid-containing structures, and it competes with cationic dyes used to stain the tissue.

Chapter 3

1. Na_2SO_4 would not neutralize acid. It would restrain swelling of inadequately fixed material immersed in water, but a poorly fixed specimen will already have been damaged by the decalcifying solution.

2. If very little bone is present, Bouin alone may decalcify the specimen. Wash thoroughly in 70% alcohol to remove picric acid, and proceed with dehydration, clearing and embedding. If more decalcification is needed, transfer the washed specimen to one of the decalcifying agents decribed in this chapter, and follow the instructions. For a Zenker-fixed specimen it is necessary to wash out all free dichromate with water (not alcohol) before immersion in formic acid, which would reduce free $Cr(VI)$ to $Cr(OH)_3$ (see Chapter 2, Section 2.4.5). If the decalcifier is nitric acid or EDTA, the washing could be done later, because these substances are not reducing agents. Mercury deposits must also be removed from Zenker-fixed material (see Chapter 4) before staining.

Chapter 4

1. Embedding in wax, nitrocellulose or plastic would extract the lipid. Frozen or cryostat sections must be cut. Gelatin embedding (Section 4.1.1) may be needed to provide support for the specimen.

2. (a) The dioxane replaces the water in the specimen. After fixation in Helly, wash in running water until all excess dichromate is removed. The specimen may be brought gradually (e.g. through 30%, 70%, 90%) to 100% dioxane, or may be transferred from water to the first of 4 changes of 100% dioxane. Dioxane mixes with molten wax, so the specimens are then put in the first of 3 changes of wax and are infiltrated and blocked out in the usual way.

(b) DMP reacts with water. The products of reaction are methanol and acetone.

3. Coating with nitrocellulose would be advisable (Section 4.2.2). The slides have to be manipulated carefully, one at a time.

4. Euparal is unsuitable for (a) and (d) because the solvent will extract some of all of the dye. It is suitable for (b) and (c) because the low refractive index makes unstained structures visible.

5. Mercury deposits look like sand spilt upon the section. The distribution of the dirt is not related to structural features, but it is clearly in the tissue, not on the surface of the slide.

6. Water or alcohol in a resinous medium appears as colourless droplets. Xylene in an aqueous medium looks much the same. Loss of dye into an inappropriate mounting medium may take a few days. The colour spreads outwards from the specimen, and the specific staining is lost.

Chapter 5

1. (a), (c) and (e) are coloured. (a) and (c) are cationic dyes.

2. The 3,4-benzpyrene molecule is hydrophobic (no atoms capable of hydrogen bonding), so it behaves like a solvent dye. It has been used as a fluorochrome for cytoplasmic lipid droplets, but is unpopular on account of its carcinogenicity. Nile red (Chapter 12) is a safe alternative.

3. Weigert's stain contains a large excess of Fe^{3+} ions even after all the haematoxylin has been oxidized to haematein and formed soluble [Fe^{3+}-haematein] complex ions or molecules. This excess of $FeCl_3$ also makes the solution acid. Free Fe^{3+} competes with the tissue for binding of the dye-iron complex. The high [H^+] breaks iron-tissue, iron-haematein and [Fe^{3+}-haematein]-tissue bonds, except at the sites of strongest binding, which are in nuclear chromatin.

With Heidenhain's method, all the iron-binding sites in the tissue are saturated and the bound iron is then complexed with haematein. Subsequent treatment with a Fe^{3+} salt removes haematein as a soluble [Fe^{3+}-haematein] complex. The acidity of the solution breaks the [complex]-tissue and Fe(III)-tissue bonds. The weakest bonds to components of the tissue (those to extracellular and cytoplasmic proteins) are the first to be broken. Bonds to membrane-rich structures (mitochondria,

myelin) follow, and the strong bonds to nuclear chromatin are the last to be broken. See also Chapter 12, Section 12.11.1.

4. (a), (b) Staining is due primarily to ionic bonding, so urea in the solution is only a weak competitor. High concentrations of urea will inhibit staining.

(c) Binding of this large molecule requires non-ionic, non-covalent bonding. Urea competes effectively with these weak forces, and prevents staining of the tissue.

(d) Urea is very hydrophilic, so it does not interfere with the attachment of a hydrophobic dye to a hydrophobic substrate.

5. (a) and (d) have coloured cations, and binding to anionic substrates is competitively inhibited by inorganic cations in the solvent. See also Chapter 11, Section 11.3.1. (b) is a reactive dye, and a salt in the solution does not prevent its combination with $-OH$, $-COOH$ and $-NH_2$. The salt may, however, promote the movement of unreacted dye into the interstices of the tissue (see Sections 5.5.1 and 5.9.3.8) (c) is anionic but is a direct cotton dye and as such it is bound mainly by non-ionic, non-covalent forces. An inorganic salt does not inhibit staining, and may even enhance the uptake of dye (Section 5.9.3.8).

6. Dye-metal binding is explained in Sections 5.5.8 and 5.9.3.4.

(a):

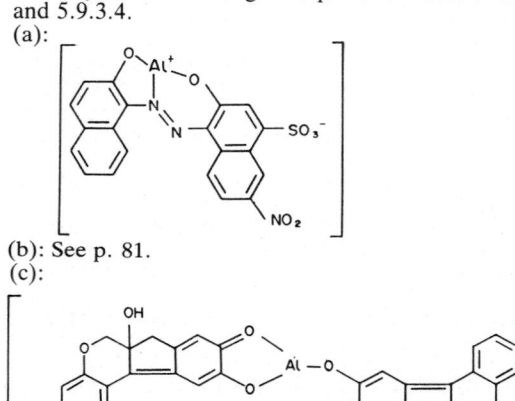

(b): See p. 81.

(c):

7. An excess of OH^- ions decolorizes the dye by changing all the coloured cations into molecules of the leucobase (see p. 75).

8. (a) Colour changes to violet.

(b) Colour changes to pale yellow.

(c) Solution becomes purple. Darkens with addition of alkali; changes to red with addition of a mineral acid.

(d) Solution becomes grey-black. Changes to deep purple with addition of alkali; changes to red with addition of a mineral acid.

The colour changes are due to the acid-base indicator properties of free haematein and of its metal complexes.

9. (a) A dark precipitate is formed. It can be collected and dissolved in alcohol. This is a salt in which the anion and cation are both dyes. When the oppositely charged parts of the ions are close together, the exposed parts of the dye molecules are not hydrophilic enough to allow solution in water.

(b) There is no precipitation, because no reaction occurs between the two cationic dyes.

(c) As with (a) an alcohol-soluble precipitate forms. Eosinates are discussed in Chapter 7.

(d) The two anionic dyes mix without precipitation.

10.

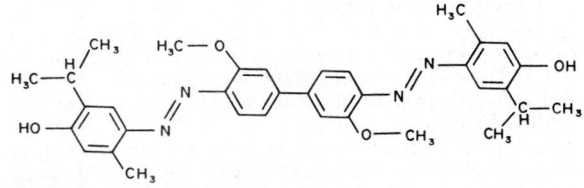

The compound has structural similarities to the Sudan dyes (few atoms able to form hydrogen bonds; not ionized) and can be used as a stain for fat.

Chapter 6

1. The Nissl substance of the neurons (cytoplasmic RNA) is strongly stained (blue) by azure A but is usually pink with H. & E. Some blue colour can be seen in Nissl substance if the alum-haematein has not been adequately differentiated.

2. Weigert's solution contains a large excess of Fe(III), which eventually oxidizes all the haematein to useless substances. Excess oxidation of the haematein in Mayer's haemalum can be brought about only by atmospheric oxygen, which is not available in large amounts in a closed bottle.

3. (a), (c) and (f) are unsuitable because they are cationic dyes and they too would stain nuclei.

(b) and (e) are suitable because their colours contrast with the blue of iron-chromoxane cyanine R and, being anionic dyes, they stain cytoplasmic and extracellular proteins.

(d) is unsuitable because although it is anionic its colour does not differ appreciably from that of the stained nuclei.

4. The formaldehyde has reacted with all the amino groups of the tissue, so there are no cationic sites left to attract eosin anions.

5. Everything is stained, including structures that are not usually basophilic, because oxidative deamination of proteins has left an abundance of carboxylic acid groups in the tissue.

6. Alum-brazilin, adequately differentiated by acid-alcohol, will give selective nuclear staining similar to that obtained with haemalum. Fast green FCF is a suitable counterstain.

7. For (a), (b) and (c), see Sections 6.1.3, 6.1.1 and 6.1.4 respectively.

Chapter 7

1. The shapes of the nuclei are clearly shown, as are the sizes and shapes of the cells. The different types of granular leukocyte cannot be identified because their cytoplasms are all stained pink.

2. The non-aqueous solvent is needed to make a concentrated solution. Water is needed to solvate the dye anions and cations, which must be present in the staining solution.

3. Formaldehyde, by reacting with amino groups of protein, reduces the number of cationic sites in the tissue. A more acid staining solution opposes this effect by promoting ionization of the remaining amino groups, thereby increasing the overall affinity of the tissue for eosin. Blood films are fixed in alcohol, which does not react with amino groups.

4. Unfixed cells flatten and spread as they settle. If the cells are fixed in suspension, they retain their spherical shapes and become physically stronger. When the fixed cells settle, they do not flatten very much. A smaller area is in contact with the glass slide, so fixed cells are more likely to be dislodged during staining than unfixed cells. Cells in smears and films are often distorted as they are dragged across the surface of the slide. When cells are centrifuged on to a slide, they flatten and spread more than cells that have settled slowly. Consequently, the apparent sizes of cells vary with the methods used to deposit them on the slide.

5. The metachromasia (see Chapter 11) of the basophil granules makes these cells (normally less than 1% of the leukocyte population) conspicuous. (Eosin alone, in alkaline solution, can be used for selective staining of eosinophils, whose granules contain much arginine. See Chapter 10 for rationale.)

Chapter 8

1. The chemistry of collagen and elastin is reviewed in Section 8.1. The staining properties of elastin are largely attributable to the abundance of hydrophobic amino acid, whereas collagen is stained by more hydrophilic dye anions. Silver staining of reticulin, which is due to the presence of oxidizable carbohydrate, is the only method in this chapter that involves chemical reactions between reagents and tissue.

2. Reticulin is a name applied to thin collagen fibres (type III) and basement membranes (type IV). A high carbohydrate content probably accounts for its staining by the classical silver methods.

3. The actions of the heteropolyacids are discussed in Section 8.2.4.

4. One such method is: Iron-haematoxylin for nuclei (black), followed by aldehyde-fuchsine for collagen (red-purple) and then picro-aniline blue for muscle and other cytoplasm (yellow) and collagen (blue). It is also possible to combine orcein (for brown elastin) with iron-chromoxane cyanine R (blue nuclei) and Gabe's trichrome (red cytoplasm, greenish collagen).

5. Aniline blue entered and stained nuclei that had been cut by the microtome knife, but orange G stained only those nuclei that were entirely contained within the section.

6. The blue colour is strongest in collagen. Stannous ions reduce the bound phosphomolybdate to molybdenum blue (see Section 8.2.4.1).

Chapter 9

1. The sections must be treated with sodium borohydride (see Chapter 10) to remove aldehyde groups derived from the fixative. If this is not done, everything in the tissue will stain with Schiff's reagent even without prior hydrolysis of nucleic acids.

2. Treat sections with the enzymes. Stain with a dilute solution of a cationic dye. After pure RNase, only the nuclei (excluding nucleoli) are stained. After pure DNase, staining is confined to nucleoli and Nissl substance (RNA in cytoplasm of neurons). A mixture of the two enzymes abolishes all basophilia in central nervous tissue. (The nucleases have no effect on the basophilia due to substances other than nucleic acids, such as that of cartilage matrix and mucus.)

3. Nucleoproteins are quite strongly basic, so they can be stained with neutral or slightly alkaline solutions of anionic dyes (see Chapter 10, Sections 10.4.1 and 10.5). More cationic sites are made available if the DNA is first extracted, chemically or enzymatically. Some of the mordant dye methods for nuclei probably demonstrate nucleoprotein (see Chapter 6).

4. Cross-linking of proteins by aldehyde fixatives makes the tissue less permeable to the large enzyme molecules.

6. In a large nucleus, the DNA is diluted, so the intensity of staining is less. A small nucleus contains a higher concentration of DNA, and is also more likely to be contained entirely in a section than a large nucleus.

Chapter 10

1. Acidophilia is due to amino groups and/or guanidino (arginine) groups. The former can be removed by deamination or blocked by acetylation or benzoylation. The latter can be blocked by reaction with benzil. If staining can be obtained at high pH, it must be due to the strongly basic guanidino group. Paneth cell granules are rich in arginine.

2. Unless the incubation is brief, the section is destroyed. Carefully controlled incubation in trypsin can remove some proteins and make others accessible to histochemical reagents such as antibodies (see Chapter 19). With animal tissues, non-protein structures are lost when the proteins around them are digested, so trypsin has little or no value as a test for protein. In plant tissues, digestion of cytoplasmic protein is likely to cause loss of such objects as lipid droplets and starch granules, even though most of the structural matrix (cellulose) is not attacked by proteolytic enzymes.

3. Methylation blocks the basophilia of the cartilage matrix by removing sulphate (leaving −OH in its place) and esterifying carboxyl groups. Ether-sulphuric acid introduces sulphate ester groups at sites of hydroxyl groups, which are widespread. Thus, all components of the tissue, including methylated cartilage matrix, will be stainable with a basic dye at pH 1.0. Hydroxyl groups of proteins cannot be demonstrated in isolation, because it is not possible to eliminate all carboxyl hydroxyls. Protein −OH is assumed to be present at sites stained by the sulphation-basic dye sequence but not stainable by any methods for carbohydrates (Chapter 11).

4. (a): With Millon's reagent or by coupling with a suitable diazonium salt (such as fast black K salt).
(b): In the hydroxynaphthaldehyde method.
(c): By reaction with hydroxynaphthoic acid hydrazide followed by coupling with a diazonium salt.

Chapter 11

1. (a) Glycoproteins; neutral polysaccharides (glycogen, starch, cellulose).

(b) Glycoproteins that contain D-glucose (i.e. collagen) or D-mannose (most secretory glycoproteins); polysaccharides that contain these sugars in α- linkage (starch, glycogen; not cellulose). Most PAS-positive substances stain with Con A; cell surfaces and basement membranes are more conspicuous with the latter. All Con A-positive substances are PAS-positive.

2. Nearly all components of the tissue become basophilic, and can be stained with strongly acidic (pH 1) solutions of cationic dyes.

3 (a): None.
(b) and (d) All that were initially PAS-positive, and also sialic acid groups that initially were acylated at C8. But see (c) below.
(c) No effect on PAS reactivity, but the reagents can cause loss of some sialic acid residues, causing diminished staining of some types of mucus.

4. (a) and (b) No effect of acetylation on alcian blue staining.
(c) Nothing stainable with alcian blue at pH 1.0 or 2.5.
(d) Restoration of stainability at pH 2.5. Nothing stainable at pH 1.0.
(e) Loss of some of the stainability at pH 2.5 (that due to sialic acids).

5. Hyaluronic acid, chondroitin −4− and −6−sulphates, dermatan sulphate, heparin, heparitin sulphate and glycoproteins containing sialic acids. (*Not* the keratan sulphates or glycoproteins that lack sialic acids.)
Amylase has no effect on binding of fluorescent aprotinin. Hyaluronidase should suppress staining in extracellular materials (cartilage matrix etc.). Neuraminidase suppresses staining of mucus and basement membranes.

6. Glycogen can be stained with PAS, Best's carmine, and the lectins of Group 1 in Table 11.1. Instructions are given in this chapter. It is important to remember that most fixatives do not react chemically with glycogen and that the polysaccharide is soluble in water.

7. Chemical blocking procedures are of low specificity. Periodate oxidation blocks lectin-binding by (a) but not by (b) (see Section 11.4.1). Before staining with the lectin it is necessary to block periodate-engendered aldehyde groups (see Chapter 10, Section 10.11.10) which would combine non-specifically with lectins or any other proteinaceous reagents. Specific binding can be inhibited by adding the appropriate sugar to the solution containing the lectin.

8. Chitin accounts for only half the material in the exoskeleton. The other half includes PAS-positive substances.

9. NANA is identifiable with specialized PAS methods (Section 11.4.2), alcian blue at pH 2.5, or some of the lectins of Group 5 in Table 11.1, and it can be extracted enzymatically or chemically. After removal of NANA, the glycoprotein would be PAS-positive only with the standard technique and it should show increased affinity for lectins in Group 2 of Table 11.1 (PNA, SBA, PHA, RCA_1). The Group 2 lectins with preference for α-linked galactosides (DBA, NPA) would not stain this substance. Prior incubation of a section in β-galactosidase would prevent all staining of the glycoprotein, and no other glycosidase would have this effect.

10. First (full-strength) HIO_4 treatment oxidizes glycols of unsubstituted hexoses and of sialic acids not acylated at C8. The aldehydes from the unsubstituted hexoses and sialic acids are destroyed by the first borohydride reduction.

Second (mild) HIO_4 treatment oxidizes glycols of sialic acids that originally were acylated at C7 and C8.

Third (full-strength) HIO_4 treatment oxidizes glycols formed by saponification of the acylated hexoses. All other glycols formed (from sialyl residues) in the saponification step are destroyed by the second oxidation and subsequent borohydride reduction. Thus, the final treatment with Schiff's reagent shows only the sites of aldehydes generated from acylated hexose sugars.

11. A solution of the neoglycoprotein is applied to sections of unfixed plant tissue. After washing off excess reagent, the section is examined by fluorescence microscopy. Control sections are stained with a mixture of the neoglycoprotein with an excess of the appropriate sugar.

Chapter 12

1. Triglycerides can be extracted with cold acetone before staining. Phosphoglycerides resist extraction by cold acetone.

2. (a) Not stained, because melting point of saturated fatty acid (stearic) is too high.

(b) Stained at higher temperature, which melts tristearin.

(c) Not stained, because no olefinic linkages.

(d) Not stained, because not a plasmalogen.

3. (a) and (b) Stained, because cis-unsaturated oleic acid is liquid at room temperature.

(c) Stained, because of the unsaturated linkage in each oleyl group.

(d) Not stained, because not a plasmalogen. Oxidation of –C=C– by air could cause a pseudoplasmal reaction.

4. Positive Schiff reaction indicates that some of the unsaturated linkages in the fatty acid chains of the myelin lipids have been oxidized by air to −CHO.

(a) Reduce pre-existing aldehydes to hydroxyls with borohydride before carrying out the plasmal reaction.

(b) Alkaline hydrolysis prior to staining lipids by any method (Sudan black B is suitable) does not extract fatty acid amides.

5. In order to take up Nile red from the oxidized Nile blue solution, the lipid droplets must be in liquid state. Requirement of a higher temperature for staining indicates a higher proportion of saturated fatty acid residues, which impart higher melting temperatures. (These considerations do not apply when Nile red is detected with much greater sensitivity by fluorescence microscopy.

Smaller amounts of the dye may then be seen in hydrophobic domains other than those of unsaturated lipid droplets).

6. The adrenal droplets can be stained by any method for hydrophobic lipids. Extraction by cold acetone excludes phospholipids. Alkaline hydrolysis removes esters (fats, cholesterol esters) but not steroid hormones (which are, however, extracted by acetone).

7. Sulphatide can be stained with cationic dyes at pH < 2, because of the strongly acid sulphate-ester group.

(a), (b) Any proteoglycans and glycoproteins that stain similarly will not be extracted by prior treatment of the sections with hot methanol-chloroform.

(c) Cerebrosides are not basophilic (no ionizable acid group).

(d) Gangliosides contain sialic acid acid residues, which are basophilic, but not stained at low pH.

(e) Neutral fats are not basophilic (no ionizable acid group).

(f) Although the acyl groups of sulphatide are largely saturated, tests of unsaturation (OsO_4, $PdCl_2$) are likely to be positive because of other lipids present in or near the deposits.

8. The paraffin inclusions are sudanophilic at temperatures approaching the melting point of wax (50–60°C). They are not extracted by alcohol or acetone (excluding fats, free fatty acids and free cholesterol), but are soluble in chloroform-methanol (excluding lipoprotein and other insoluble hydrophobic materials). They do not reduce osmium tetroxide (excluding almost all lipids except cholesterol esters), are unaffected by alkaline hydrolysis (excluding cholesterol esters), and cannot be stained with basic dyes or by the PAS method (excluding glycolipids).

Chapter 13

1. Na^+ occurs as soluble salts, mainly in extracellular fluids. The few reagents that precipitate sodium (e.g. potassium pyroantimonate, zinc uranyl acetate) penetrate and react too slowly to ensure accurate localization.

2. Immerse tiny fragments of tissue or sections of unfixed freeze-dried tissue in $AgNO_3$ solution. Wash in water. Expose to light. Accuracy of localization is poor for the same reasons given for sodium above. Specificity is also poor, because of binding of Ag^+ to proteins (see Chapter 18, Section 18.4.1.1).

3. See Section 13.4. Test for masked iron if Perls' reaction is negative.

4. Precipitation of insoluble, coloured salts; acid hydrolysis to release metal ions ("unmasking") from insoluble compounds in tissue; formation of insoluble, coloured chelates; use of chelating agents to "mask" (extract) coloured chelates of metals one does not want to stain. When an ion is released for reaction with a precipitant, it diffuses away from its site of origin and may bind to other components of tissue.

5. It is the free base of a cationic azo dye. Metal binding involves the phenolic hydroxyl and the azo nitrogen attached to the pyridine ring, to form a six-membered chelate ring.

6. Calcified deposits are in the basement membranes of the renal tubules (also PAS-positive). GBHA also stains nuclei, which bind traces of Ca^{2+} derived from extracellular fluids of the tissue.

Chapter 14

1. Diffusion of the enzyme during fixation or staining; diffusion of intermediate or final reaction products; binding of intermediate or (more commonly) final reaction product to structures other than those containing the enzyme. Hydrophobic final products are particularly prone to artifactual false localization.

2. See answer to Question 1 above for artifactual binding of final reaction product to ribosomes, secretory granules etc. An enzyme uniformly distributed in the cytosol is likely to form a final product that consists of solid particles, and these may be confused with mitochondria in light microscopy.

3. The substrate is the physiological one. The products of enzymatic reaction are the oxidized substrate, electrons and hydrogen ions. The trapping agent is the tetrazolium salt, which is reduced by the released electrons to a coloured formazan. This is an over-simplification. See Chapter 16 for more.

4. (a) 85 ions.
(b) 22,780 ions.

Chapter 15

1. General principles are explained in Chapter 14.
(a) Often (as in the case of esterases), the hydrolysis of an artificial substrate is catalysed by more than one enzyme.
(b) See answer to Question 1 of Chapter 14.
2. A reversible inhibitor competes with the substrate. It must be included in the incubation medium. An irreversible inhibitor binds firmly to the active site of the enzyme, so sections can be pre-incubated, and then incubated in substrate without added inhibitor.
3. (a) Arylesterase or acetylesterase (could be distinguished using PCMB after E600).
(b) Cholinesterase (= pseudocholinesterase).
(c) Cholinesterase (= pseudocholinesterase). AChE would give stronger reaction with AThCh, and not inhibited by ethopropazine or very low concentration of DFP.
Thus, the cells contain at least two esterases: ChE, and also arylesterase and/or acetylesterase.
4. The incubation medium would contain TPP (substrate), cysteine (to inhibit interfering alkaline phosphatase), $CaCl_2$ and a buffer at about pH 8 (sufficiently basic for precipitation of calcium phosphate). After incubation, the sections are treated with an aqueous solution of a Co^{2+} salt, washed and then placed in dilute aqueous ammonium sulphide, which converts the cobalt phosphate to the black sulphide. They can then be dehydrated, cleared and covered. For reasons given in Chapter 4 (Section 4.3.1.1), Canada balsam is the mounting medium of first choice.
5. Stain the substrate film by the Feulgen method, or with a cationic dye. Sites of hydrolysis of DNA appear as clear areas. Controls should be stained with an anionic dye, to make sure that the loss of DNA is not due to proteolytic enzymes acting on the gelatin.

Chapter 16

1. (a) Sodium ions are reduced to the metal. Oxidation occurs at the site from which electrons enter the cathode (for example, within a chemical battery used as source of current).
(b) Cupric ions are reduced to metallic copper; iron is oxidized to ferrous ions.
(c) Hydrogen gas is oxidized; the carbon atoms of the olefin are reduced (oxidation number changes from -1 to -2).
2. The long conjugated chains, which include chromophoric azo ($N = N$) and imine ($C = N$) linkages and two $-NO_2$ groups, ensure absorption of visible light. The formazan is insoluble and is not ionized, so it is a pigment, not a dye.
3. One of the molecules of hydrogen peroxide is the oxidizing agent; the other is the reducing agent.
4. A compound with an E' of $+1.0$ V would be a much stronger oxidizing agent than FAD (see Table 16.1), so it would be expected to accept electrons from $FADH_2$ ($E' = -0.22$ V), the reduced prosthetic group of SDH. However, such a strong oxidizing agent would also accept electrons from all the cytochromes ($E' = +0.22$ to $+0.82$ V) and from non-enzymatic reducing groups (such as $-SH$) in the tissue. Although the oxidation-reduction potentials are only a crude indicator of the properties of a tetrazolium salt (see Section 16.4.1), an E' of $+1.0$ is far too high (see Table 16.2), and such a compound could not interrupt the flow of electrons from $FADH_2$ to oxygen.
5. For (a) and (b), the medium must contain the natural substrate of the enzyme and a tetrazolium salt. For (a), it is also necessary to include NAD. For other ingredients, see Sections 16.4.4 and 16.4.5. Controls include omission of the substrate and, for NAD-linked dehydrogenases, omission of the coenzyme to exclude an NADP-linked enzyme or a flavoprotein that catalyses oxidation of the same substrate. If a specific inhibitor of the enzyme is available, this should also be used. If an intermediate electron carrier such as PMS is not included in the incubation medium for (a), it is necessary to ensure that the sections contain NADH diaphorase.
6. (i) In mitochondria of type B cells.
(ii) In mitochondria of both cell-types.
An enzyme that does not require NAD^+, presumably a flavoprotein dehydrogenase, has been detected in the type A cells.
7. Citrate ions form a complex with cupric ions, so that $[Cu^{2+}]$ is too low to allow precipitation of the sparingly soluble copper ferricyanide. The reduced coenzyme or flavin nucleotide formed by the action of the enzyme reduces $Fe(CN)_6^{3-}$ to $Fe(CN)_6^{4-}$. Copper ferrocyanide (Hatchett's brown) is extremely insoluble, so it is precipitated even in the presence of citrate, and is the electron-dense final reaction product.
8. This is the classical NADI reaction (Section 16.5.1). It does not occur in the presence of azide or cyanide ions, which inhibit oxidase.
9. In mammals, enzymatic activity is seen in melanocytes, at the junction of dermis with epidermis and commonly also within the dermis. These cells secrete their pigment granules, which are immediately taken up by the

basal cells of the epidermis (pigment donation). Consequently, most of the melanin in the skin is in the epidermis. In amphibian skin, the melanophores synthesize and store melanin, so the pigment and the monophenol oxygenase are found in the same cells.

Chapter 17

1. The fluorescence method can detect approximately 1,781,000 molecules of NA, which are contained in 81 chromaffin granules.

2. Proteins containing tyrosine, histidine and tryptophan (see Chapter 10, Section 10.9.2).

3. Histamine can be demonstrated with o-phthaldialdehyde (low specificity) or immunohistochemically. The sulphated proteoglycan can be stained by alcian blue, pH 1.0 or other methods described in Chapter 11. Serotonin in mast cells can be demonstrated by formaldehyde-induced fluorescence or by an azo-coupling method.

4. (a) No. Heavy metals like Cr quench fluorescence.

(b) Yes, if the amines are also phenols.

(c) Yes, because of the chromium atoms in the reineckate.

5. Serotonin is detectable in some of the enteroendocrine cells of the intestinal epithelium, and in connective tissue mast cells.

6. Use an aqueous fixative (SUSA recommended). Cut paraffin sections. Axons can be stained by Holmes's method; neurosecretory material by an oxidation-basic dye method. Myelin stains show that there are no myelinated fibres in the pituitary gland.

7. The treatment with dichromate can make the specimens brittle, causing difficulty cutting sections.

8. With the glutaraldehyde-osmium tetroxide method, lipid droplets in the cells of the adrenal cortex are blackened. They can be extracted from the sections (see Chapter 12) before treating with OsO_4.

Chapter 18

1. (a) Nissl methods.

(b) Golgi methods.

(c) Either luxol fast blue or iron-chromoxane cyanine R (for myelin), with a red Nissl counterstain.

(d) Cajal's gold-sublimate or immunohistochemical staining for GFAP.

2. Sudan IV stains hydrophobic lipids more strongly than hydrophilic lipids (see Chapter 12). Myelin lipids (phosphoglycerides and sphingomyelins) are hydrophilic.

3. The SO_3^{2-} ions withdraw unreduced silver from the sections, as a soluble complex. This extracted silver makes the solution a physical developer. Sulphite also reacts with the oxidation product of hydroquinone to form a colourless, soluble product.

4. Immerse the film in a physical developer.

5. Silver methods and vital methylene blue stain axons of all types but can be unreliable, especially for fibres of small diameter. Histochemical methods for choline esterases (Chapter 15) and biogenic amines (Chapter 17), and immunohistochemical methods for various enzymes and peptides are valuable for axons that contain these substances.

Chapter 19

1. The eye can detect smaller quantities of a fluorescent substance (see Chapter 1).

2. Coagulant fixatives distort the shapes of protein molecules, exposing more of the antigenic sequences of amino acids (see Chapter 2). Picric acid also suppresses some autofluorescence. By cross-linking proteins, aldehydes can mask epitopes. Aldehyde fixatives also induce fluorescence non-specifically in some tissues.

3. Treat sections with rabbit anti-HRP serum, followed by fluorescently labelled goat anti-rabbit immunoglobulin. For controls: (i) omit the primary antiserum; (ii) substitute non-immune rabbit serum for the primary antiserum; (iii) mix the primary antiserum with purified HRP to absorb out the antibodies. The enzyme is also easily demonstrated by direct histochemical staining for peroxidase activity.

4. Sensitivity is much less than with indirect methods. The reason is the same as that for direct fluorescent antibody methods (Section 19.5).

5. The reasons are explained in Sections 19.5.3 and 19.6.

6. Apply (a), (b) and (c) sequentially to the section, with washes between treatments. This forms the sandwich:

(Antigen A)—(Anti-A)—(Fluorescent Antigen A)

With a monoclonal IgG antibody, only one fluorescent antigen molecule is bound at each antigenic site. If Anti-A is a polyclonal antiserum, several molecules can bind to each molecule of Antigen A by attaching to different epitopes. The method is then more sensitive.

7. When glutaraldehyde cross-links proteins, many free aldehyde groups are bound to all parts of the tissue (see Chapter 2). These free aldehydes will bind any soluble proteins, including immunohistochemical reagents, that are applied to the section. Non-immune goat serum binds to the aldehyde groups first, and thereby prevents the non-specific binding of primary and secondary antisera and of PAP.

8. It is important to provide secondary antibody molecules in numbers large enough to bind monovalently to the F_c segments of all the attached molecules of the primary antibody. This provides an excess of free F_{ab} segments (of the secondary antibody molecules) to bind the subsequently applied PAP.

9. The rabbit PAP could not attach to goat anti-dog immunoglobulin. It would be necessary to use dog PAP if the primary antiserum had been raised in a dog. Dog PAP is unlikely to be available, so the correct procedure is to use staphylococcal protein A instead of a secondary antiserum in stage (b) of the procedure. Protein A binds to the F_c segments of most IgG molecules, thus building the sandwich:

(Dog anti-hormone)—(Protein A)—(PAP)

10. Obtain a cDNA probe that recognizes an mRNA for some part of the F_c segment of the type of immunoglobulin detected in the lesions. *In situ* hybridization will then reveal sites in which the gene for that type of immunoglobulin is being transcribed.

INDEX

Salts are indexed under their histochemically significant ions, *e.g.* Permanganate (*not* Potassium permanganate); Picric acid (*not* Acid, picric).

Numbers or single letters that begin chemical names are omitted. Thus, *N*-acetylneuraminic acid, *n*-butanol, *N,N*-dimethylformamide, 2-hydroxynaphthoic acid hydrazide and α-naphthol are indexed as Acetylneuraminic acid, Butanol, Dimethylformamide, Hydroxynaphthoic acid hydrazide and Naphthol, respectively.